OXFORD MATHEMATICS

Intermediate GCSE for AQA

Linear Specification

Course Editors
Peter McGuire   Ken Smith

Course Consultants
Keith Gordon
Head of Mathematics at Wath Comprehensive School, Rotherham;
Chief Examiner

Trevor Senior
Head of Mathematics at Wales High School, Rotherham;
Principal Examiner

Spirals grow according to a simple numerical sequence.

You can find out more about sequences in section 2.

OXFORD
UNIVERSITY PRESS

OXFORD
UNIVERSITY PRESS
Great Clarendon Street, Oxford OX2 6DP

Oxford University Press is a department of the University of Oxford.
It furthers the University's objective of excellence in research, scholarship,
and education by publishing worldwide in

Oxford New York
Athens Auckland Bangkok Bogotá Buenos Aires Cape Town
Chennai Dar es Salaam Delhi Florence Hong Kong Istanbul Karachi
Kolkata Kuala Lumpur Madrid Melbourne Mexico City Mumbai Nairobi
Paris São Paulo Shanghai Singapore Taipei Tokyo Toronto Warsaw
with associated companies in Berlin Ibadan

Oxford is a registered trade mark of Oxford University Press
in the UK and in certain other countries

10 9 8 7 6 5 4 3 2 1

Many thanks to the original authors of this series for the use of their material:
Sue Briggs Peter McGuire Derek Philpott Susan Shilton Ken Smith

The authors would like to thank Paul Metcalf for his authoratitive coursework guidance.

Database right Oxford University Press (maker)

First published 2001

British Cataloguing in Publication Data

Data available

ISBN 0 19 914808 2

The publishers would like to thank AQA for their kind permission to reproduce past paper questions. AQA except no responsibility for the answers to the past paper questions which are the sole responsibility of the publishers.

The publishers and authors are grateful to the following:

Illustrators	Gecko, Oxford Illustrators
Photographers	Mike Dudley, Martin Sookias, Andrew Ward
Photographic Libraries	Ancient Art and Architecture Collection Ltd, Architectural Association, Mary Evans Picture Library
Suppliers	Cornish Seal Sanctuary, Eurostar
Cover artwork	Image Colour Library

Typeset in Great Britain by Mathematical Composition Setters Ltd, Hilltop Business Park, Devizes Road, Salisbury, Wiltshire.
Printed and bound in Great Britain by Butler and Tanner Ltd., Frome and London

About this book

This book is designed to help you achieve your best possible grade in the **AQA Specification A** Intermediate GCSE examination. The book is divided into two parts: a full-colour part containing the GCSE content that you need to learn, and a black-and-white part containing plenty of exam practice and answers.

This is more than just a book of questions: it is a learning package that will help you to make the most of your mathematical talents and expertise. You can be confident that you will be well prepared for your AQA examination.

Wordfinder

As well as a detailed **Contents** list and an **index**, the book contains a **Wordfinder** on pages 10 and 11. This provides an alphabetical list of the mathematical terms used in the book, and tells you where to find them.

Starting points

The content of the book is divided into sections, each covering a particular topic. Each section begins with **Starting points**, which lists the facts and techniques that you should already know. There are some questions for you to try out before starting the section.

Sections

Colour is used in the sections in the following ways:

- Yellow panels contain explanations and worked examples.
- The more difficult questions in an exercise are numbered in blue - for example, on page 84:
  7 Show that $2(n + 2) + 2n$ and $4n + 4$ are equivalent expressions.
- Blue text is used at times to stress important points – for example, on page 98, the digits 4 and 0 are emphasized in the number 0.00403.
- Words in the margin that link with the main text are coloured red. For example, on page 69: quadrilateral is defined in the margin and referred to in the main text.
  There are also plenty of exercises to consolidate your learning.
  A **non-calculator icon** is used to identify exercises where you should not use a calculator.

End points

Each section finishes with **Endpoints**, where the main work of the section is summarized. These are excellent for revision purposes.

Skills break

At certain points throughout the book there are **Skills breaks**. Each one provides a variety of questions all linked to the same data, and allows you to draw on the skills and techniques that you have learned.

In focus

Towards the end of the book are **In focus** pages which relate to a particular section, and you can use these for in-depth revision.

Coursework

There is a separate **Coursework Guidance** section, which details what you need to do for your investigative task and your statistical task. It contains sample tasks, with **Moderator comments** to help you get better marks.

Exam questions

The **Exam questions** section contains hundreds of past paper questions from AQA to help you prepare for your exam. There are also two **practice exam papers**, one calculator and one non-calculator, so that you can gain familiarity and confidence with the new AQA examination style.

Answers

Numerical **answers** are given at the end of the book.

Good luck in your exams!

CONTENTS

Note for students taking the modular GCSE course:

If you are following **AQA Specification B (modular)**, these symbols will tell you which module the topic is assessed in.

■ Module 1
■ ■ Module 3
■ ■ ■ Module 5

The coursework component (comprising Modules 2 and 4) is covered in the Coursework Guidance section.

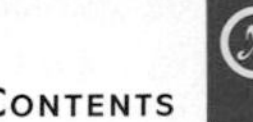

A note on accuracy

Make sure your answer is given to any degree of accuracy stated in the question, for example 2 dp or 1 sf. Where it is not stated, choose a sensible degree of accuracy for your answer, and make sure you work to a greater degree of accuracy through the problem. For example, if you choose to give an answer to 3 sf, work to at least 4 sf through the problem, then round your final answer to 3 sf.

Examination groups differ in their approach to accuracy. Some say that you should not give your final answer to a greater degree of accuracy than that used for the data in the question, but others state answers should be given to 3 sf.

If you are in any doubt, check with your examination group.

Metric and imperial units

	Metric	Imperial	Some approximate conversions
Length	millimetres (mm) centimetres (cm) metres (m) kilometres (km) 1 cm = 10 mm 1 m = 100 cm 1 km = 1000 m	inches (in) feet (ft) yards (yd) miles 1 ft = 12 in 1 yd = 3 ft 1 mile = 1760 yd	1 inch = 2.54 cm 1 foot ≈ 30.5cm 1 metre ≈ 39.4 in 1 mile ≈ 1.61 km
Mass	grams (g) kilograms (kg) tonnes 1 kg = 1000 g 1 tonne = 1000 kg	ounces (oz) pounds (lb) stones 1 lb = 16 oz 1 stone = 14 lb	1 pound ≈ 454 g 1 kilogram ≈ 2.2 lb
Capacity	millilitres (ml) centilitres (cl) litres 1 cl = 10 ml 1 litre = 100 cl = 1000 ml	pints (pt) gallons 1 gallon = 8 pints	1 gallon ≈ 4.55 litres 1 litre ≈ 1.76 pints ≈ 0.22 gallons

WORDFINDER

Starting points

You need to know about ... ... so try these questions.

A Negative numbers

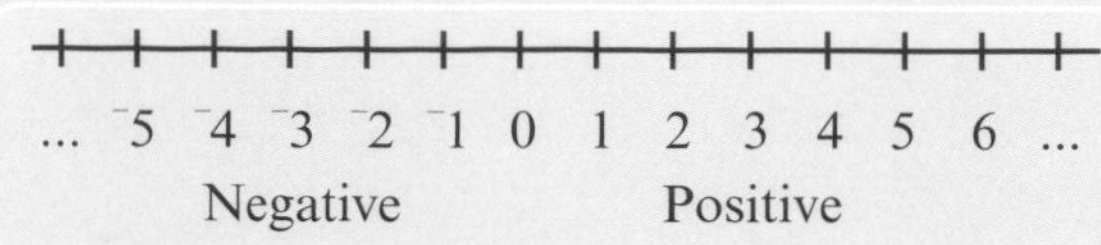

- To **add** a negative number:
  - subtract the corresponding positive number.

$7 + {}^-4$ $= 7 - 4$ $= 3$	$2 + {}^-5$ $= 2 - 5$ $= {}^-3$	${}^-4 + {}^-3$ $= {}^-4 - 3$ $= {}^-7$

- To **subtract** a negative number:
  - add the corresponding positive number.

$7 - {}^-4$ $= 7 + 4$ $= 11$	$2 - {}^-5$ $= 2 + 5$ $= 7$	${}^-4 - {}^-3$ $= {}^-4 + 3$ $= {}^-1$

A1 Calculate these missing numbers.

a $1 + {}^-3 = \square$ **b** ${}^-5 + {}^-2 = \square$
c $9 + {}^-5 = \square$ **d** $4 - {}^-6 = \square$
e ${}^-7 - {}^-3 = \square$ **f** ${}^-2 - {}^-7 = \square$

A2 Calculate these missing numbers.

a $3 + {}^-5 = \square$ **b** ${}^-5 + 3 = \square$
c $7 - {}^-2 = \square$ **d** ${}^-2 - 7 = \square$
e $6 + \square = 3$ **f** $7 - \square = 5$
g ${}^-3 - \square = 1$ **h** ${}^-4 + \square = {}^-9$
i ${}^-5 + \square = {}^-1$ **j** ${}^-6 - \square = 0$

B Multiples, factors and primes

- The **multiples** of a number can be divided by the number without leaving a remainder.
- The **common multiples** of two numbers are numbers that are multiples of both the numbers.

Multiples of 4:
4, 8, 12, 16, 20, 24, 28, 32, ...
Multiples of 6:
6, 12, 18, 24, 30, 36, 42, ...
Common multiples of 4 and 6:
12, 24, 36, ...

- The **factors** of a number are whole numbers that divide into it without leaving a remainder.
- The **common factors** of two numbers are numbers that are factors of both the numbers.

Factors of 18:
1, 2, 3, 6, 9, 18
Factors of 42:
1, 2, 3, 6, 7, 14, 21, 42
Common factors of 18 and 42:
1, 2, 3, 6

- A number is **prime** if it has only two different factors.

5 is a prime number: it has factors 1 and 5.

9 is not prime: it has factors 1, 3 and 9.

B1 Give seven multiples of:
a 5 **b** 8 **c** 12

B2 Give three common multiples of:
a 5 and 8 **b** 8 and 12

B3 Give all the factors of:
a 24 **b** 30
c 25 **d** 17

B4 Give the common factors of:
a 24 and 30 **b** 25 and 30

B5 Explain why 1 is not prime.

B6 Give all the prime numbers less than 20.

C Writing powers using index notation

$3 \times 3 \times 3 \times 3 = 3^4$
$1.6 \times 1.6 = 1.6^2$

4 is the **index**.
4 and 2 are **indices**.

We say 3^4 is:
3 **to the power** 4, or
3 **raised to the power** 4.

C1 Write in index notation:
a 5×5 **b** $2 \times 2 \times 2 \times 2 \times 2$

C2 Evaluate:
a 4^3 **b** 3^4 **c** 1.5^2 **d** 1.1^5

D Writing a number as a product of primes

- The factors of a number which are prime are called **prime factors**.
- A multiplication of prime factors is called a **product of primes**.
- To write a number as a product of primes:
  - break the number down into pairs of factors until all the factors are prime
  - use index notation to write the powers of each prime.

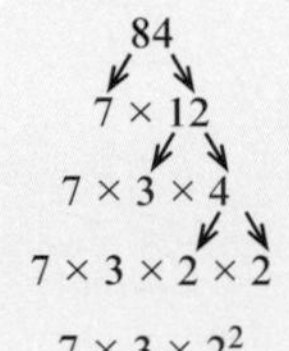

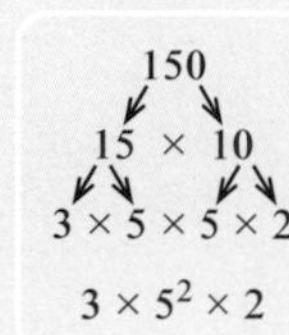

D1 Give the prime factors of:
a 84 **b** 154

D2 Write these as a product of primes.
a 120 **b** 350

E Types of fraction

- There are several ways to think of a fraction such as $\frac{3}{5}$.

This shows that **multiplying by $\frac{1}{5}$** has the same effect as **dividing by 5**.

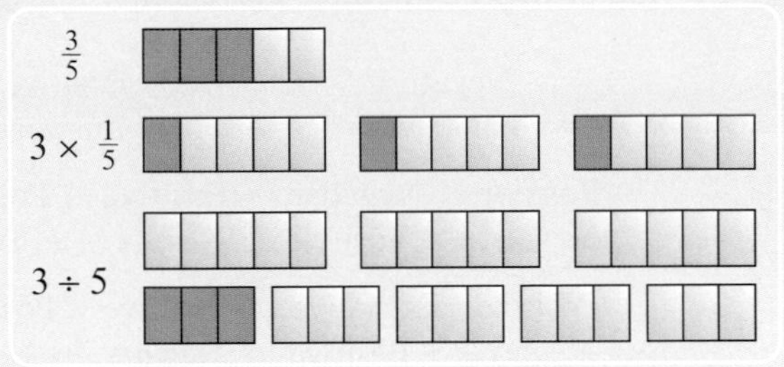

- An **improper fraction** is one where the numerator is larger than the denominator. $\frac{9}{7}$
- An improper fraction is greater than 1, so it can be written as a **mixed number**. $\frac{9}{7} = \frac{7}{7} + \frac{2}{7} = 1\frac{2}{7}$
- A **whole number** can be written as an improper fraction. $3 = \frac{3}{1}$

E1 Draw a diagram to show that $\frac{2}{3}$ is the same as $2 \times \frac{1}{3}$ and $2 \div 3$.

E2 Write as a mixed number:

a $\frac{4}{3}$ **b** $\frac{13}{6}$ **c** $\frac{23}{4}$

E3 Write as an improper fraction:

a $1\frac{2}{5}$ **b** $3\frac{1}{3}$ **c** 5

d $2\frac{11}{13}$ **e** 11

F Equivalent fractions

- To write an equivalent fraction:
  - multiply or divide the numerator or denominator by the same number.

$\frac{3}{4} \xrightarrow{\times 5} \frac{15}{20}$ $\qquad$ $\frac{27}{42} \xrightarrow{\div 3} \frac{9}{14}$

$$\frac{3}{4} = \frac{15}{20} \quad (\times 5) \qquad \frac{27}{42} = \frac{9}{14} \quad (\div 3)$$

F1 Write three equivalent fractions for:

a $\frac{2}{5}$ **b** $\frac{4}{12}$ **c** $\frac{20}{30}$

G Writing fractions as decimals

- To write a fraction as a decimal:
  - divide the numerator by the denominator.

$$\frac{3}{8} = 3 \div 8 = \mathbf{0.375}$$

Some fractions give **recurring decimals**.

$$\frac{2}{3} = 0.\dot{6} \qquad \frac{5}{6} = 0.8\dot{3} \qquad \frac{59}{270} = 0.2\dot{1}8\dot{5}$$

G1 Write these fractions as decimals.

a $\frac{7}{16}$ **b** $\frac{5}{9}$

c $\frac{7}{11}$ **d** $\frac{23}{54}$

H Calculating with fractions

- Multiplying by a fraction less than 1 gives a smaller amount.
- To calculate a fraction of an amount: either
  - divide the amount by the denominator and multiply by the numerator

  or
  - write the fraction as a decimal and multiply by the amount.

$\frac{3}{8}$ of **£112**

$112 \div 8 \times 3 =$ **£42**

$0.375 \times 112 =$ **£42**

- Dividing by a fraction less than 1 gives a larger amount.

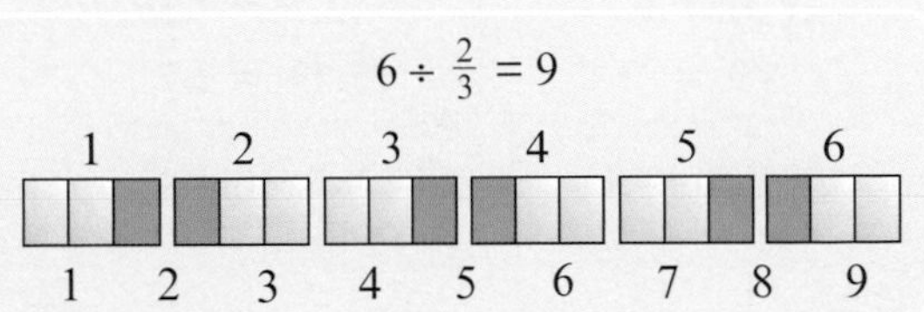

H1 Calculate:

a $\frac{5}{8}$ of £144

b $\frac{2}{7}$ of 322 litres

c $\frac{9}{14}$ of 154 m

H2 Draw a diagram to show that $6 \div \frac{3}{4} = 8$

Multiplying and dividing negative numbers

- Multiplying or dividing numbers give either a negative or a positive answer.

Negative answer	Positive answer
Positive × Negative Negative × Positive	Positive × Positive Negative × Negative
Positive ÷ Negative Negative ÷ Positive	Positive ÷ Positive Negative ÷ Negative

Exercise 1.1
Multiplying and dividing negative numbers

Do not use a calculator for Exercise 1.1.

1 Which of these calculations give a negative answer?

A $4 \times {}^{-}5$ B ${}^{-}2.5 \times 6$ C ${}^{-}3 \div {}^{-}2.8$ D ${}^{-}7.1 \div 4.6$ E ${}^{-}6.9 \times {}^{-}5$

F ${}^{-}3.7 \times 4.2$ G ${}^{-}3 \div {}^{-}4$ H ${}^{-}1.8 \times 6$ I $6.2 \div {}^{-}3.9$ J $12.5 \div 3$

2 Copy and complete these calculations.

a $3 \times {}^{-}5 = \square$ **b** ${}^{-}2 \times 6 = \square$ **c** $5 \times {}^{-}4 = \square$ **d** ${}^{-}3 \times {}^{-}6 = \square$
e ${}^{-}12 \div {}^{-}3 = \square$ **f** ${}^{-}8 \div 4 = \square$ **g** $24 \div {}^{-}6 = \square$ **h** ${}^{-}40 \div 5 = \square$

3 Copy and complete these calculations.

a $4 \times \square = {}^{-}28$ **b** ${}^{-}15 \div \square = 3$ **c** ${}^{-}4 \times \square = {}^{-}32$ **d** $\square \times {}^{-}6 = 18$
e $\square \div {}^{-}7 = 5$ **f** $\square \div {}^{-}6 = {}^{-}6$ **g** ${}^{-}3 \times \square = 36$ **h** $\square \div 7 = 2$

4 ${}^{-}2.3 \times 4.1 = {}^{-}9.43$ ${}^{-}12.6 \div {}^{-}1.75 = 7.2$

Use these two calculations to answer:

a 2.3×4.1 **b** ${}^{-}2.3 \times {}^{-}4.1$ **c** ${}^{-}4.1 \times 2.3$
d ${}^{-}12.6 \div 1.75$ **e** $12.6 \div 1.75$ **f** $12.6 \div {}^{-}1.75$

5 ${}^{-}1.9 \times 7.4 \times {}^{-}5.8$ ${}^{-}3.4 \times 5.7 \times 1.2 \times {}^{-}5.3 \times {}^{-}2.6$

Can you predict if these calculations have a negative or a positive answer? If so, explain how.

Squares and square roots

- To **square** a number is to multiply the number by itself.

3 × 3 = 9
3 rows of 3 dots gives 9 dots.

9 is the **square** of 3
3 is a **square root** of 9

$\mathbf{3^2 = 9}$
$\mathbf{\sqrt{9} = 3}$

- You can also square negative numbers and decimal numbers.

${}^{-}3 \times {}^{-}3 = 9$	$2.7 \times 2.7 = 7.29$	${}^{-}2.7 \times {}^{-}2.7 = 7.29$
$\mathbf{({}^{-}3)^2 = 9}$	$\mathbf{2.7^2 = 7.29}$	$\mathbf{({}^{-}2.7)^2 = 7.29}$
$\mathbf{\sqrt{9} = {}^{-}3}$	$\mathbf{\sqrt{7.29} = 2.7}$	$\mathbf{\sqrt{7.29} = {}^{-}2.7}$
${}^{-}3$ is a square root of 9	2.7 is a square root of 7.29	${}^{-}2.7$ is a square root of 7.29

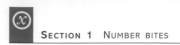

> This type of curve is called a **parabola**.

- The square roots of 9 are **3** and **$^{-}$3**.
- The square roots of 7.29 are **2.7** and **$^{-}$2.7**
- Each positive number has two square roots:
  - a **positive** one
  - a **negative** one.

Square roots graph: numbers 0 to 100

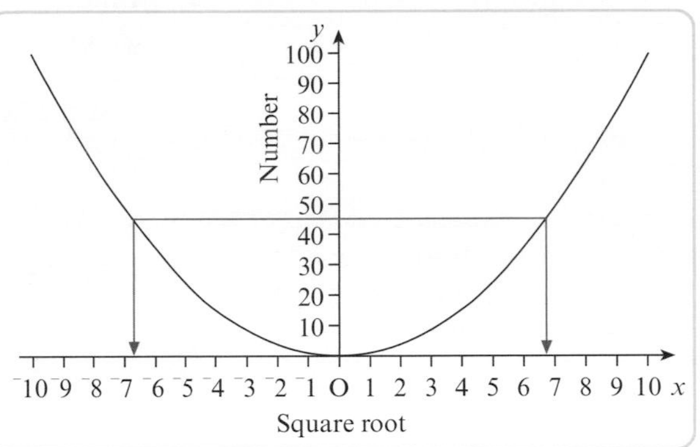

The square roots of 45 are about **$^{-}$6.7** and **6.7**.

Exercise 1.2
Square roots

1 Draw the square roots graph above.

2 Use your graph to estimate the values of each of these square roots.

a $\sqrt{30}$ **b** $\sqrt{74}$ **c** $\sqrt{12}$ **d** $\sqrt{45}$ **e** $\sqrt{60}$ **f** $\sqrt{95}$

3 Use your calculator to find the values to 2 dp of each of these square roots.

> Using the square root key on a calculator only gives the positive square root.

a $\sqrt{48}$ **b** $\sqrt{72}$ **c** $\sqrt{39.1}$ **d** $\sqrt{26.82}$ **e** $\sqrt{900}$

f $\sqrt{0.01}$

4 Without a calculator find each of these square roots.

a $\sqrt{169}$ **b** $\sqrt{196}$ **c** $\sqrt{100}$ **d** $\sqrt{81}$ **e** $\sqrt{225}$

Cubes and cube roots

- To **cube** a number is to multiply the number by itself twice.

> The symbol for a cube root is $\sqrt[3]{\ }$.

4 × 4 × 4 = 64
4 layers of 4 rows of
4 cubes gives 64 cubes.

64 is the cube of 4
4 is the cube root of 64

$4^3 = 64$
$\sqrt[3]{64} = 4$

- You can also cube negative numbers and decimal numbers.

$^{-}4 \times {^{-}4} \times {^{-}4} = {^{-}64}$
$(^{-}4)^3 = {^{-}64}$
$\sqrt[3]{^{-}64} = {^{-}4}$

2.7 × 2.7 × 2.7 = 19.683
$2.7^3 = 19.683$
$\sqrt[3]{19.683} = 2.7$

- All numbers, positive and negative, have only one cube root.

Exercise 1.3
Cubes and cube roots

1 List the key presses on your calculator to find 6^3.

2 Find:

a 9^3 **b** 30^3 **c** 3.5^3 **d** $(^-9)^3$ **e** $(^-1.1)^3$ **f** 0.4^3

3 List the key presses on your calculator to find $\sqrt[3]{68.921}$.

4 Find:

a $\sqrt[3]{8}$ **b** $\sqrt[3]{64}$ **c** $\sqrt[3]{1000}$ **d** $\sqrt[3]{27}$ **e** $\sqrt[3]{729}$

f $\sqrt[3]{343}$ **g** $\sqrt[3]{15.625}$ **h** $\sqrt[3]{8000}$ **i** $\sqrt[3]{125}$ **j** $\sqrt[3]{0.343}$

An integer is a positive or negative whole number. 0 is also an integer.

5 **a** Find the cube of each integer from $^-5$ to 5.
b Draw a graph to show these integers and their cubes.

Reciprocals

Zero does not have a reciprocal as ÷ 0 is not defined

- Two numbers which multiply together to equal 1 are **reciprocals** of each other.

  $0.8 \times 1.25 = 1$

  0.8 is the reciprocal of 1.25.
  1.25 is the reciprocal of 0.8.

- The reciprocal of any number n is: $\frac{1}{n}$ or $1 \div n$.

  $\frac{1}{0.8} = 1.25$ $\frac{1}{1.25} = 0.8$

- The reciprocal of any fraction $\frac{p}{q}$ is: $\frac{q}{p}$.

  Writing 0.8 and 1.25 as fractions gives: $\frac{4}{5} \times \frac{5}{4} = 1$

  $\frac{4}{5}$ is the reciprocal of $\frac{5}{4}$.
  $\frac{5}{4}$ is the reciprocal of $\frac{4}{5}$.

Exercise 1.4
Reciprocals

1 Find the reciprocal of:

a 4 **b** 0.1 **c** 2.5 **d** 3 **e** $0.\dot{6}$ **f** $0.8\dot{3}$ **g** 0

2 Write as a fraction the reciprocal of:

a $\frac{3}{5}$ **b** $\frac{7}{4}$ **c** $1\frac{1}{2}$ **d** $3\frac{1}{3}$ **e** 1.2

3 **a** Copy these axes.
b Complete the graph by plotting the whole numbers 1 to 10 against their reciprocal.

4 Investigate the relationship between negative numbers and their reciprocals.

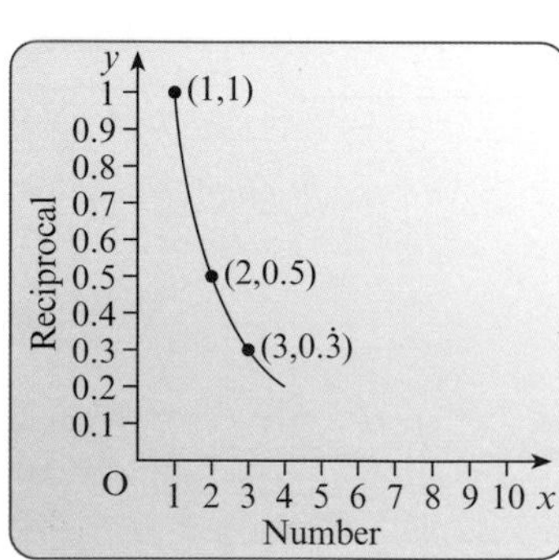

Standard form

- A number can be written in several ways using powers of 10. For example:

10^{-3}	10^{-2}	10^{-1}	10^{0}	10^{1}	10^{2}	10^{3}	10^{4}
0.001	0.01	0.1	1	10	100	1000	10 000

3600

$0.36 \times 10\,000$
3.6×1000
36×100
360×10
3600×1
$36\,000 \times 0.1$
$360\,000 \times 0.01$
$3\,600\,000 \times 0.001$

0.729

$0.000\,729 \times 1000$
$0.007\,29 \times 100$
0.0729×10
0.729×1
7.29×0.1
72.9×0.01
729×0.001
7290×0.0001

Standard form is also called standard index form.

- One way of writing a number is called standard form. 3600 and 0.729 in standard form are:
  - the first part is a number greater than or equal to 1 and less than 10

  3.6×10^3
  7.29×10^{-1}

  - the second part is a power of 10 written in index notation.

Include any zeros between non-zero digits. For example: $40\,900\,000 = 4.09 \times 10^7$

- To write a number in standard form:
  - use all the non-zero digits for the first part

  43 900 000 → 4.39 | 0.000 58 → 5.8

  - decide what power of 10 the first part must be multiplied by to give the number

  $43\,900\,000 = 4.39 \times 10\,000\,000$ | $0.000\,58 = 5.8 \times 0.0001$

  - write the power of 10 in index notation.

  $43\,900\,000 = 4.39 \times 10^7$ | $0.000\,58 = 5.8 \times 10^{-4}$

- To rewrite a number in standard form as an ordinary number:
  - write the power of 10 in full and multiply by the first part.

  $2.4 \times 10^6 = 2.4 \times 1\,000\,000 = 2\,400\,000$

  $3.152 \times 10^{-3} = 3.152 \times 0.001 = 0.003\,152$

Exercise 1.5
Standard form

1 Write these numbers in standard form.
  a 380 000 **b** 45 100 000 **c** 0.000 92 **d** 0.000 0262

2 Write these as ordinary numbers.
  a 9.42×10^4 **b** 2.5×10^{-3} **c** 7.414×10^8 **d** 6.27×10^{-7}

3 **P** 13×10^4 **Q** 5.7×1000 **R** $6.42 \div 10^3$ **S** 0.58×10^{-4}
  a Explain why each of these numbers is not written in standard form.
  b Write each number in standard form.

4 List the key presses on your calculator to find $60 \times (6.19 \times 10^{-11})$.

Give your answers in standard form.

5 Use your calculator to find:

a $365 \times (2.4\times10^4)$ b $(8.84\times10^6) \div 52$
c $(5.5\times10^{-4}) \times (3.6\times10^7)$ d $(8.2\times10^8) \div (1.04\times10^{-2})$
e $(6.8\times10^{-16}) + (7\times10^{-15})$ f $(4\times10^4) - (2.8\times10^2)$

Rules of indices

- To multiply powers of the same number:
  - add the indices.

$3^2 \times 3^3 = 3^{2+3} = 3^5$ $(3\times3) \times (3\times3\times3) = 3\times3\times3\times3\times3 = 243$

- To divide powers of the same number:
  - subtract the indices.

$3^5 \div 3^2 = 3^{5-2} = 3^3$ $(3\times3\times3\times3\times3) \div (3\times3) = 243 \div 9 = 27$

- To raise a power of a number to another power:
  - multiply the indices.

$(3^4)^2 = 3^{4\times2} = 3^8$ $(3\times3\times3\times3) \times (3\times3\times3\times3) = 81 \times 81 = 6561$

Exercise 1.6
Rules of indices

1 Give the answer to these using index notation.

a $3^2 \times 3^5$ b $4^{-3} \times 4^5$ c $2^7 \div 2^4$ d $7^{-4} \times 7^{-2}$ e $6^3 \div 6^5$
f $7^{-4} \div 7^2$ g $5^{-1} \times 5^1$ h $2^4 \div 2^{-1}$ i $6^{-3} \div 6^{-2}$ j $7^{-3} \times 7^0$

2 Copy and complete these calculations.

a $2^3 \times 2^{\square} = 2^7$ b $4^5 \times 4^{\square} = 4^3$ c $3^5 \div 3^{\square} = 3^2$ d $7^{\square} \times 7^{-3} = 7^4$
e $8^4 \div 8^{\square} = 8^{-2}$ f $5^{\square} \div 5^2 = 5^2$ g $2^{\square} \times 2^5 = 2^{-1}$ h $3^4 \div 3^{\square} = 3^3$

3 Give the answer to these using index notation.

a $(3^2)^4$ b $(5^3)^3$ c $(4^5)^{-2}$ d $(4^{-2})^5$ e $(2^4)^{-4}$ f $(7^3)^0$

4 Copy and complete these calculations.

a $2^{\square} \times 2^2 = 32$ b $3^6 \times 3^{-3} = \square$ c $2^8 \div 2^{\square} = 64$ d $4^2 \div 4^{-1} = \square$
e $(3^2)^{\square} = 81$ f $(5^3)^{\square} = 125$ g $(2^{-3})^{\square} = 0.125$ h $(5^{\square})^2 = 0.04$

Standard form without a calculator

- To multiply: $(6 \times 10^7) \times (3 \times 10^{-2})$
  - multiply the first parts
  - multiply the second parts
  - combine the two answers
  - write in standard form.

$6 \times 3 = 18$
$10^7 \times 10^{-2} = 10^5$
18×10^5
1.8×10^6

- To divide: $(4 \times 10^{-3}) \div (8 \times 10^5)$
  - divide the first parts
  - divide the second parts
  - combine the two answers
  - write in standard form.

$4 \div 8 = 0.5$
$10^{-3} \div 10^5 = 10^{-8}$
0.5×10^{-8}
5×10^{-9}

Exercise 1.7
Standard form without a calculator

1 Give the answer to these in standard form.

a $(6\times10^{-4}) \times (4\times10^7)$ b $(6\times10^{-2}) \div (3\times10^4)$
c $(5\times10^{-9}) \div (2.5\times10^{-12})$ d $(4\times10^{-3}) \times (1.5\times10^{-8})$

Least common multiples

- To find the least common multiple (LCM) of 24 and 45:
  - write each number as a product of primes — $24 = 2^3 \times 3$, $45 = 3^2 \times 5$
  - take the highest power of each prime factor to give a new product of primes — $2^3 \times 3^2 \times 5$
  - evaluate the new product of primes. — $8 \times 9 \times 5$ **= 360**

Exercise 1.8
Least common multiples

1 Find the LCM of:

a 12 and 18 b 20 and 42 c 28 and 30 d 150 and 315
e 10, 18 and 35 f 15, 27 and 70 g 66, 135 and 275

2 The LCM of two numbers is 630. One number is 42.
Give the numbers the other one could be.

Adding and subtracting fractions

- You can only add or subtract fractions when the denominators are the same.
- To add or subtract fractions: $\frac{3}{4} + \frac{1}{6}$ $\quad$ $\frac{7}{8} - \frac{1}{3}$
  - find the LCM of the denominators — 12 $\quad$ 24
  - find equivalent fractions with the LCM as the new denominator — $\frac{9}{12} + \frac{2}{12}$ $\quad$ $\frac{21}{24} - \frac{8}{24}$
  - add or subtract the numerators. — $\frac{11}{12}$ $\quad$ $\frac{13}{24}$

Any common multiple can be used as the new denominator.

Using the LCM will usually give an answer in its lowest terms.

Exercise 1.9
Adding and subtracting fractions

1 Evaluate:

a $\frac{1}{2} + \frac{1}{3}$ b $\frac{2}{5} + \frac{1}{4}$ c $\frac{1}{2} - \frac{1}{3}$ d $\frac{3}{4} - \frac{1}{5}$
e $\frac{2}{5} + \frac{3}{8}$ f $\frac{1}{3} + \frac{1}{4} + \frac{1}{6}$ g $\frac{2}{3} - \frac{1}{6}$ h $\frac{3}{5} + \frac{1}{2} - \frac{1}{4}$

2 Write the answer to each of these as a mixed number.

a $\frac{2}{3} + \frac{1}{2}$ b $\frac{1}{4} + \frac{4}{5}$ c $\frac{1}{2} + \frac{1}{3} + \frac{1}{4}$ d $\frac{3}{8} + \frac{3}{4} + \frac{1}{2}$

3 Evaluate:

a $1\frac{1}{2} + \frac{4}{5}$ b $1\frac{4}{5} - \frac{1}{2}$ c $1\frac{1}{3} + 2\frac{1}{2}$
d $2\frac{2}{5} - 1\frac{1}{2}$ e $2\frac{5}{8} + 1\frac{1}{3}$ f $2\frac{1}{4} - 1\frac{3}{5}$

Write any mixed numbers as improper fractions before adding or subtracting.

4 Fractions with a numerator of 1 are called **unit fractions**.

Investigate fractions like this which can be written as the sum of two unit fractions.

$$\frac{2}{5} = \frac{1}{3} + \frac{1}{15}$$

Thinking ahead to ... multiplying and dividing fractions

A Calculate:

a $10 \div 1.25$ **b** 10×0.8
c $16 \div 1.25$ **d** 16×0.8

$1.25 = \frac{5}{4}$ $0.8 = \frac{4}{5}$

B What do your answers to Question **A** tell you about dividing by a fraction?

Multiplying and dividing fractions

- To multiply fractions:
  - multiply the numerators
  - multiply the denominators
  - write the answer in its lowest terms.

$\frac{4}{5} \times \frac{3}{8} = \frac{12}{40} = \mathbf{\frac{3}{10}}$

The answer may already be in its lowest terms.

- Dividing by a fraction has the same effect as multiplying by its reciprocal.
- To divide fractions:
  - write the division as a multiplication
  - multiply the fractions.

$\frac{1}{6} \div \frac{2}{3}$
$= \frac{1}{6} \times \frac{3}{2} = \frac{3}{12} = \mathbf{\frac{1}{4}}$

Exercise 1.10 Multiplying and dividing fractions

1 Evaluate:

a $\frac{2}{3} \times \frac{5}{8}$ **b** $\frac{3}{5} \times \frac{5}{6}$ **c** $\frac{4}{7} \times \frac{3}{5}$ **d** $\frac{9}{14} \times \frac{2}{3}$
e $\frac{1}{4} \div \frac{5}{8}$ **f** $\frac{2}{3} \div \frac{6}{7}$ **g** $\frac{2}{5} \div \frac{3}{4}$ **h** $\frac{1}{4} \div \frac{3}{5}$

2 Write the answer to each of these as a mixed number or a whole number.

a $6 \times \frac{2}{5}$ **b** $8 \times \frac{3}{4}$ **c** $10 \div \frac{5}{6}$ **d** $7 \div \frac{2}{3}$
e $12 \times 1\frac{3}{4}$ **f** $9 \times 3\frac{1}{2}$ **g** $5 \div 1\frac{1}{6}$ **h** $11 \div 2\frac{3}{5}$

Write any mixed numbers or whole numbers as improper fractions before multiplying or dividing.

3 Use the fraction key on your calculator to evaluate:

a $2\frac{1}{2} \times 1\frac{4}{5}$ **b** $1\frac{1}{6} \div \frac{2}{5}$ **c** $4\frac{1}{2} \div 1\frac{3}{4}$ **d** $2\frac{3}{4} \times 2\frac{2}{3}$
e $\frac{4}{5} \times 1\frac{1}{4}$ **f** $2\frac{1}{2} \div 2\frac{1}{4}$ **g** $1\frac{3}{5} \times 3\frac{3}{4}$ **h** $\frac{5}{6} \div 1\frac{3}{8}$

Highest common factors

- To find the highest common factor (HCF) of 120 and 252:
  - write each number as a product of primes
  - take the lowest power of each common prime factor to give a new product of primes
  - evaluate the new product of primes.

$120 = 2^3 \times 3 \times 5$

$252 = 2^2 \times 3^2 \times 7$

$2^2 \times 3$

4×3
$= \mathbf{12}$

Exercise 1.11 Highest common factors

1 Use products of primes to find the HCF of:

a 252 and 360 **b** 120 and 315 **c** 693 and 1078
d 48, 66 and 225

End points

You should be able to ...	... so try these questions.
A Multiply and divide negative numbers	**A1** Copy and complete these calculations. **a** $7 \times {}^{-}2 = \square$ **b** ${}^{-}3 \times {}^{-}5 = \square$ **c** ${}^{-}12 \div 4 = \square$ **d** $20 \div {}^{-}5 = \square$ **e** $\square \times {}^{-}3 = {}^{-}18$ **f** $\square \div {}^{-}2 = 8$ **g** ${}^{-}4 \times \square = 36$ **h** $\square \div 7 = {}^{-}3$
B Find roots, cubes and reciprocals	**B1** Use your calculator to find: **a** $\sqrt{1.44}$ **b** 3.3^3 **c** $\sqrt[3]{216}$ **d** $\sqrt[3]{3.375}$ **B2** Find the reciprocal of: **a** 8 **b** 0.2 **c** $0.\dot{5}$ **d** $\frac{2}{5}$ **e** $1\frac{1}{4}$
C Understand and use the rules of indices	**C1** Give the answer to these using index notation: **a** $3^3 \times 3^4$ **b** $4^5 \div 4^2$ **c** $(7^2)^3$ **C2** Copy and complete these calculations: **a** $2^3 \times 2^{\square} = 128$ **b** $3^5 \times 3^{\square} = 27$ **c** $(4^2)^{\square} = 0.0625$ **C3** Give the answer to these using index notation. **a** $2^5 \times 2^{-3}$ **b** $3^{-2} \div 3^7$ **c** $(5^2)^5$ **d** $(4^{-2})^3$ **C4** Copy and complete these calculations. **a** $3^4 \times 3^{\square} = 3^{10}$ **b** $4^3 \div 4^{\square} = 4^7$ **c** $7^{\square} \times 7^7 = 7^5$ **d** $(6^{-2})^0 = 6^{\square}$
D Understand and use standard form	**D1** Write these numbers in standard form. **a** 70 620 000 **b** 0.000 003 75 **D2** Write these as ordinary numbers. **a** 6.9×10^{8} **b** 1.03×10^{10} **D3** Use your calculator to find: **a** $52 \times (1.92 \times 10^{13})$ **b** $(6.58 \times 10^{-7}) \div 7$ **c** $(4.1 \times 10^{27}) + (7 \times 10^{26})$ **D4** Without using a calculator, give the answer to these in standard form. **a** $(7 \times 10^5) \times (6 \times 10^{-9})$ **b** $(8 \times 10^{-2}) \div (4 \times 10^{-6})$
E Calculate with fractions	**E1** Evaluate: **a** $\frac{2}{3} + \frac{1}{5}$ **b** $\frac{5}{6} - \frac{2}{5}$ **c** $2\frac{3}{8} - 1\frac{3}{4}$ **d** $1\frac{5}{7} + 2\frac{1}{6}$ **E2** Evaluate: **a** $\frac{3}{4} \times \frac{2}{5}$ **b** $\frac{1}{6} \div \frac{2}{3}$ **c** $3\frac{1}{3} \times 1\frac{4}{5}$ **d** $2\frac{1}{2} \div 1\frac{3}{4}$
F Find least common multiples and highest common factors	**F1** Find the LCM of: **a** 12 and 21 **b** 24 and 90 **F2** Find the HCF of: **a** 14 and 35 **b** 105 and 350

Some points to remember

- Each positive number has two square roots: a positive one and a negative one.
- All numbers, positive and negative, have only one cube root.
- A number to a negative power is a fraction.
- Fractions can only be added or subtracted when the denominators are the same.

Starting points

You need to know about ... | ... so try these questions.

A Finding the value of an expression

- Examples of expressions are:

$3n + 1 = (3 \times n) + 1$ $\quad 9 - 4n = 9 - (4 \times n)$ $\quad 2a - b = (2 \times a) - b$

$k^2 + 1 = (k \times k) + 1$ $\quad 2y^2 = 2 \times y \times y$ $\quad g(g + 1) = g \times (g + 1)$

- The value of an expression depends on the value of the letters.

For example:
- when $n = 5$, $3n + 1 = (3 \times n) + 1 = (3 \times 5) + 1 = 16$
- when $y = 3$, $2y^2 = 2 \times y \times y = 2 \times 3 \times 3 = 18$

A1 Find the value of these expressions when $n = 4$.
a $5n + 1$ **b** $7 - n$
c $2n - 8$ **d** $3(2 + n)$

A2 Find the value of these expressions when $a = 3$.
a $a^2 + 10$ **b** $4a^2$
c $a^2 - 5$ **d** $a(a + 1)$

B Some types of number sequences

- Even numbers: 2, 4, 6, 8, 10, ...
- Odd numbers: 1, 3, 5, 7, 9, ...
- Square numbers: 1, 4, 9, 16, ...

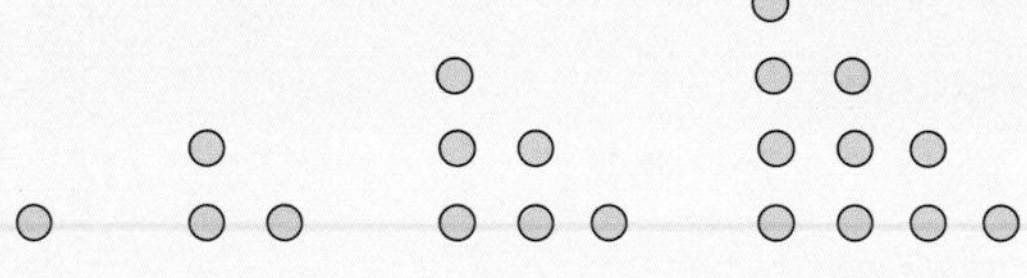

- Triangle numbers: 1, 3, 6, 10, ...
- Powers of 2: 2, 4, 8, 16, 32, ...
- Powers of 3: 3, 9, 27, 81, 243, ...
- Fibonacci numbers: 1, 1, 2, 3, 5, 8, 13, ...

The sequence of Fibonacci numbers begins 1, 1, ...
Further numbers in the sequence are found by adding the previous two numbers together. For example, the next Fibonacci number is $8 + 13 = 21$.

Start with two different numbers to find other sequences like this, for example: 3, 6, 9, 15, 24, ...

B1 Draw a pattern of dots to show that 25 is a square number.

B2 Find four odd square numbers.

B3 Write down the first six triangle numbers.

B4 List the first four powers of 5.

B5 Which of these are Fibonacci numbers:
30 34 62 90?

B6 Each number in this sequence is found by adding the previous two numbers together.

1, 3, 4, 7, 11, 18, ...

Find the next four numbers in this sequence.

C Continuing a sequence

- A **sequence** of numbers usually follows a pattern or rule.
- Each number in a sequence is called a **term**.

  For example,
  in the sequence: 3, 5, 7, 9, 11, ... the 1st term is 3
  the 2nd term is 5
  the 3rd term is 7.

- A sequence can often be continued by finding a pattern in the **differences**.

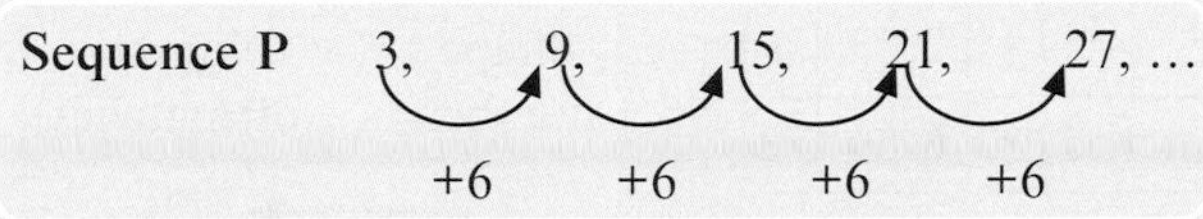

❖ In Sequence P, the **first difference** is 6 each time.
So continue the sequence by adding 6.

C1 Find the 6th and 7th term in sequence P.

C2 What is the 10th term in this sequence?

4, 7, 10, 13, 16, ...

C3 For each sequence, find the next three terms.

a 6, 7, 10, 15, 22, ...
b 2, 8, 14, 20, 26, ...
c 2, 5, 11, 20, 32, ...
d 2, 3, 9, 20, 36, ...

D A rule for a sequence

These are the first three matchstick patterns in a sequence.

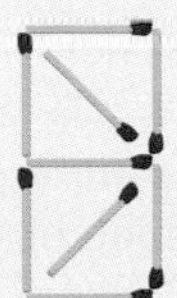
Pattern 1

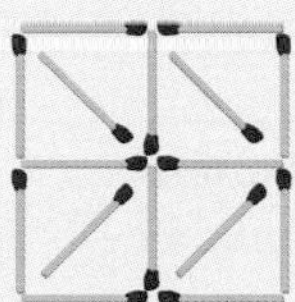
Pattern 2

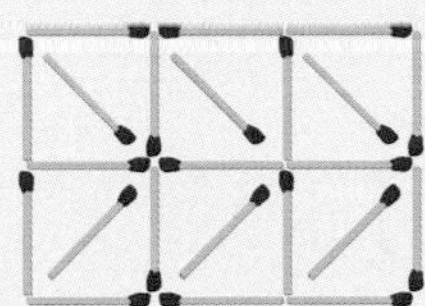
Pattern 3

- The number of matches in each pattern can be shown in a table.

Pattern number (n)	Number of matches (m)
1	9
2	16
3	23
4	30

(Pattern number: +1, +1, +1; Number of matches: +7, +7, +7)

❖ The pattern number goes up by 1 each time.

❖ The number of matches goes up by 7 each time.

❖ So a rule that links the number of matches (m) with the pattern number (n) begins $m = 7n$ ...

❖ A rule that fits all the results in the table is $m = 7n + 2$

- This rule can be used to calculate the number of matches in any pattern,
  for example: in Pattern 10 there are $(7 \times 10) + 2 = 72$ matches.

D1 These are the first three matchstick patterns in a sequence.

Pattern 1

Pattern 2
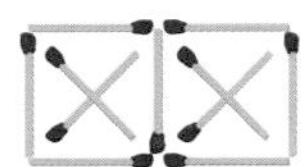

Pattern 3
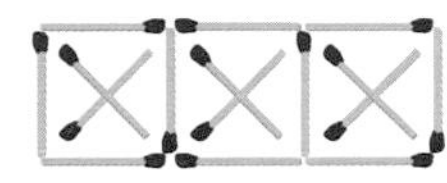

a Draw pattern 4 and pattern 5 in this sequence.

b How many matches are in:
i pattern 3 **ii** pattern 5?

c Make a table for the first five patterns in this sequence.

d Which of these rules fits the results in your table?

$m = 4n + 2$ $m = 5n + 1$ $m = 6n - 1$

Sequences and mappings

These equilateral triangles are the first three in a sequence of shapes.

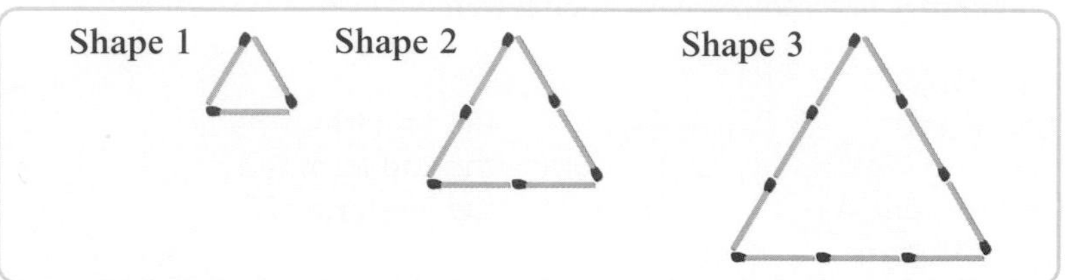

- Data for this sequence can be shown in a table.

Shape number	1	2	3	4
Number of matches	3	6	9	12

- The data can also be shown in a **mapping diagram** like this.

Shape number		Number of matches
1	$\longrightarrow$	3
2	$\longrightarrow$	6
3	$\longrightarrow$	9
4	$\longrightarrow$	12 ...

You can choose any letter to stand for the shape number.

For example:
the rule
$n \longrightarrow 3n$
can be written as
$s \longrightarrow 3s$
or
$p \longrightarrow 3p$
or ...

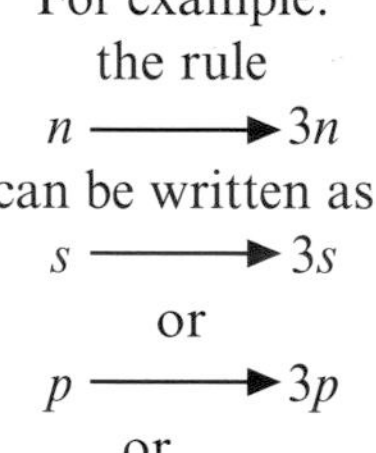

- The total number of matches is 3 times the shape number. For example, the 50th shape in the sequence uses 150 matches.

  Using n to stand for the shape number, the rule for the mapping diagram can be written: $n \longrightarrow 3n$

Exercise 2.1
Sequences and mappings

1 These are the first three patterns in a sequence.

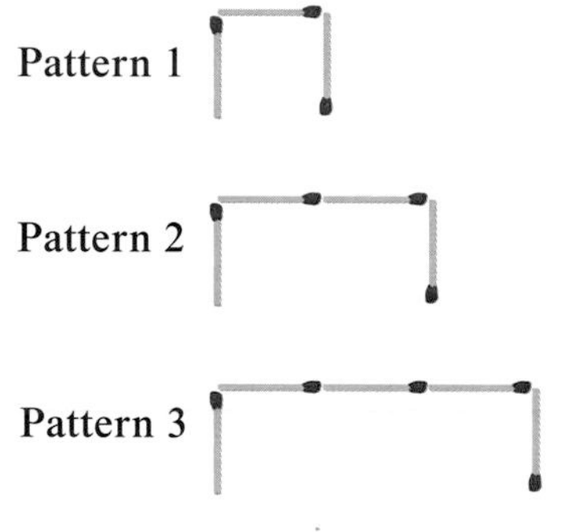

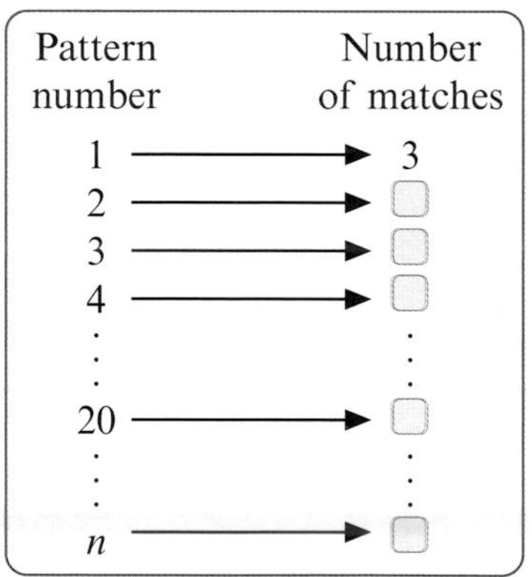

Copy and complete the mapping diagram for the sequence.

2 These patterns of touching squares are the first four in a sequence.

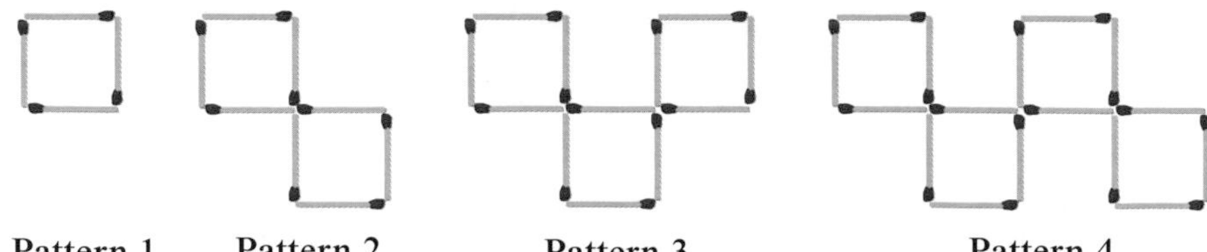

a Draw a mapping diagram for the first six patterns of touching squares.
b Find a rule for the sequence in the form $n \longrightarrow \ldots$, where n is the pattern number.
c Use your rule to calculate the number of matches in the 100th pattern.

Thinking ahead to ...
finding rules

A These are the first three patterns in sequence A.

Sequence A

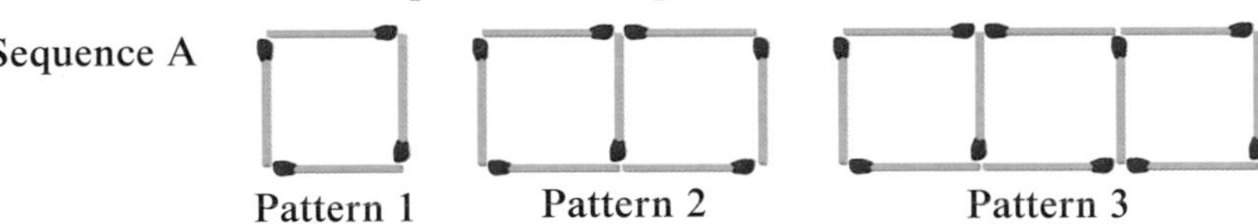

How many matchsticks are in the 100th pattern?
Explain how you worked it out.

Finding rules

To find a rule for the number of matches (m) in the nth pattern in sequence A above

- **Method 1** Look at how the patterns are made.

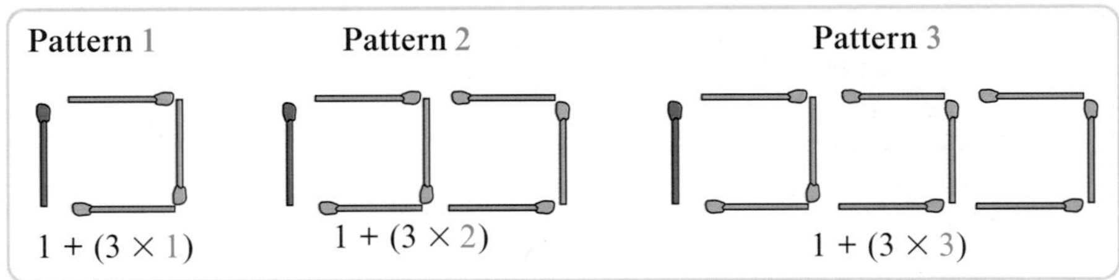

  - So a rule for the number of matches (m) in the nth pattern is $\boldsymbol{m = 1 + 3n}$

- **Method 2** Look at differences.

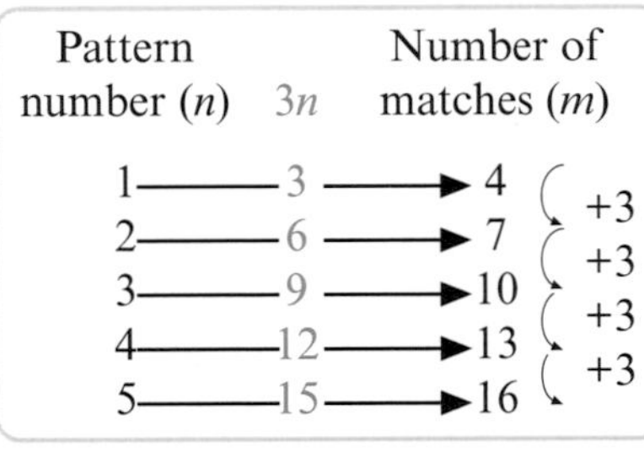

Pattern number (n)	$3n$	Number of matches (m)	
1	3	4	
2	6	7	+3
3	9	10	+3
4	12	13	+3
5	15	16	+3

  - The pattern number goes up by 1 each time.
  - The number of matches goes up by 3 each time so there is a linear rule that begins $m = 3n$ ... .

> Examples of linear rules are:
>
> $m = 4n + 3$
> $y = 2 - 5x$
> $a = 3b - 1$

  - Compare $3n$ with the number of matches.
  - The number of matches is 1 more than $3n$ each time.
  - So a rule for the number of matches (m) in the nth pattern is $\boldsymbol{m = 3n + 1}$.

Exercise 2.2
Finding rules

1 These triangle patterns are the first three in sequence B.

Sequence B

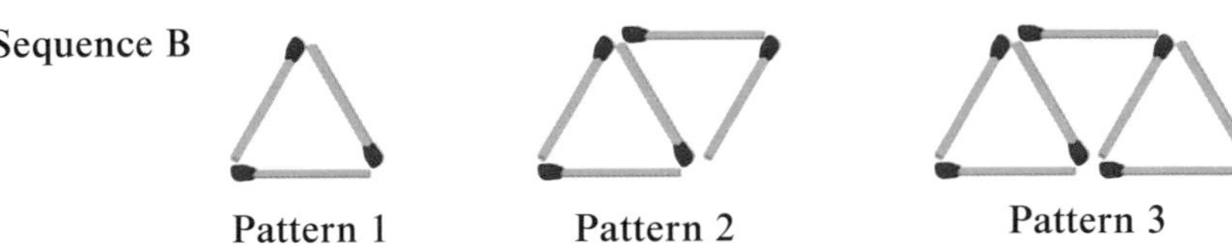

a Find a rule for the number of matches (m) in the nth triangle pattern. Explain your method.
b Use your rule to find the number of matches in the 8th pattern.
c Check your answer by drawing the 8th pattern and counting the matches.

2 **Sequence C**

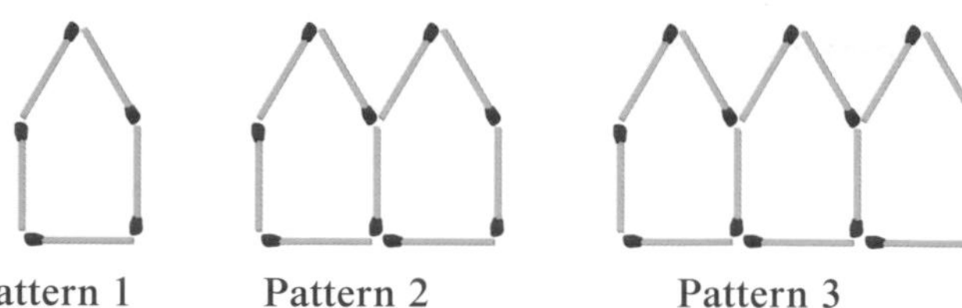

Pattern 1 Pattern 2 Pattern 3

a For sequence C, find a rule for the number of matches (m) in the nth pattern.
Explain your method.

b Calculate the number of matches in the 40th pattern.

c Which pattern uses exactly 129 matches?

3 **Sequence D**

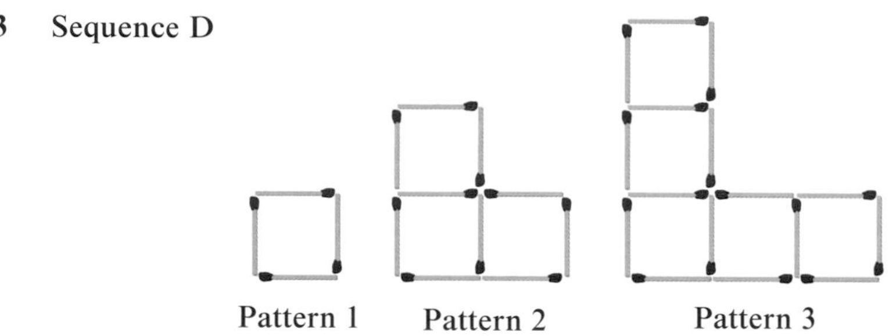

Pattern 1 Pattern 2 Pattern 3

a For sequence D, find a rule for the number of matches (m) in the nth pattern.
Explain your method.

b How many of matches are in the 100th pattern?

4 This mapping diagram fits a sequence of matchstick patterns.

Pattern number (n)		Number of matches (m)
1	→	6
2	→	10
3	→	14
4	→	18
5	→	22

a Draw a sequence of matchstick patterns that fits this mapping diagram.

b Find a rule for the number of matches (m) in the nth pattern.

5 Copy and complete each mapping diagram.

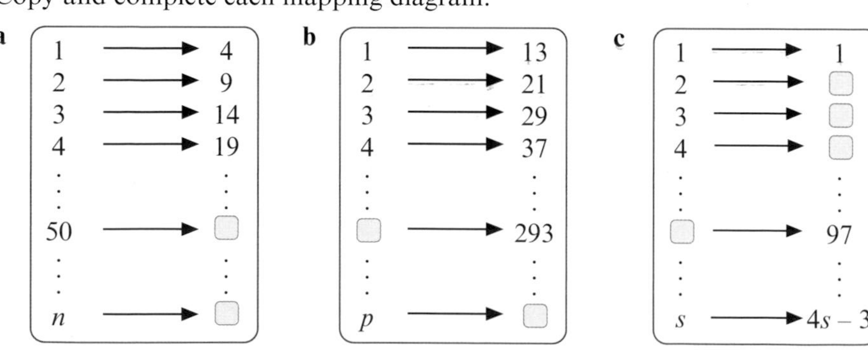

Thinking ahead to ...
finding the nth term

A 7, 9, 11, 13, 15, 17, ...

a Write the next two terms in this sequence.
b Find the 12th term.
c What is the 50th term?

The nth term of a sequence

To find an expression for the nth term in the sequence 7, 9, 11, 13, 15, ...

The 1st term is 7
The 2nd term is 9
The 3rd term is 11
The 4th term is 13
The 5th term is 15
. . .

- These results can be shown in a mapping diagram.

Term number (n)	$2n$	Term
1	2	7
2	4	9
3	6	11
4	8	13
5	10	15

- The first difference for the terms is 2 each time.
So there is a linear expression for the nth term that begins $2n$ ... .

- Each term is 5 more than $2n$.
So an expression for the nth term is $2n + 5$.

Exercise 2.3
Finding the nth term

1 A 6, 9, 12, 15, 18, ... B 1, 6, 11, 16, 21, ...
C 13, 23, 33, 43, 53, ... D 2, 10, 18, 26, 34, ...

For each of the sequences A to D:
a find an expression for the nth term
b use your expression to calculate the 50th term.

2 A student has tried to find the nth term of this sequence.

5, 8, 11, 14, 17, ...
nth term is $n + 3$ ✗

a Explain the mistake you think he has made.
b Find a correct expression for the nth term of this sequence.

3 The 2nd term of a sequence is 7.
Which of these could not be an expression for the nth term?

$3n + 1$ $11 - 2n$ $n + 5$ $n + 7$ $5n - 3$

4 Find an expression for the nth term of the sequence: 20, 18, 16, 14, 12, ...

Extending number patterns

Exercise 2.4
Extending patterns

1 Morag finds an expression for the nth number in this sequence.

4, 10, 18, 28, 40, ...

This is her working:

1st number	4	$= 1 \times 4$
2nd number	10	$= 2 \times 5$
3rd number	18	$= 3 \times 6$
4th number	28	$= 4 \times 7$ ...
So nth number		$= n \times (n + 3)$

a Show Morag's line of working for the 5th number.
b Find the 10th number in this sequence.
c Explain how Morag's working helps to find an expression for the nth number in this sequence.

2 These are the first five triangle numbers:

1, 3, 6, 10, 15, ...

The triangle numbers follow this pattern:

1st triangle number	$1 = \frac{1 \times 2}{2}$
2nd triangle number	$3 = \frac{2 \times 3}{2}$
3rd triangle number	$6 = \frac{3 \times 4}{2}$
4th triangle number	$10 = \frac{4 \times 5}{2}$

a What is the next line in this pattern?
b Use the pattern to find the 12th triangle number.
c Find an expression for the nth triangle number.

3 These are the first five powers of 2:

2, 4, 8, 16, 32, ...

Powers of 2 follow this pattern:

1st power	2		$= 2^1$
2nd power	4	$= 2 \times 2$	$= 2^2$
3rd power	8	$= 2 \times 2 \times 2$	$= 2^3$
4th power	16	$= 2 \times 2 \times 2 \times 2$	$= 2^4$

a What is the next line in this pattern?
b Use the pattern to find the 8th power of 2.
c Write an expression for the nth power of 2.

Writing rules in different ways

Exercise 2.5
Different ways to write rules

1 These hollow square tile designs are the first three in a sequence.

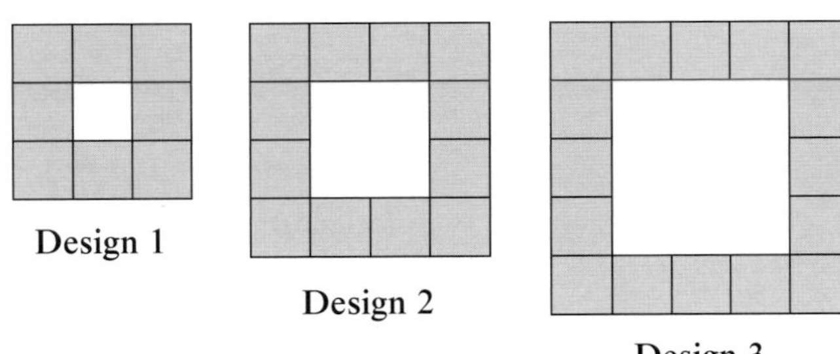

Three students find a rule for the number of tiles in the nth design.
All three students are correct.

Andrew
Number of tiles
= 4(n + 1)
Aisha
Number of tiles
= 4n + 4
Fiona
Number of tiles
= 2(n + 2) + 2n

The way each student wrote their rule shows how they found it.

Andrew drew this diagram to show how he found his rule.

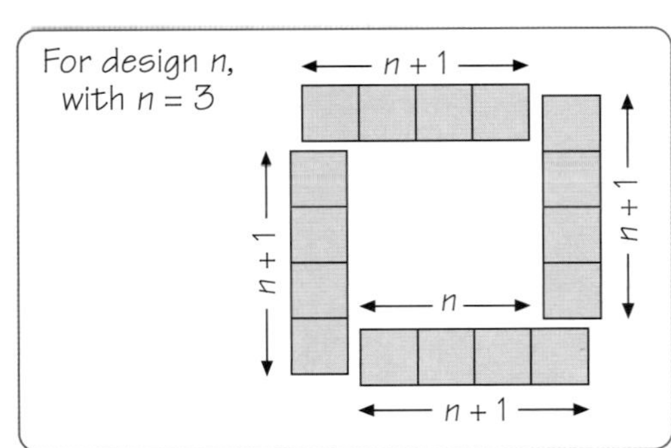

a Draw a diagram for Aisha's rule.
b Draw a diagram for Fiona's rule.
c Show how you can calculate the number of tiles in the 50th design using:
i Andrew's rule **ii** Aisha's rule **iii** Fiona's rule
d Which design has 324 tiles?
Explain how you found your result.

Fiona's rule is not as easy to use as the other two rules.

Using Fiona's rule to find the number of tiles in the 10th design gives:

Number of tiles
$= 2 \times (10 + 2) + (2 \times 10)$
$= 24 + 20$
$= 44$

2 These designs are the first three in a sequence.

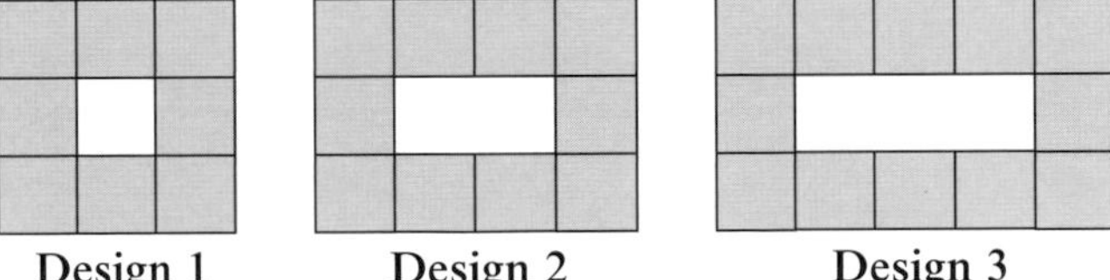

a Find two different ways of writing a rule for the number of tiles in the nth design.
b Explain how you found your results.

End points

You should be able to ... | **... so try these questions**

A Find a rule that fits a sequence of patterns

A1 These are the first three matchstick patterns in a sequence.

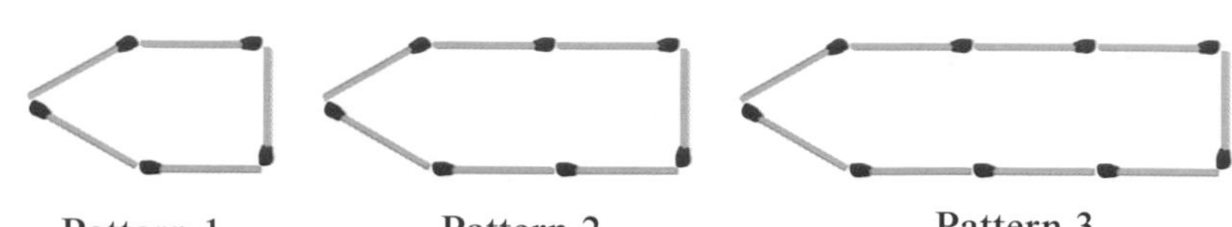

Pattern 1 Pattern 2 Pattern 3

a Find a rule for the number of matches (m) in the nth pattern. Explain your method.
b Use your rule to find the number of matches in the 100th pattern.

B Find rules for mapping diagrams

B1 Copy and complete these mapping diagrams.

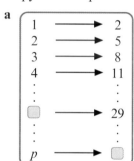

a

1	→	2
2	→	5
3	→	8
4	→	11
⋮		⋮
□	→	29
⋮		⋮
p	→	□

b

1	→	7
2	→	12
3	→	17
4	→	22
⋮		⋮
20	→	□
⋮		⋮
n	→	□

C Find an expression for the nth term in a sequence

C1 A 7, 10, 13, 16, 19, ... B 2, 7, 12, 17, 22, ...
C 5, 9, 13, 17, ... D 6, 7, 8, 9, 10, ...
E 9, 12, 15, 18, 21, ... F 11, 18, 25, 32, ...

For each of the sequences A to F:

a find an expression for the nth term
b use your expression to calculate the 20th term.

Some points to remember

- It is often possible to find a rule for the nth pattern in a sequence of patterns by looking at how each pattern can be made.
- In a sequence: ❖ If the first differences are k each time, there is a simple linear expression for the nth term that begins kn ... .

Starting points

You need to know about ... ... so try these questions

A Naming angles and triangles

- Any angle less than 90° is an **acute angle**.
- Any angle equal to 90° is a **right angle**.
- Any angle between 90° and 180° is an **obtuse angle**.
- Any angle between 180° and 360° is a **reflex angle**.
- Any triangle which has:
  - three sides of equal length
  - three equal angles (60°)

  is an **equilateral triangle**.
- Any triangle which has:
  - two sides of equal length
  - two equal angles

  is an **isosceles triangle**.

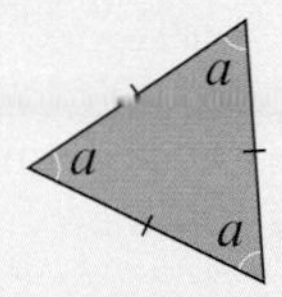

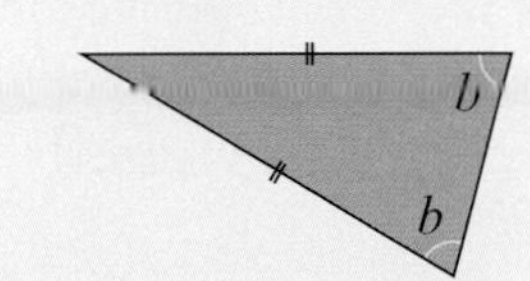

- Any triangle which has no sides of equal length and no equal angles is a **scalene triangle**.
- Any triangle which has one right angle is a **right-angled triangle**.

A1

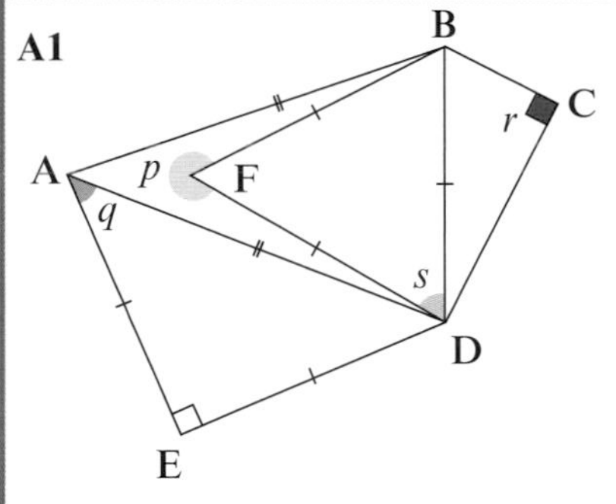

a What type of angle is:
i p **ii** q **iii** r?
b What type of triangle is:
i BFD **ii** BCD?
c Which triangles are isosceles?

B Angle sums

- Angles at a point on a straight line add up to 180°.

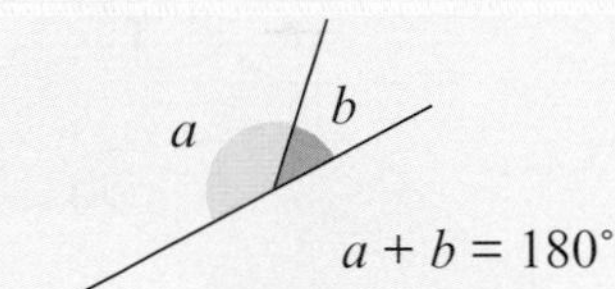

- Angles round a point add up to 360°.

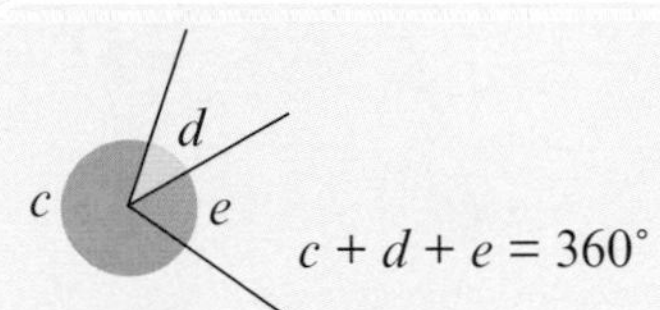

- Vertically opposite angles are equal.

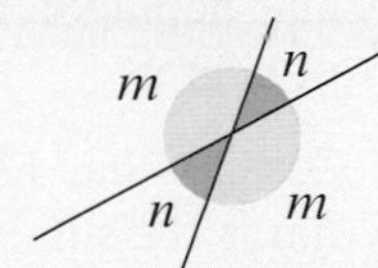

- Angles in a triangle add up to 180°.

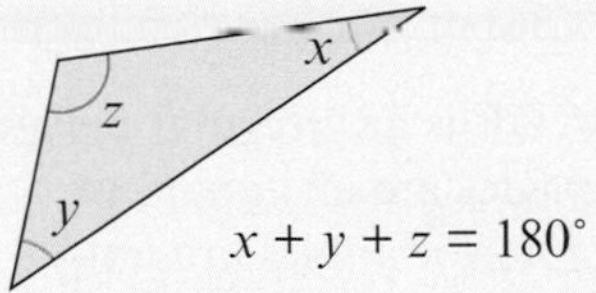

B1 In the diagram above calculate:
a the size of angle s
b angle p
c angle $A\hat{D}E$.

B2 On this diagram, angles marked with the same letter are equal in size.

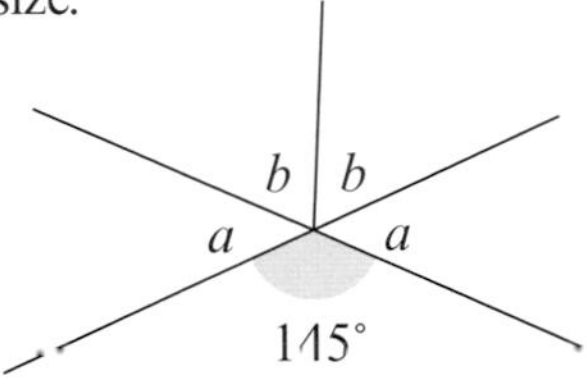

a Work out angles a and b.
b Explain why a triangle can only have one obtuse angle.

C Parallel lines

At each point where a straight line crosses a set of parallel lines there are two pairs of vertically opposite angles.

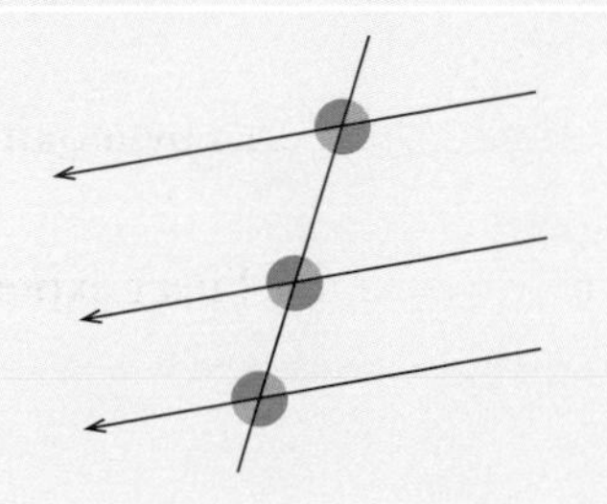

Parallel lines are marked with arrows.

Here equal angles are marked with the same colour.

C1

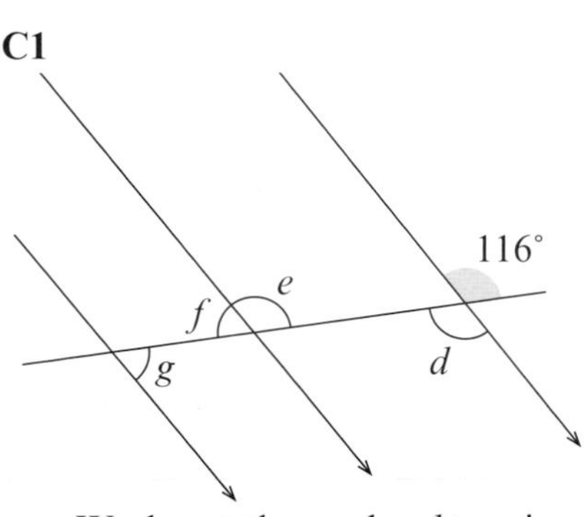

Work out the angles d to g in this diagram.

D Quadrilaterals

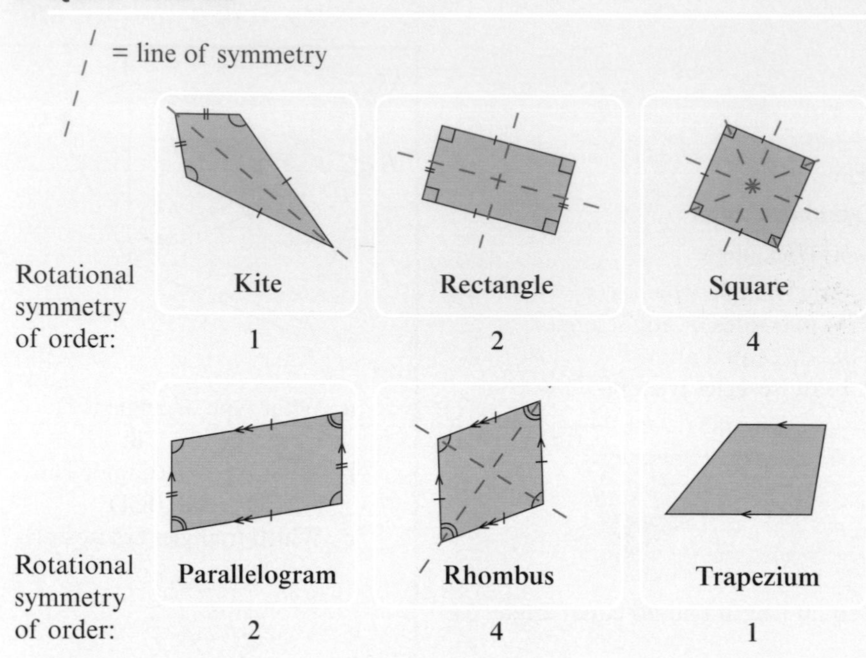

Rotational symmetry of order:

Parallelogram	Rhombus	Trapezium
2	4	1

D1 Name all the quadrilaterals that fit each of these labels.

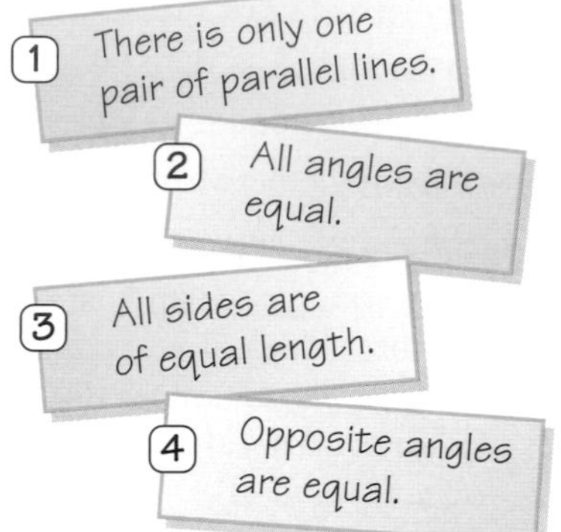

1 There is only one pair of parallel lines.
2 All angles are equal.
3 All sides are of equal length.
4 Opposite angles are equal.

D2 Draw a trapezium with one line of symmetry.

E Polygons

- In ABCDE:
  - the **interior angles** are marked in red
  - the angles marked in blue are not interior angles.

- The **sum of the interior angles** of a polygon with n sides is: $(n - 2) \times 180°$
  So for ABCDE the sum of interior angles is: $(5 - 2) \times 180°$
  $= 3 \times 180°$
  $= 540°$

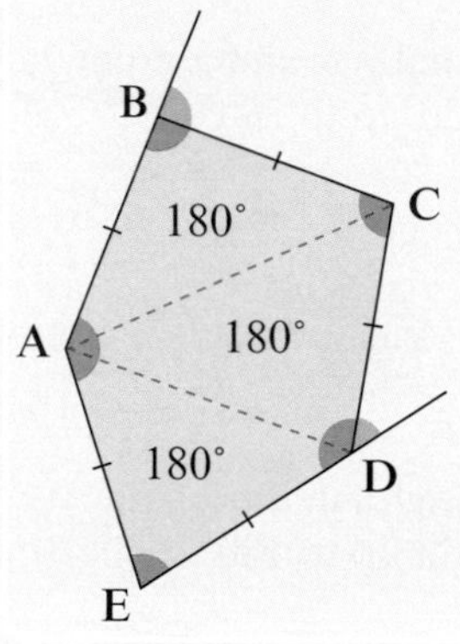

- In a **regular polygon** all the sides are equal and all the interior angles are equal.

- ABCDE is an **irregular polygon** The sides are all equal but the interior angles are not.

Name of polygon	Number of sides	Sum of interior angles	Interior angle of a regular polygon
Triangle	3	180° —— ÷3 ——▶	60°
Quadrilateral	4	360° —— ÷4 ——▶	90°
Pentagon	5	540° —— ÷5 ——▶	108°
Hexagon	6	720° —— ÷6 ——▶	120°
Heptagon	7		
Octagon	8		
Nonagon	9		
Decagon	10		

Another expression for the **sum of the interior angles** of a polygon with n sides is: $(180° \times n) - 360°$

E1 What is the sum of the interior angles of an octagon?

E2 Calculate the angle a in this pentagon.

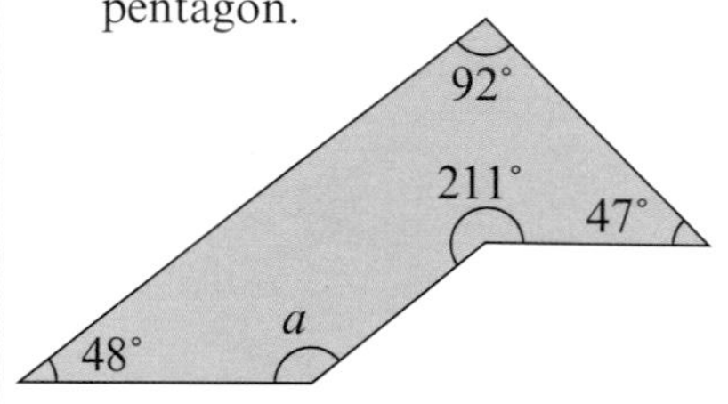

E3 Calculate the interior angle of a regular heptagon to the nearest degree.

E4 A dodecagon has 12 sides.
- **a** What is the sum of the interior angles of a dodecagon?
- **b** Calculate the interior angle of a regular dodecagon.

Angles in triangles

To calculate an angle you may need to work out some other angles first.

Example

Calculate the angle $E\hat{G}F$.

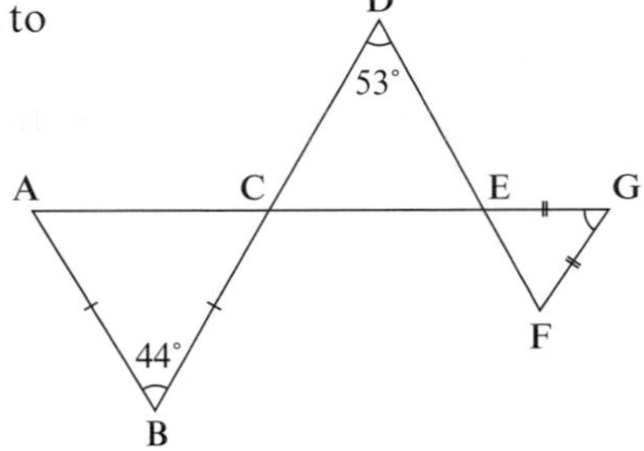

◆ You should sketch a diagram and label each angle that you calculate.

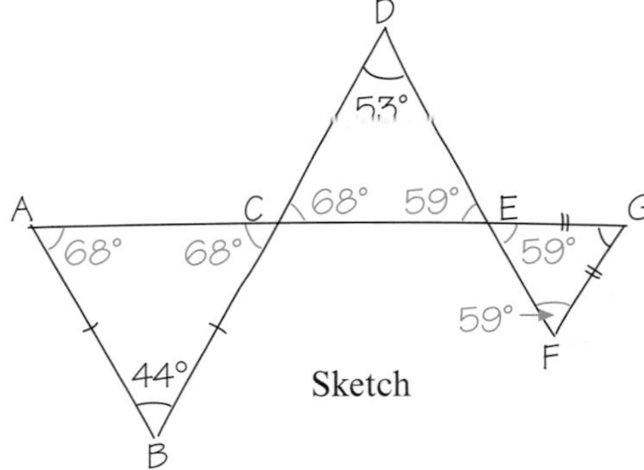

Sketch

Δ stands for triangle

An angle can be written in different ways.

For example:
$D\hat{C}E$ is the same angle as $D\hat{C}G$ and $E\hat{C}D$
$D\hat{E}C$ is the same angle as $D\hat{E}A$ and $C\hat{E}D$.

You may not need to calculate all the intermediate angles.

To calculate $E\hat{G}F$

Calculation …	**… Reason**
$A\hat{C}B = C\hat{A}B$… $= (180° - 44°) \div 2$ $= 68°$	…ABC is an isosceles Δ
$D\hat{C}E = 68°$…	…Vertically opposite ACB
$D\hat{E}C = 180° - (68° + 53°)$… $= 59°$	…Angle sum of Δ
$F\hat{E}G = 59°$…	…Vertically opposite DEC
$G\hat{E}F = E\hat{F}G$…	…EFG is an isosceles Δ
$E\hat{G}F = 180° - (59° \times 2)$…	…Angle sum of Δ

So the angle $E\hat{G}F = 62°$

Exercise 3.1
Angles in triangles

1 **a** Which is the easiest angle to calculate in this diagram?
b Calculate the angles a to f in this diagram.
c In what order did you calculate the angles? Explain why.

106°, a, b, c, d, e, f

2

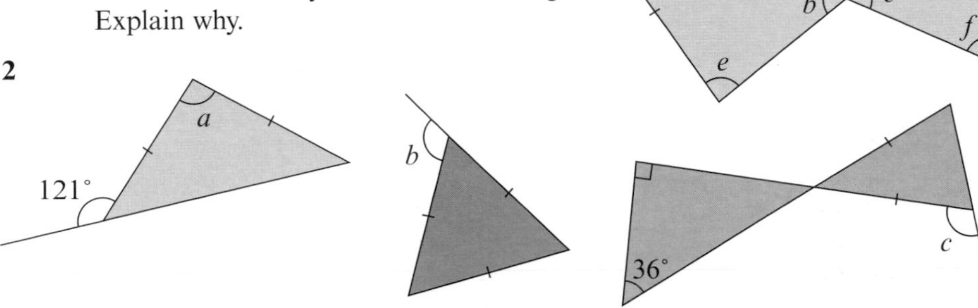

You will need to work out some other angles first.

Calculate the angles a, b and c in these diagrams.
Give a reason for each calculation that you do.

Parallel lines

In each of these diagrams a straight line crosses two parallel lines.

- In each diagram a pair of **corresponding angles** is labelled. Corresponding angles are equal.

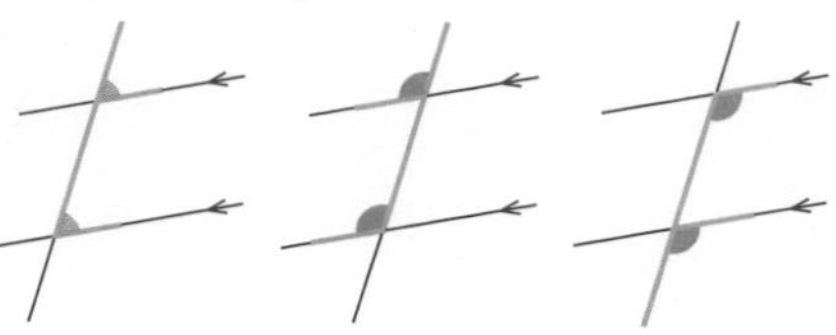

- In each diagram a pair of **alternate angles** is labelled. Alternate angles are equal.

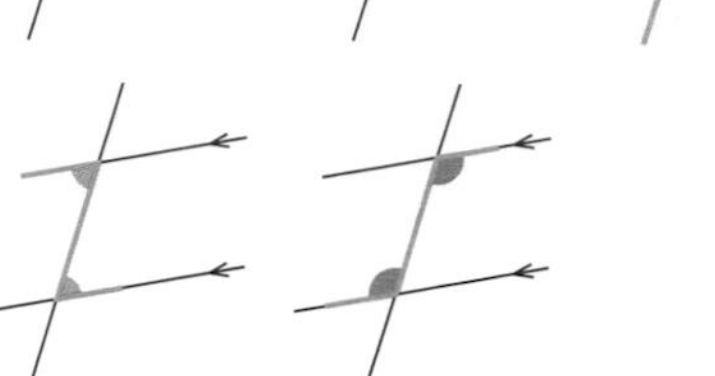

Exercise 3.2
Angles in parallel lines

> To find corresponding angles in a diagram you could look for an F shape which may be upside down and/or back to front.
>
> To find alternate angles in a diagram you could look for a Z shape which may be back to front.

1 **a** List five pairs of corresponding angles in this diagram.
b List three pairs of alternate angles.

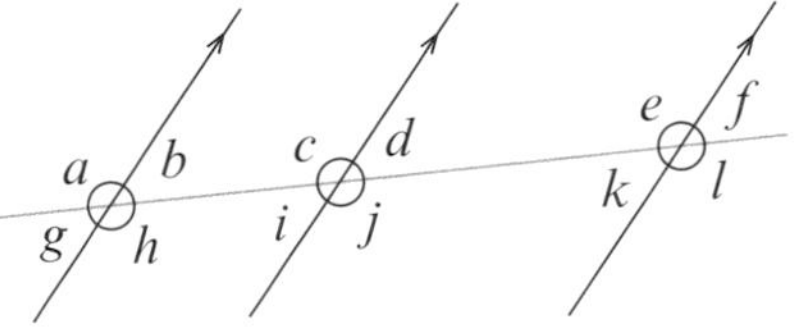

2 Sketch these diagrams.
Work out the angles a, b and c.
You may need to calculate some other angles first.

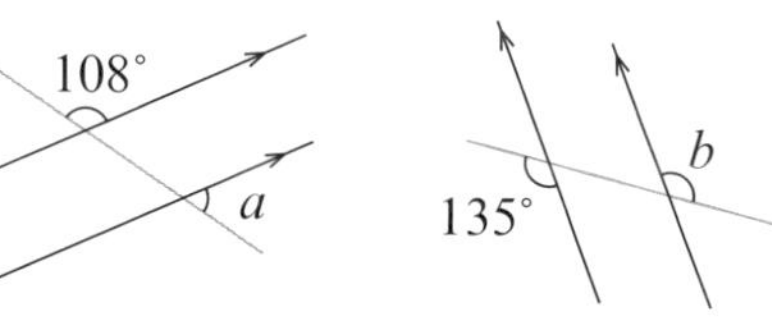

3 In this diagram AS, BR and NQ intersect to make the triangle DPO.
CE, FI, JM and NQ are parallel.

a List three pairs of corresponding angles along the line:
i AS **ii** BR
b Explain why $\hat{DGH}$ and $\hat{DHI}$ are not corresponding angles.
c List three pairs of alternate angles in this diagram.
d Calculate each of these angles.

> Give a reason for each calculation that you do.

i $\hat{HDG}$ **ii** $\hat{ADC}$
iii $\hat{GKL}$ **iv** $\hat{HLK}$
v $\hat{QPL}$ **vi** $\hat{NOR}$

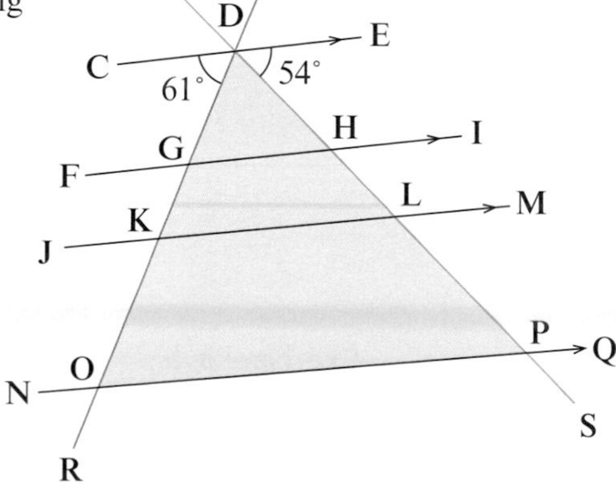

4 In each of these diagrams there is one pair of parallel lines.

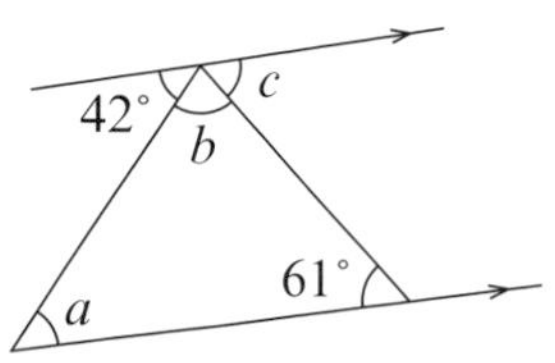

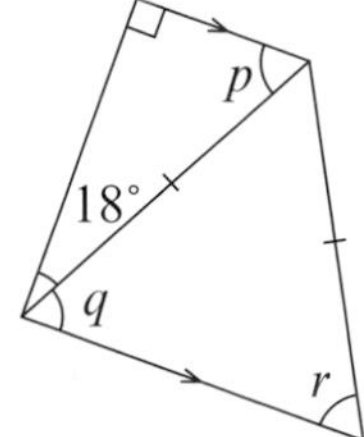

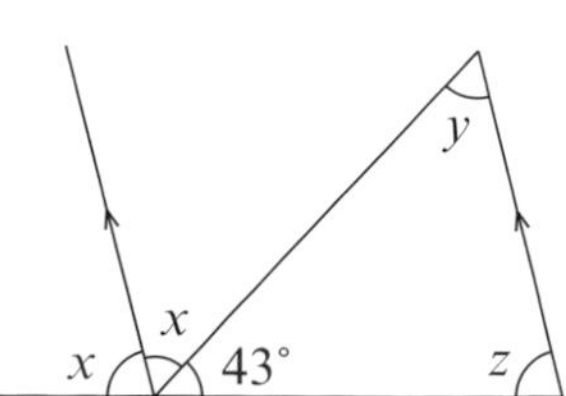

Sketch these diagrams and calculate each of the angles marked with a letter.

Angles in polygons

- At each vertex of a polygon the angle between an extended side and the adjacent side is called **an exterior angle**.

  In ABCDE:
  - the exterior angles are marked in orange
  - the interior angles are marked in blue.

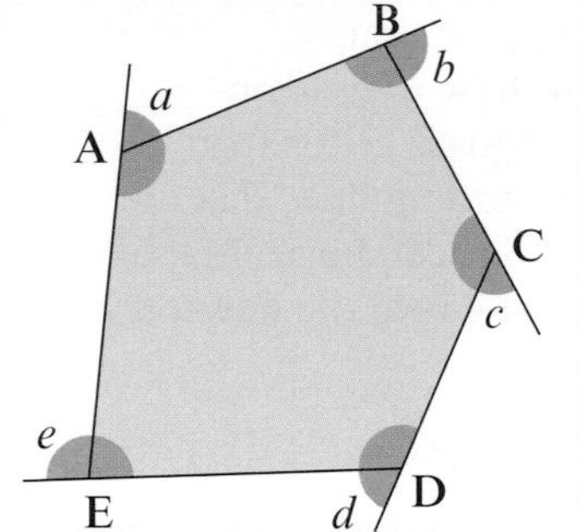

- The sum of the exterior angles of any polygon is 360°.

  In ABCDE:
  $a + b + c + d + e = 360°$

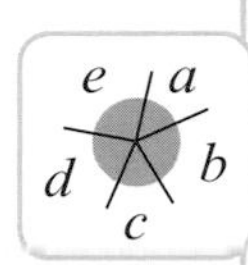

  You can show this by tracing the angles and fitting them together round a point.

- At each vertex the sum of the interior angle and exterior angle is 180°.

Exercise 3.3
Exterior angles of polygons

1 For this polygon:
  a calculate each exterior angle
  b check that the total of the exterior angles is 360°.

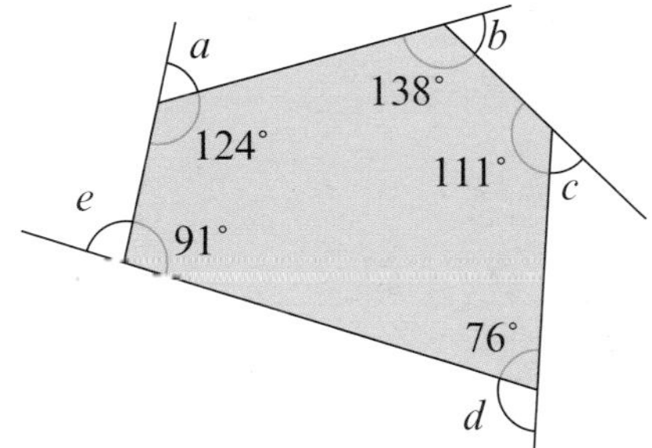

2 A dodecagon has 12 sides.
In this regular dodecagon one side is extended to form the angle p.
  a Explain why the exterior angles of a regular dodecagon are all equal to 360° ÷ 12.
  b Calculate the angle p.
  c Calculate the interior angle of a regular dodecagon.

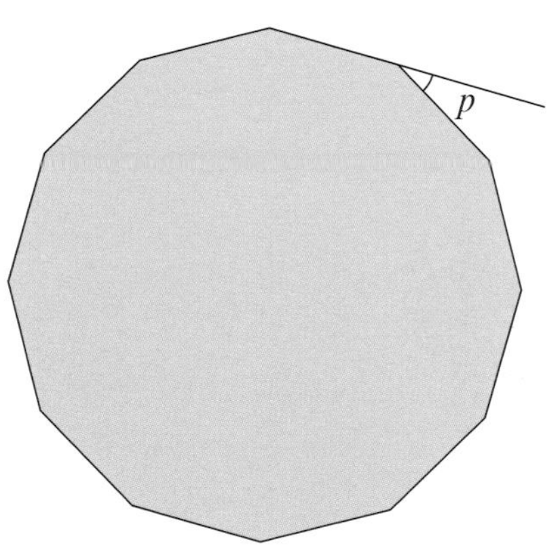

3 This is part of regular nonagon drawn inside a circle with centre C.
One side of the nonagon is extended to form the angle e.
Calculate the angles e to h.

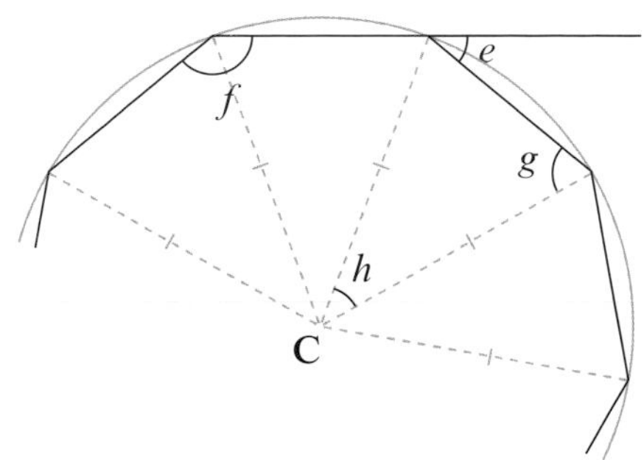

Exercise 3.4
Triangles investigation

You can mark eight points that are equally spaced on the circumference of a circle if you:
- draw a circle on square grid paper
- mark in lines that are vertical, horizontal and at 45° to the horizontal.

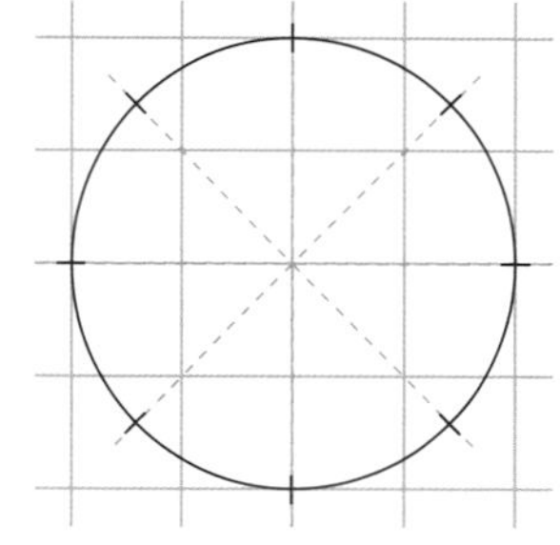

1 This eight-point circle has the points A to H equally spaced on the circumference.
Δ ABD is drawn by joining three of the points.

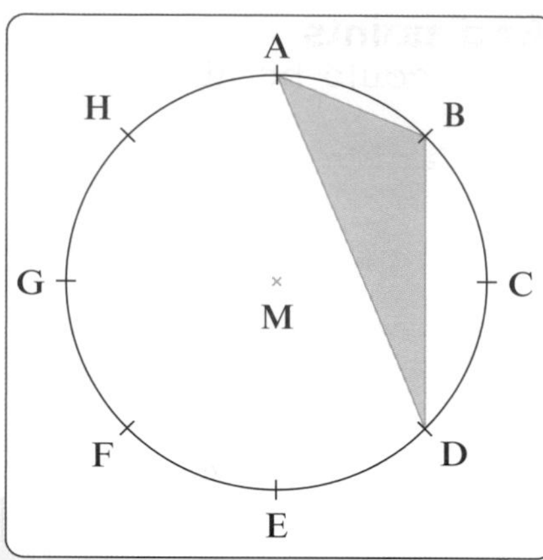

This is how a student calculated the exterior angle at A for Δ ABD.

To find the exterior angle at A

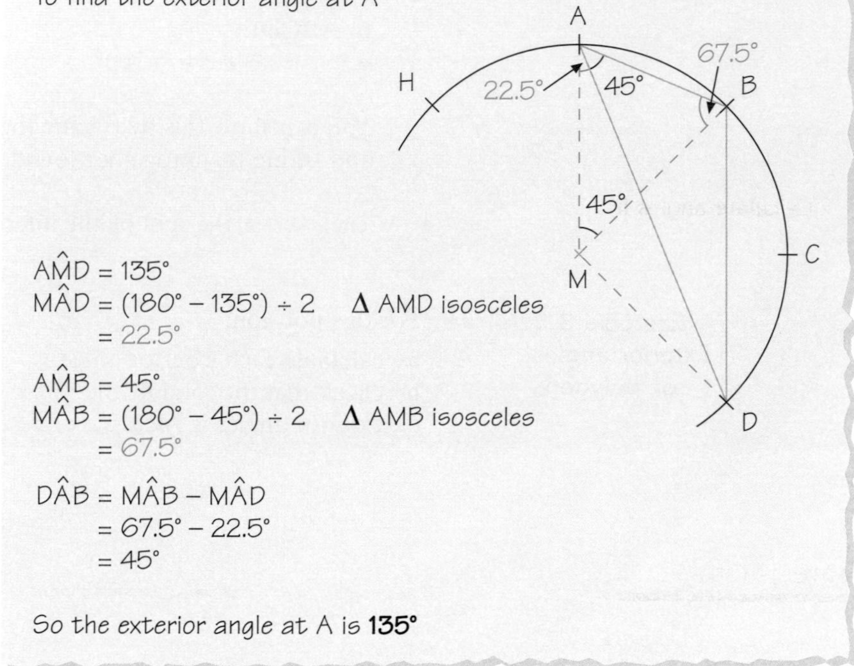

$A\hat{M}D = 135°$
$M\hat{A}D = (180° - 135°) \div 2$ Δ AMD isosceles
$= 22.5°$

$A\hat{M}B = 45°$
$M\hat{A}B = (180° - 45°) \div 2$ Δ AMB isosceles
$= 67.5°$

$D\hat{A}B = M\hat{A}B - M\hat{A}D$
$= 67.5° - 22.5°$
$= 45°$

So the exterior angle at A is **135°**

a Explain why $A\hat{M}D$ is 135°.
b For triangle ABD:
  i calculate the exterior angles at B and D
  ii check that the total of the exterior angles is 360°.

2 Triangle ACF is also drawn on an eight-point circle.

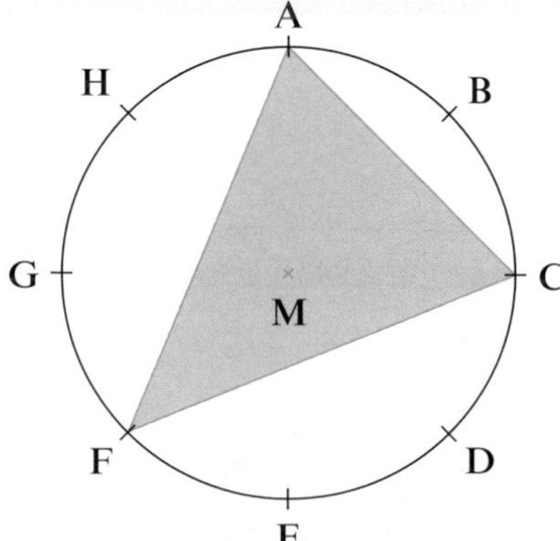

a For triangle ACF:
  i calculate each interior angle
  ii calculate each exterior angle.
b How many different triangles is it possible to draw in an eight-point circle?
c What different exterior angles are possible for triangles drawn on an eight-point circle?

Do not count any that are reflections or rotations of another polygon.

End points

You should be able to ...	... so try these questions
A Calculate angles in parallel lines	**A1** Calculate the angles a, b and c in this diagram.
B Use the properties of polygons	**B1** Polygons A to E are drawn on an equilateral grid. Which of these polygons: **a** is a regular polygon **b** has only one obtuse angle?
C Calculate angles in polygons	**C1** For each of the polygons B and E: **a** calculate the interior angles **b** calculate the exterior angles. **C2** What is the exterior angle of a regular octagon?

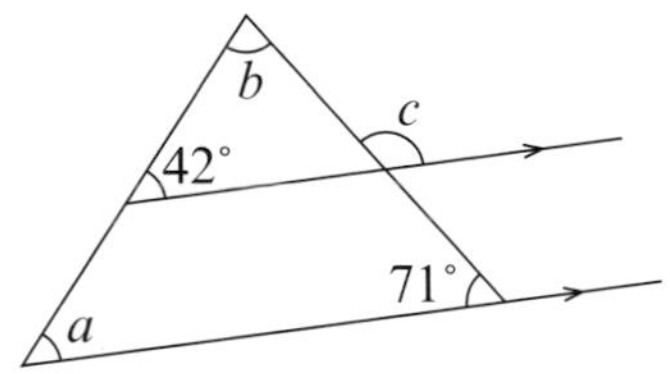

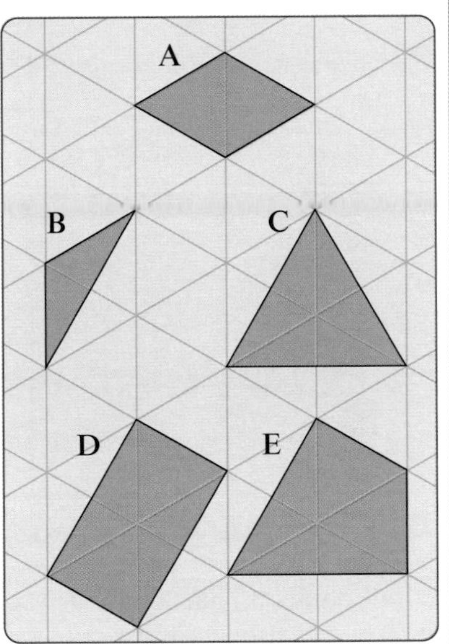

Some points to remember

- For a polygon with n sides:
  - the sum of the interior angles is $(n - 2) \times 180°$
  - the sum of the exterior angles is 360°
  - each exterior angle of a regular polygon is $360° \div n$.
- Examples of quadrilaterals

	Square	Rectangle	Kite	Rhombus	Parallelogram
The diagonals:					
◆ bisect the interior angles	✓	✗	✗	✓	✗
◆ bisect each other	✓	✓	✗	✓	✓
◆ intersect at 90°.	✓	✗	✓	✓	✗
Number of lines of symmetry	4	2	1	2	0
Order of rotational symmetry	4	2	1	2	2

Starting points

You need to know about ... / ... so try these questions

A Graphs of vertical and horizontal lines

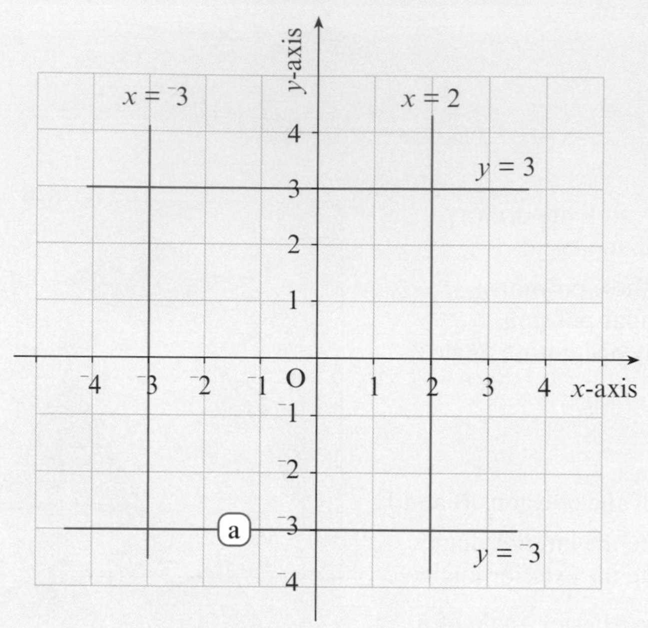

The equation of **line a** is $y = {}^{-}3$ as:
all points on **line a** have a y-coordinate of $^{-}3$.
The equation of the x-axis is $y = 0$ and
the equation of the y-axis is $x = 0$.

A1 **a** Draw a pair of axes.
b Label the x-axis.
c Label the y-axis.
d Draw and label the line $y = 2$ and the line $x = {}^{-}1$
e Give the coordinates of where lines $y = 2$ and $x = {}^{-}1$ cross.

A2 **a** Give the coordinates of where the lines $x = 0$ and $y = 2$ cross.
b Will the line $y = 2$ cross the line $y = {}^{-}2$?
Explain your answer.
c What can you say about all the points on the line $x = 0$?

A3 Will each of these lines be vertical or horizontal?

a $y = 4$ **b** $x = {}^{-}2$
c $x = 3$ **d** $y = {}^{-}8$
e $y = 0$ **f** $x = 0$

B A graph from a table of values

This table of values shows how values of x and y are linked by the equation $y = x + 1$

x	$^{-}1$	0	1	1.5	2
y	0	1	2	2.5	3

From the table of values:
the points

($^{-}1$, 0)
(0, 1)
(1, 2)
(1.5, 2.5)
(2, 3)

can be plotted and joined for the graph.

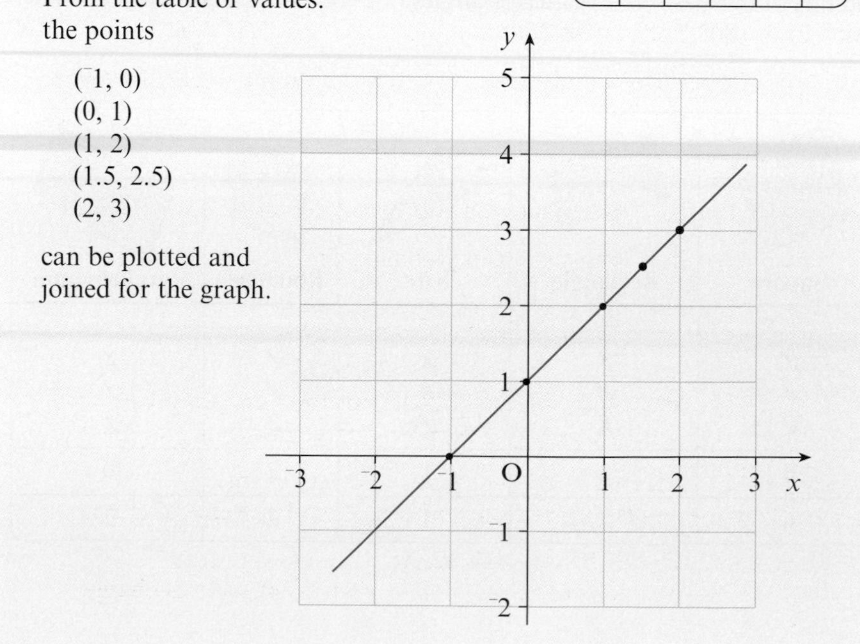

B1

x	$^{-}2$	$^{-}1$	0	1	2	3
y						

Copy and complete this table of values for each equation.

a $y = 3x - 1$
b $y = x + 2$
c $y = x$
d $y + 3 = x$
e $y = 1.5x + 1$

B2 For each equation in **B1** draw and label a graph.

B3 Do not draw a graph.
Which of these points lie on the line $y = 2x - 3$?

a ($^{-}2$, $^{-}7$) **b** (0, 3)
c (2, 1) **d** (3, 3)
e (0, 1.5) **f** ($^{-}15$, $^{-}27$)
g ($^{-}0.25$, $^{-}3.5$) **h** (4, 4)

Linear graphs

Exercise 4.1
Interpreting linear graphs

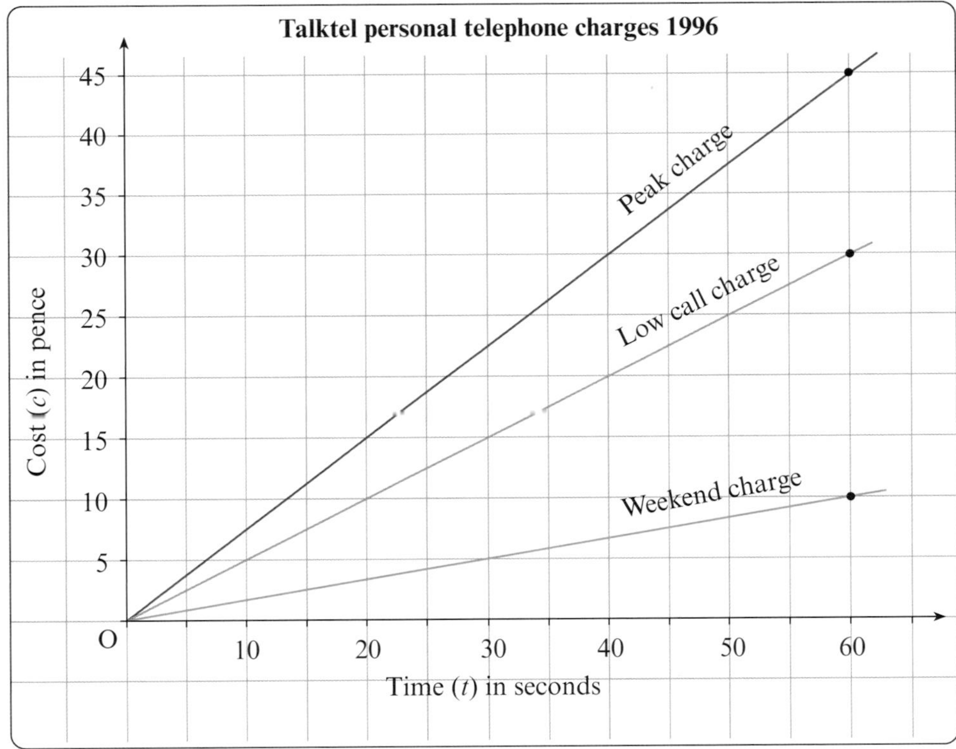

Linear is another word for straight.

A linear graph is a straight line graph.

Each linear graph shows a different charge made by Talktel.
Each call is timed (in seconds), and the cost is calculated for the customer bill.

1 Talktel's peak charge is 45 pence per minute.
- **a** From the graph, what is the low call charge per minute?
- **b** What is the weekend charge per minute?

2
- **a** Does a 35 second call at peak charge cost more or less than 25 pence?
- **b** Estimate the cost of this call.

3 Estimate the cost of a 45 second call at peak charge.

4 Which is cheaper: 15 seconds at peak or 20 seconds at low call charge? Explain your answer.

5 Jess paid 7.5 pence for a 15 second call. Which charge was used for this?

6 At the weekend charge, estimate the cost of a 2 minute 35 second call.

7 At peak charge, how long a call can you make for 75 pence?

8
- **a** Copy the graph for the Talktel charges in 1996.
- **b** Add a line to show the Infotel charge of 32 pence per minute.
- **c** The line graph for which charge is the steepest?
- **d** List the charges in order of the steepness of their line graphs.
- **e** Describe any link you can spot between charge rates and steepness.

In the formula $c = t \div 6$:

c is the cost in pence

t is the time in seconds.

9 Talktel use the formula $c = t \div 6$ for their weekend charge.
Use the formula to find the cost of a 96 second call at weekend charge.

10
- **a** Write a formula for the low call charge.
- **b** Use your formula to find the cost of $1\frac{1}{2}$ minutes at low call charge.

The gradient of a linear graph

- The gradient of a linear graph is a measure of how steep the line is.
- The gradient of a linear graph is the same for any part of the line.
- The gradient of a linear graph is given by: $\frac{\text{Change along the } y\text{-axis}}{\text{Change along the } x\text{-axis}}$

The change along an axis can be an **increase** or a **decrease**.

For a gradient we need to look at what happens for an **increase** along the x-axis.

When:
an **increase** along the x-axis
gives
an **increase** along the y-axis

We say the gradient of the line is **positive**.

For example:
To find the gradient of this linear graph:

- choose two points on the line, e.g. A and B
- along the y-axis the change is 6 units (from 8 to 14)
- along the x-axis the change is 2 units (from 2 to 4).

The gradient is given by:

$$\frac{6}{2} = 3$$

The gradient of the line is 3.

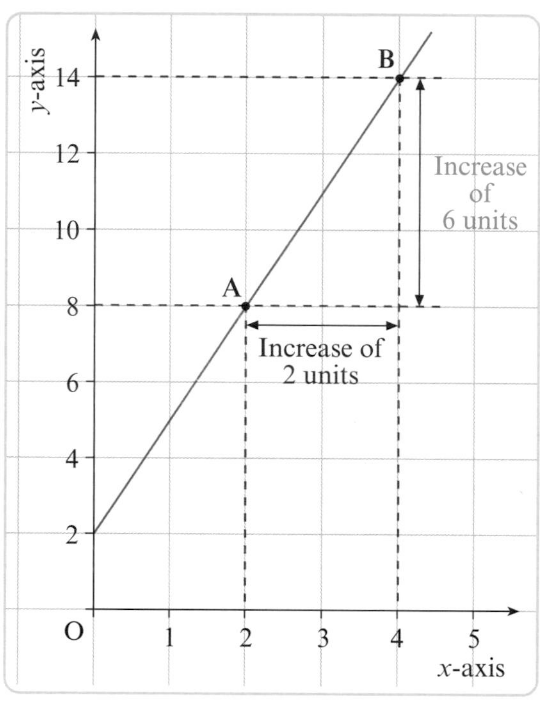

Exercise 4.2
Gradients of linear graphs

1 Calculate the gradients of lines **a** to **h**.

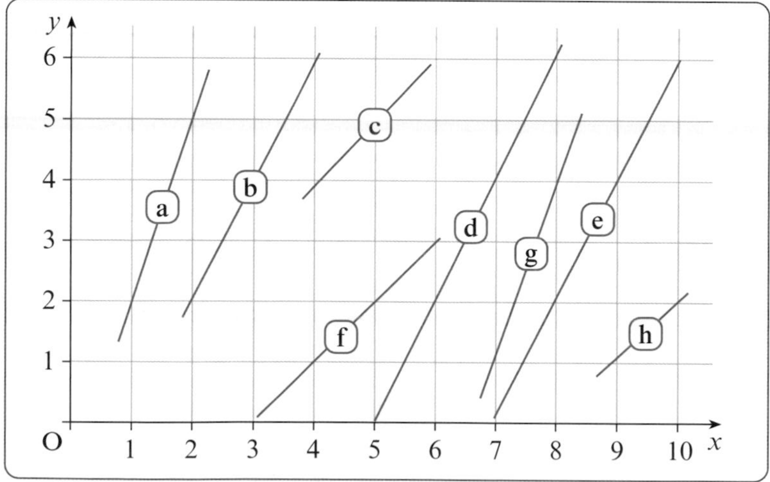

2 **a** Draw a pair of axes: x-axis from 0 to 10, y-axis from 0 to 10.
b On your axes draw and label lines with each of these gradients:
5 6 4 1 2 0.5

3 How are the gradient of a linear graph and its steepness linked?

The gradient of a linear graph has a value that can be given:

- as a whole number
- or as a fraction (in its lowest terms)
- or as a decimal.

For example:

The gradient of a linear graph is given by:

$$\frac{\text{Change in } y\text{-coordinates}}{\text{Change in } x\text{-coordinates}}$$

The linear graph **a** has a gradient given by:

$\frac{3}{1}$ (increase of 3) (increase of 1)

$= 3$ (whole number)

The linear graph **b** has a gradient given by:

$$\frac{5}{2}$$

$\frac{5}{2}$ is a fraction in its lowest terms

but $\frac{5}{2} = 2\frac{1}{2}$

The linear graph **c** has a gradient given by:

$\frac{2}{4}$ or 0.5 (decimal)

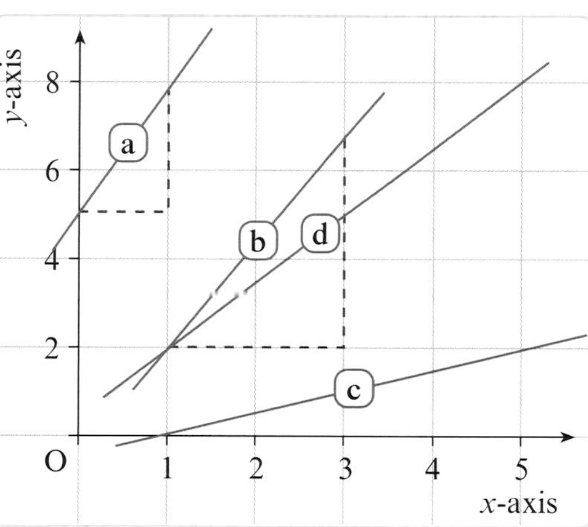

The linear graph **d** has a gradient of:

$\frac{6}{4} = \frac{3}{2}$ (in its lowest terms)

Exercise 4.3
Calculating gradients

1

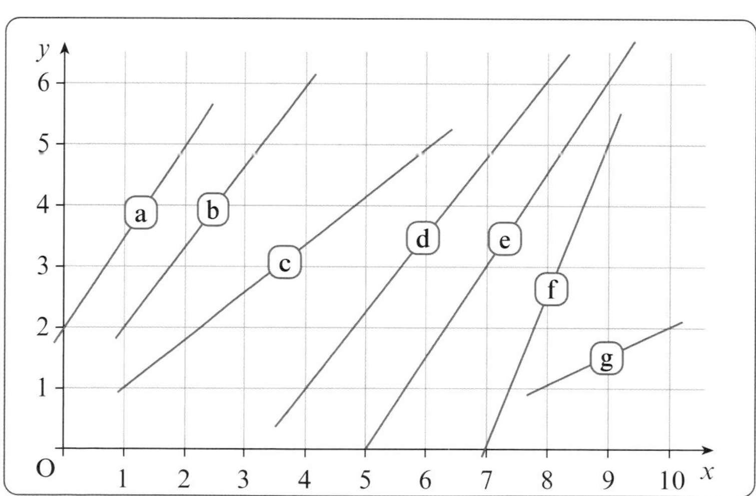

a Give the gradient of each linear graph **a** to **g**, as:
i a fraction in its lowest terms
ii a decimal value.

b List the lines with their gradients in order of steepness. Start with the steepest.

2 What can you say about the gradients of lines which are parallel? Explain your answer with an example and a diagram.

Drawing linear graphs from their equation

As the equation of a line tells us about:

- the gradient of the line
- the y-intercept

it is possible to draw a linear graph just by using the data in the equation.

Example

To draw a linear graph of the equation $y = 3x - 4$

We can tell that the gradient of the line is 3 and the y-intercept is $^-4$.

To draw the graph of $y = 3x - 4$:

- Draw a pair of axes.

- **Step 1** Mark the y-intercept, $^-4$.
- **Step 2** From the y-intercept draw in the gradient 3.
- **Step 3** Join the points with a line and label.

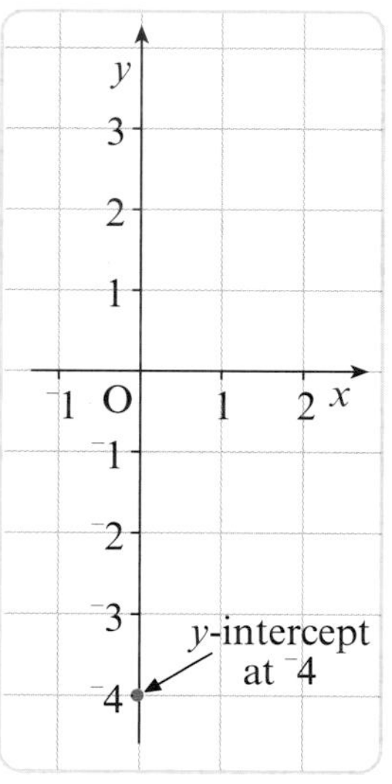

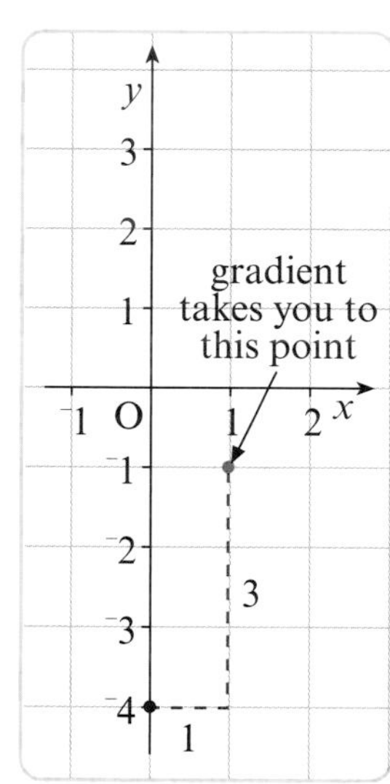

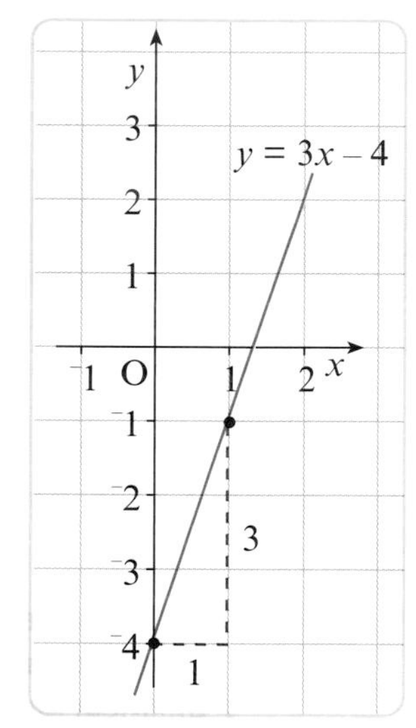

Exercise 4.8
Drawing linear graphs from equations

1 **a** Draw a pair of axes. Use x from $^-1$ to 2, and y from $^-5$ to 5.
b On your axes draw and label graphs of these lines:
$y = 4x - 3$ $\quad y = 2x + 1$ $\quad y = x - 1$

2 **a** Draw a pair of axes. Use x from $^-1$ to 4, and y from $^-4$ to 6.
b On your axes draw graphs of these lines:
$y = 3x - 3$ $\quad y = x + 2$ $\quad y = 3x$
c Which two of these lines go through the point (1, 3)?
d Which of these lines go through the point (3, 5)?

3 **a** What is the gradient of the line $y = 2 - 3x$? What is the y-intercept?
b Draw the graph of $y = 2 - 3x$.

Exercise 4.9
Building equations from graphs

1 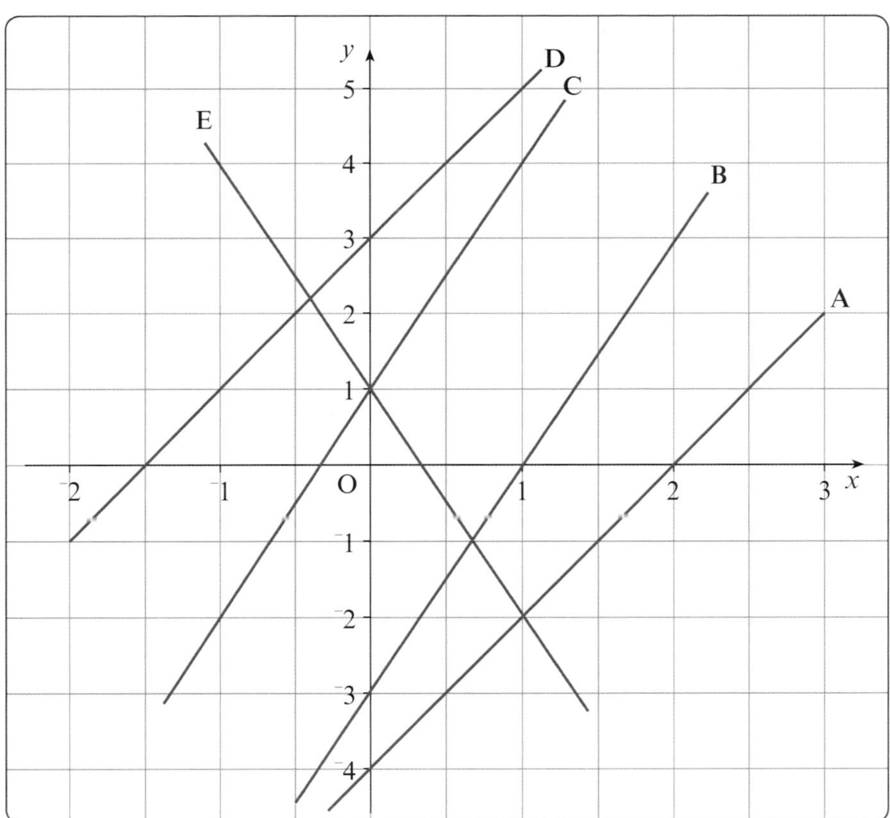

a What is the gradient of line A?
b What is the y-intercept for line A?
c Write an equation for line A.

2 **a** What is the y-intercept for line B?
b Write an equation for line B.

3 Write an equation for each of the lines C, D, and E.

Writing equations in the form $y = mx + c$

'Rewrite an equation' is not the same as 'change an equation'.

An equation, and its rewritten form give the same data, we say they are equivalent.

When the equation of a linear graph starts $y = \ldots$, it is easy to draw a graph. Sometimes you will have to rewrite an equation to make it read $y = \ldots$
One way to rewrite an equation is to divide each term by the same number.

Examples

To rewrite the equation: $3y = 6x - 3$

divide each term by 3

$$y = 2x - 1$$

as $3y \div 3 = y$, $6x \div 3 = 2x$
and $^-3 \div 3 = {}^-1$

To rewrite the equation: $5y = 3x + 10$

divide each term by 5

$$y = \frac{3}{5}x + 2$$

as $5y \div 5 = y$, $3x \div 5 = \frac{3}{5}x$
and $10 \div 5 = 2$

Exercise 4.10
Rewriting equations

1 **a** To rewrite the equation $4y = 8x - 20$, to make it read $y = \ldots$, what number will you divide each term by?
b Rewrite $4y = 8x - 20$, to make it read $y = \ldots$.

2 a To rewrite $2y = 3x + 4$ in the form $y = \dots$, what will you divide each term by?
b Rewrite $2y = 3x + 4$ in the form $y = \dots$.

3 Rewrite each of these equations in the form $y = \dots$.

a $2y = 4x + 8$ b $3y = 4x - 9$ c $3y = 3x - 6$ d $5y = 3x$
e $5y = 2x + 5$ f $2y = 6x - 4$ g $2y = 3 - 4x$ h $4y = x$
i $4y = 4 - 4x$ j $6y = 12 + 12x$ k $5y = x + 1$ l $2x = 3y$

4 Which of these equations can be rewritten as $y = 2x - 3$?

a $2y = 3x - 6$ b $4y = 8x - 16$ c $3y = 6x - 9$ d $2y = 4x$

5 Write three different equations that can be rewritten as $y = 3x - 1$.

Rewriting equations and drawing graphs

When you divide your equation, the gradient can work out as a fraction.

A fraction is easier to work with for the gradient than a decimal value.

Example

Draw the graph of $3y = 4x - 6$
One way is with the equation in the form $y = \dots$.

- Rewrite the equation
  $3y = 4x - 6$
  Divide each term by 3
  $y = \frac{4}{3}x - 2$
- Draw the graph from the equation:
  gradient $\frac{4}{3}$
  y-intercept, $^{-}2$
- Label the graph with its equation $3y = 4x - 6$.

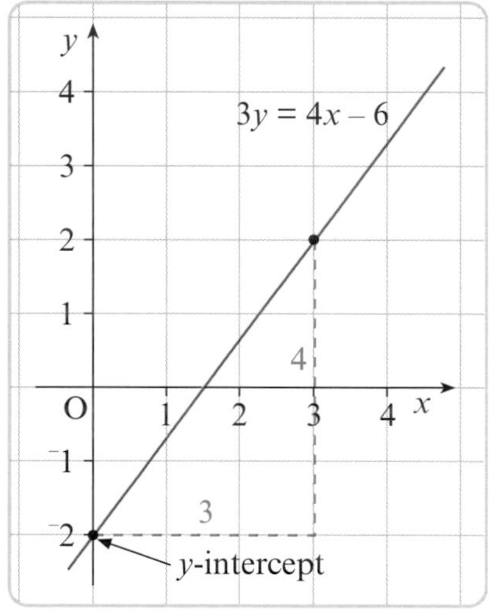

Exercise 4.11
Drawing graphs

1 For the graph of $2y = 3x + 2$:
a Rewrite the equation in the form $y = \dots$.
b What is the gradient of the line?
c What is the y-intercept?
d Draw a pair of axes: use x from $^{-}1$ to 3 and y from $^{-}1$ to 5.
e Draw the graph of $2y = 3x + 2$.

2 For the graph of $3y = 2x - 3$:
a What is the gradient of the line?
b What is the y-intercept?
c Draw a graph of $3y = 2x - 3$.

3 Draw a graph of $4y = 8x - 12$.

4 a Draw a pair of axes, use x from $^{-}1$ to 5 and y from $^{-}4$ to 4.
b On your axes draw a graph for each of these:
$2y = x$ $4y = x + 4$ $4y = 5x - 12$
c Give the coordinates of a point where all three lines cross.

5 Draw a graph of $3y = 4x - 2$.

6 Is the point (2, 3) on the graph of $2y = 5x - 4$?
Explain your answer.

Linear graphs that cross

Any linear graph is part of a straight line drawn through every point that fits its equation.

Where two graphs cross, the point must be on both lines.

For example, at A the graphs of:

$$y = 2x - 2$$

and

$$2y = x + 2$$

both have a point of (2, 2)

We can say that at (2, 2) the equations must both be true for these values of x and y.

You can test this by using the values of the x- and y-coordinates in each equation.

At A: $x = 2$ and $y = 2$

$y = 2x - 2$ is true as:

$y = 2$ and $2x - 2 = 2$

$2y = x + 2$ is true as

$2y = 4$ and $x + 2 = 4$

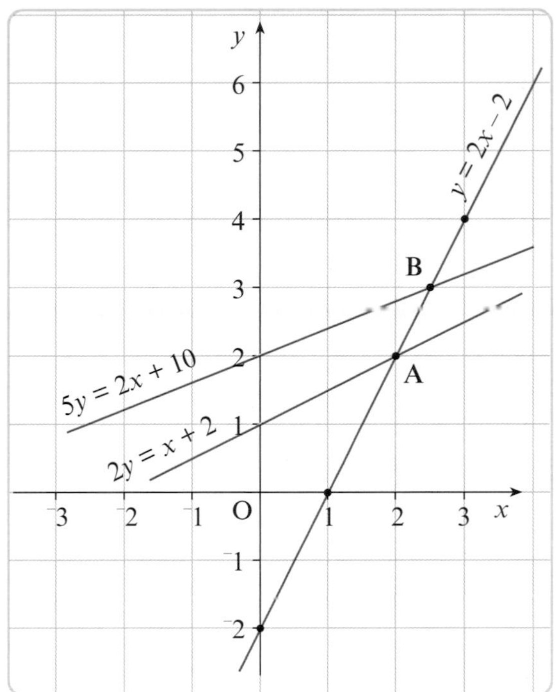

Exercise 4.12
Crossing points

1 a Give the coordinates of point B.
  b Give the equations of two lines that go through B.
  c Test that the values for x and y at B are true for the equations.

2 a On a pair of axes draw the graphs of:
  $y = x + 1$ and $y = 2x - 1$
  b Give the coordinates of a point that both lines are drawn through.
  c Test that the equations are true at this point.

3 As lines C and D cross, they must have one point that is the same.

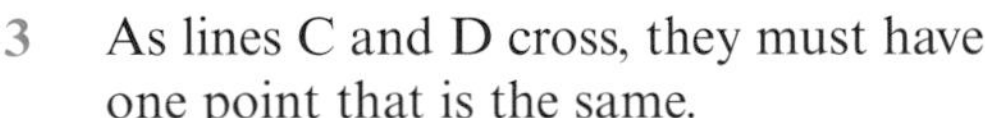

  a Estimate the coordinates where lines C and D cross.
  b Test these values on each equation.
  c Try other estimates to see if you can find the exact coordinates.

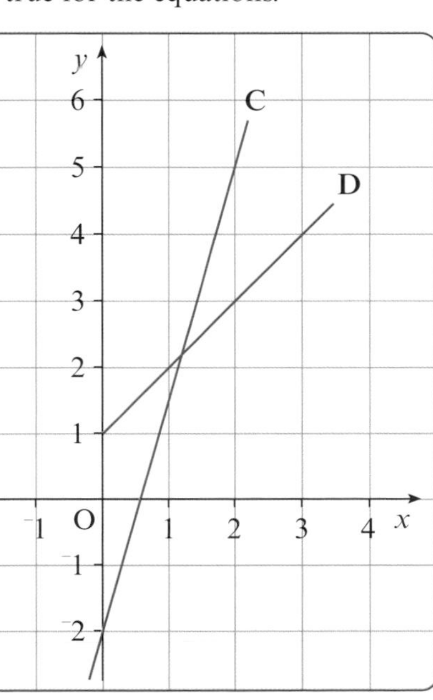

End points

You should be able to … … **so try these questions**

A Explain gradients of linear graphs

A1 **a** In words, and with a diagram, describe a positive gradient.
b In words, and with a diagram, describe a negative gradient.
c What does the gradient of a line tell you:
- the length of the line
- the colour of the line
- something else?

Explain your answer.

B Give the value of the gradient of a linear graph

B1 Give the gradient of line D.

B2 Give the gradient of line A
a as a fraction
b as a decimal.

B3 Give the gradients of lines B and C.

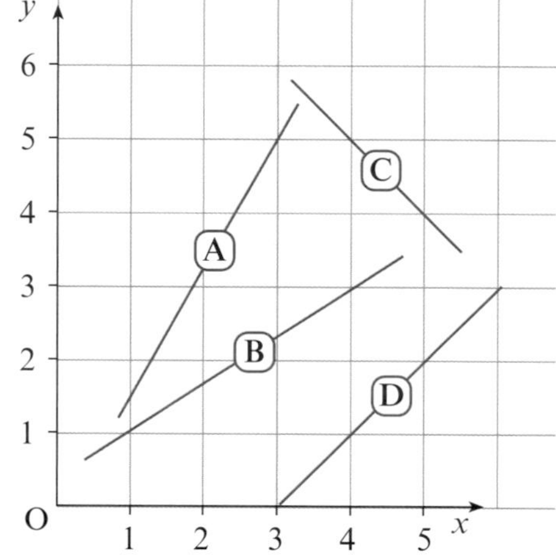

C Use a table of values

C1 Draw up a table of values for the graph of $y = 3x - 4$
Use values of x from $^{-}1$ to 3.

C2 Draw the graph of $y = 3x - 4$.

D Rewrite equations to read $y =$

D1 Change the equation $4y = 3x + 6$ to read $y = \ldots$.

E Use $y = mx + c$

E1 For this line, give the gradient and the y-intercept.

$$y = 2x - 5$$

E2 Draw the graph of $y = \frac{1}{2}x + 1$.

E3 Draw a graph of $3y = 5x - 6$.

F Test to see if a line goes through a point

F1 Which of these lines pass through the point (3, 4)?

$3y = 2x + 6$ $\qquad y = 2x - 2$ $\qquad 3y = 4x + 1$

Some points to remember

- $y = mx + c$ is the general form of a linear equation where, m is the gradient and c the y-intercept.
- Rewriting a linear equation is not the same as changing the equation.

Starting points

You need to know about ... ... so try these questions

A Types of data

- There are two main types of data:
  - data that is divided into **categories**, such as:
    make of car,
    colour of car.
  - data that is **numerical**, such as:
    number of people in car,
    length of car.

TRAFFIC SURVEY

Car	Make of car	Number of people in car
A	Peugeot	1
B	Ford	3
C	Vauxhall	1
D	Ford	2
E	Rover	2

NUMBER OF CHILDREN IN CAR

1 0 1 2 1 1 0 0
4 1 0 1 0 0 2 0
2 0 3 0 0 2 0 1

COLOUR OF CAR

Red	Blue	Green	White
Silver	Red	White	Black
Black	Green	Red	Blue
Black	Blue	Blue	Silver
Red	White	Blue	Brown
Blue	Black	Brown	Red

B Presenting data in frequency tables

Make of car	Tally	Number of cars
Ford	~~||||~~ ||||	9
Rover	~~||||~~		6
Vauxhall	||	2
Others	~~||||~~ ||	7

Number of people in car	Frequency
1	3
2	10
3	6
4	3
5	2

B1 Use a tally to present the number-of-children-in-car data in a frequency table.

B2 Present the colour-of-car data in a frequency table.

C Diagrams that present one set of data

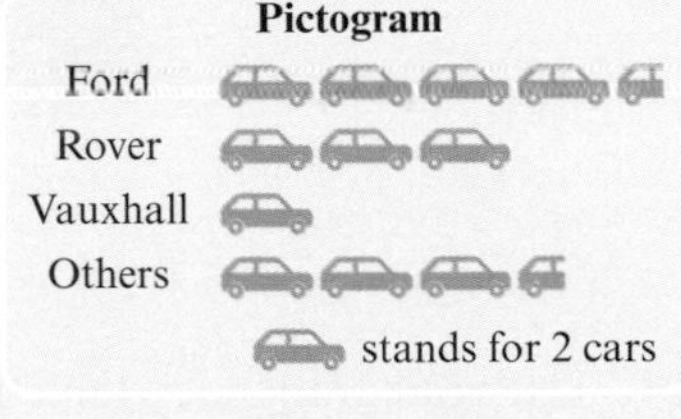

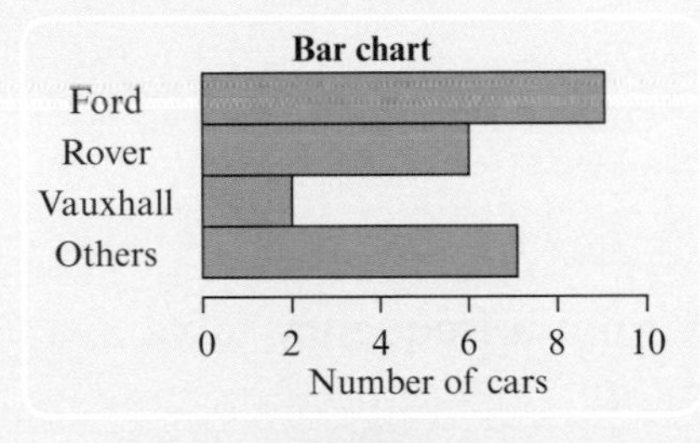

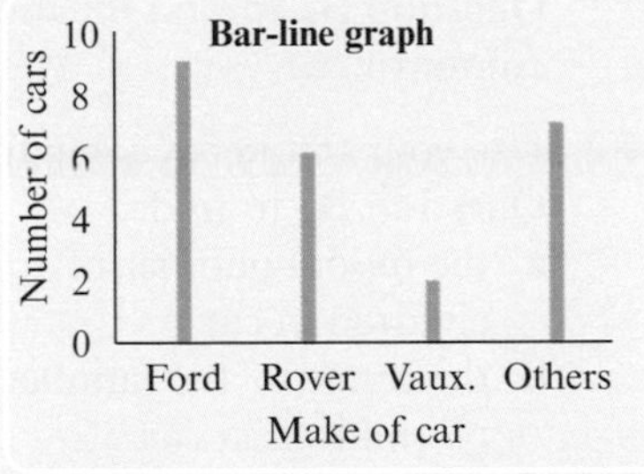

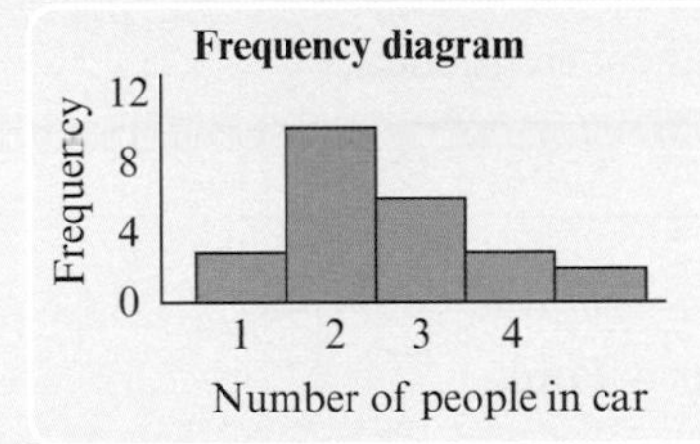

C1 Draw a bar-line graph to show the colour-of-car data.

C2 Draw a frequency diagram to show the number-of-children-in-car data.

C3 Draw a pictogram to show the colour-of-car data.

C4 Draw a bar chart to show the colour-of-car data.

D Constructing a pie chart

- To calculate the percentage for each category.
  For Vauxhall, there are 2 cars out of 24 so
  $100 \div 24 \times 2 = 8.3\%$ (to 1 dp).

	No. of cars	%
Ford	9	37.5
Rover	6	25
Vauxhall	2	8.3
Others	7	29.2
Totals	24	100

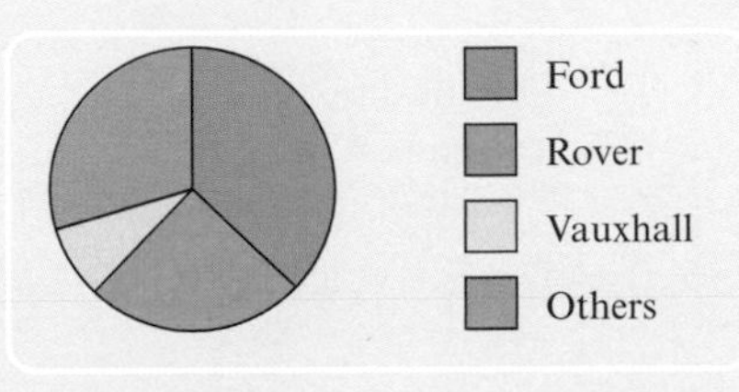

D1 For the colour-of-car data, calculate the percentage for each colour.

D2 Draw a pie chart to show the colour-of-car data.

E Finding averages and the range

- For data in categories, you can only find one average:
  - the **mode** (the **modal** category is the most common)

Red Blue Yellow Blue Green Black Red Blue

The modal colour is **Blue**.

- For numerical data, you can find several averages:
  - the **mode** is the most common value (or values)
  - the **median** is the middle value when the data is in order (for an **even** number of values, take the median as halfway between the middle pair of values)
  - the **mean** is the total of all the values divided by the number of values.
- A measure of how spread out the data is can also be found:
  - the **range** is the difference between the highest and lowest values.

46 27 82 46 102 27 60

Mode = **27** and **46**
Median = **46** (27 27 46 [46] 60 82 102)
Mean = **55.7** (to 1 dp) $\left(\frac{46 + 27 + 82 + 46 + 102 + 27 + 60}{7}\right)$
Range = **75** (102 – 27)

87 43 101 56 87 67

Mode = **87**
Median = **77** (43 56 [67 87] 87 101)
Mean = **73.5** (441 ÷ 6)
Range = **58** (101 – 43)

E1

15 42 33 37 84 42 50
81 29 26 67 15 19 55

For this set of data, find:
a the mode
b the median
c the mean
d the range.

E2

38 25 106 78 44 62
13 90 25 31 25

For this set of data, find:
a the mode
b the median
c the mean
d the range.

E3

Mode 3 and 7
Median 6
Mean 6

These are averages for a set of data with 8 values.
List what the values might be.

F Finding the mode and the range from a frequency table

- For data in categories:
  - the **mode** is the category with the highest frequency.

Make of car	Ford	Rover	Vaux.	Other
Number of cars	9	2	6	7

The modal make of car is **Ford**.

- For numerical data:
  - the **mode** is the value (or values) with the highest frequency
  - the **range** is the difference between the highest and lowest values.

Number of people in car	**Frequency**
1	3
2	10
3	6
4	3
5	2

The mode is **2** people.
The range is **4** people (5 – 1).

F1 Use your frequency table from Question **B2** to find the modal colour of car.

F2 Use your frequency table from Question **B1** to find:
a the modal number of children in car
b the range of the number of children in car.

G Deciding which average to use

- The **mean** is the most widely used average.
  It is not a sensible average to use for data with **extreme values**, values which are much smaller or much greater than the others.

G.T. Small & Son – Monthly Salaries (£)					
870	870	870	870	1050	1050
1050	1210	1210	1210	**2080**	**2330**

Mean = £1222.50
Median = £1050
Mode = £870

10 of the 12 salaries are smaller than the mean, so the **median** is the more sensible average to use.

- The **mode** can also be a poor choice of average.
  For this data, it is a poor choice because it is the lowest salary.

H Diagrams that present two or more sets of data

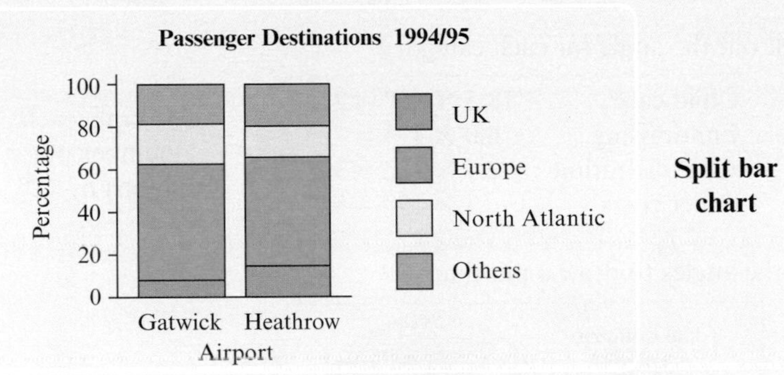

Split bar chart

Line graph

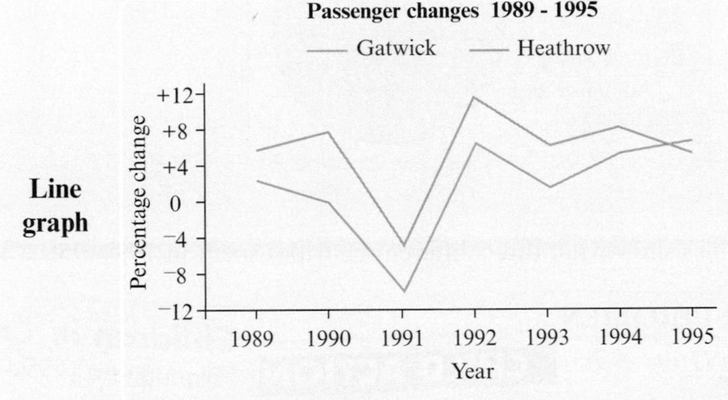

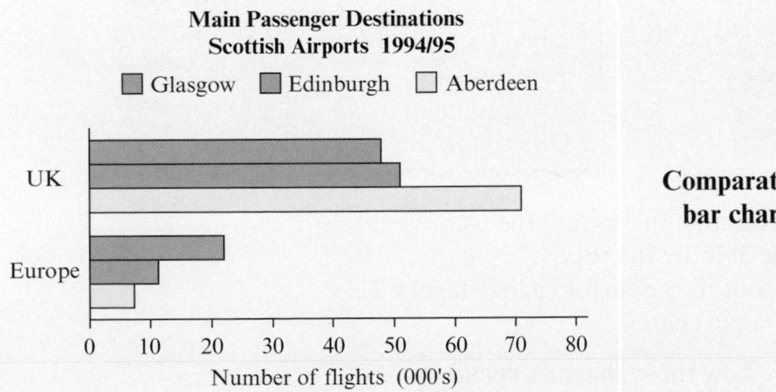

Comparative bar chart

G1

S. Fry & Partners – **Weekly Wages (£)**				
112	285	285	340	340
340	372	372	388	

Find
a the modal wage
b the median wage
c the mean wage
d For each average wage:
  i decide whether it is sensible to use or not
  ii if you think it is not sensible, explain why.

H1

Southampton to Channel Islands *1994/95*	Number of passengers (000's)	Number of flights (000's)
Jersey	156	4.0
Guernsey	100	3.4
Alderney	31	2.9
Totals	287	10.3

Calculate the percentage of passengers going to:
a Jersey **b** Guernsey
c Alderney

H2 Calculate the percentage of flights going to each of the three islands.

H3 Draw a split bar chart to show the two sets of data.

H4

% change in no. of passengers	1990	1991	1992	1993	1994
Glasgow	11.0	⁻3.1	12.4	7.4	8.8
Edinburgh	5.3	⁻6.1	8.4	7.2	10.3

Draw a line graph to show this data for Glasgow and Edinburgh.

H5

NUMBER OF FLIGHTS 1994/95 (000's)	**Heathrow**	**Gatwick**
UK	75	33
Europe	252	111
North Atlantic	40	19
Others	47	21

Draw a comparative bar chart to show this data.

Constructing a pie chart using degrees

£m stands for millions of pounds.

- A set of data in categories can be shown on a pie chart.
- To construct a pie chart using degrees:

CHILD CONCERN		£m		£m
Expenditure	Child care	18.5	Administration	2.5
(1994/95)	Fundraising	6.0	Other costs	3

- add the amounts to find the total

  18.5 + 6.0 + 2.5 + 3 = **30**

£28.8m is shared between the 360° in a circle: each £1m is given 12°

There are 360° at the centre of a circle.

360°

- divide 360° by the total

  360° ÷ 30 = **12°**

- work out the angle for each category

Child care	18.5 × 12° = 222°
Fundraising	6.0 × 12° = 72°
Administration	2.5 × 12° = 30°
Other costs	3 × 12° = 36°

Multiply each number of millions by 12°

- use the angles to draw a pie chart.

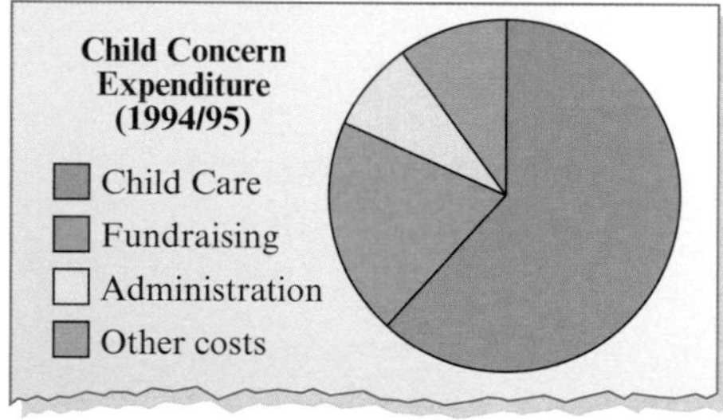

Exercise 5.1
Presenting data divided into categories

Use the same method for data given in %: divide 360° by 100.

1 These tables show what three different charities spent in 1994–95.

CARING FOR CHILDREN *Expenditure (1994–95)*	£m
Child Care	61.2
Fundraising	6.4
Administration	1.6
Other costs	2.8

CHILD ACTION Expenditure (1994–95)	£m
Child Care	33.6
Fundraising	8.8
Administration	2.4
Other costs	3.2

Children in Crisis Expenditure (1994–95)	%
Child Care	70
Fundraising	15
Administration	10
Other costs	5

For each charity:

a add the amounts to find the total
b divide 360° by the total
c work out the angle for each category
d draw a pie chart.

2 Compare how these charities spend money.

Thinking ahead to ... finding the median

A Archers use a 10-ring target.
Give the ring score for:

a the outer red ring
b the inner black ring.

Finding the median of a frequency distribution

10-ring target

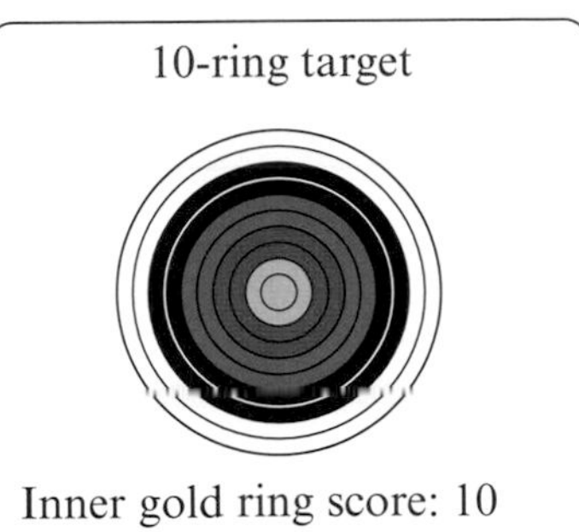

Inner gold ring score: 10
Outer gold ring score: 9
Inner red ring score: 8
and so on.

The frequency here is the number of arrows.

- A set of data presented in a frequency table is called a **frequency distribution**.
- To find the median by listing the data:

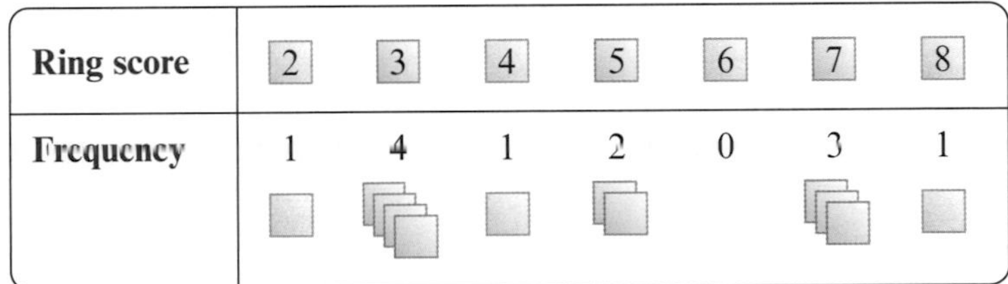

Ring score	2	3	4	5	6	7	8
Frequency	1	4	1	2	0	3	1

- list the data in order

2 3 3 3 3 4 5 5 7 7 7 8

- find the middle value.

2 3 3 3 3 [4 5] 5 7 7 7 8
Median score = **4.5**

Exercise 5.2 Finding the median by listing the data

1

Ring score	5	6	7	8	9
Frequency	4	2	1	2	1

Ravi

a How many of Ravi's arrows hit the 8 ring?
b List Ravi's scores for his ten arrows.
c Find Ravi's median score.

2

Ring score	5	6	7	8	9	10
Frequency	3	2	4	0	3	3

Sally

a How many arrows did Sally fire in total?
b List all Sally's scores.
c Find Sally's median score.

3

Ring score	3	4	5	6	7	8	9
Frequency	1	5	6	2	1	1	2

Peta

Peta hit seven different rings: 3 4 5 6 7 8 9
The middle ring of these is the 6 ring.

a Explain why 6 is not Peta's median score.
b Find Peta's median score.

4 Karl fires 14 arrows: his lowest score is 2,
his highest score is 10,
his median score is 6.5

Write a possible frequency distribution for Karl's scores.

Thinking ahead to ...
calculating the mean of a frequency distribution

A

Ring score	1	2	3	4	5	6	7
Frequency	1	0	1	3	5	4	3

Saul

What is the total score for Saul's arrows that hit the 6 ring?

Calculating the mean of a frequency distribution

- To calculate the mean of a frequency distribution:

Ring score	2	3	4	5	6	7	8
Frequency	1	4	1	2	0	3	1

- calculate the total of all the values

Ring score	**Frequency**	**Total score in ring**	
2	1	2×1	2
3	4	3×4	12
4	1	4×1	4
5	2	5×2	10
6	0	6×0	0
7	3	7×3	21
8	1	8×1	8
Totals	12		57

1 arrow hit the 2 ring: a total score of 2.

2 arrows hit the 5 ring: a total score of 10.

12 arrows scored 57 in total.

- divide the total of all the values by the **total frequency**.

Mean score $= \frac{57}{12} = \mathbf{4.8}$ (to 1 dp)

Exercise 5.3
Calculating the mean

Give your answers to 1 dp.

1

Ring score	3	4	5	6	7	8
Frequency	2	5	2	2	4	1

Jodie

Geeta

Ring score	Frequency
6	2
7	6
8	7
9	4
10	1

For each archer:

a calculate the total of all their scores
b calculate the total frequency
c calculate their mean score.

2

Ring score	6	7	8	9	10
Frequency	4	2	1	2	1

The calculation at the side is wrong.

a Explain the mistakes.
b Calculate the correct mean score.

Mean score $= \frac{6+7+8+9+10}{5}$

$= \frac{40}{5}$

$= 8$ ✗

Comparing sets of data

A measure of spread measures how spread out data is.

The simplest measure of spread is the range.

- One way to compare sets of data is to compare two types of value:
  A – an average
  B – a measure of spread.

Example

Compare Kate's and Bob's scores using the median and the range.

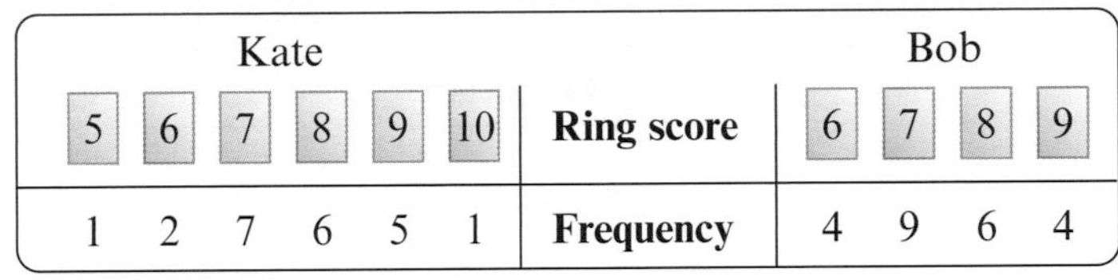

Kate							Bob			
5	6	7	8	9	10	**Ring score**	6	7	8	9
1	2	7	6	5	1	**Frequency**	4	9	6	4

Kate 5 6 6 7 7 7 7 7 7 7 8 8 8 8 8 8 9 9 9 9 9 10

Median = **8**
Range = 10 – 5 = **5**

Bob 6 6 6 6 7 7 7 7 7 7 7 7 7 8 8 8 8 8 8 9 9 9 9

Median = **7**
Range = 9 – 6 = **3**

A – Kate's scores are higher on average.
B – Bob is more consistent because his scores are less spread out.

Exercise 5.4 Comparing sets of data

1

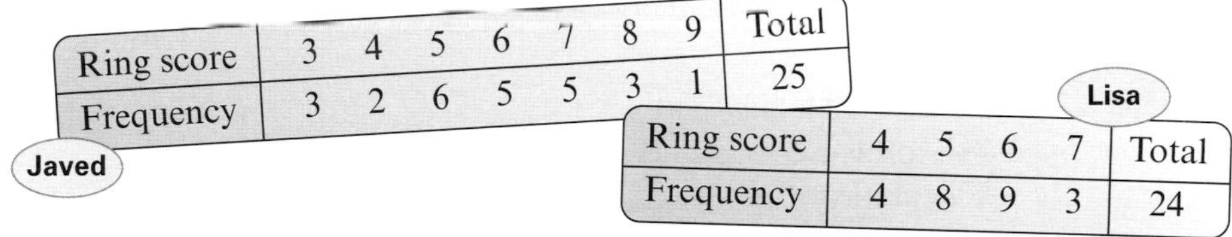

Javed

Ring score	3	4	5	6	7	8	9	Total
Frequency	3	2	6	5	5	3	1	25

Lisa

Ring score	4	5	6	7	Total
Frequency	4	8	9	3	24

a Find the median score for each frequency distribution.
b Calculate the range of the scores for each distribution.
c Use the median and range to compare Javed's and Lisa's scores.

2

Amy

Ring score	4	5	6	7	8	Total
Frequency	3	5	4	3	2	17

Paul

Ring score	3	4	5	6	7	8	Total
Frequency	4	2	0	7	5	2	20

a Calculate the mean and the range of each distribution.
b Compare Amy's and Paul's scores using the mean and the range.

3

Imran

Ring score	1	2	3	4	5	6	7	8	9	Total
Frequency	1	1	1	2	3	2	1	2	2	15

Viv

Ring score	1	2	3	4	5	6	7	8	9	Total
Frequency	1	0	0	2	1	3	4	0	1	12

a Compare Imran's and Viv's scores using the mode and the range.
b Explain why the range is not a sensible measure of spread for Viv's scores.

Calculating the interquartile range

- The **interquartile range** of a set of data measures how spread out the middle 50% of the data is.
- To calculate the interquartile range:

Ring score	1	2	3	4	5	6	7	8	9
Frequency	1	0	0	2	1	3	4	0	1

❖ list the data in order

1 4 4 5 6 6 6 7 7 7 7 9

❖ divide the data into four quarters

1 4 4 ¦ 5 6 6 ¦ 6 7 7 ¦ 7 7 9

❖ find the end values of the middle 50% of the data

1 4 4 ¦ 5 6 6 ¦ 6 7 7 ¦ 7 7 9

4.5 **Lower quartile** — **7** **Upper quartile**

You can divide the data into four quarters by first dividing it into two halves.

The end values are the medians of the two halves.

❖ calculate the difference between the upper and lower quartiles.

Interquartile range = 7 – 4.5
= 2.5

Exercise 5.5
Calculating the interquartile range

1

William

Ring score	2	3	4	5	6	7	8	Total
Frequency	1	0	3	2	5	3	2	16

Bryony

Ring score	4	5	6	7	8	9	Total
Frequency	2	4	3	3	0	2	14

Daniel

Ring score	2	3	4	5	6	7	8	9	10	Total
Frequency	1	1	2	3	5	3	0	3	2	20

For each archer:

a list the data
b find the upper and lower quartiles
c calculate the interquartile range of their scores.

2

Sheera

Ring score	1	2	3	4	5	6	7	8	9	Total
Frequency	1	2	5	4	2	1	0	2	1	18

Dave

Ring score	1	2	3	4	5	6	7	Total
Frequency	2	4	4	3	4	5	2	24

Is Sheera or Dave the more consistent archer?
Give reasons for your answer.

Calculate some figures to support your explanation.

Thinking ahead to ... constructing a cumulative frequency table

A

Ring score	2	3	4	5	6	7	8	Total
Frequency	1	4	1	2	0	3	1	12

Calculate how many arrows in total scored:

a 2 or 3
b less than or equal to 5
c less than or equal to 6.

Constructing a cumulative frequency table

- The total of frequencies up to a particular value in a set of data is called the **cumulative frequency**.
- To construct a cumulative frequency table from a frequency table.

Frequency table

Ring score	2	3	4	5	6	7	8	Total
Frequency	1	4	1	2	0	3	1	12

Cumulative frequency table

Ring score	Cumulative Frequency
≤2	1
≤3	5
≤4	6
≤5	8
≤6	8
≤7	11
≤8	12

6 arrows hit the 2, 3 or 4 rings: 1 + 4 + 1

11 arrows hit the 2, 3, 4, 5, 6 or 7 rings: 1 + 4 + 1 + 2 + 0 + 3

≤ stands for 'is less than or equal to'

- You can find the cumulative frequency for each score by adding its frequency to the cumulative frequency for the previous score.

Ring score	2	3	4	5	6	7	8
Frequency	1	4	1	2	0	3	1
Cumulative frequency	1	5	6	8	8	11	12

Exercise 5.6 Constructing cumulative frequency tables

1

Ring score	5	6	7	8	9	Total
Frequency	2	5	12	7	5	31

André

Ring score	Cumulative Frequency
≤5	
≤6	
≤7	
≤8	
≤9	

a Copy this cumulative frequency table.
b Calculate the cumulative frequencies for André's scores.

2

Ring score	2	3	4	5	6	7	8
Frequency	3	7	12	19	13	9	4
Cumulative frequency							

Simone

a Copy this table.
b Calculate the cumulative frequencies for Simone's scores.

3 This distribution shows the ages of archers at a junior event.

Age	10	11	12	13	14	15	16	Total
Number of archers	8	7	10	11	8	9	11	64

How many archers can enter these age group competitions?
a 11 and under **b** 13 and under **c** 16 and under.

4 **a** Copy this cumulative frequency table for the data in question 3.
b Calculate the cumulative frequency for each age group.

Age	Cumulative frequency
10 and under	
11 and under	
12 and under	
13 and under	
14 and under	
15 and under	
16 and under	

5 Dani fired 20 arrows at the target from each of seven different distances.

Distance to target in metres	10	15	20	25	30	35	40
Number of arrows hitting target	20	19	17	16	15	13	11

Construct a cumulative frequency table for this distribution.

6 This table shows Jade's scores in the April, May and June competitions.

	Ring score										
	1	2	3	4	5	6	7	8	9	10	Total
April	4	7	7	8	5	6	2	3	5	1	48
May	6	5	4	9	5	3	8	2	2	4	48
June	3	6	6	4	5	4	8	4	2	6	48

a Construct a cumulative frequency table like this.

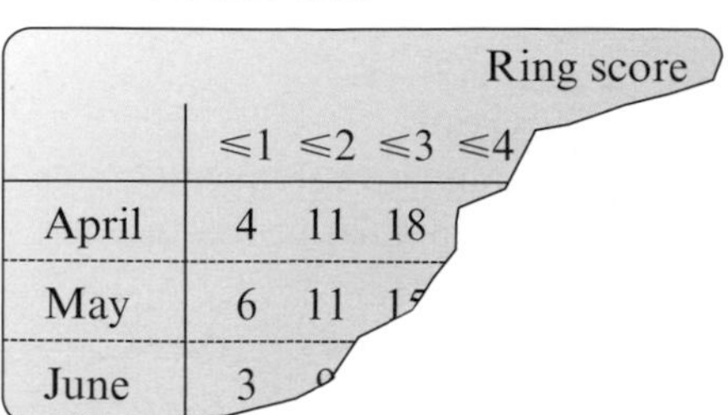

	Ring score			
	≤1	≤2	≤3	≤4
April	4	11	18	
May	6	11	1[illegible]	
June	3	[illegible]		

b Do you think Jade is improving as an archer or getting worse? Explain why.

Thinking ahead to ...
finding the median

A

Ring score	5	6	7	8	9	Total
Frequency	7	8	13	18	14	60

Emily

a List all Emily's scores.
b Find Emily's median score.

Using cumulative frequencies to find the median of a frequency distribution

You can find the median for a small set of data by listing it.

This method is quicker for a distribution with a large total frequency.

- The median can be found without listing all the data.
- To find the median using cumulative frequencies:
  - construct a cumulative frequency table
  - use the total frequency to decide where the median is
  - find the median.

Ring score	Frequency	Cumulative Frequency
5	7	7
6	8	15
7	13	28
8	18	46
9	14	60

The total frequency is 60, so the median score is halfway between the 30th largest and the 31st largest.

The 28th largest score is 7.

All the scores between the 29th and 46th largest are 8.

The 30th and 31st largest scores are both 8, so
Median score = **8**

Exercise 5.7
Finding the median using cumulative frequencies

1

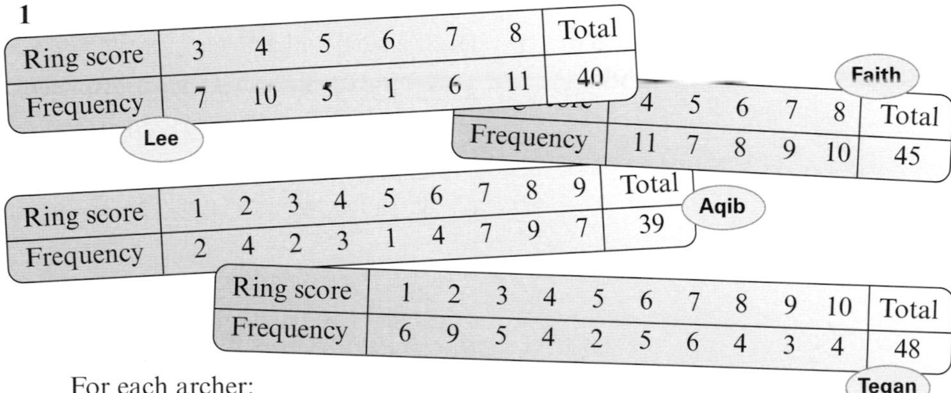

Lee

Ring score	3	4	5	6	7	8	Total
Frequency	7	10	5	1	6	11	40

Faith

Ring score	4	5	6	7	8	Total
Frequency	11	7	8	9	10	45

Aqib

Ring score	1	2	3	4	5	6	7	8	9	Total
Frequency	2	4	2	3	1	4	7	9	7	39

Tegan

Ring score	1	2	3	4	5	6	7	8	9	10	Total
Frequency	6	9	5	4	2	5	6	4	3	4	48

For each archer:
a construct a cumulative frequency table
b decide where the median is
d find the median score.

2

Ring score	4	5	6	7	8	9	10
Cumulative Frequency	7	18	31	40	48	62	80

Jake

a Explain why Jake's median score is 7.5.
b Explain why the interquartile range of this distribution is 3.

Drawing and using a stem and leaf plot

- The stem and leaf plot is a frequency diagram.
- The actual data is displayed together with its frequency.
- Part of the value of each piece of data is used to fix the data class ie the **stem**.

  Part of the value is listed in the diagram ie the **leaves**

Example

Show this set of data with a stem and leaf plot

6, 13, 23, 42, 30, 32, 36, 27, 24, 32, 40, 45, 8

As the data is listed and shown in order on a stem and leaf plot the median value can be easily found. Here the median value is 6.

stem	leaves
4	0 2 5
3	0 2 2 6
2	3 4 7
1	3
0	6 8

Exercise 5.8
Stem and leaf plots

1 Draw a stem and leaf plot for each data set.

a 23, 12, 14, 26, 38, 46, 33, 32, 19, 7, 24, 12, 40, 20
b 15, 22, 24, 20, 10, 12, 31, 30, 24, 26, 26, 24, 32, 40
c 42, 16, 9, 7, 6, 18, 25, 6, 24, 30, 36, 25, 30, 41, 16, 44
d 38, 14, 22, 56, 18, 24, 28, 37, 40, 30, 32, 28, 46, 40, 28.

2 The data gives the number of minutes waited by people to fly on the 'London Eye'.

44, 56, 50, 52, 42, 40, 38, 44, 52, 54, 53, 38, 30
41, 40, 51, 59, 62, 34, 40, 47, 50, 54, 50, 57

a Draw a stem and leaf plot to show this data.
b Find the median value for the data set.

Thinking ahead to ...
box-and-whisker plots

A A survey of three fast food restaurants recorded the number of chips in a serving. A sample of 80 servings was taken from each restaurant.

	Number of chips													
Restaurant	32	33	34	35	36	37	38	39	40	41	42	43	44	Total
X	1	2	3	5	9	13	14	12	8	5	4	3	1	80
Y	4	6	11	14	13	11	8	6	4	3	0	0	0	80
Z	1	1	2	6	9	13	15	13	10	6	2	1	1	80

a **i** Which restaurant do you think gives more chips in a serving?
ii Explain your answer.

b **i** Find the median number of chips for each restaurant.
ii Calculate the interquartile range for each distribution.

Drawing box-and-whisker plots

- A **box-and-whisker** plot shows a frequency distribution by using:
  - the lowest and highest values
  - the median
  - the lower and upper quartiles.

Example

Draw a box-and whisker plot for these chips from restaurant W.

Number of chips	34	35	36	37	38	39	40	41	42	43
Frequency	1	2	4	7	12	14	16	15	7	2
Cumulative frequency	1	3	7	14	26	40	56	71	78	80

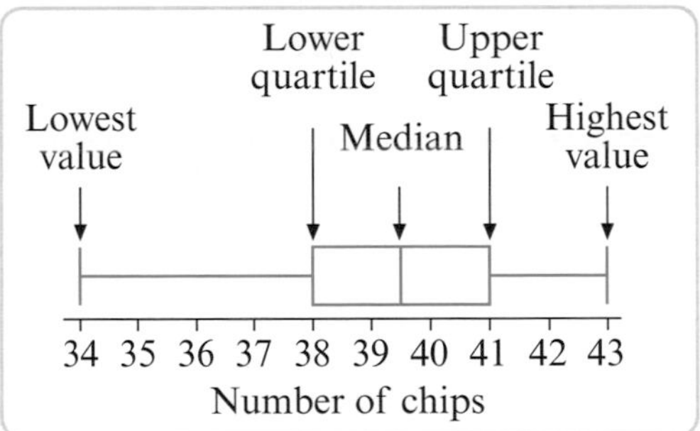

Exercise 5.9
Drawing box-and-whisker plots

1 Copy this box-and-whisker diagram on squared paper. Use the data from Question **A** to draw a box-and-whisker plot for each restaurant.

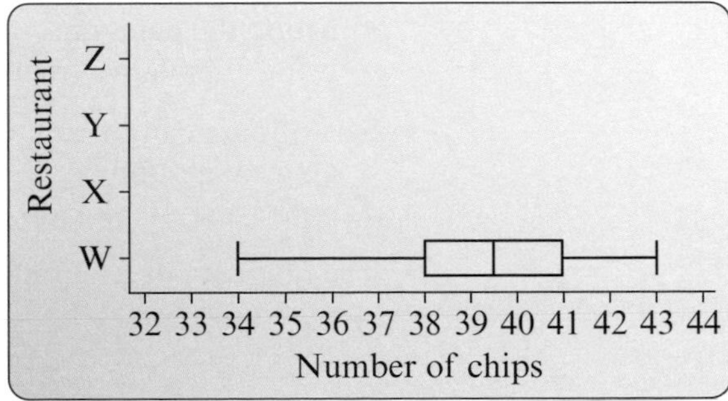

2 Write a short report that compares the number of chips per serving from the four fast food restaurants.

End points

You should be able to ... ... so try these questions.

A Construct a pie chart using degrees

A1

SAVING CHILDREN

Expenditure (1994/95)		£m		£m
	Child care	9.2	Administration	1.0
	Fundraising	3.4	Other costs	0.8

Draw a pie chart to show how Saving Children spend their money.

Manoj

Ring score	2	3	4	5	6	7	Total
Frequency	1	0	3	5	7	2	18

Pat

Ring score	1	2	3	4	5	6	7	8	Total
Frequency	2	1	5	8	7	5	3	2	33

B Find the median of a frequency distribution

B1 **a** Find Manoj's median score by listing the data.
b **i** Construct a cumulative frequency table for Pat's scores.
ii Find Pat's median score.

C Calculate the mean of a frequency distribution

C1 Calculate the mean of:
a Manoj's scores **b** Pat's scores
Give your answers to 1 dp.

D Compare sets of data

D1 Compare Manoj's and Pat's scores using the mean and the range.

Kim

Ring score	2	3	4	5	6	7	8	Total
Frequency	3	2	1	3	4	5	2	20

Stuart

Ring score	3	4	5	6	7	8	9	10	Total
Frequency	2	0	3	6	5	2	3	1	22

D2 Compare Kim's and Stuart's scores using the mode and the range.

E Draw a box-and-whisker plot and a stem and leaf plot

E1 Draw box-and-whisker plots on the same diagram to show
a Kim's scores **b** Stuart's scores

Some points to remember

- The median of a frequency distribution can be found by:
  - listing the data, or
  - using cumulative frequencies.
- Sets of data can be compared using:
  - an average, and
  - a measure of spread.
- The range is not a sensible measure of spread to use when there are extreme values in the data.

Starting points

You need to know about … … so try these questions

A Perimeter and circumference

The perimeter of a shape is the distance around the edges of the shape.

Example

Find the perimeter of ABCDE.

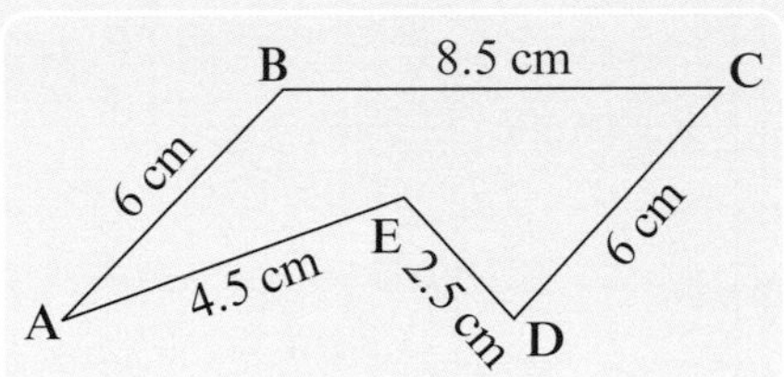

The perimeter (p) of ABCDE is given by:

$$p = 6 + 8.5 + 6 + 2.5 + 4.5$$
$$= 27.5$$

Perimeter of ABCDE is 27.5 cm.

The perimeter of a circle is called the circumference.
The formula for the circumference (c) of a circle is:

$c = \pi D$ (where D is the diameter of the circle)

Example

The circumference (c) of this circle is given by:

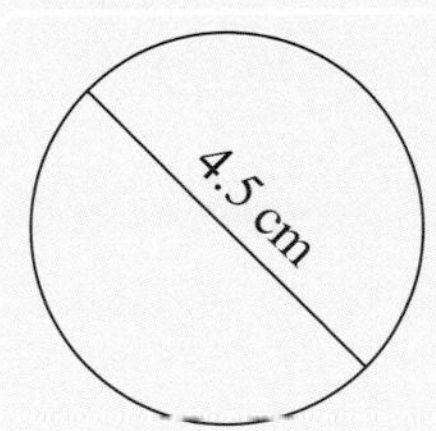

$c = \pi \times 4.5$
$c = 14.1$ (to 1 dp)

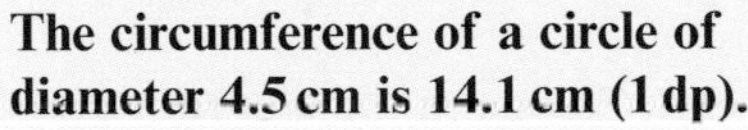
The circumference of a circle of diameter 4.5 cm is 14.1 cm (1 dp).

A1 Calculate the perimeter of a square of side 5.8 cm.

A2 Find the perimeter of RSTVW.

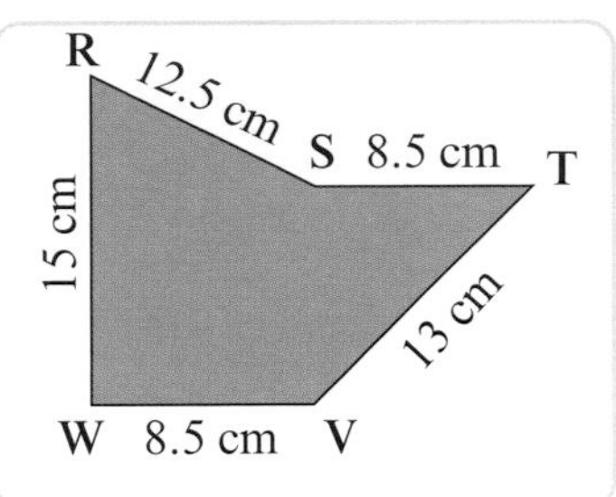

A3 A rectangle has:
long sides of 12.4 cm
and short sides of 9.8 cm
Calculate its perimeter.

A4 Calculate the circumference of each of these circles (to 1 dp):
- **a** circle with diameter 5.2 cm
- **b** circle with diameter 0.7 cm
- **c** circle with radius 2.4 cm.

B Area

The area of a rectangle is given by: Base × Height

The area of a parallelogram is given by: Base × Height

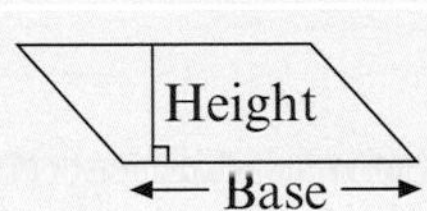

The area of a triangle is given by: 0.5 × Base × Height

or $\dfrac{\text{Base} \times \text{Height}}{2}$

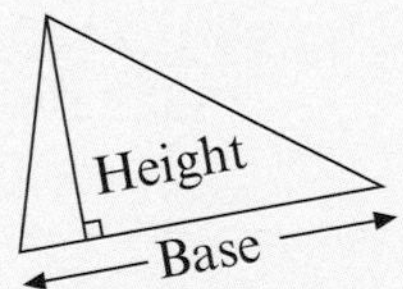

The area of a circle is given by: πr^2 (where r is the radius)

Example

The area (A) of this circle is given by:

5.4 cm radius

$A = \pi \times 5.4^2$ ($\pi \times 5.4 \times 5.4$)
$A = 91.6$ (to 1 dp)

The area of a circle of radius 5.4 cm is 91.6 cm² (1 dp).

B1 A rectangle has:
long sides of 15.2 cm
and short sides of 8 cm.
Calculate its area.

B2 Calculate the area of CDEF.

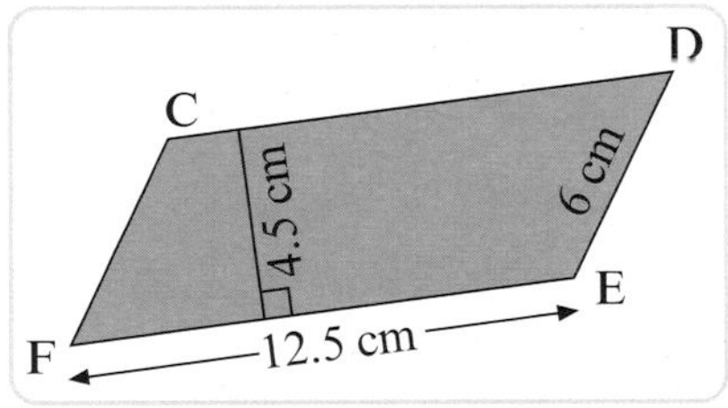

B3 Find the area of triangle JKL.

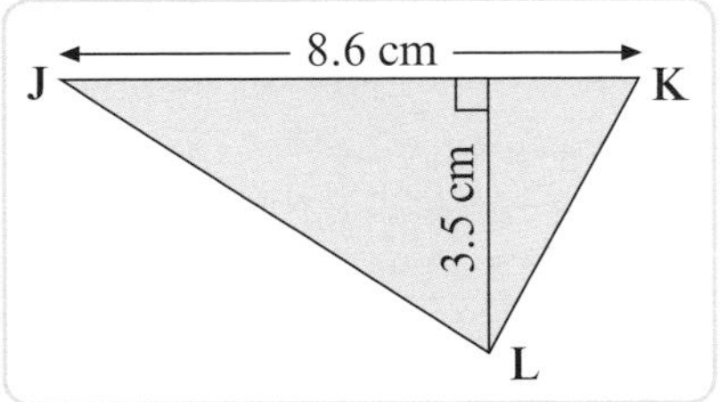

B4 Find the area of a circle of radius:
a 4.3 cm **b** 7.5 cm **c** 12.4 cm

C Splitting up a shape to find its area

Often you can find the area of a complicated shape by splitting it into simple shapes that you can calculate the area for.

Example

Find the area of the whole shape. One way to split the shape is to make two rectangles and one triangle as shown.

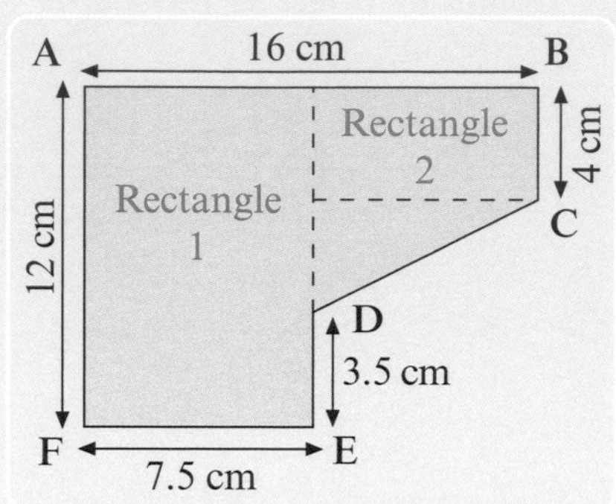

The area (A) of ABCDEF is given by:

A = Area of rectangle 1 + Area of rectangle 2 + Area of triangle

Area Rectangle 1 given by:	Area Rectangle 2 given by:	Area Triangle given by:
7.5×12 $= 90$	8.5×4 $= 36$	$0.5 \times 8.5 \times 4.5$ $= 19.125$

So: $A = 90 + 36 + 19.125 = 145.125$

Area of whole shape = 145.1 cm² (1 dp)

D Pythagoras' rule

Pythagoras' rule is:

In any **right-angled triangle**, the area of the square on the hypotenuse is equal to the sum of the areas of the squares on the other two sides.

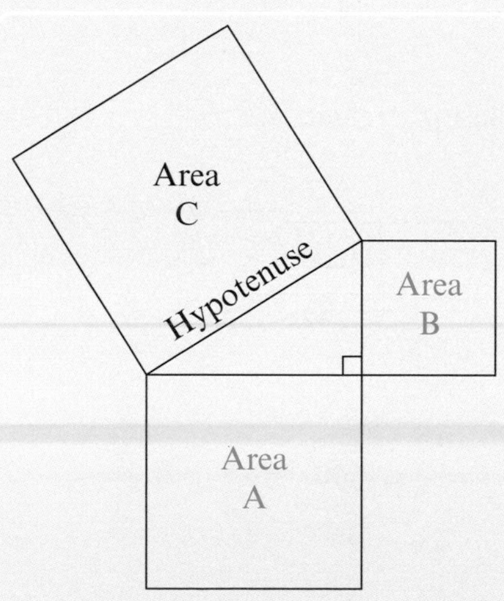

For this triangle:

Area C = Area A + Area B

From this we can also say:

Area A = Area C – Area B

and

Area B = Area C – Area A

For RST: to find the length of RS

$RS^2 = RT^2 - ST^2$
$RS^2 = 6.4^2 - 3.7^2$
$RS^2 = 40.96 - 13.69 = 27.27$
$RS = 5.2$ (1 dp)

RS is 5.2 cm long (1 dp).

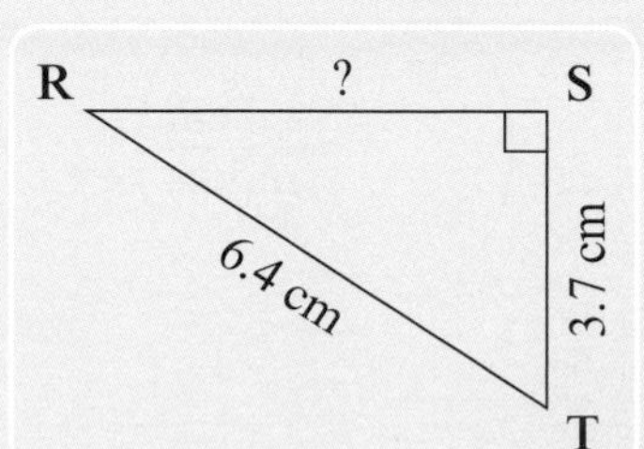

C1 Calculate the area of each shape.

a

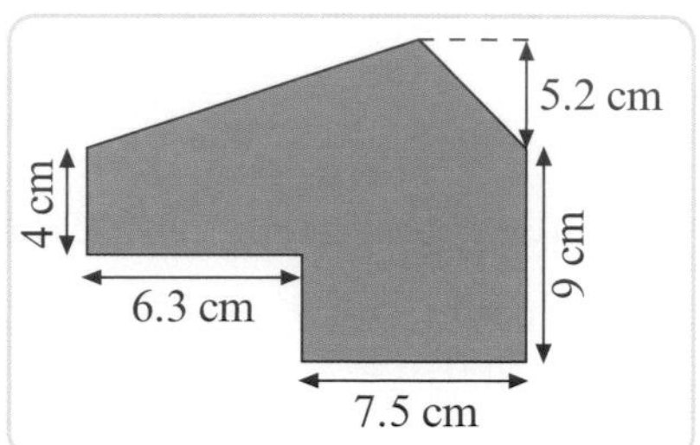

b

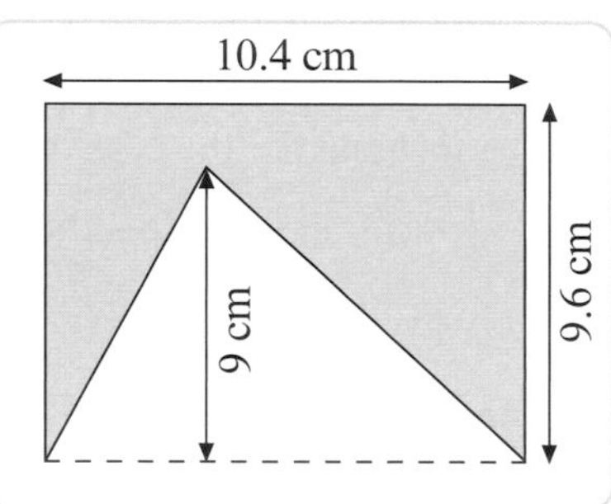

D1 Calculate the length marked x in each triangle.

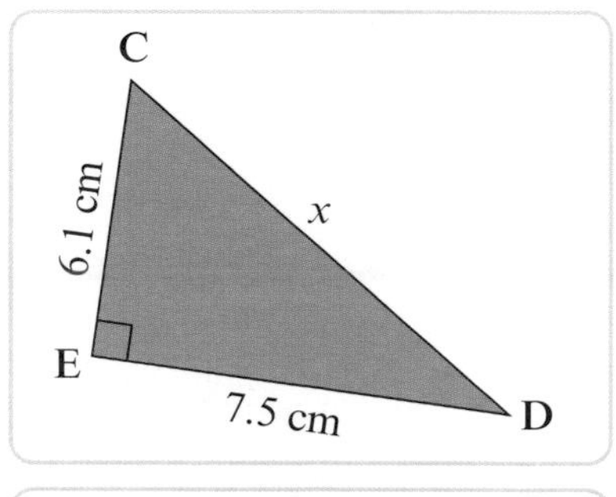

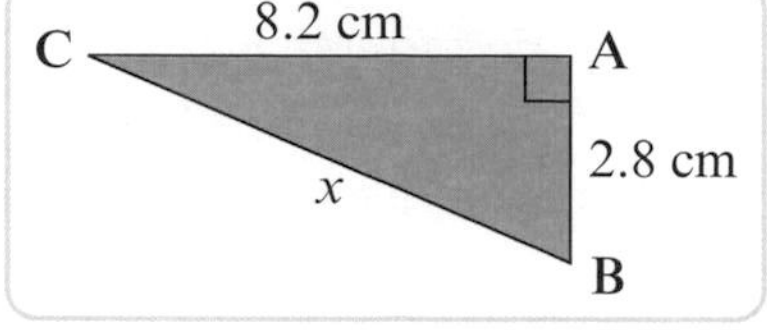

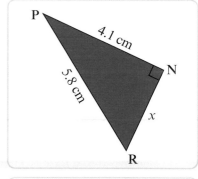

The area of a trapezium

A quadrilateral is a shape with four straight edges.

A trapezium is a quadrilateral with one pair of opposite sides that are parallel.

ABCD is a trapezium as:

- it is a quadrilateral
- AB and CD are parallel.

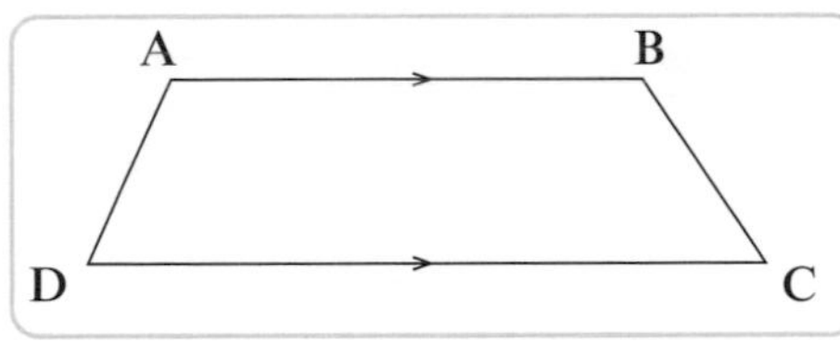

One way to find the area of ABCD is to split it into two triangles and a rectangle, then calculate the total area.

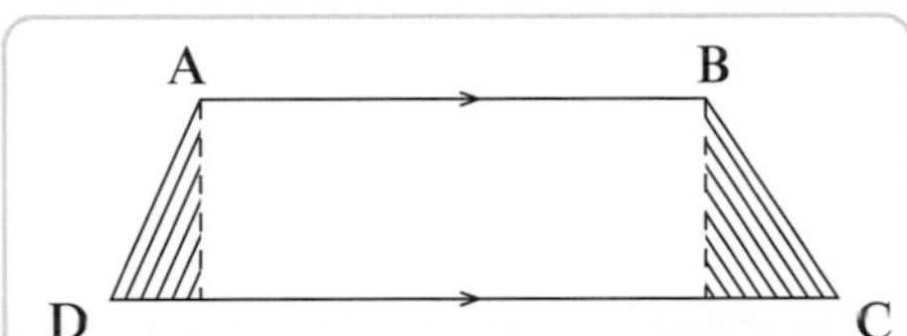

Another way to find the area is to use the formula for the area of a trapezium.

Start with ABCD, then add to it a rotation of ABCD to make a parallelogram.

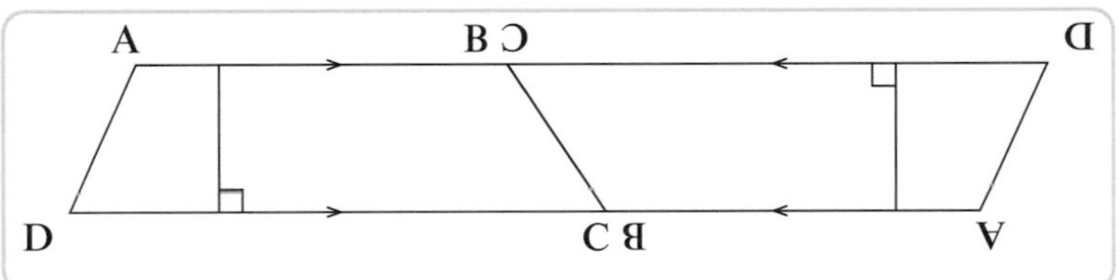

The formula for the area (A) of a parallelogram is: A = Base × Height

The area of this parallelogram is: (DC + BA) × Height
Base

The area of the trapezium ABCD is half the area of the parallelogram
So, the area (A) of trapezium ABCD is given by:

$$A = 0.5 \times (AB + CD) \times \text{Height}$$

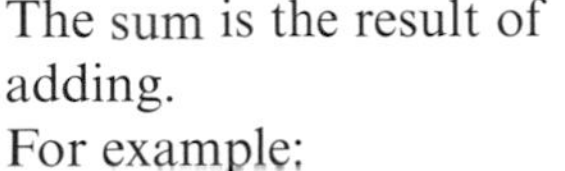

The sum is the result of adding.
For example:
the sum of 3 and 5 is 8.

In general, the formula for the area (A) of a trapezium is:

$$A = 0.5 \times (\text{sum of the parallel sides}) \times \text{Height}$$

In calculations with decimals you should either:

- give your answer to the degree of accuracy asked for e.g. 1 dp, or 2 sf

or

- give your answer to the same degree of accuracy as is used in the question.

Example

Calculate the area of STVW.

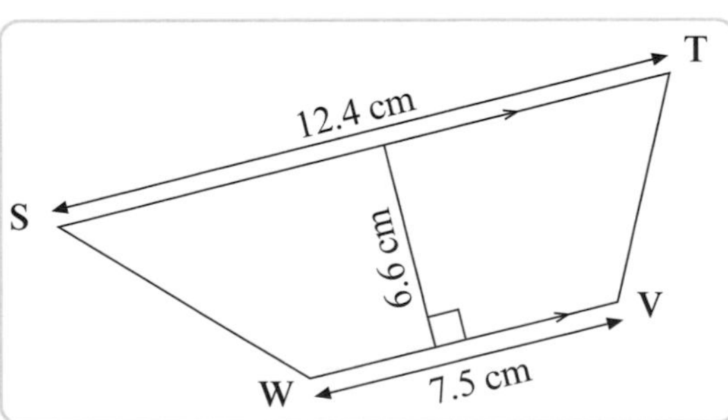

Area (A) given by
$A = 0.5 \times (ST + VW) \times 6.6$
$= 0.5 \times (12.4 + 7.5) \times 6.6$
$= 0.5 \times 19.9 \times 6.6$
$= 65.7$ (to 1 dp)

Area of STVW is 65.7 cm².

Exercise 6.1
Using the formula for the area of a trapezium

1 Calculate the area of STVW:

a when ST = 15.2 cm, VW = 9.5 cm, and the height is 8.3 cm
b when the height is 1.7 cm, VW = 2.1 cm, and ST = 4.5 cm.

2 Calculate the area of each shape:

a

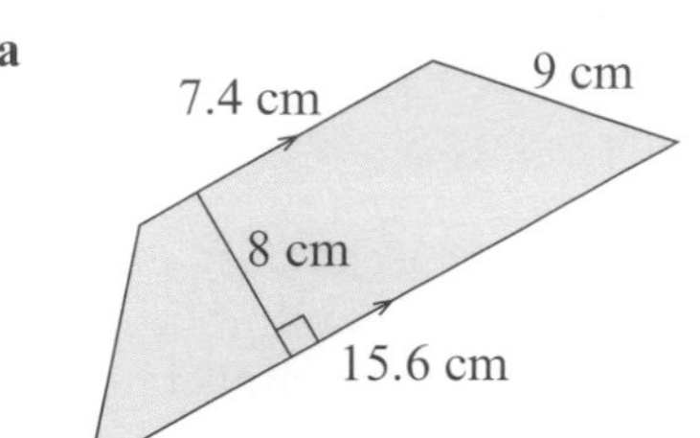

b

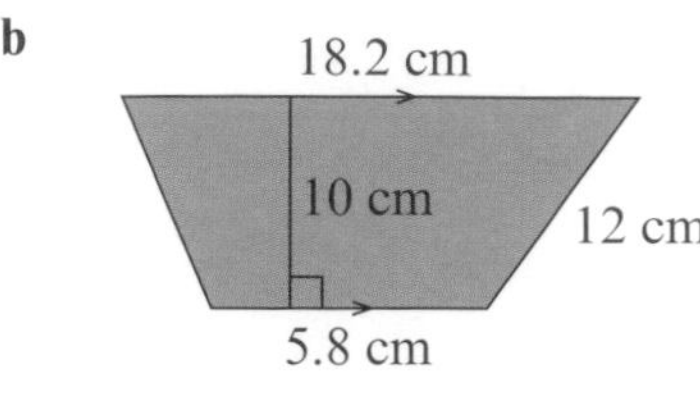

c

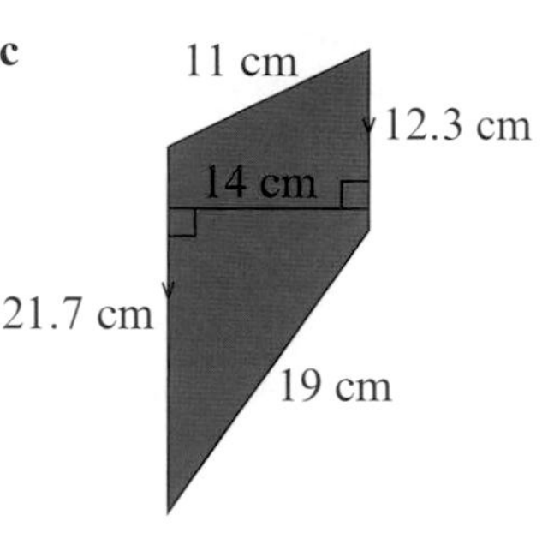

d

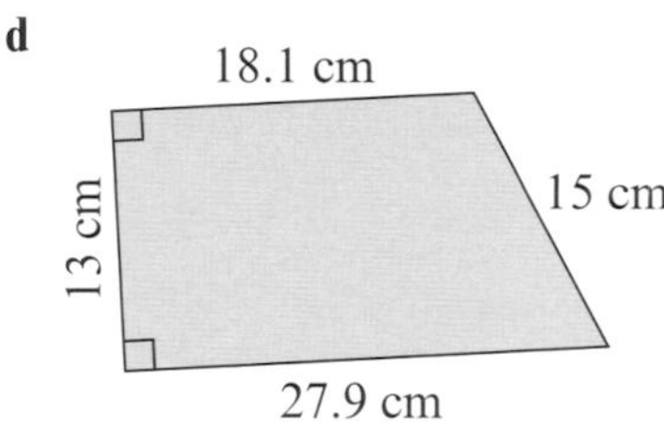

e

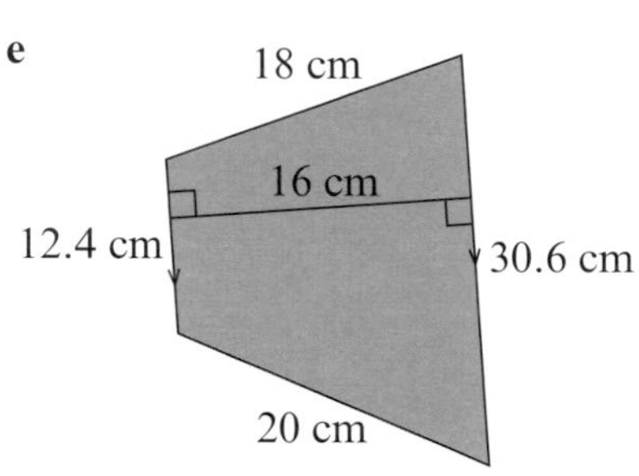

f

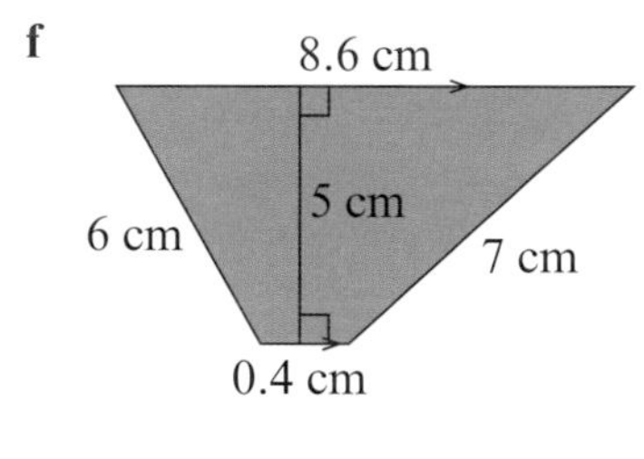

g

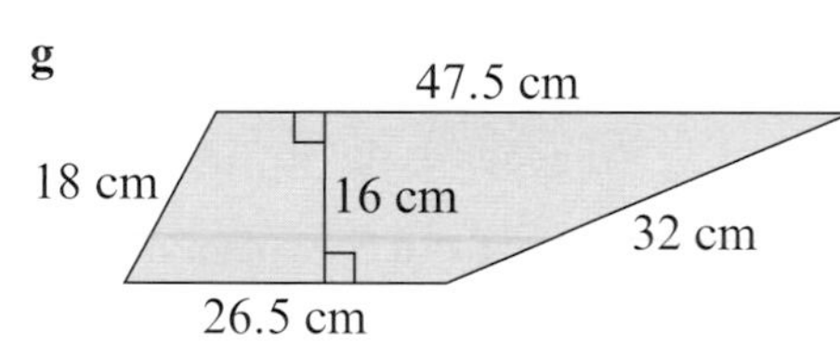

h

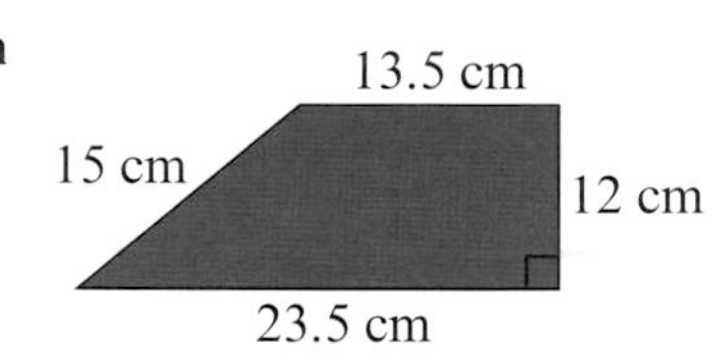

Calculate the shaded area of these shapes:

i

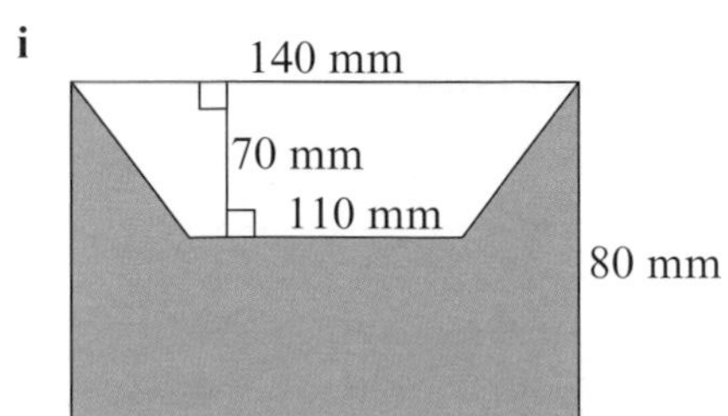

j

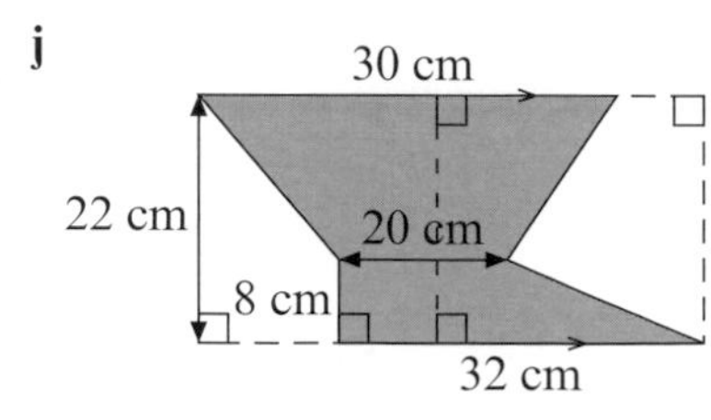

3 This logo uses the same shape trapezium three times.

a Calculate the area of CDEF.
b What is the area of the complete logo?

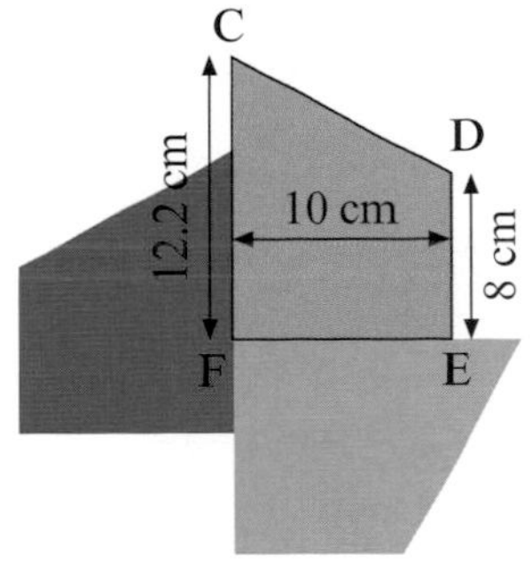

4 The diagram shows the door of a cupboard under some stairs. The door is 76.4 cm wide.

a Draw a diagram of the door and show all the distances you know.
b What shape is the front of the door?
c Calculate the area of the front of the door.

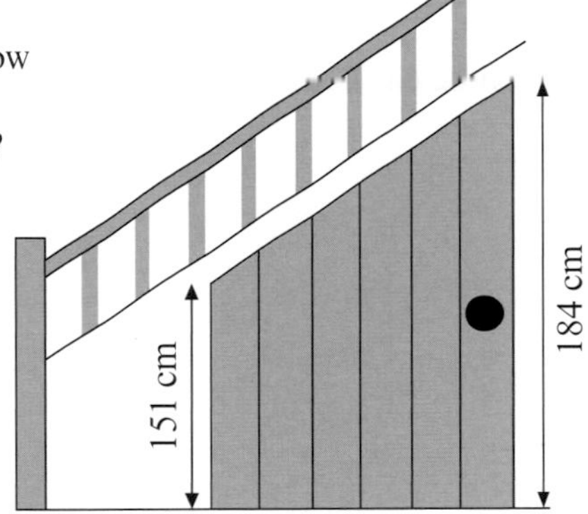

5 The diagram shows the baffles on the front of a spotlight. All four baffles are the same size.

a Draw a diagram of one baffle, and label all the dimensions you know.
b Calculate the area of one baffle.

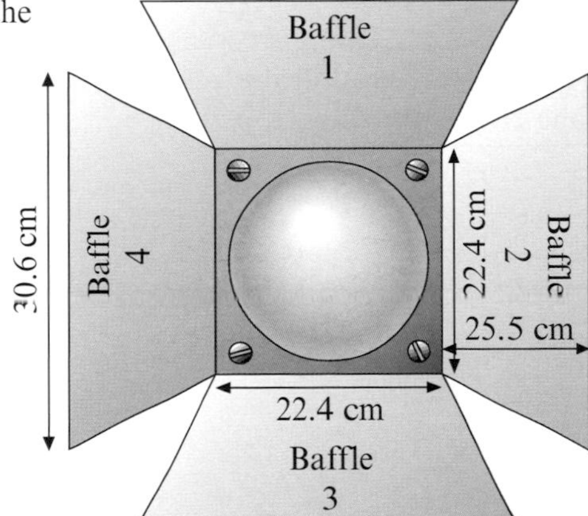

6 The diagram shows the vertical end of a feed bin.

Calculate the area of the end of the bin.

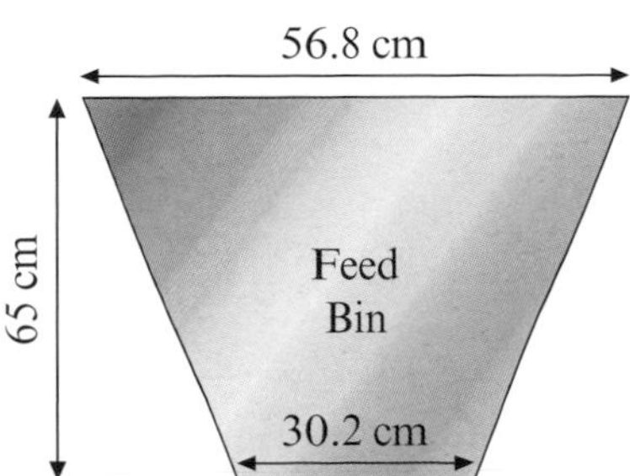

Composite shapes

Composite shapes are made up from more than one shape.
Sometimes you can find the area by adding areas, at other times it is easier to subtract.

Example 1

Find the area of ABCDEF.

The area (A) of ABCDEF can be given by:

$A = (12.4 \times 4.5) + (4.6 \times 4.8)$
$= 55.8 + 22.08$
$= 77.9$ (1 dp)

The area of ABCDEF is $77.9\,\text{cm}^2$

or

The area (A) of ABCDEF is given by:

$A =$ Area of large rectangle – Area cut out for L-shape
$A = (12.4 \times 9.3) - (4.8 \times 7.8)$
$A = 115.32 - 37.44$
$\mathbf{A = 77.9}$ **(1 dp)**

It can be cut out in this way.

Example 2

Find the shaded area.

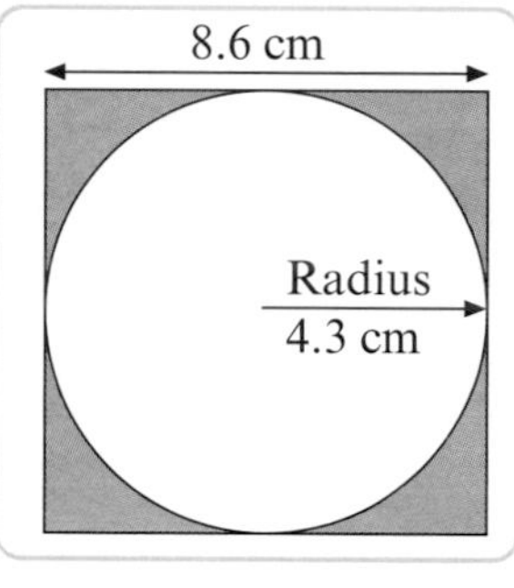

The area of the shaded part (A) is given by:

$A =$ Area of square – Area of circle
$A = (8.6 \times 8.6) - (\pi \times 4.3^2)$
$A = 73.96 - 58.09\ldots$
$\mathbf{A = 15.9}$ **(1 dp)**

The shaded area is $15.9\,\text{cm}^2$

The answer is rounded at the end of the calculation.

Exercise 6.2
Working with composite shapes

1 a Calculate the area of a circle of radius 7.2 cm.
b A circle has a diameter of 12.4 cm. Find its area.
c A circle has a diameter of 3 cm.
Show that its area can be written as 2.25π.
d The radius of a circle is given as $\sqrt{8}$.
Write an expression for the area of the circle in terms of π.
e The area of a circle is given as 53π.
What is the radius of the circle?

2 This shows how milk bottle tops are cut from a strip of foil.

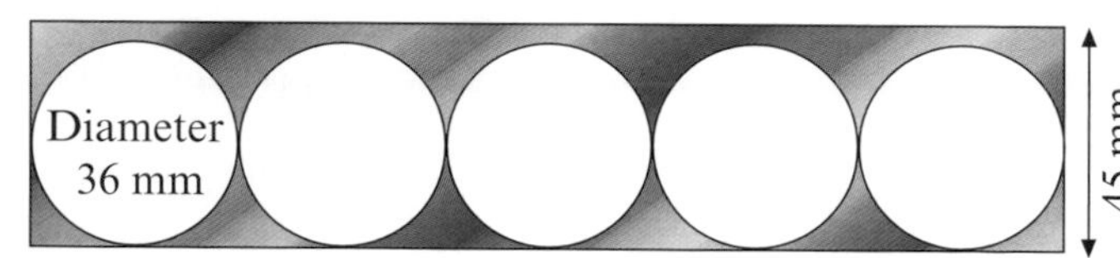

a What is the radius of a foil bottle top?
b Calculate the area of a foil bottle top.
c How long is this strip of foil?
d Calculate the area of foil wasted by cutting out these five tops.

3 Tops are cut from a 7.2 m strip of foil. Calculate the area of wasted foil.

4 A rectangle of card 7.4 cm by 8.3 cm had a piece cut from it to leave shape R.

a What shape was cut from the rectangle?
b What was the area of the rectangle before the cut?
c Calculate the area of shape R.

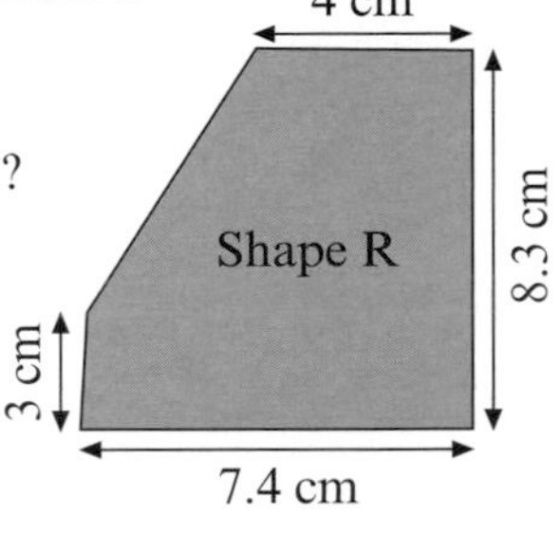

5 This card is used to frame photographs.
A circle of radius 4 cm is cut from the card.
Gold lines are printed 1 cm from the edge of the circle and the card as shown.

a Calculate the area of card after the cut out.
b Is the area of photograph showing more or less than $\frac{1}{4}$ of the blue card?

Explain your answer.

c Calculate the length of gold line on the card.

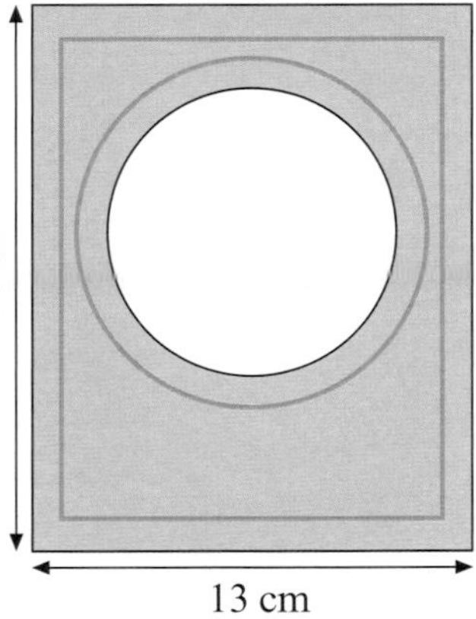

6 This diagram shows the net of a box for playing cards.
The net is cut from a rectangle of card.

For the rectangle of card:

a what is its length?
b how wide is it?
c Calculate the area of the glue flap.
d The box opens at either end.

Find the area of an opening end.

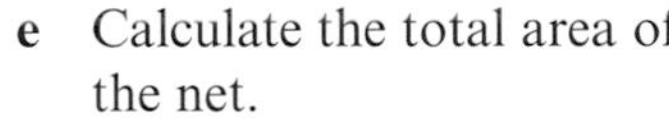

e Calculate the total area of the net.
f Calculate the area of waste card.
g Roughly, what fraction of the card is wasted? Explain how you decided.

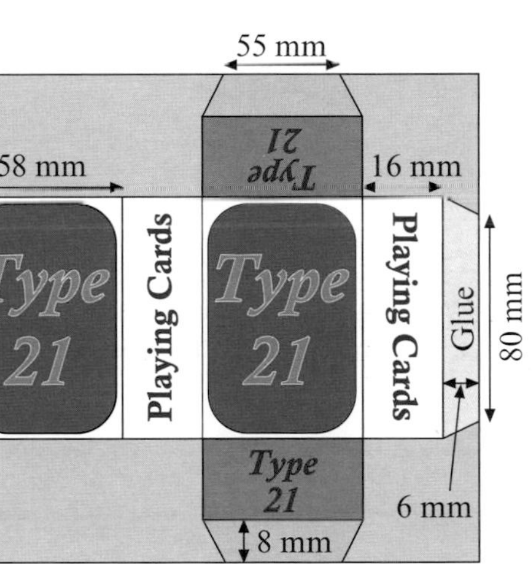

7 This logo design uses two parts of a circle in a red square.

a What is the radius of the circle?
b What fraction of the circle stands for the visor (the blue part)?
c Calculate the area of the visor in the logo.
d Find the area of the helmet in the logo (the yellow part).
e Calculate the area of the logo that is red.

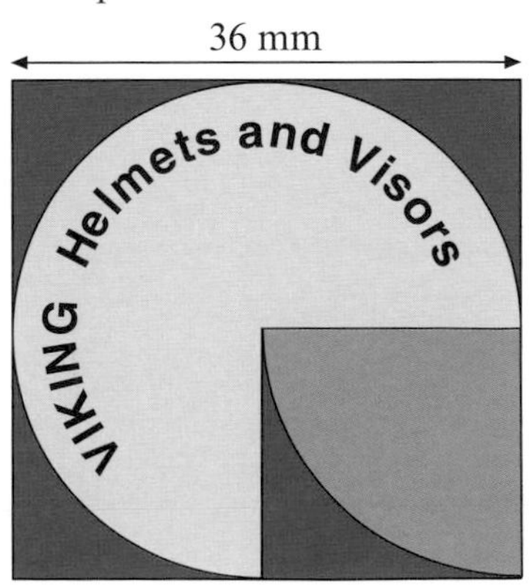

Involving Pythagoras

Sometimes to find the area of a shape you have to calculate a distance before you can calculate the area.

Example

Calculate the area of ABC.

> The area of ABC is given by:
> $0.5 \times$ Base $\times$ Height
>
> Think of AC as the base and DB as the height.

The area of ABC (A) can be given by:

$$A = 0.5 \times AC \times DB$$

but we first have to calculate the length of AC:

$$AC = AD + DC$$

To find AD. In triangle ADB:

$$AD^2 = AB^2 - DB^2$$
$$AD^2 = 3.5^2 - 2.8^2 = 12.25 - 7.84 = 4.41$$
$$AD = \sqrt{4.41} = 2.1$$

To find DC. In triangle DBC:

$$DC^2 = BC^2 - DB^2$$
$$DC^2 = 5.5^2 - 2.8^2 = 30.25 - 7.84 = 22.41$$
$$DC = \sqrt{22.41} = 4.73 \ldots$$

> The answer is rounded at the end of the calculation.

So $AC = 2.1 + 4.73 \ldots = 6.83 \ldots$

Area of ABC (A) $= 0.5 \times AC \times DB = 0.5 \times 6.83 \ldots \times 2.8 = 9.6$ (1 dp)

The area of ABC is 9.6 cm²

Exercise 6.3
Finding areas involving Pythagoras

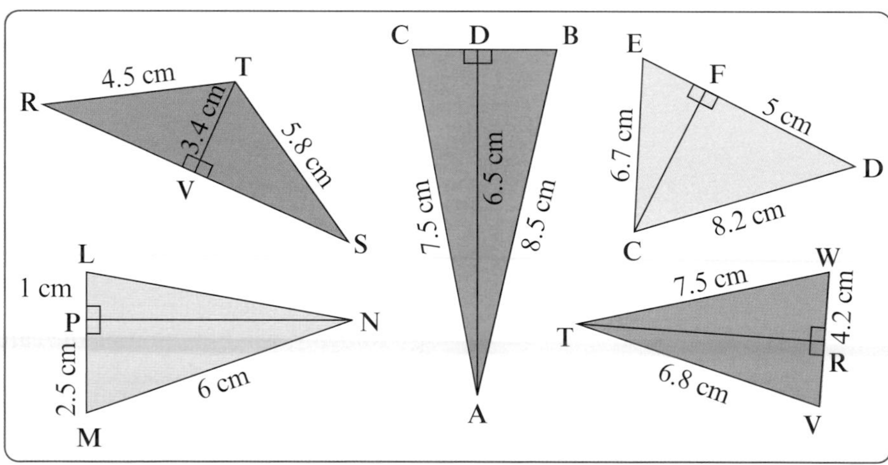

1 In triangle RST:
  - a Calculate the length of RV.
  - b Calculate the length of VS.
  - c Calculate the area of triangle RST.

2 Calculate the area of triangle ABC.

3 a Calculate the height of triangle CDE.
  b Calculate the area of triangle CDE.

4 a Calculate the area of LMN.
  b Calculate the area of TVW.

> Each face of this box is a triangle. This 3-D shape is a **tetrahedron**.

5 The diagram shows the net for this paperclip box.

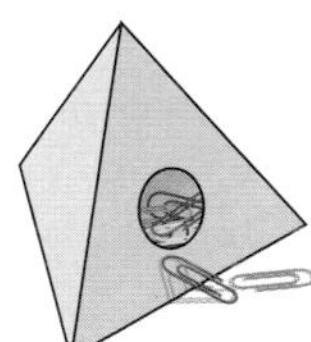

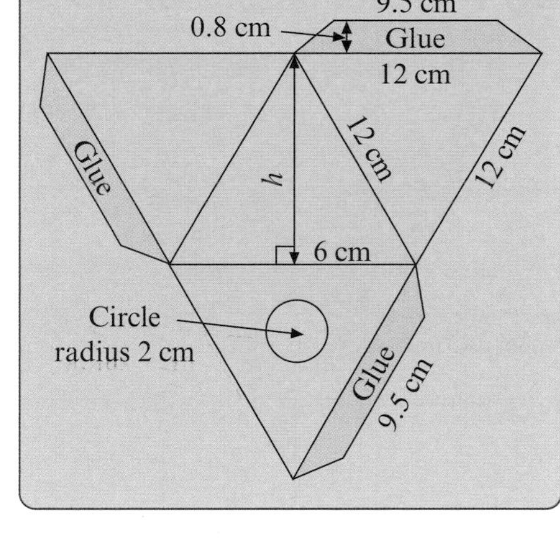

Calculate the length marked h.

6 Find the area of the bottom of the box.

7 Calculate the area of a glue tab.

8 Find the area of the face with the circle cut out.

9 Calculate the total area of the net.

10 Find the area wasted, when the net is cut from a 26 cm square of card.

11 Calculate the perimeter of the net (including the glue flaps).

Using circumference

Exercise 6.4
Using circumference

1 **a** Find the circumference of a circle of radius 9.5 cm.
b A circle has a diameter of 24.5 cm. Find its circumference.
c Show that the circumference of a circle of radius 4.8 cm can be written as 9.6π.
d The circumference of a circle is given as 15π. What is the radius of the circle?
e The semicircle has a diameter of 21.5 cm. Find the perimeter of the shape.

> Remember
>
> Diameter = 2 × radius.

2 The larger of these crop circles has a diameter of 31.5 metres, and the smaller a radius of 5.25 metres.

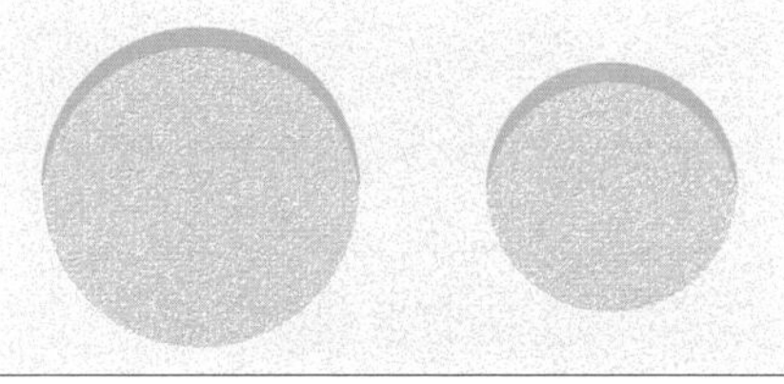

Calculate the circumference of:

a the larger circle
b the smaller circle.

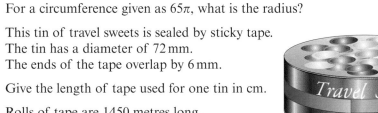

3 For a circumference given as 65π, what is the radius?

4 This tin of travel sweets is sealed by sticky tape. The tin has a diameter of 72 mm. The ends of the tape overlap by 6 mm.

Give the length of tape used for one tin in cm.

5 Rolls of tape are 1450 metres long, and 1 metre is wasted.

How many tins can be sealed with one roll of tape?

2D representations of 3D shapes

3D shapes can be shown in 2D in several ways.

For example:

- A drawing or sketch.

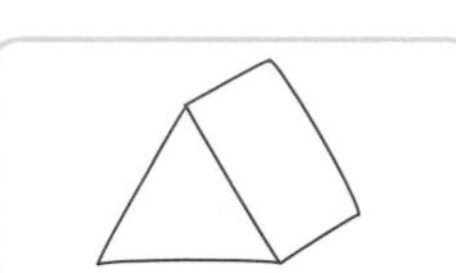

An isometric grid is like this:

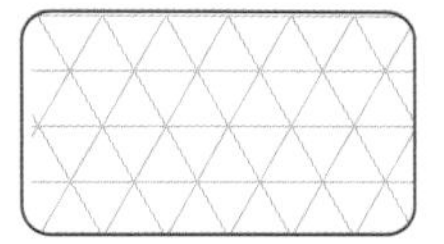

this way up

- A diagram on isometric grid paper.

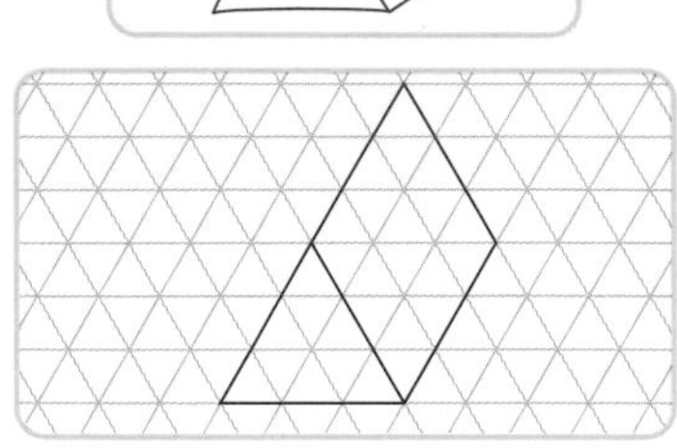

- A drawing of a net of the shape.

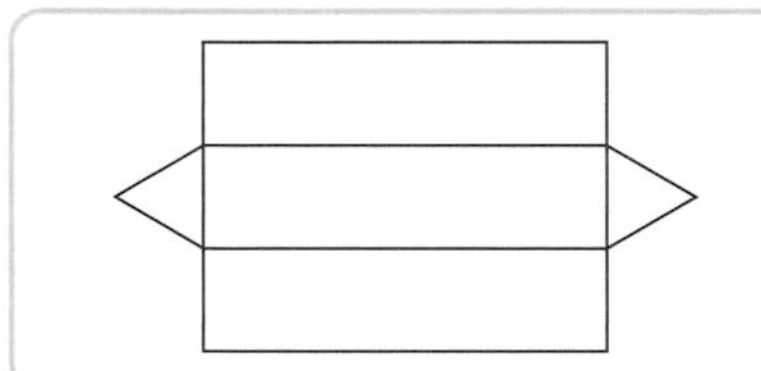

Exercise 6.5
2D/3D representation

1 This is a drawing of a box for a double CD.

Draw a diagram of a box like this on isometric paper.

2 Draw a net for a box of this shape. (Do not include glue flaps.)

3 This drawing is of a box for sweets.

What shape is:

a an end of the box
b a side of the box
c the bottom of the box?

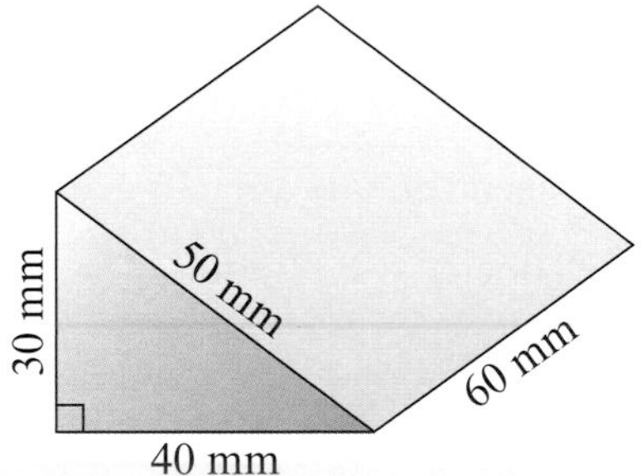

An accurate drawing has angles and lengths drawn using measuring instruments.

4 Make an accurate drawing of one end of the box.

5 Calculate the total area of all five faces of the box.

6 Draw an accurate net for the box.

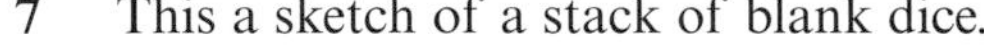

7 This a sketch of a stack of blank dice.

a Draw this stack on isometric paper.
b How many dice are in this stack?

8 A stack of fifteen 3 cm dice is packed in a box.

a Sketch a possible box you think can be used.
b Draw an accurate net for your box.

End points

You should be able to ...   ... so try these questions

A Use the formula for the area of a trapezium

A1

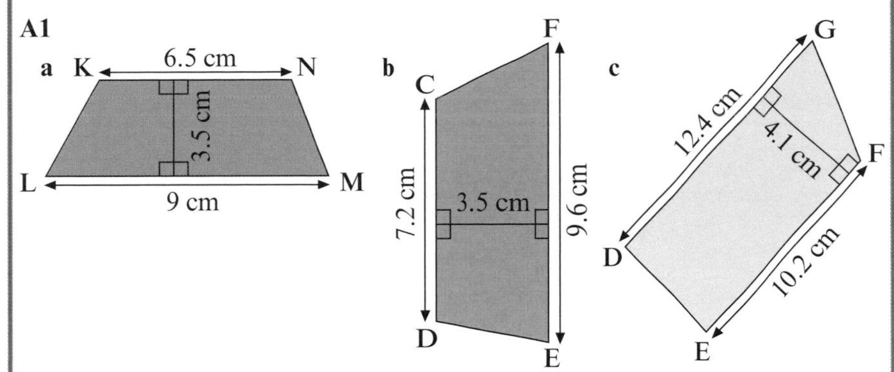

Calculate the area of each shape.

B Work with composite shapes

B1 Explain, with a diagram, what is meant by a composite shape.

B2 **a** Calculate the area of shape P.

Shape P is cut from a 16 cm square of card.

b Calculate the area of waste card.

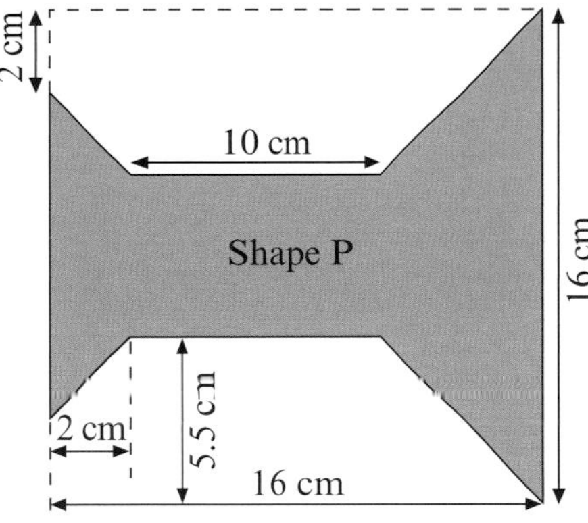

B3 Calculate the area of RST.

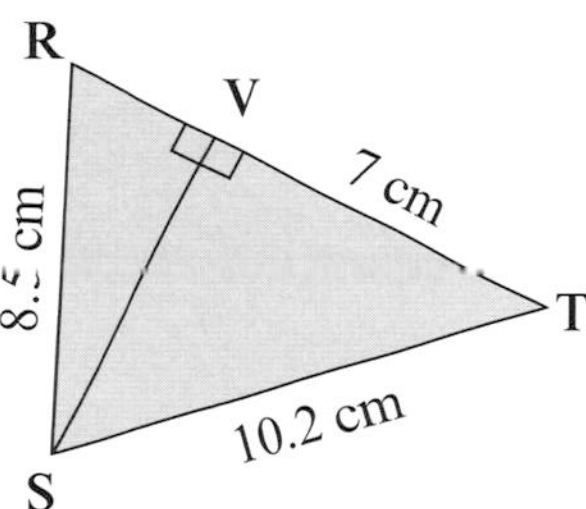

C Calculate perimeters

C1 This hoax crop circle was made by two people with a piece of rope 3.2 metres long.
They simply made two half-circles.

Calculate the perimeter of the crop circle.

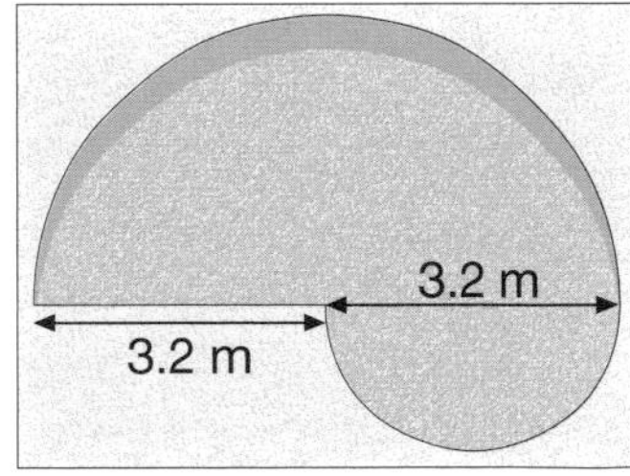

D Work with 2D representations of 3D objects

D1 **a** Make a sketch of the two parts of a matchbox.

b Draw the two parts of a matchbox on isometric grid paper.

About the Cornish Seal Sanctuary

It was founded in 1957.
In 1975, it moved to a site of about 40 acres in Gweek.
It has a fully equipped seal hospital and 10 outdoor pools.

Opening hours
Open every day except Christmas
09 00 to 18 30

Feeding times
11 00 • 13 30 • 16 00

Cafe prices

Coffee		60p
Tea		40p
Cola		50p
Rolls:	Bacon	£1.50
	Sausage	£1.25
	Egg	£1.10
Chips		70p
Beans		40p
Salad		55p

Seals rescued in Cornwall (1990–91)

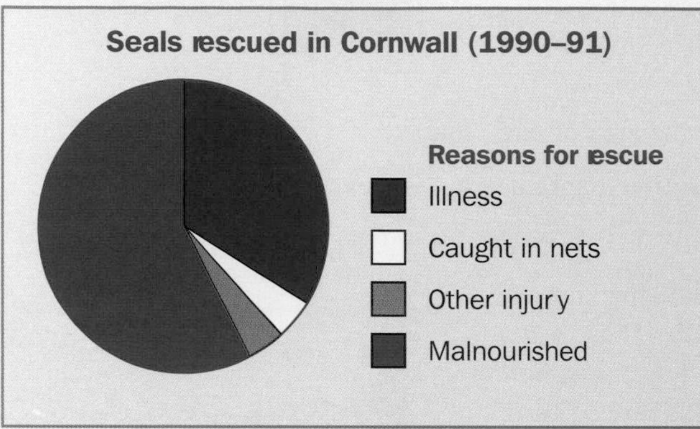

Ratio of male: female seals rescued between Land's End and Porthleven

Year	Ratio
1992–93	1:1
1993–94	1:2
1994–95	2:1

About the grey seals

Weights of pups rescued in December 1992

Name	Weight at rescue (kg)	Weight at release (kg)
Bill	23.4	65.0
Ben	16.8	56.0
Mandy	12.3	66.0
Tony	24.0	98.0
Rory	18.5	85.0

Seals are usually released when they weigh about 60 kg.

Seals at the sanctuary eat over a tonne of fish per week.

Food weight chart (hospital)

Weight of food per day (kg): 0.5, 0.6, 0.7, 0.8, 0.9, 1.0, 1.1
Weight of seal (kg): 10, 12, 14, 16, 18, 20

Major breeding areas for grey seals

Location	Pups born (1989)	Pups born (1990)	Total population in 1990 (to nearest 100)
Inner Hebrides	2051	2256	7800
Outer Hebrides	9537	9823	34000
Orkney	7038	7319	25400
Isle of May	933	1185	4100
Farne Islands	892	1004	3500

The world population of grey seals is estimated at 120000.

About two thirds of them live around the British coastline.

In water, seals can reach speeds of up to 20 km per hour.

1 How many years is it since the Seal Sanctuary was founded?

2 How long is it open each day?

3 In the cafe, Pritpal orders a cola, bacon roll and chips.
How much does this cost?

4 What was the most common reason for a seal being rescued in Cornwall in 1990–91?

5 What was the weight of the lightest pup rescued in December 1992?

6 About how many kilograms of fish do the seals at the Sanctuary eat each week?

7 What weight of food will be given to a 10 kg seal pup during a day in hospital?

8 What percentage of seals rescued between Land's End and Porthleven in 1992–93 were male?

9 How many pups were born in Orkney in 1990?

10 How long has the Seal Sanctuary been at Gweek?

11 Pritpal gets there at 8.25 am. How many minutes is it till the Sanctuary opens?

12 About what fraction of the seals rescued in 1990–91 were ill?

13 About what percentage of the seals rescued in 1990–91 were malnourished?

14 24 seals were rescued in Cornwall in 1990–91. About how many of them were ill?

15 What fraction of the seals rescued between Land's End and Porthleven in 1994–95 were female?

16 About what percentage of the seals rescued in Cornwall in 1990–91 were caught in nets?

17 4 seals were rescued between Land's End and Porthleven in 1992–93.
How many were male?

18 3 seals were rescued between Land's End and Porthleven in 1993–94.
How many were female?

19 In 24 hours in hospital, what weight of food is given to a 15 kg seal pup?

20 In a day, what weight of food in grams will be given to a pup that weighs 14 kg?

21 In a day, a pup in hospital is given 800 grams of food. Give the weight of this pup in kilograms.

22 Which of the pups rescued in December 1992 gained the most weight before being released?

23 Write Tony's rescue weight as a percentage of his release weight.

24 For the rescue weights, find:
a the mean **b** the median
c the range

25 Write each rescue weight rounded to the nearest kilogram.

26 Write each release weight in grams.

27 About how many kilograms of fish do the seals at the Sanctuary eat each day?

28 Out of the major breeding areas, which location had the highest total population of grey seals in 1990?

29 For each location in 1989, write down how many pups were born, to the nearest hundred.

30 For each location in 1990, write down how many pups were born, to the nearest thousand.

31 Draw a graph to show the number of pups born in 1989 and 1990 for these locations.

32 About how many grey seals live around the British coastline?

33 At 20 km per hour:
a How far could a seal travel in 2 hours?
b How long would it take to travel 35 km?
c How long would it take to travel 1 km?
d How far could a seal travel in 1 minute?

34 In June about 800 people visit the Seal Sanctuary each day.
In total, about how many people visit in June?

35 Susan arrives at the Sanctuary at 3.15 pm.
How long has she to wait for feeding time?

36 In the cafe, Susan asks her aunt for a hot drink and a roll.
If her aunt chooses a hot drink and roll at random, what is the probability of her choosing:
a a coffee and an egg roll?
b a tea and a bacon roll?

37 If cafe prices are increased by 15%, what is the new price of:
a a tea? **b** a sausage roll?
c chips?

Dale Valley Railway Spring 1997

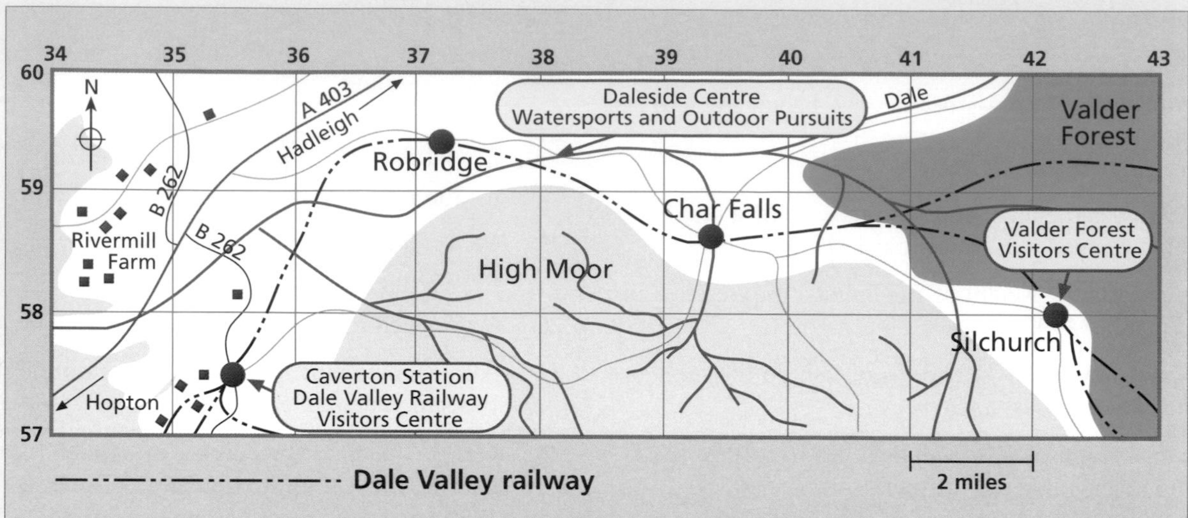

Dale Valley Railway

Timetable 1 May to 30 September

	Depart	Depart	Depart
Caverton Station	1000	1200	1430
Robridge	1012	1212	1445
Char Falls	1025	1225	1505
Silchurch	*a* 1040	*a* 1240	*a* 1520

	Depart	Depart	Depart
Silchurch	1105	1330	1610
Char Falls	1120	1345	1628
Robridge	1135	1400	1645
Caverton Station	*a* 1145	*a* 1410	*a* 1700

a – arrival time

** Note – in May and September there is no 1200 departure

Fares

Adult	£5.50
Child (under 14 years)	£3.00
Senior Citizen	£3.50

* Note all fares are return no single tickets are for sale

Festival Special

Adult	£12.50
Child (under 14 years)	£6.50
Senior Citizen	£9.50

* Note all fares include lunch at the Railway Arms in Silchurch

Information card

Weights and Measures

Locomotive	55 tons
Type A coach	27 tons
Type B coach	37 tons
Coal	1.5 tons
Water	500 gallons

1 gallon of water weighs 10 lb
1 ton = 2240 lb

	Tally for Festival Special
Adult	// ///////////////////////////
Child	///////////////////
Senior Citizen	////////////////////////// /////////////////////////////////////// ////////////////////////////

1 Estimate how far it is by rail from Caverton Station to Silchurch.

2 What area is shown by one square on the map?

3 Estimate the area of High Moor on the map.

4 Estimate the area of Valder Forest on the map.

5 When you travel from Robridge to Char Falls by road, is the railway on your left or right?

6 Which station is at the eastern end of the Dale Valley railway?

7 **a** Which station is roughly north-east of Caverton?
b Give the bearing of Robridge from Silchurch.

8 Give a 6-figure grid reference for:
a the Daleside Centre
b Rivermill Farm
c places where the railway crosses the River Dale
d where the B262 crosses the River Dale
e the Valder Forest Visitors Centre.

9 As a very good guide:
100 miles is equivalent to 160 kilometres.
Draw a conversion graph for miles to kilometres.

10 Estimate these distances in kilometres:

a 30 miles	**b** 55 miles	**c** 80 miles
d 25 miles	**e** 130 miles	**f** 285 miles

11 Estimate these distances in miles:

a 55 km	**b** 96 km	**c** 40 km
d 120 km	**e** 235 km	**f** 375 km

12 1 km is roughly what fraction of a mile?
Explain your answer.

13 In kilometres, roughly how far is it by rail from Char Falls to Caverton Station?

14 What is the mean journey time for a journey between Caverton and Silchurch stations.
Give your answer to the nearest minute.

15 For a return journey, the same day, what is:
a the shortest time between departure and return?
b the longest time?

16 Between 10 00 and 17 00, how long in total does the train spend in Silchurch?

17 Calculate the shortest journey time between Robridge and Char Falls.

18 **a** For how many days does the timetable run?
b How many return trips will the train make during the season?

Next year the Dale Valley Railway estimate that they will carry 25 000 passengers.
They expect the ratio of adults to children to be 5 : 3.

19 **a** How many adults do they expect to carry?
b How many children?

20 Will £120 000 be a good estimate of ticket sales? Explain your answer.

21 Calculate the mean number of passengers expected per day (1 May – 30 Sept).

22 DVR expect about 35% of the adults to be Senior Citizens.

How many Senior Citizens are expected?

23 The yearly running costs of the railway are estimated at £7000 per mile between Caverton and Silchurch.

Do you expect the railway to make a profit next year? Explain your answer.

24 To build a cafe, Dale Valley Railway plan to increase the ticket charges by 6%, then to round the new charge to the nearest 10 pence.

Give the new charge for each type of ticket.

25 The Festival Special ran on 1 May 1992.
Draw up a frequency table to show the number of tickets sold.

26 What was the total amount taken in ticket sales for the Festival Special?

27 To the nearest penny, what was the mean price of a ticket on the Festival Special?

28 Draw a pie chart to show the numbers of tickets sold for the Festival Special.

29 What is the weight of 500 gallons of water?

30 Which is heavier:
500 gallons of water or 1.5 tons of coal?

31 At the start of the day, the locomotive is loaded with 500 gallons of water and 1.5 tons of coal.
What is the total weight of the locomotive?

32 An eight coach train is made up of:
p type A coaches, and n type B coaches.
Write an equation with p and n for the number of coaches in the train.

33 **a** Write an expression in p and n, for the weight of the coaches in the train.
b Write an expression for the weight of coaches and loco.

SECTION 7 SOLVING EQUATIONS

Starting points

You need to know about ...

... so try these questions

A Finding the value of linear expressions

- Examples of linear expressions are: $2m + 4$ $6p - 4t$
- The value of an expression depends on the value of the letters.

When $p = 2$ and $t = 1$,

$$\begin{aligned} 6p - 4t &= (6 \times p) - (4 \times t) \\ &= (6 \times 2) - (4 \times 1) \\ &= 12 - 4 = 8 \end{aligned}$$

A1 Find the value of these expressions when $t = 5$.

a $5t - 8$ **b** $11 - t$
c $2(t + 7)$ **d** $3(t - 6)$

A2 Evaluate these expressions when $a = 2$ and $b = 5$.

a $4a + b$ **b** $3b - 4a$
c $7(a + b)$ **d** $2a - 3b$

B Adding like terms

- In the expression $2x + 4y + 3x + 5y$:
  - $2x$, $4y$, $3x$ and $5y$ are called **terms**.
  - $2x$ and $3x$ are called **like terms** as both give the number of xs.
- The expression can be simplified by adding like terms:

$$\begin{aligned} &2x + 4y + 3x + 5y \\ &= 2x + 3x + 4y + 5y \\ &= 5x + 9y \end{aligned}$$

$2x + 4y + 3x + 5y$ and $5x + 9y$ are **equivalent expressions**.

B1 Which of these is equivalent to $5x + 2y + 3x + y$?

A $15x + 2y$ B $8x + 2y$
C $8x + 3y$ D $15x + 3y$

B2 Simplify each of these by adding like terms.

a $5t + 3t + t$
b $7s + 3s + 5k + 4k$
c $2c + 6b + c + 2b$
d $6x + 2 + 5x$
e $5v + 6 + 2v + 10$

C Solving linear equations

- Solving an equation is finding the possible values for each letter.
- With the value of one letter to find, add, subtract, multiply or divide **both** sides of the equation by equal amounts.

Example

Solve $6n - 2 = 2n + 8$

$$\begin{aligned} 6n - 2 &= 2n + 8 && (+2) \\ 6n &= 2n + 10 && (-2n) \\ 4n &= 10 && (\div 4) \\ n &= 2.5 \end{aligned}$$

The **solution** of this equation is $\boldsymbol{n = 2.5}$.

C1 Which of these is the solution of: $4x - 5 = 2x + 7$?

A $x = 1$ B $x = 2$ C $x = 6$

C2 Solve:

a $6z + 1 = 10$
b $5y - 2 = 13$
c $3x + 14 = 8$
d $2w + 5 = 3w + 1$
e $5 + v = 2v - 3$
f $6t + 8 = t + 3$
g $5s - 1 = 7s - 5$

D Linear graphs

- An example of an equation of a straight line is $y = 3x + 5$.
- For any x-coordinate, you can calculate the y-coordinate.

When $x = {}^{-}1$,

$$\begin{aligned} y &= (3 \times {}^{-}1) + 5 \\ &= {}^{-}3 + 5 = 2 \end{aligned}$$

So the line $y = 3x + 5$ goes through the point $({}^{-}1, 2)$.

D1 Which of these points is on the line $y = 5x - 1$?

A $(1, 4)$ B $(2, 11)$ C $({}^{-}1, {}^{-}6)$

D2

x	$^{-}2$	$^{-}1$	0	1	2
y					

Copy and fill this table for:

a $y = 2x + 5$ **b** $y + x = 6$

Thinking ahead to ... simplifying linear expressions

(P) $10 + (1 \times 0.6)$ (Q) $10 + (2 \times 0.6) - (3 \times 0.6)$ (R) $10 - (5 \times 0.6)$

(S) $10 - (1 \times 0.6)$ (T) $10 - (2 \times 0.6) - (3 \times 0.6)$

(U) $10 - (2 \times 0.6) + (3 \times 0.6)$

A Sort these calculations into pairs with the same value.

Simplifying linear expressions

To simplify an expression
- reorder the terms if you need to
- add or subtract like terms

Examples

$6y - x - 3x$
$= 6y - 4x$

$5p + 6 - 3p - 1$
$= 5p - 3p + 6 - 1$
$= 2p + 5$

$7a - 3b - 4a + 2b - a$
$= 7a - 4a - a - 3b + 2b$
$= 2a - b$

Exercise 7.1 Adding and subtracting like terms

1 Simplify each of these expressions:

a $4t + 7t - 3t$ **b** $6x - 5 - x + 5$ **c** $6 + 4h - 3 + h$
d $3a - 2b + 2b - 2a$ **e** $2 - y + 6 - 4y - 7$ **f** $10 - 5f - 4 + 2f$
g $3x + 4y - 5 - 2x$ **h** $3x - 5y - x$ **i** $5x - 4 - 3x + 2y$
j $2ax + 3x + 4ax$ **k** $xy + 2x + y + xy$ **l** $4x - 3 - 2y - 7$
m $3x + 4y + x + y - 3$ **n** $5y - x - 2y - 3x$ **o** $4 - 5y - x - 2y$
p $12 - 7y - 5x - 8y$ **q** $7x - 3 - 8x + 2y$ **r** $ax + ay + 3ax + y$

2 Some expressions are arranged in a square.

$3x - y$	$x - 2y$	$2x$
x	$2x - y$	$3x - 2y$
$2x - 2y$	$3x$	$x - y$

In a magic square, the numbers in each row, each column and each diagonal add to give the same total.

a Find the value of each expression in the square when $x = 5$ and $y = 2$.
b Draw the square with these values. Is it a magic square?
c For each row, column and diagonal, find the total of the three expressions.
d Explain why your totals show that any values for x and y will give a magic square.

Thinking ahead to ...
using brackets

A Sort these calculations into pairs with the same value.

(A) $200 - 30$ (B) $5 \times (100 + 6)$ (C) $5 \times (100 + 30)$

(D) $5 \times (100 - 30)$ (E) $500 + 30$ (F) $2 \times (100 - 15)$

(G) $(5 \times 100) + (5 \times 30)$ (H) $(5 \times 100) - (5 \times 30)$

Using brackets in linear expressions

To multiply out brackets:

- Multiply **every** term inside the brackets.

Examples

$$2(n + 3) = 2 \times (n + 3) = (2 \times n) + (2 \times 3) = 2n + 6$$

$2(n + 3)$ and $2n + 6$ are **equivalent expressions**

$$5(3n - 2) = 5 \times (3n - 2) = (5 \times 3n) - (5 \times 2) = 15n - 10$$

$5(3n - 2)$ and $15n - 10$ are equivalent expressions

Exercise 7.2
Using brackets

1 For $n = 5$, find the value of $2(n + 3)$ and $2n + 6$.

2 For $n = 2$, find the value of $5(n - 2)$ and $5n - 10$.

3 Sort these into pairs of equivalent expressions.

$2(a + 4)$	$8a - 12$	$2(a + 2)$	$2a + 8$
$4(2a - 3)$	$2a - 12$	$2a + 4$	$2(a - 6)$

4 A student has tried to multiply out the brackets from $3(4 + x)$.

$3(4 + x) = 4 + 3x$ ✗

a Explain the mistake you think she has made.
b Multiply out the brackets from $3(4 + x)$.

5 Two students find expressions for the nth term in a sequence.

Sue: nth term = $5(2n - 1)$

Ahmet: nth term = $10n - 5$

Show that the two expressions are equivalent.

6 Multiply out the brackets from:

a $4(n + 1)$ **b** $5(m - 3)$ **c** $6(c - 9)$ **d** $2(3p + 2)$
e $8(2s - 5)$ **f** $4(3 + t)$ **g** $3(5 - k)$ **h** $2(7n - 5)$
i $30(2f - 13)$ **j** $180(5 + 3h)$ **k** $10(3 - 7q)$ **l** $5(2y + 3z)$

7 Show that $2(n + 2) + 2n$ and $4n + 4$ are equivalent expressions.

8 Expand and simplify:

a $5(p + 2) + 3p$ b $4(2q - 1) - 3q$ c $3(x + 2) - 4(x - 5)$
d $4r + 10(3 + 5r)$ e $3(2s - t) + 5t$ f $2(x - 3) + 5(x + 1)$
g $3(x - 4) - 5(x - 2)$ h $2(x - 3) + 3(x - 1)$ i $5(2x - 3) - 4(3x + 2)$

Solving linear equations

You can check by finding the value of both sides of the equation for your solution. The values should be equal.

For linear equations with brackets you can start by multiplying out the brackets.

Examples

- Solve $2(p + 4) = 8p - 1$

$$2(p + 4) = 8p - 1$$
$$2p + 8 = 8p - 1 \quad (-2p)$$
$$8 = 6p - 1 \quad (+1)$$
$$9 = 6p \quad (\div 6)$$
$$1.5 = p$$

- Solve $3(4 - t) + 5 = 13 - 5t$

$$3(4 - t) + 5 = 13 - 5t$$
$$12 - 3t + 5 = 13 - 5t$$
$$17 - 3t = 13 - 5t \quad (+5t)$$
$$17 + 2t = 13 \quad (-17)$$
$$2t = {}^{-}4 \quad (\div 2)$$
$$t = {}^{-}2$$

Exercise 7.3
Solving linear equations

1 Solve these equations.

a $3x + 5 = 17$	**b** $4y - 3 = 21$	**c** $5x + 8 = 43$
d $2(w + 8) = 52$	**e** $3(2v - 1) = 21$	**f** $6(3 + 5u) = 54$
g $4(t + 1) = 33$	**h** $4(3s - 1) = 4s + 39$	**i** $5(2r + 3) = 7r + 42$
j $2(q - 3) + 3q = 3$	**k** $10 - p = 4p$	**l** $14 - 4n = 2(1 + 2n)$
m $3m - 5 = 11 - m$	**n** $2(5l - 3) = 48 - 5l$	**o** $5(2k - 3) = 15 - 2k$
p $4j - 7 = 3(18 - j) + 9$	**q** $17 - 2h = 22 - 3h$	**r** $2(9 - 2g) = 25 - 6g$

Each equation has a negative number as its solution.

2 Solve these equations.

a $2z + 12 = 3z + 17$	**b** $4(y + 6) = 20$	**c** $5(2x + 7) = 15$
d $5(w + 5) = 2(w + 8)$	**e** $v + 17 = 9 - v$	**f** $3u + 35 = 5 - 2u$
g $2(3 - t) = 4t + 9$	**h** $8 - 5s = 2 - 10s$	**i** $2r + 29 = 1 - 6r$

$\frac{2x + 1}{4}$ stands for $(2x + 1) \div 4$

3 Solve these equations.

a $4z + 1 = 10z - 1$	**b** $2(3y - 1) = 12y - 7$	**c** $\frac{2x + 1}{4} = 2$
d $\frac{1}{4}(3w + 1) = 4$	**e** $\frac{1}{3}(2v - 1) = v - 1$	**f** $\frac{5u + 6}{2} = 4$

Forming and solving linear equations

For some problems, you can form a linear equation and then solve it.

Example

The lengths of the sides of a triangle are y cm, $2y$ cm and $(y + 3)$ cm.

If the perimeter of the triangle is 63 cm, what is the length of each side?

2y

y

y + 3

The perimeter is the total distance around the outside edge of a shape.

- The perimeter, in terms of y, is: $y + 2y + y + 3 = 4y + 3$
- The perimeter is 63 cm so: $4y + 3 = 63$
- Find the value of y: $4y = 60$, $y = 15$
- Find the value of $2y$ and $y + 3$: $2y = 30$ and $y + 3 = 18$
- So the lengths of the sides are: **15 cm, 30 cm and 18 cm.**

Exercise 7.4 Forming and solving linear equations

1 The lengths of the sides of a triangle are x cm, $2x$ cm and $(3x - 4)$ cm.

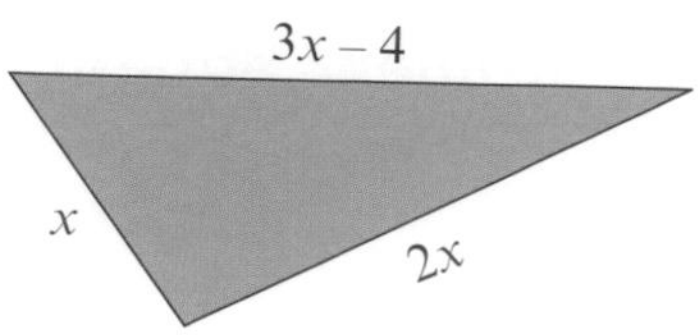

An expression in terms of x does not include any letters other than x.

a List the lengths of the sides when $x = 5$.
b What is the perimeter when $x = 4$?
c What is the perimeter of the triangle in terms of x?
d If the perimeter is 104 cm, form an equation in x and solve it to find the length of each side.

2 A square has sides of length $2p$ metres.
A rectangle has width p metres and length $(p + 5)$ metres.

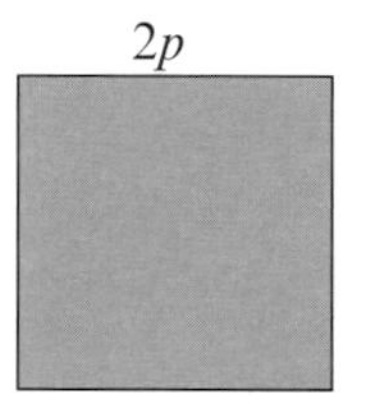

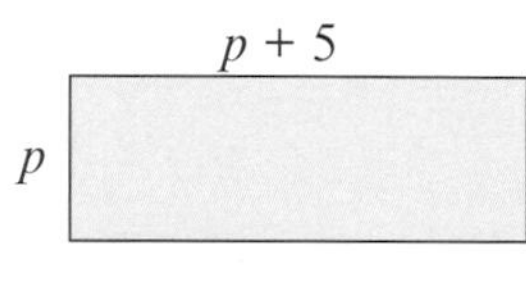

a What is the perimeter of each shape when $p = 10$?
b Find the perimeter of the square in terms of p.
c Find the perimeter of the rectangle in terms of p.
d If the perimeter of the rectangle is 90 m, how long is each of its sides?
e Find a value of p so that the perimeters of the square and rectangle are equal.

3 An equilateral triangle has sides of length $(t + 1)$ cm.
A rectangle has width t cm and length $2t$ cm.

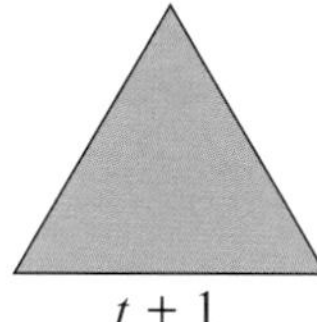

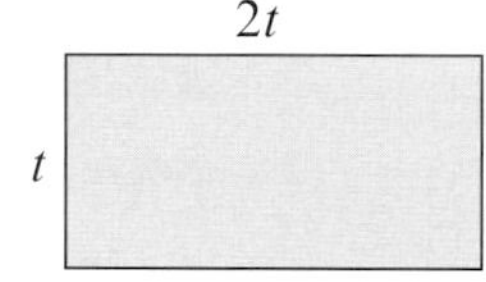

Find a value of t so that the perimeter of the triangle and rectangle are equal.

4 These expressions are arranged in a magic square.

$2x$	$x - 2$	$3x - 1$
$3x - 2$	$2x - 1$	x
$x - 1$	$3x$	$2x - 2$

The magic total is the total of the numbers in each row, column or diagonal.

a What value for x gives 21 at the centre of the magic square?
b Draw the complete magic square.
c Find a value for x that gives a magic square with a magic total of 117.
d i Make a magic square where the four numbers in the corner squares add to give 100.
ii Explain how you decided on a value for x.
e With these expressions, explain why the total of the numbers in the corner squares will always be four times the number in the centre square.

Some number puzzles can be solved using equations.

Example

I think of a number, add 1, and then double. I get the same answer if I subtract my number from 14. What is my number?

> The left-hand side of the equation is for 'add 1 and then double' $2(n + 1)$ is equivalent to $(n + 1) \times 2$.
>
> The right-hand side is for 'subtract ... from 14' $14 - n$ is **not** equivalent to n 14.

- Choose a letter to stand for the number: n is the number
- Write an equation for the puzzle: $2(n + 1) = 14 - n$
- Solve the equation to find the number: $2(n + 1) = 14 - n$

$$2n + 2 = 14 - n \quad (+n)$$
$$3n + 2 = 14 \quad (-2)$$
$$3n = 12 \quad (\div 3)$$
$$n = 4$$

- So the number is **4**.

Exercise 7.5
Solving number puzzles

1 I think of a number, subtract 2 and then double. I get the same answer if I multiply my number by 3 and subtract from 21. What is my number?

Which of these equations fits the number puzzle?

A $2(n - 2) = 3n - 21$

B $2n - 2 = 21 - 3n$

C $2n - 2 = 21 - 3n$

D $2(n - 2) = 21 - 3n$

E $n - 4 = 21 - 3n$

> Use n to stand for the number each time.

2 For each puzzle A to E, write an equation and solve it to find the number.

A I think of a number, multiply it by 3, and add 5. I get the same answer if I subtract my number from 13. What is my number?

B I think of a number, subtract 2, and multiply by 4. I get the same answer if I subtract 3 and multiply by 5. What is my number?

C I think of a number, subtract 1, and multiply by 3. I get the same answer if I subtract my number from 8 and multiply by 4. What is my number?

D I think of a number, multiply it by 6, and subtract from 10. I get the same answer if I multiply my number by 3 and subtract from 7. What is my number?

E I think of a number, double it, and subtract 11. I get the same answer if I double my number and subtract from 15. What is my number?

3 **a** Write a number puzzle for the equation $3(n - 5) = 20 - 2n$
b Solve the equation to find the value of n.

4 Make up some number puzzles for someone else to solve.

Using graphs to solve problems with two values to find

For some problems, you can form two equations and use graphs to solve them.

Example

Two families visit a funfair.
One family buys 1 adult ticket and 2 child tickets. The total is £10.
The other family buys 2 adult tickets and 1 child ticket. The total is £14.

How much is each type of ticket?

- Choose letters to stand for each type of ticket:
  Use a to stand for the cost in pounds of an adult ticket.
  Use c to stand for the cost in pounds of a child ticket.
- Write an equation for each family: $a + 2c = 10$
  $2a + c = 14$
- Draw up a table of values for each equation:

Negative values for a and c are not included as the cost of a ticket cannot be negative.

$a + 2c = 10$

a	0	1	2	3	4	5
c	5	4.5	4	3.5	3	2.5

$2a + c = 14$

a	0	1	2	3	4	5
c	14	12	10	8	6	4

- Draw a graph for each equation, so that the lines cross at a point:

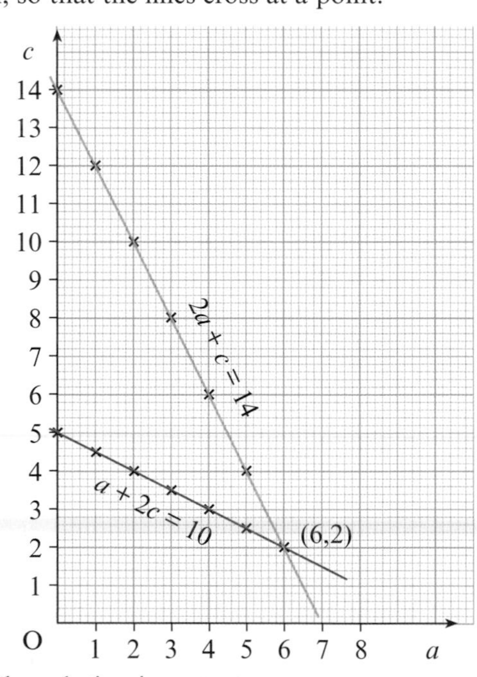

- Where the lines cross, the values of a and c fit both equations, so they give the cost of each ticket:

  The lines cross when $a = 6$ and $c = 2$,

  So an adult ticket costs £6 and a child ticket costs £2

- Check your solution:

  When $a = 6$ and $c = 2$,

  $$a + 2c = 6 + (2 \times 2) = 10$$

  $$2a + c = (2 \times 6) + 2 = 14$$

- These are the correct values so the solution is correct.

Exercise 7.6
Using equations and graphs

£1.30 = 130p

1 Two people visit a cafe.
Rita pays 70p for 1 tea and 2 biscuits.
Simon pays £1.30 for 3 teas and 1 biscuit.

a Write an equation for each person, using t for the cost of a tea in pence and b for the cost of a biscuit in pence.

b Draw up a table of values like this for each equation for values of t from 0 to 40.

t	0	10	20	30
b				

c Draw a graph for each equation so that the lines cross.

d From your graphs, how much does each item cost?

In Questions **2**, **3** and **4** the letters can stand for any type of number, including decimals and negative numbers.

2 Equation 1 $x + y = 6$

Equation 2 $3x + y = 9$

- **a** For each equation, what is the value of y when $x = 4$?
- **b** Draw up a table of values for each equation with these values of x: 0, 1, 2, 3, 4, 5.
- **c** On one set of axes, draw a graph for each equation.
- **d** What are the values of x and y at the point where the lines cross?
- **e** Check that these values fit Equations 1 and 2.

3 Equation 1 $y = x + 4$

Equation 2 $y = 2x + 5$

- **a** For each equation, what is the value of y when $x = {}^{-}2$?
- **b** Draw up a table of values for each equation with these values of x: $^{-}4$, $^{-}2$, 0, 2, 4.
- **c** On one set of axes, draw a graph for each equation.
- **d** Use your graphs to find the values of x and y that fit both equations.

4 A $y = x - 5$, $y = 3x - 11$ B $y = 2x + 6$, $y = 4x + 11$ C $y = 2x$, $y = 7x - 9$

For each pair of equations:

- **a** Draw up a table of values for each equation with these values of x: $^{-}4$, $^{-}2$, 0, 2, 4.
- **b** On one set of axes, draw a graph for each equation.
- **c** Use your graphs to find the values of x and y that fit both equations.

Solving problems without using graphs

Exercise 7.7
Solving problems without using graphs

1 In a cafe, 2 teas and 4 coffees cost £4.60.
From this information, which of these can you find the cost of?

A 4 teas and 8 coffees B 1 tea and 4 coffees

C 1 tea and 2 coffees D 1 coffee E 6 teas and 8 coffees

2 In a shop:
1 cola and 3 bags of crisps cost £1.52
2 colas and 4 bags of crisps cost £2.48

Find the cost of:

- **a** 2 colas and 6 bags of crisps
- **b** 1 cola and 2 bags of crisps
- **c** 3 colas and 7 bags of crisps
- **d** 1 cola and 1 bag of crisps
- **e** 1 bag of crisps
- **f** 1 cola

3 A woman has twins.
The twins are the same height.
Adding the height of the woman to the height of one twin gives 280 cm.
The total height of the woman and her twins is 400 cm.

- **a** What is the height of each person?
- **b** Make up a problem like this for someone else to solve.

Thinking ahead to ...
using algebra

A A value for p and a value for q fit both these equations.

$3p + 4q = 23$

$p + 2q = 11$

a Copy and complete these equations for p and q.
i $3p + 6q = \square$ **ii** $2p + \square = 22$ **iii** $4p + 6q = \square$
iv $2p + 2q = \square$ **v** $p + q = \square$

b Find the values of p and q.

Using algebra to solve problems with two values to find

Values that fit two equations can be found using algebra.

Example

Solve these equations to find the values of a and b.

$$2a + b = 19$$
$$3a + 4b = 26$$

- Label the equations (1) and (2): $2a + b = 19 \ldots (1)$; $3a + 4b = 26 \ldots (2)$
- Multiply both sides of equation (1) by 4 to give two equations with '+ $4b$': (1) × 4 ... $8a + 4b = 76 \ldots (3)$; $3a + 4b = 26 \ldots (2)$
- Subtract: (3) – (2) ... $(8a - 3a) + (4b - 4b) = 76 - 26$; $5a = 50$
- Find the value of a: $a = 10$
- Substitute the value of a in one equation to find the value of b: $2a + b = 19 \ldots (1)$; $(2 \times 10) + b = 19$; $20 + b = 19$; $b = {}^-1$
- **So the solution is $a = 10$, $b = {}^-1$.**

Subtract to remove '+ $4b$' from each equation:
$4b - 4b = 0$

Equation (2) could also be used to find the value of b:

$$3a + 4b = 26$$
$$(3 \times 10) + 4b = 26$$
$$30 + 4b = 26$$
$$4b = {}^-4$$
$$b = {}^-1$$

Exercise 7.8
Using algebra

1 For each pair of equations, use algebra to find the values of x and y.

a $x + 4y = 42$, $2x + 5y = 57$ **b** $11x + 3y = 91$, $3x + y = 25$ **c** $5x + 7y = 32$, $x + 3y = 12$

2 For which pair of equations is it true that $s = 5$ and $t = 2$?

A $s + 3t = 11$, $5s + t = 32$

B $3s + t = 17$, $4s + 5t = 30$

C $3s + 2t = 20$, $s + 2t = 9$

3 A value for m and a value for n fit both these equations.

$2m + 3n = 28$ ... (1)
$3m + 4n = 37$ ... (2)

Multiply **both** equations to give two equations with '$6m$'.

a Multiply equation (1) by 3.
b Multiply equation (2) by 2.
c Subtract to find the value of n that fits both equations.
d Substitute in one of the equations to find the value of m.

4 For each pair of equations, use algebra to find the values of v and w.

a $2v + 3w = 40$, $5v + 2w = 34$
b $3v + 2w = 3$, $6v + 10w = 24$
c $4v + 2w = 9$, $3v + 7w = 4$

Sometimes it is simpler to **add** the equations.

Example

Solve these equations to find the values of x and y.

$$6x - 2y = 18$$
$$5x + 3y = 1$$

- Label the equations (1) and (2): $6x - 2y = 18$ (1), $5x + 3y = 1$ (2)
- Multiply equation (1) by 3 and equation (2) by 2 to give '$-6y$' and '$+6y$':
  (1) × 3: $18x - 6y = 54$ (3)
  (2) × 2: $10x + 6y = 2$ (4)
- Add: (3) + (4): $(18x + 10x) + (-6y + 6y) = 54 + 2$
  $28x = 56$
- Find the value of x: $x = 2$
- Substitute the value of x in one equation to find the value of y:
  $5x + 3y = 1$ (2)
  $(5 \times 2) + 3y = 1$
  $10 + 3y = 1$
  $3y = {}^{-}9$
  $y = {}^{-}3$
- **So the solution is $x = 2$, $y = {}^{-}3$.**

This is one way to use algebra to solve this problem. There are other ways.

Add to remove '$-6y$' and '$+6y$': $-6y + 6y = 0$

Exercise 7.9 Using algebra

1 For each pair of equations, use algebra to find the values of a and b.

a $a - b = 8$, $4a + b = 42$
b $5b + 2a = 29$, $b - 2a = 1$
c $3a - b = 15$, $4a + 2b = 25$
d $b + 3a = 2$, $3b - a = 26$
e $5a + 2b = 17$, $2a - 3b = 3$
f $7a + 5b = 27$, $3a - 2b = 24$

2 Two numbers m and n fit both these equations.

$5m - n = 15$
$3m - n = 5$

Subtract to remove '$-n$' from each equation: $(-n) - (-n) = 0$

a Subtract to find the value of m that fits both equations.
b Substitute in one of the equations to find the value of n.

3 Find the values of p and q that fit: $6p - 2q = 16$
and $p - 2q = 1$

4 Two numbers m and n fit both these equations:

$5m - 2n = 28$ (1)
$7m - 5n = 37$ (2)

Here we multiply **both** equations to give two equations with '$-10n$'.

a Multiply equation (1) by 5.
b Multiply equation (2) by 2.
c Subtract to find the value of m that fits both equations.
d Substitute in one of the equations to find the value of n.

5 For each pair of equations, use algebra to find the values of x and y.

a $5x - 2y = 16$, $2x - 3y = 2$ **b** $4y - x = 17$, $3y - 4x = 3$ **c** $5x - 3y = 7$, $2x - 4y = 0$

d $5y + 3x = 15$, $5y + 7x = 25$ **e** $4x + 3y = 17$, $5x - 7y = 32$ **f** $3x + 2y = 13$, $5x - 6y = 59$

Using algebra to solve word problems

Example

A father's age and his son's age add to give 54.
The father is 30 years older than his son.

How old is his son?

- Use f to stand for the father's age and s to stand for the son's age:
- The ages add to give 54, so: $f + s = 54$ ... (1)
- The father is 30 years older than the son, so: $f - s = 30$ ... (2)
- Add: (1) + (2) ... $(f + f) + (s + -s) = 54 + 30$

$2f = 84$

- Find the value of f: $f = 42$
- Substitute the value of f in one equation to find the value of s: $f + s = 54$ ... (1)

$42 + s = 54$
$s = 12$

- **So the son is 12 years old.**

Exercise 7.10
Solving word problems

1 Susan's age and her brother's age add to give 73.
Susan is 3 years older than her brother.
Find the ages of Susan and her brother.

2 A bag contains a mixture of large and small marbles.
Each small marble weighs 2 g.
Each large marble weighs 5 g.
The total weight of the marbles in the bag is 256 g.
Altogether there were 89 marbles in the bag.

Use x to stand for the number of small marbles.
Use y to stand for the number of large marbles.

a Use this information to write two equations in x and y.
b Solve these equations to find the number of each type of marble.

End points

You should be able to ...	... so try these questions

A Add and subtract like terms in linear expressions

A1 Simplify:

- **a** $6k + 8m - 4k + m$
- **b** $5p - 1 - 3p + 9$

B Form and solve linear equations

B1 Solve:

- **a** $2z - 1 = 12$
- **b** $4y + 1 = 2y + 7$
- **c** $6(x - 1) = 3$
- **d** $15 - 3w = w + 11$
- **e** $3(2v + 7) = v + 6$
- **f** $2(t + 3) = 18 - 6t$

B2 The lengths of the sides of a triangle are x cm, $(x + 8)$ cm and $(x - 6)$ cm.

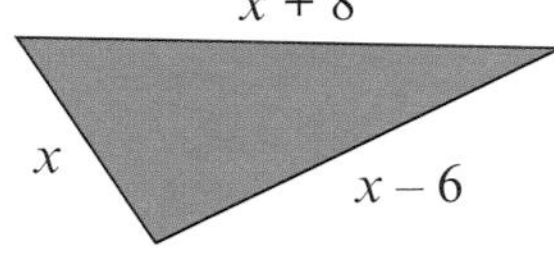

- **a** What is the perimeter of the triangle in terms of x?
- **b** The triangle has a perimeter of 56 cm. Write down an equation in x, and solve it to find the length of each side of the triangle.

B3

> I think of a number, double it and subtract 1. I get the same answer if I multiply my number by 4 and subtract from 14. What is my number?

Write an equation for this number puzzle and solve it to find the number.

C Solve problems using graphs

C1 Equation 1 $y = 2x - 3$

Equation 2 $y = 4x - 4$

- **a** Draw up a table of values for each equation with these values of x: $^{-}4$, $^{-}2$, 0, 2, 4.
- **b** On one set of axes, draw a graph for each equation.
- **c** Use your graphs to find the values of x and y that fit both equations.

D Solve problems using algebra

D1 For each pair of equations, use algebra to find the values of m and n.

- **a** $3m + 2n = 5$, $5m + 2n = 8$
- **b** $2m + 3n = 5$, $3m + 4n = 6$
- **c** $2m - n = 22$, $3m + 2n = 40$

D2 A slot machine takes only 20p and 50p coins.
It contains a total of 140 coins.
The value of the coins is £45.10.

Use x to stand for the number of 20p coins.
Use y to stand for the number of 50p coins.

- **a** Show that $20x + 50y = 4510$.
- **b** Write down a different equation in x and y.
- **c** Use your equations to find the number of each type of coin.

Some points to remember

- When solving an equation, add, subtract, multiply or divide **both** sides by equal amounts.
- Problems where two values have to be found can be solved using graphs or algebra. The most accurate answer can be found using algebra.

Starting points

You need to know about ... / ... so try these questions

A Rounding numbers

Rounding is a way to approximate numbers when an exact value is not needed.

Whole numbers are usually rounded to the nearest 10, 100, 1000 and so on.

Example

3837.6 rounded to the nearest whole number is 3838
to the nearest 10 is 3840
to the nearest 100 is 3800
to the nearest 1000 is 4000
to the nearest 10 000 is 0

Decimals are rounded to a given number of decimal places (dp).

Example

3.6748 rounded to 1 dp is 3.7
to 2 dp is 3.67
to 3 dp is 3.675

Halfway numbers can be rounded either way but are usually rounded up.

Example

735 to the nearest ten is 740
56.75 to 1 dp is 56.8

A1 Round 2175.6 to the nearest:
a thousand **b** hundred
c ten **d** whole number

A2 Round 45.638 to:
a 2 dp **b** 1 dp

A3 Round these numbers.
a 34.597 to 2 dp
b 2.501 to the nearest whole number
c 38.45 to 1 dp
d 3496 to the nearest ten

A4 Which numbers are not 2.56 when rounded to 2 dp?
a 2.5555 **b** 2.5648
c 2.5651 **d** 2.550 99
e 2.5500 **f** 2.5666

B Adding and subtracting decimals

When you add or subtract decimal numbers you may find it easier to arrange the digits in columns.

Example 1

45.346 + 8.6 + 237

1000	100	10	U		$\frac{1}{10}$	$\frac{1}{100}$	$\frac{1}{100}$	
		4	5	.	3	4	6	
			8	.	6			+
	2	3	7	.				
	2	9	0	.	9	4	6	

Example 2

34.6 – 2.784

3	4	.	6	0	0	–
	2	.	7	8	4	
3	1	.	8	1	6	

B1 Add these decimal numbers:
a 45.73 + 8.423 + 123.6
b 14 + 0.563 + 28.9
c 0.004 + 0.03 + 0.95 + 3
d 17.8 + 3425 + 0.0895

B2 Subtract these numbers:
a 34.78 – 6.8
b 14 – 2.83
c 154.36 – 4.5
d 256 – 23.764

B3 What mistake has been made here?
34.5 + 2.34 = 57.9

B4 Explain the mistake made in this calculation.
34.6 – 2.278 = 32.478

Rounding up or down?

People make estimates every day.
They often base an estimate on a calculation they do in their head.

Sometimes it is best to **overestimate** so they **round up** their answer.
At other times it is best to **underestimate** and **round down**.

We should be able to cycle about 78 miles each day so how far apart do we want the hostels to be?

In this case it might be better to **round down** the 78 to say 60 miles a day just in case they felt tired or had an accident.

I've worked out that we need 4756 bricks for the extension. How many should I order?

Here, it would be better to **round up** the number of bricks. An order of say 5000 bricks would allow for breakage or error. Bricks ordered later might not be exactly the same colour.

Exercise 8.1
Rounding up or down?

1 In each of these situations do you think it is better to round up, round down, or not to round at all. Explain why.

a You have to draw out some money from the bank.
You work out that you need £8.35 for your trip.

b The seat number on your concert ticket is 213.
You must decide where to sit.

c You calculate that you need 11 rolls of wallpaper for your room.
You go into the shop to buy the paper.

d Don lives at 26 Hayward Road. You decide to pay him a call.

e You think you may earn £360 from your holiday job.
You look at hi-fis you think you will be able to afford.

2 Describe a new situation for each of these:

a when it would be a good idea to round up

b when it would be best to round down

c when it would be silly to do any rounding.

3 For each of these situations decide if it is better to round up or down.
Say what number you would use, and explain why.

a You calculate that the gap for a desk is 98 cm wide.
You have to decide what width of desk to ask for.

b The milometer in your sister's car reads 67 673 miles.
You advertise the car with the mileage it has done.

c You expect 74 people for the school-leavers meal.
You have to hire some glasses.

d You calculate that you need 15.3 metres of wood for some shelves.
You have to buy the wood.

Significant figures

Exercise 8.2
Rounding

1 In this extract some numbers are given to a greater accuracy than they need to be.

Practical Green Keeping

March edition

Crew measure up while the games are on

THE MAINTENANCE CREW arrived at the stadium at 10:43 while the athletics events were taking place. The 71 934 crowd was already seated when measuring up started. The perimeter railings were measured as 63 479.6 cm long and the supports as 19.6 cm thick. At one point the crowd rose to its feet as Mary Taylor set a new European record of 10.84 sec for the 100 metres sprint. From their calculations the crew estimated the area of grass which needed re-seeding was 1452.56 metres^2. The measuring was completed in about 56 minutes with little disruption to the 492 or so competitors.

The maintenance contract with the sports committee expires in 2003

'Appropriate' means 'sensible'.

'Significant' in this case means 'important'.

a List the numbers which you think are more accurate than is appropriate. Write what you think each one should be rounded to.

b Which numbers should not be rounded? Explain why.

c In both 63 479.6 and in 19.6 the last digit stands for $\frac{6}{10}$.
In which of these numbers do you think this 6 is more significant? Explain why.

2 Draw a line on your page.
Measure its length as accurately as you can.
How many digits are in the number you have written?

3 Estimate how far it is from Land's End to John O'Groats.
How many of the digits in your answer are not zero?

4 The digit 4 is in both of these numbers: **24.6** **10 243**

a In which number does the 4 have the greater actual value?
b In which number do you think the 4 is more significant?

5 To what accuracy do you think a 100 metre running track must be measured when it is marked out?

Significant figures means the most important digits in a number.

In 6351.2 the 6 and 3 are the two most significant figures because they show the largest numbers 6000 and 300.

Significant figures can be written as **sf**.

Ways of approximating include rounding a number to the nearest ten or to a set number of decimal places.

Another way is to round a number is to a set number of significant figures.

63 479.6 rounded to **2 significant figures** is 63 000

The three zeros are added to keep the value of the number about the same.

63 479.6 rounded to **3 significant figures** is 63 500

Note how the 4 has rounded up to 5 because the next digit 7 is above halfway.

63 479.6 rounded to **4 sf** is 63 480

63 479.6 rounded to **5 sf** is 63 480

Exercise 8.3
Rounding using significant figures

1 Round each of these numbers to 3 sf:

a 1452.56 **b** 21 675 **c** 142.51 **d** 2 134 518.4 **e** 149 625

2 Round 71 934 to:

a 1 sf **b** 2 sf **c** 3 sf **d** 4 sf

3 When 63 479.6 is rounded to 4 sf or 5 sf the answer is the same.
Why do you think this is?

4 A number, rounded to 2 sf, is 3200.
Give three numbers it could be.

5 Copy and complete this table.

Number	45 287	2395	302 604.32	14.823
to 2 sf				
to 3 sf				
to 4 sf				

6 The grass that needed re-seeding in the stadium was a rectangle this size:

54.2 m

26.8 m

Area = 1452.56 m^2

Why do you think it is not sensible to use all the digits for the area?
What would you round the area to?

In calculations you should either:

- give your answer to the degree of accuracy asked for e.g. 1 dp or 2 sf, or
- give your answer to the same degree of accuracy as is used in the question.

7 A number, when rounded to 2 sf or to 3 sf is 420 000.
Give an example of what the number might be.

8 To how many significant figures could a number be rounded to give 164 000?

Numbers less than 1 can also be rounded using significant figures.

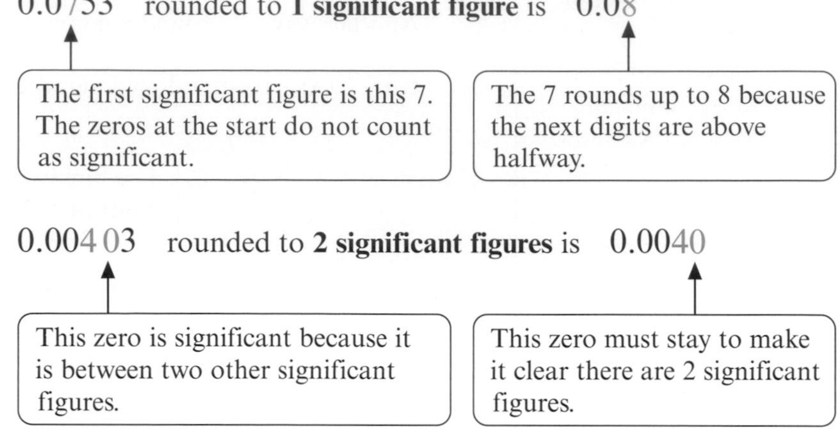

0.0753 rounded to **1 significant figure** is 0.08
The first significant figure is this 7. The zeros at the start do not count as significant.
The 7 rounds up to 8 because the next digits are above halfway.
0.004 03 rounded to **2 significant figures** is 0.0040
This zero is significant because it is between two other significant figures.
This zero must stay to make it clear there are 2 significant figures.

Exercise 8.4
Significant figures

1 The width of a human hair found at the scene of a crime was 0.007 64 cm. Round this number to 1 sf.

2 When 0.000 524 6 is rounded to 2 sf it becomes 0.000 52. 0.000 520 0 is not really correct. Explain why.

3 For these conversions round the blue numbers.

Metric/Imperial conversions

	to 1 sf	to 2 sf	to 3 sf
1 inch = 0.0254 metres	0.03	0.025	0.0254
1 yard = 0.009 144 kilometres			
1 millimetre = 0.0394 inches			
1 kilometre = 0.6214 miles			
1 millilitre = 0.001 76 pints			
1 centimetre^3 = 0.061 023 74 inches^3			
1 foot^3 = 0.0283 metres^3			
1 pound = 0.004 535 9237 tonnes			

4 The table shows land areas of some countries. Which country has a land area of four hundred thousand km^2 when rounded to 1 sf?

5 Why might it not be helpful for a book to give all the areas to 1 sf?

6 Give the following land areas:

a United Kingdom to 5 sf
b Andorra to 2 sf
c Japan to 3 sf
d China to 4 sf
e Monaco to 1 sf.

7 For two countries the area stays the same when rounded to 3 sf, 4 sf or 5 sf. Which countries are these?

COUNTRY	AREA (km^2)
Andorra	464
Argentina	2758 829
Bangladesh	143 998
Cambodia	181 035
Cameroon	475 499
China	9560 948
Congo	348 999
Denmark	43 030
Japan	369 698
Monaco	1.6
United Kingdom	244 019

Using significant figures to estimate answers

One way to check if a calculation gives an answer of about the right size is to round each of the numbers to 1 sf.

For example, to estimate an answer:

> Here are some answers given by four students when they had to calculate the value of **43 183.5** × **184.23** without using a calculator.
>
> **a** 795 569.6 **b** 79 556 962 **c** 7955 696 **d** 79 557
>
> Which answer is likely to be most accurate?

A value which is of the correct order of magnitude is about the right size.

To 1 sf these numbers become $40\,000 \times 200$

which is $40\,000 \times 2 \times 100$

$= 80\,000 \times 100$

$= \underline{8\,000\,000}$

Answer **c**. 7955 696 is about the same order of magnitude as 8 000 000 so it is most likely to be most accurate.

Exercise 8.5
Estimating answers

1 Work out estimates of the answers to each of these.
Show all the stages you use.

a 31.2 × 41.45 **b** 677 × 3.764
c 856 × 83.42 **d** 542 × 52
e 56 234 ÷ 82.5 **f** 62 381.23 ÷ 478.23
g 452 ÷ 2.34 **h** 28 536 ÷ 0.9623

2 A theatre sells 562 tickets at £28.50 each.

a Roughly what is their income from ticket sales?
b Why does rounding both numbers to 1 sf give too large an estimate?

Population density is the average (mean) number of people to each square kilometre.

3 France has a land area of 549 619 km^2.
In 1990 the population was 56 304 000.
Estimate the population density in people per km^2.

4 When the numbers in 341.2 × 14.25 are rounded to 1 sf and then multiplied the estimate is much smaller than the true answer.
For the problem 156 ÷ 34.7 the estimate is much larger.
Explain why.

5 For each of these problems, say if rounding all numbers to 1 sf makes estimates too large, too small, or about the right size.

a 56.5 × 1763.2 **b** 184 ÷ 19.6
c 491.432 × 2061.4 **d** 445 × 84 632
e 2265 ÷ 27.7 **f** 6834 ÷ 14.23
g 453 782 + 242 565 **h** 7452.3 – 2837.324

6 For the problem 342 561 + 453, why is rounding to 1 sf not helpful?

Thinking ahead to ...
working with numbers less than 1

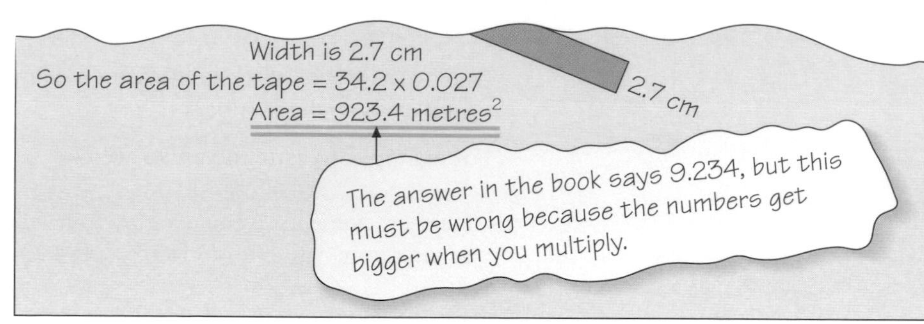

Width is 2.7 cm
So the area of the tape = 34.2 x 0.027
Area = 923.4 metres^2
2.7 cm
The answer in the book says 9.234, but this must be wrong because the numbers get bigger when you multiply.

A Work out 342 × 0.027 with a calculator.
Is the answer larger or smaller than 342?

B Work out 342 ÷ 0.027. Is this answer larger or smaller than 342?

C What can you say about the answer when you:

a multiply by a number less than 1
b divide by a number less than 1?

You may need to do some more calculations to decide this.

Working with numbers less than 1

Example 1 Estimate the value of 342 × 0.052.

Approximating to 1 sf, this becomes 300 × 0.05.

The answer to this estimate will be smaller than 300.
One way to work out the value is to look for patterns.

$300 \times 5 = 1500$

$300 \times 0.5 = 150$

$300 \times 0.05 = 15$

So the estimate is 15.

Example 2 Estimate for the value of 26 ÷ 0.0056.

To 1 sf this is 30 ÷ 0.006.

$30 \div 6 = 5$

$30 \div 0.6 = 50$

$30 \div 0.06 = 500$

$30 \div 0.006 = 5000$

So the estimate is 5000.

Exercise 8.6
Calculating and estimating answers

1 **a** Calculate 342 × 52 without a calculator.
b Use the example above to help decide what 342 × 0.052 is.

2 Estimate the value of 45 × 0.0023.
Show all the stages you use to find the estimate.
Calculate the exact answer and check it with your estimate.

3 Estimate, then calculate, the exact values of these.

a 346.3 ÷ 0.04 **b** 26.23 × 0.6 **c** 876.2 × 0.0002
d 2.448 ÷ 0.0018 **e** 35 465 × 0.000 12 **f** 53 546.2 ÷ 0.045

4

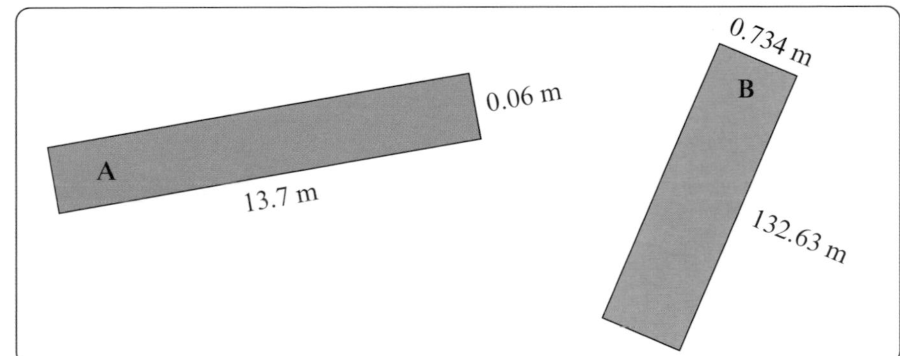

Remember.
When you are dealing with length and area you must be clear about the units you give in your answer.

For example
cm for length
cm^2 for an area.

a Estimate the areas of shapes A and B.
Show the stages in your working of the estimate.

b Calculate the areas without using a calculator.

5

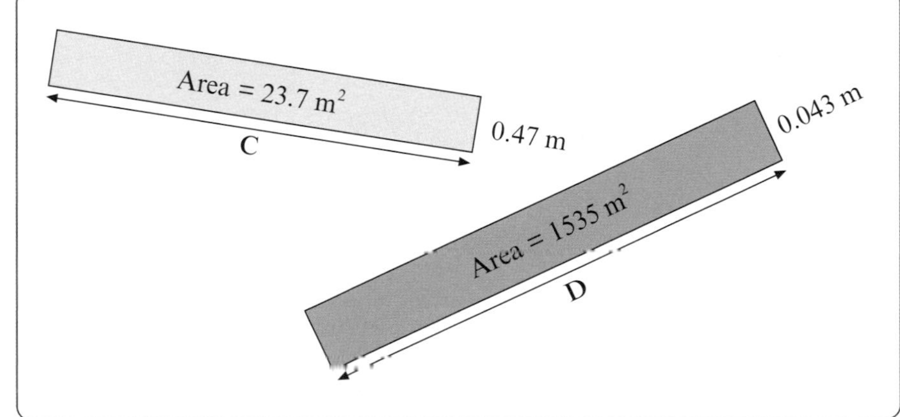

Estimate the dimensions C and D.
Show the stages in your working of the estimate.

6

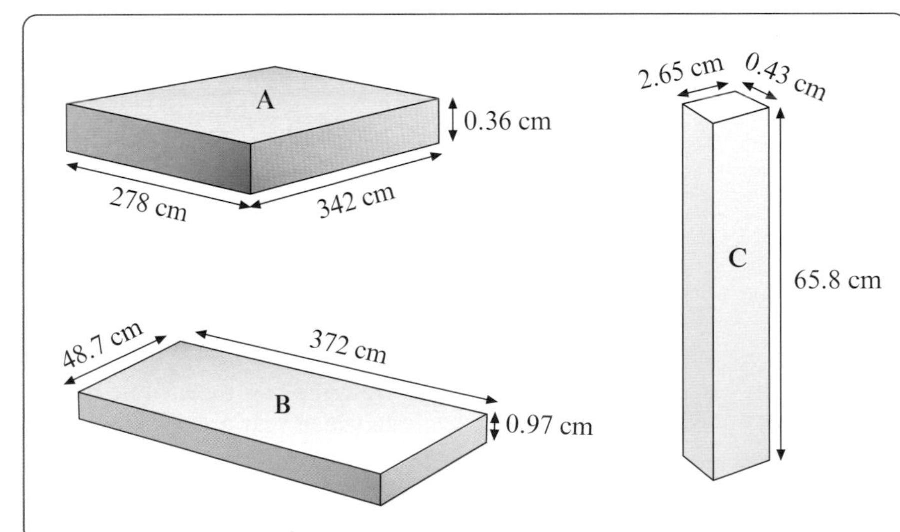

The units for volume are cubic, for example cm^3

Volume of a cuboid = Length × Width × Depth

Estimate the volumes of cuboids A to C.
Explain your working.

7 A cuboid is 113.6 cm long, 0.42 cm deep and has a volume of $18.42\,cm^3$.
Estimate its width.

When should you round?

What is the area of the £20 note? Round your answer to the nearest cm^2.

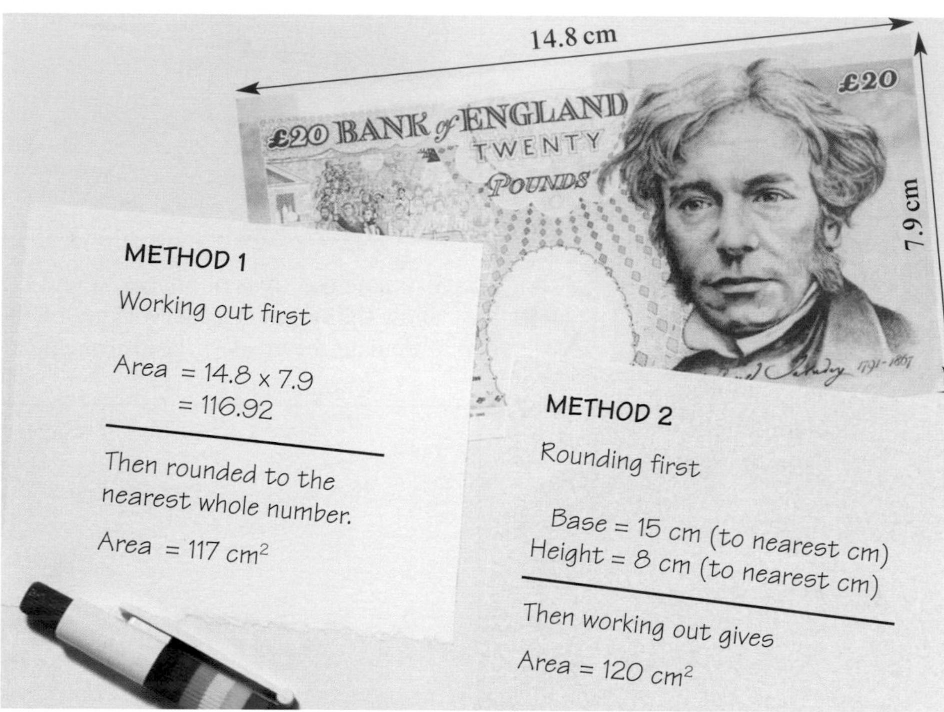

14.8 cm
7.9 cm
£20 BANK of ENGLAND
TWENTY
POUNDS
£20
METHOD 1
Working out first
Area = 14.8 x 7.9
= 116.92
Then rounded to the nearest whole number.
Area = 117 cm^2
METHOD 2
Rounding first
Base = 15 cm (to nearest cm)
Height = 8 cm (to nearest cm)
Then working out gives
Area = 120 cm^2

Alison uses Method 1. She calculates then rounds the answer.
Mark uses Method 2. He rounds all the numbers, then calculates.

Exercise 8.7
The effect of rounding

1 The £5 note has changed in size since 1900.

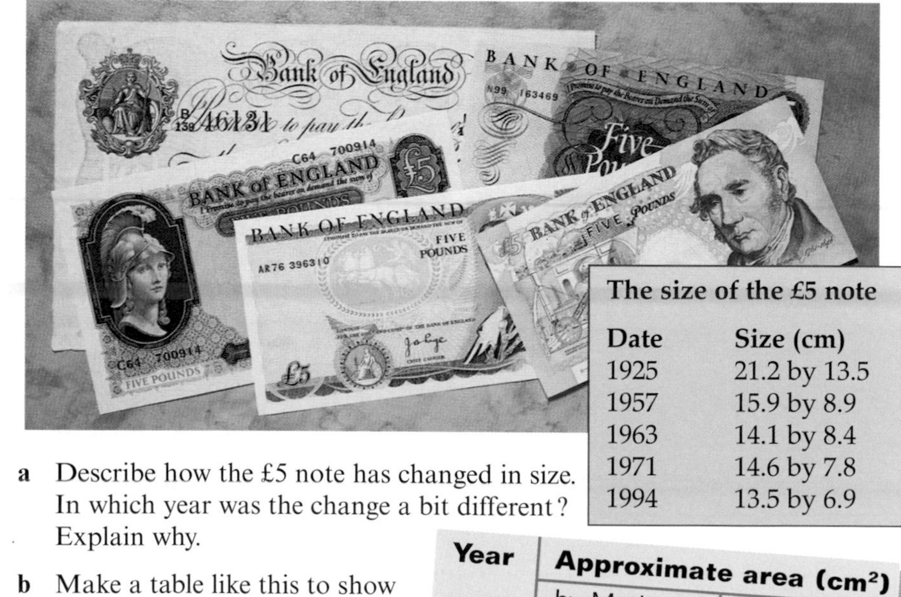

The size of the £5 note

Date	Size (cm)
1925	21.2 by 13.5
1957	15.9 by 8.9
1963	14.1 by 8.4
1971	14.6 by 7.8
1994	13.5 by 6.9

a Describe how the £5 note has changed in size. In which year was the change a bit different? Explain why.

b Make a table like this to show the areas by both methods.

Year	Approximate area (cm^2)	
	by Method 1	by Method 2
1925		
1957		

c Which method do you find easier to use?

d Which method gives the more accurate answer? Explain why.

End points

You should be able to ... ... so try these questions

A Decide what rounding is appropriate for the situation

A1 Alez tuned into her favourite station, Channel 162 on the infrawave. She knew the slot lasted for about 91 minutes so she would need a compulsory meal before the end. She dined on 27 of her favourite food pills with 785 ml of ice-cold isophoric delight. Jeq materialised in 11 minutes raving about some antique maths book with about 416 pages that he'd found and dated as 1997.

Rewrite this extract and round numbers where it is appropriate.

A2 For each of these situations would you round up, down or not at all? Explain why.

a You find the wall area of your bedroom is 330 square feet. You go to buy paint.

b The manual says your car can pull a trailer with a maximum weight of 470 kg. You are loading your camping gear.

B Round a number to a given number of significant figures

B1 Round 345.683 to:

a 3 sf **b** 5 sf **c** 1 sf

B2 Round each number to 3 sf.

a 34.673 **b** 1.974 **c** 194 638.2
d 0.003 186 **e** 143.5 **f** 6.987 36

C Estimate then calculate the answers to problems and know when to round

C1 For the cuboid:

a Estimate its volume.

b Calculate the volume and give your answer correct to 2 dp.

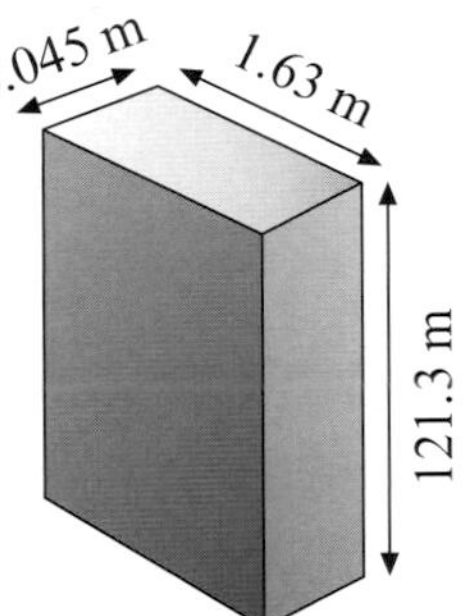

Some points to remember

- Give your final answers to an appropriate degree of accuracy for the situation.
- In exams, round your answers when asked to do so, and give any units.
- One way to estimate an answer is to round all the numbers to 1 sf.
- It is a good idea to estimate the order of magnitude of your answers before you calculate.
- Do not round at the start of, or during, a calculation if you want an accurate answer.
- When you multiply a number by a number less than 1, the answer is smaller.
- When you divide a number by a number less than 1, the answer is larger.

Starting points

You need to know about...

A Probability from equally likely outcomes

If outcomes are equally likely, then you can calculate the probability that something will happen by counting the outcomes.

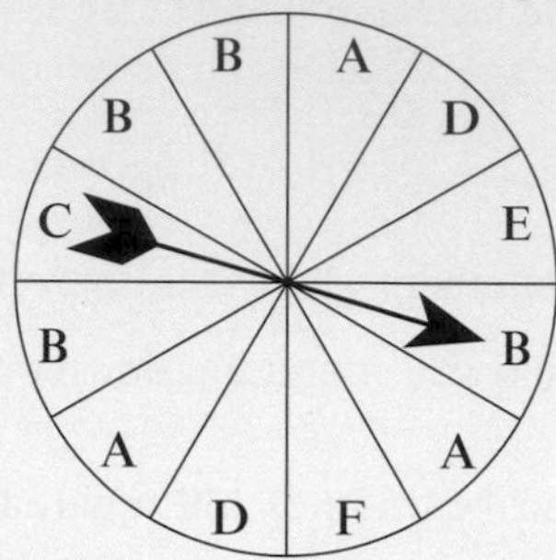

For example: There are twelve sections of equal size.
Four sections have B on them.
So the probability that the spinner stops on B is $\frac{4}{12} = \frac{1}{3}$.

Three sections have A on them and four sections have B.
So the probability that it stops on either A or B is $\frac{7}{12}$.

Probabilities can be shown on a probability scale from 0 to 1.

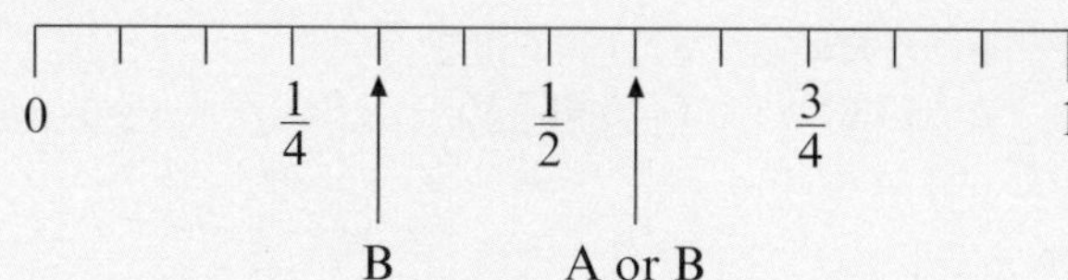

B The probability of a non-event

If you know the probability of something happening, then you can also calculate the probability of it **not** happening.

The probability of **getting B** on the spinner above is $\frac{1}{3}$.

The probability of **not getting B** is $1 - \frac{1}{3} = \frac{2}{3}$.

C Multiplication of fractions

In probability, sometimes you need to multiply fractions.

You can think of $\frac{3}{4} \times \frac{2}{5}$ like this:

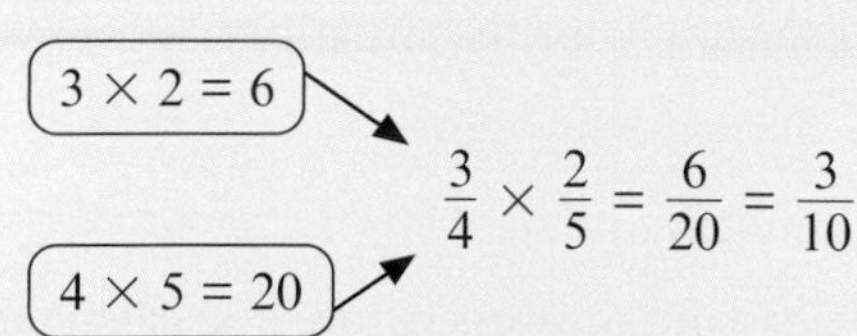

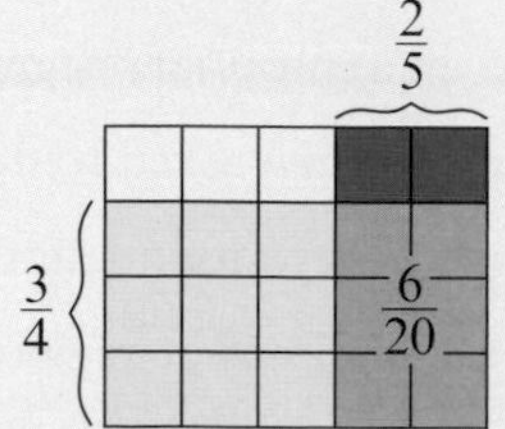

- Multiply the numerators.
- Multiply the denominators.
- Then reduce to the simplest terms if you need to.

...so try these questions

A1 What is the probability that the wheel will stop on:
a A **b** E **c** C
d D **e** F?

A2 What is the probability that the wheel will stop on:
- **a** either B or F
- **b** either E or A
- **c** either A, B or C
- **d** a letter after D in the alphabet
- **e** a letter of the alphabet
- **f** the letter N?

A3 Draw a probability scale and show the probabilities of the wheel stopping on each of A, B, C, D, E and F.

A4 For a 1 to 6 dice what is the probability that for one roll you will get:
- **a** the number 5
- **b** an even number
- **c** a number less than 3?

B1 The probability of getting a red colour on a spinner is $\frac{4}{5}$.
What is the probability of not getting red?

C1 Multiply these fractions.
- **a** $\frac{3}{4} \times \frac{1}{4}$
- **b** $\frac{5}{8} \times \frac{1}{2}$
- **c** $\frac{2}{3} \times \frac{3}{7}$
- **d** $\frac{1}{8} \times \frac{3}{4}$

D Sample space diagrams

A sample space diagram can be used to show the outcomes from two events which are not linked (independent events).

For example, when a coin and a dice are thrown there are twelve different pairs of outcomes.

Outcome of coin						
H	H1	H2	H3	H4	H5	H6
T	T1	T2	T3	(T4)	(T5)	(T6)
	1	2	3	4	5	6

Outcome of dice

These diagrams can be used to show the probability of two things happening. For example, to find the probability of a tail on the coin and a number more than 3 on the dice.

The three pairs which match have been circled in red **so the probability is $\frac{3}{12} = \frac{1}{4}$.**

E Tree diagrams

A tree diagram has branches showing different events. The probabilities of different events can be shown on the branches. This is a tree diagram for the spin of a coin, then Spinner A from the top of the page.

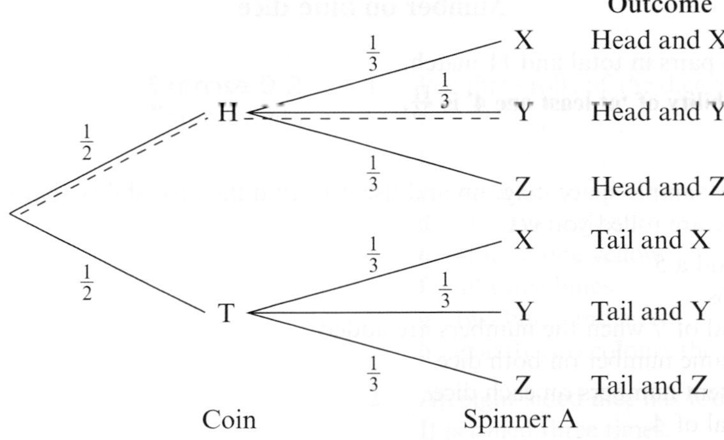

To find the probability of a particular outcome you can multiply the probabilities on the branches that lead to it.

For example, to find the probability of a head and Y you multiply the probabilities along the dotted branches.

Probability of a head and Y is $\frac{1}{2} \times \frac{1}{3} = \frac{1}{6}$.

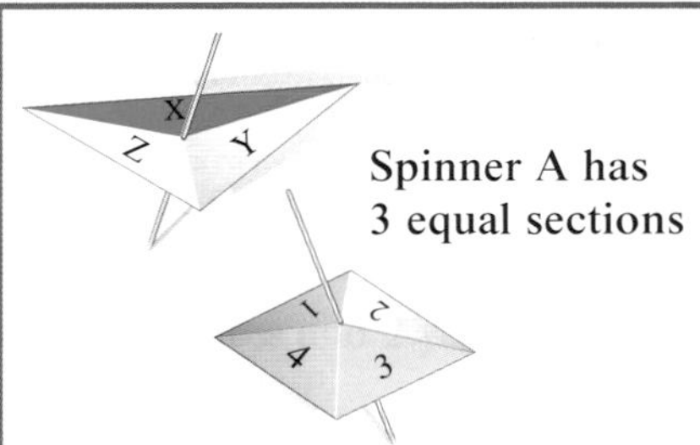

Spinner A has 3 equal sections

Spinner B has 4 equal sections

D1 Draw a sample space diagram to show the outcomes of letters and numbers for the two spinners, A and B.

D2 From your diagram calculate the probability that with a spin of each you get:

a a 4 and a Y

b a number less than 3 and an X.

E1 Draw a tree diagram to show the spinning of spinner B then a coin.

E2 From your tree diagram calculate the probability of a 3 on the spinner and a head on the coin.

Starting points

You need to know about ... | ... so try these questions

A Multiplying out brackets

- To multiply out brackets, multiply every term inside the bracket by the term outside.

Example
$$2(a-8) = 2 \times (a-8)$$
$$= (2 \times a) - (2 \times 8)$$
$$= 2a - 16$$

A1 Multiply out the brackets from:
a $3(a+4)$ **b** $2(b-6)$
c $3(2c+5)$ **d** $4(2d-2)$
e $4(8-2e)$ **f** $5(3+6f)$

B Collecting like terms

- In the expression $7a + 4b - 3a + 6b - 2a$
  - $7a$, $3a$ and $2a$ are like terms as they all give the number of *as*
  - $4b$ and $6b$ are like terms as they all give the number of *bs*.
- The expression can be simplified by collecting like terms.

$$7a + 4b - 3a + 6b - 2a$$
$$= 7a - 3a - 2a + 4b + 6b$$
$$= 2a + 10b$$

B1 Simplify each of these.
a $8a - 3a + 4b - 6$
b $7p + 2q - 3p + 4q - 10$

B2 Simplify these.
a $2(a-4) + 3(2a+6)$
b $4(2a+3) + 2(3a-1)$

C Solving linear equations

- To **solve an equation** find the possible values for each letter.

Example Solve $5(2a-1) = 2(3+4a)$

To solve this equation
- simplify each expression
- add, subtract, multiply or divide both sides of the equation by equal amounts.

$$5(2a-1) = 2(3+4a)$$
$$10a - 5 = 6 + 8a \quad (+5)$$
$$10a = 11 + 8a \quad (-8a)$$
$$2a = 11 \quad (\div 2)$$
$$a = 5.5$$

The **solution** to this equation is $a = 5.5$.

C1 Solve these equations.
a $2p + 5 = 3p + 2$
b $16 + 4q = 6q - 4$
c $10 - 5t = 3t + 4$
d $5(2x-6) = 6x + 45$
e $2(x-1) = 2(3x+8)$

D Indices

- Indices are used as shorthand for multiplication
  - n^2 stands for $n \times n$
  - $2n^2$ stands for $2 \times n^2$ or $2 \times n \times n$
  - $2mn^2$ stands for $2 \times m \times n^2$ or $2 \times m \times n \times n$

D1 Find 3 pairs of equivalent terms.

A ab^2 B a^2 C a^2b

D $a \times b \times a$ E $a \times b \times b$

F $a \times a$ G $a \times a \times b \times b$

E Evaluating an expression

- The value of an expression depends on the value of each letter.

Example Evaluate $2a^2 + 3ab + 8$ when $a = 6.4$ and $b = 2.1$

$$2a^2 + 3ab + 8 = (2 \times 6.4^2) + (3 \times 6.4 \times 2.1) + 8$$
$$= (2 \times 40.96) + (3 \times 6.4 \times 2.1) + 8$$
$$= 81.92 + 40.32 + 8$$
$$= 130.24$$

E1 Evaluate these expressions when $p = 3.4$ and $q = 1.8$
a $2p + 4pq$
b $5p + 2q - 8$
c $pq - 2$
d $8p - 3q + pq$
e $p^2 + 3p + 8$
f $2pq + 2p^2 + 3q^2$

Calculating missing dimensions

Exercise 10.1
Missing dimensions

1 This is a plan of a garden. There is a fence round the garden and edging between the lawn and the flower beds.

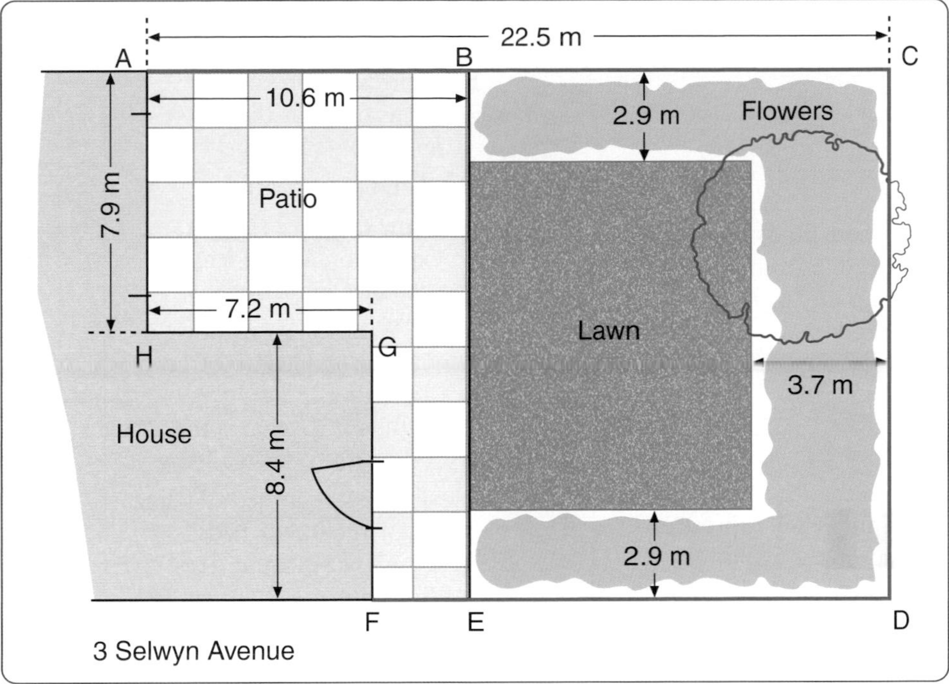

It may help to make a sketch of the garden and mark the lengths on it.

a Calculate the length of:
I BE II EF III DE

b The fence is shown in orange on the plan.
Calculate the total length of the fence.

c Calculate the dimensions of the lawn.

d The edging for the lawn is shown in green.
What is the total length of edging used for the lawn?

e What is the total area of the flower bed?

f Calculate the area of the whole garden.

The area of the garden includes the patio.

2 The fence and edging for the lawn are shown in the same way on this plan.

a Calculate the total length of the fence.

b What length of edging is used for the lawn?

c What is the total area of the flower bed?

d Calculate the area of the whole garden.

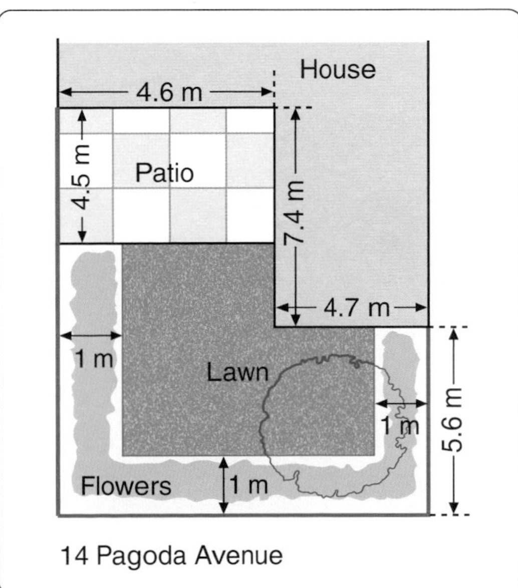

Writing expressions

Exercise 10.2
Writing expressions

On these plans the fence is marked in orange and the edging for the lawn is marked in green.

Simplify each expression by multiplying out any brackets and collecting like terms.

1 On this plan the flower beds are w metres wide.

a Explain why $8.1 + w$ is an expression for the length of BC in metres.

b Write an expression in terms of w for the length of:

i DF **ii** HI

c Show that $43 + 2w$ is an expression for the length of the fence in metres.

d **i** Write an expression for the total length of lawn edging.

ii If $w = 2$, what is the total length of lawn edging?

e **i** Show that the total area of the flower beds is $32.4w$ square metres.

ii Write an expression for the area of the lawn.

f What is the value of w when the area of the lawn is the same as the total area of the flower bed?

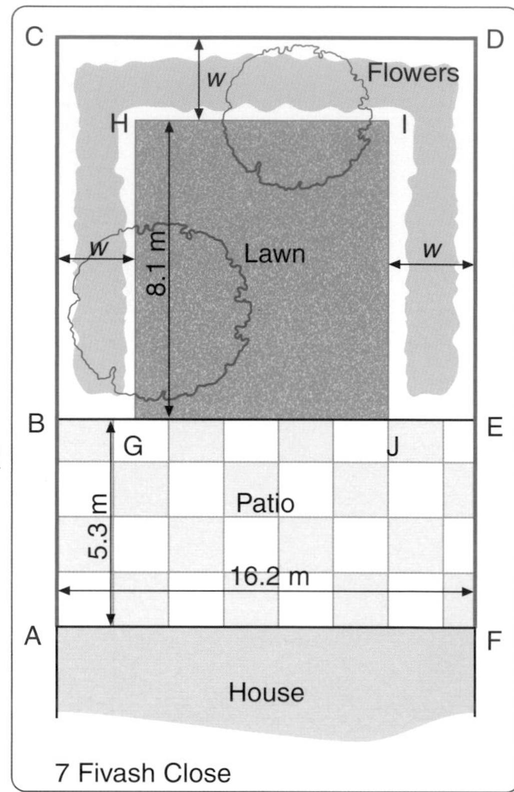

2 On this plan the flower beds are p metres wide.

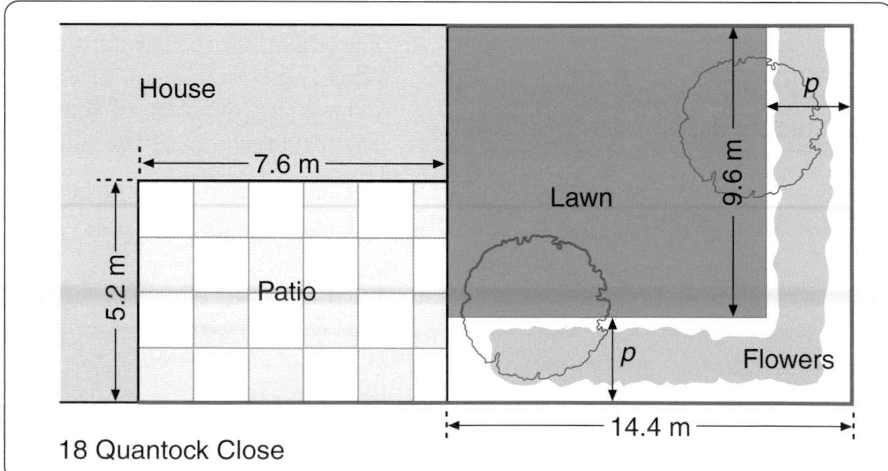

a Write an expression in terms of p for:

i the total length of the fence

ii the total length of the lawn edging

iii the area of the lawn

iv the total area of the flower bed.

b Calculate the total length of the fence if $p = 2$.

c What value of p makes the area of the lawn twice the total area of the flower bed?

Using brackets

- You can use brackets to write an expression for the shaded area in each of these rectangles.

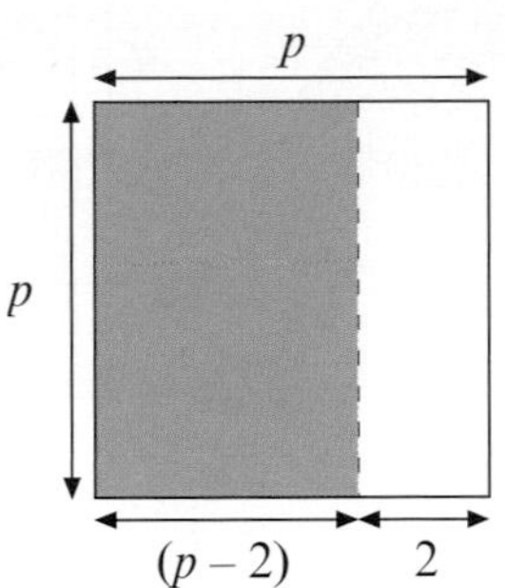

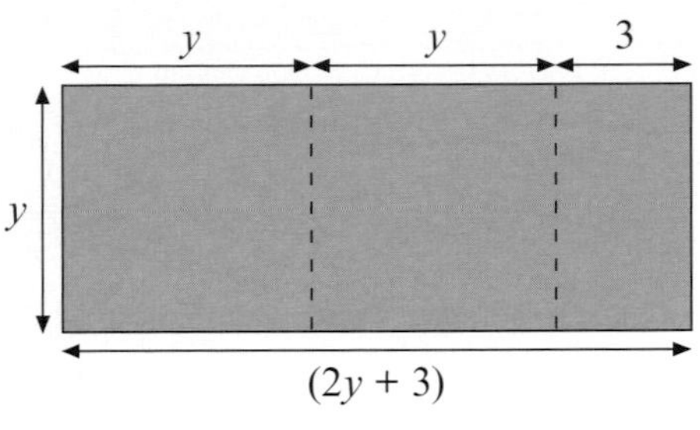

Shaded area = $p(p - 2)$

Shaded area = $y(2y + 3)$

- To multiply out a bracket it may help to use a table.

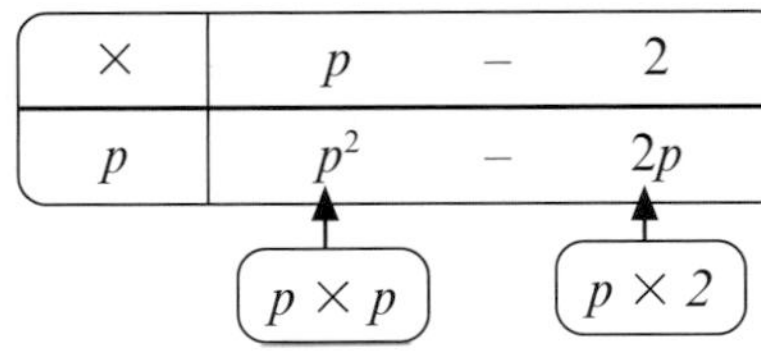

$\times$	p	$-$	2
p	p^2	$-$	$2p$

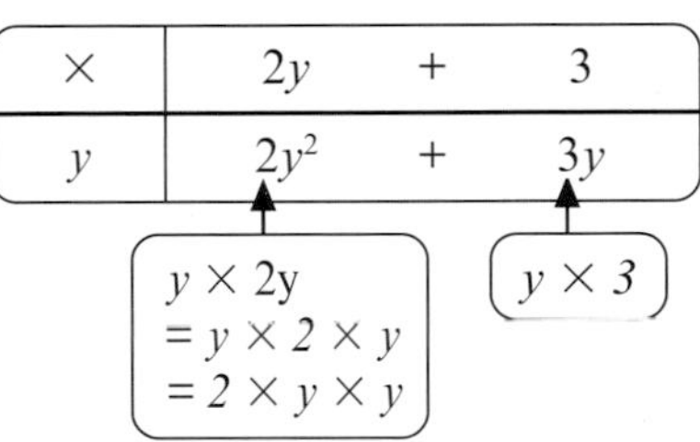

$\times$	$2y$	$+$	3
y	$2y^2$	$+$	$3y$

So $p(p - 2) = p^2 - 2p$

So $y(2y + 3) = 2y^2 + 3y$

Exercise 10.3
Using brackets

1 Multiply out each of these.

a $2(a + 4)$ **b** $3(b - 5)$ **c** $4(2c + 3)$ **d** $d(d + 7)$
e $e(6 - e)$ **f** $f(4 + f)$ **g** $p(2p + 9)$ **h** $r(4r - 2)$

2 For each of these rectangles write an expression for the shaded area:
a with brackets **b** without brackets.

For some of the shaded rectangles you will need to find an expression for the length or the width.

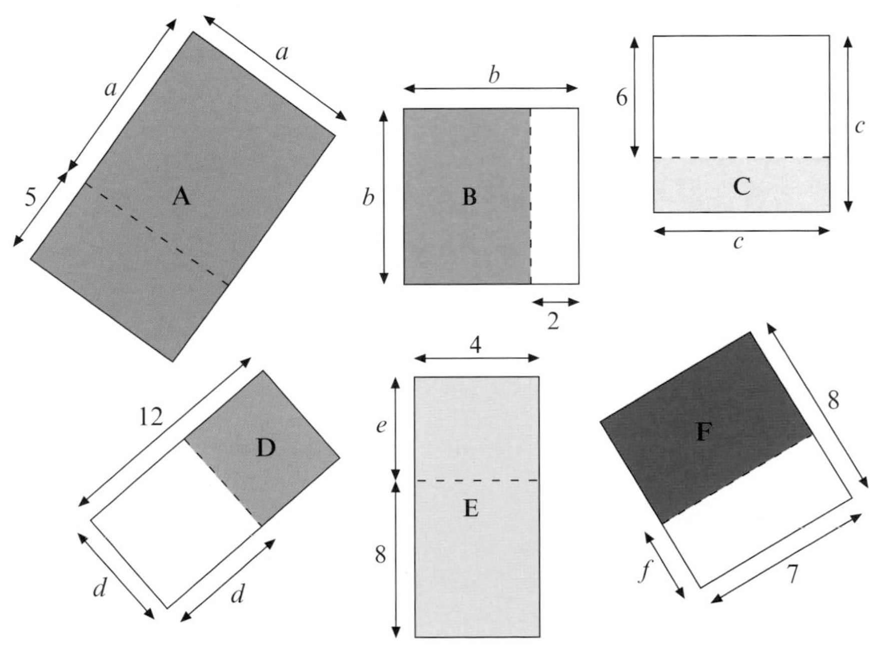

3 For each of these rectangles P, Q and R write an expression for the shaded area:

a with brackets **b** without brackets.

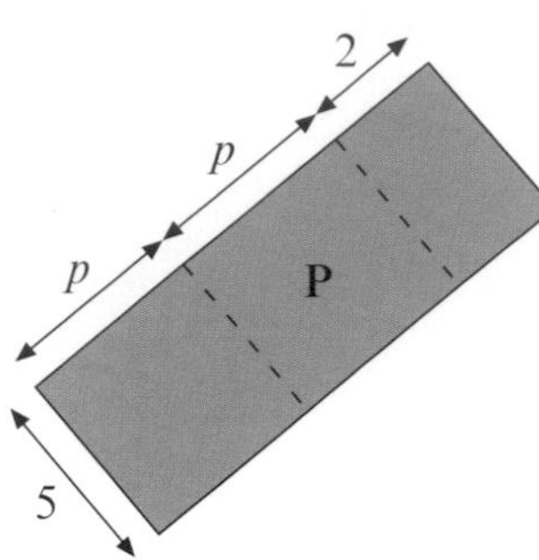

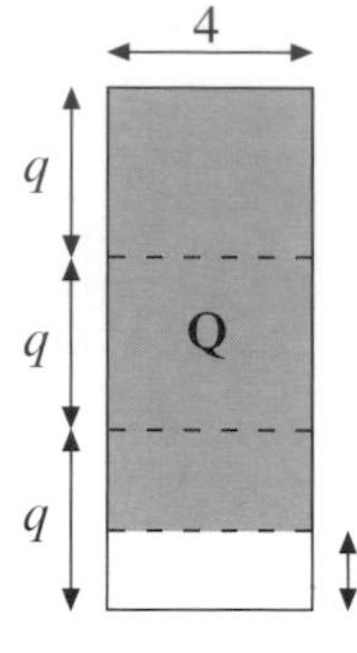

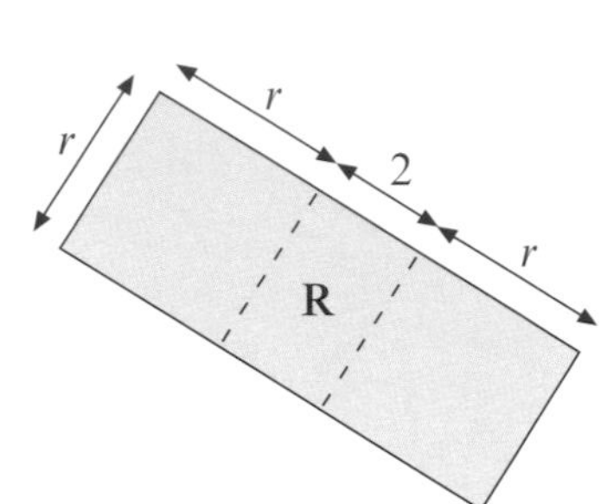

4 Write down the widths of rectangles A, B and C.

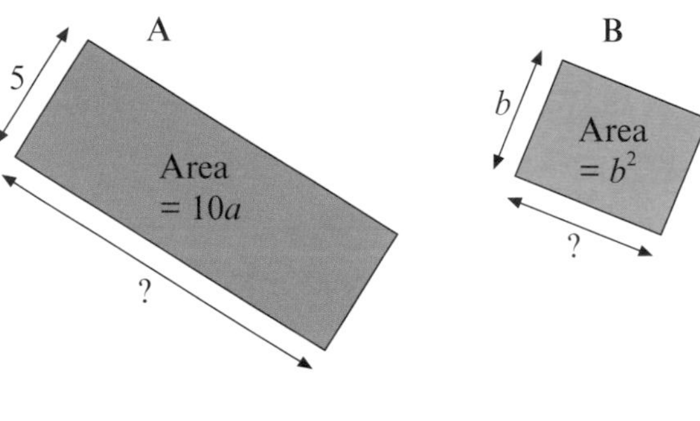

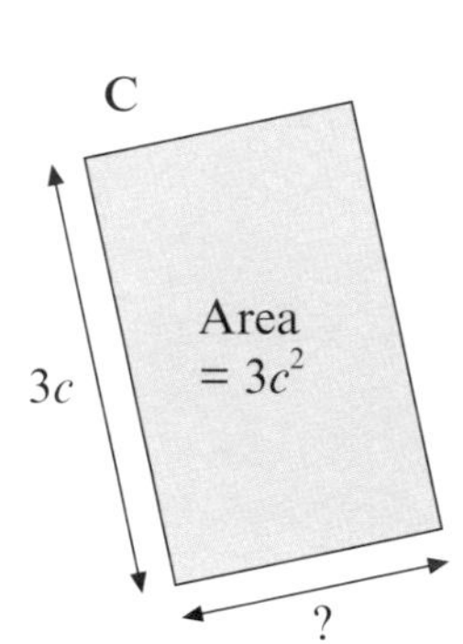

5 Write an expression for the widths of rectangles D to G.

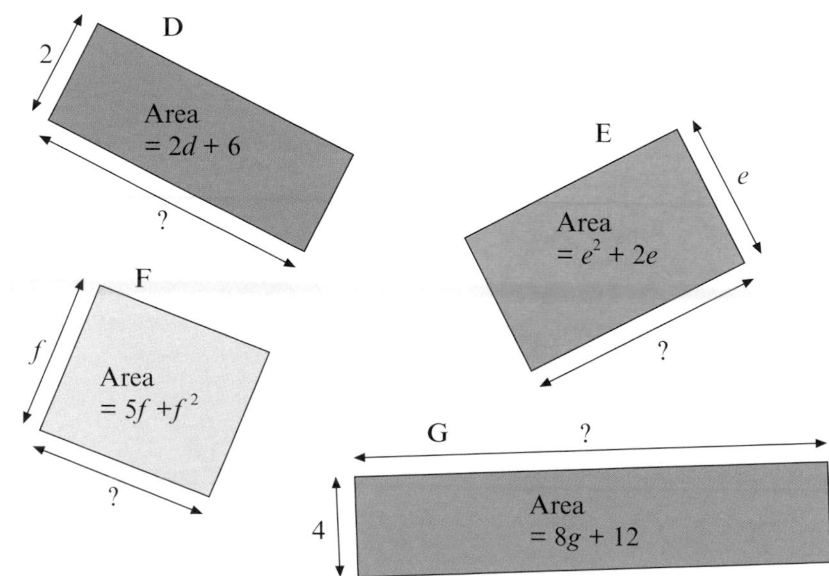

When you complete an expression using brackets check that the two expressions are equivalent.

For example:

$4a - a^2 = a(\square - \square)$

$4a = a \times 4$ $a^2 = a \times a$

$\times$	4	$-$	a
a	$4a$	$-$	a^2

So $4a - a^2 = a(4 - a)$

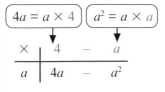

6 Copy and complete these expressions using brackets.

a $4x + 12 = 4(\square + \square)$ **b** $6p - 4 = 2(\square - \square)$
c $2a + 8 = \square(a + \square)$ **d** $10 - 5q = \square(2 - \square)$
e $b^2 - 6b = b(\square - \square)$ **f** $y^2 + 4y = y(\square + \square)$
g $2r^2 + 3r = r(\square + \square)$ **h** $3d^2 - 5d = d(\square - \square)$
i $3t^2 + 4t = \square(\square + \square)$ **j** $6s^2 + 7s = \square(\square + \square)$

Writing expressions to solve problems written in words

Example

The length of a rectangle is 4 m greater than the width.
The perimeter is 36 m. What is the area of the rectangle?

To find the length of this rectangle:

- Choose a letter to stand for the width.
  Let the width be w metres.
- Write the length in terms of this letter.
  Length $= w + 4$
- Draw and label a diagram.

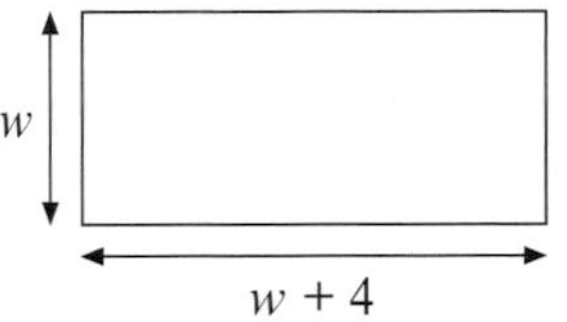

- Write and solve an equation for the information given.

$$\begin{aligned} \text{Perimeter} &= 2w + 2(w + 4) \\ &= 2w + 2w + 8 \\ &= 4w + 8 \end{aligned}$$

$$\begin{aligned} \text{Perimeter} &= 36 \text{ metres} \\ \text{So } 4w + 8 &= 36 \\ 4w &= 28 \\ w &= 7 \end{aligned}$$

$$\begin{aligned} \text{So Width} &= 7 \text{ metres} \\ \text{Length} &= 7 + 4 \text{ metres} \\ &= 11 \text{ metres} \end{aligned}$$

- Answer the problem.

$$\text{Area} = 7 \times 11 \text{ m}^2$$

So the area of the rectangle is 77 m².

Exercise 10.4
Solving problems written in words

1 The length of a rectangle is 8 centimetres more than its width.
   a If the width is a centimetres, write an expression for the length of the rectangle.
   b Write an expression for the perimeter of the rectangle in terms of a.
   c If the perimeter of the rectangle is 60 centimetres what is the area of the rectangle?

2 In this triangle AB is twice the length of BC.

B
A C

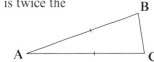

   a If BC is d centimetres, write an expression for the perimeter of the triangle in terms of d.
   b If the perimeter is 58 cm, what is the length of AB?

3 The perimeter of a square is 60 metres.
What is the area of the square?

4 The length of a rectangle is twice its width.
The perimeter is 39 centimetres.
What is the area?

Multiplying terms and simplifying

In any term the letters are usually written in alphabetical order.

For example:

$2ba$ is usually written as $2ab$

$4n^2m$ is usually written as $4mn^2$.

- You can **multiply any terms** by grouping together the numbers and each of the letters.
  For example:

$2m \times 3n$	$2ab \times a$	$2p \times 3p^2$
$= 2 \times m \times 3 \times n$	$= 2 \times a \times b \times a$	$= 2 \times p \times 3 \times p \times p$
$= 2 \times 3 \times m \times n$	$= 2 \times a \times a \times b$	$= 2 \times 3 \times p \times p \times p$
$= 6mn$	$= 2a^2b$	$= 6p^3$

The letters in each term are usually written in alphabetical order.

- **Like terms** must have exactly the same letters in them.
  For example:

$2p^2q = 2 \times p \times p \times q$	$3pq^2 = 3 \times p \times q \times q$
$8p^2q = 8 \times p \times p \times q$	$2p^2q = 2 \times p \times p \times q$
So $2p^2q$ and $8p^2q$ are like terms.	So $2p^2q$ and $3pq^2$ are not like terms.

- To **simplify an expression** collect together any **like terms**.
  For example:

These expressions **can be simplified** by collecting together like terms.

$2a^2b + 3ab^2 + 4a^2b = 6a^2b + 3ab^2$

$2x^2 + 2x + 3x^2 - x + 4 = 5x^2 + x + 4$

These expressions **cannot be simplified** because there are no like terms.

$2a^2b + 3ab^2$

$2x^2 + 4x + 3$

Exercise 10.5
Multiplying terms and simplifying

1 Multiply these terms.

a $3a \times 2b$	**b** $p \times 3q$	**c** $4y \times 5x$
d $5q \times 6p$	**e** $x \times 2x$	**f** $ab \times a$
g $2xy \times y$	**h** $2ab \times 3a$	**i** $2p^2 \times 3q$
j $a^3 \times a^2$	**k** $2b^2 \times 3b$	**l** $5c^3 \times 2b$

2 Find four pairs of equivalent terms.

A $(2b^2)^3$	B $6b^5$	C $3b^2 \times 2b^4$
D $5b^5$	E $8b^6$	F $3b^2 \times 2b^3$
G $6(b^4)^2$	H $6b^6$	I $6b^8$

To multiply out each bracket multiply each pair of terms.

$\times$	m	$+$	$3n$
$2n$	$2mn$	$+$	$6n^2$

$2n \times m$
$= 2 \times n \times m$

$2n \times 3n$
$= 2 \times n \times 3 \times n$
$= 2 \times 3 \times n \times n$

So $2n(m + 3n) = 2mn + 6n^2$

3 Multiply out these brackets.

a $a(b + 4)$	**b** $x(y + z)$	**c** $m(2n + 3p)$
d $2x(3y + 2z)$	**e** $c(a + c)$	**f** $p(p - q)$
g $3b(c + b)$	**h** $4a(a - b)$	**i** $p(3p - 4)$
j $a(2b - 4c)$	**k** $2a(3a + 4b)$	**l** $4p(2q - 3p)$
m $2pq(3p + 2q)$	**n** $4xy(x - 2y)$	**o** $3ab(x^2 - y^2)$

4 Simplify these where possible.

a $5a - 3b + 4a + 25b$	**b** $4x^2 + x - 2x^2$
c $5a + 2ab - a + 3ab$	**d** $x^2 + x^3 - 2x$
e $4a - 3b + 7a + 5b$	**f** $7x^2 + xy - x^2 + 3xy$
g $4p^2 - pq + 6q + pq$	**h** $ab + 2a - ab + 4a$

5 Multiply out these brackets and simplify.

a $2(2a + 3b) + 5(a + 4b)$ **b** $2(2x + 4y) + 3(2x - y)$
c $x(x - 3) + x(x + 4)$ **d** $2x(3x + 2y) + x(4x - y)$
e $ab(a + b) + ab(a - b)$ **f** $3a(ab + b) + 2b(ab - b)$

6 Simplify these expressions.

a $2a(3a - b) + 4b(2a + 3b)$ **b** $5xy(2x + 4y) + 3x(2xy - 2y)$
c $pq(2p - 3q) + 2p(3pq - 2q^2)$ **d** $2mn(3m - 4n) + 4m^2(2n - 3m)$

Factorising

When you look for **common factors** in algebra the factor might be numerical (a number) or algebraic (a letter).

Example

Factorise this expression: $6a + 3bx - 9y$

There is a common factor of 3 in each term:

$$\overset{2\times 3}{6a} + \overset{3}{3bx} - \overset{3\times 3}{9y}$$

So we can take the factor of 3 outside a bracket.
So $6a + 3bx - 9y$ **factorised** is $3(2a + bx - 3y)$

You can check if you have factorised correctly by multiplying out the bracket. You should get the expression you started with, for example
$3(2a + bx - 3y)$
$= 6a + 3bx - 9y$

Example

Factorise this expression: $3ax + 6ay - 9ac$

There is a common factor of $3a$ in each term so we can take the $3a$ outside a bracket so $3ax + 6ay - 9ac$ factorised is $3a(x + 2y - 3c)$

Exercise 10.6
Common factors

1 Factorise each expression.

a $5ax + 10y$ **b** $6ay - 15bc$
c $9xy - 12a$ **d** $4xy + 10x$
e $21xy + 35c$ **f** $36ab - 45y$
g $6ax + 12y + 15x$ **h** $8ac + 20cy - 24ax$
i $9x + 15ay - 21xy$ **j** $30kx + 18y + 42ac$
k $28xy + 56y - 42c$ **l** $144ac - 60y + 108x$

2 Factorise each expression.

a $3ax - 5ay$ **b** $6xy + 7ax$
c $9xy - 5ay$ **d** $6ax - 17xy$
e $3abc + 5ax$ **f** $9ab + 4b$
g $7xy - 9y$ **h** $6ax + 7axy$
i $7abc - 8c$ **j** $3ax + 5xy - 7cx$
k $5xy - 7ay + 6y$ **l** $12xy - 19ay + 15y$
m $6abc + 5bc + 7c$ **n** $8x + 7axy + 9bxy$
o $7abc + ab$ **p** $16xy + 25y - 13axy$
q $21abx - 15cbx$

You may be able to factorise by finding more than one common factor.

Example Factorise fully $6axy - 9abc$

Each term has a common factor of 3.
(6 can be written as 2×3 and 9 as 3×3)
Each term has a common factor of a.
So we can take $3a$ outside a bracket.
So $6axy - 9abc$ fully factorised is $3a(2xy - 3bc)$

Check the answer.
$3a(2xy - 3bc)$
gives
$6axy - 9abc$

3 Factorise these fully.

a $2ax - 8ay$ **b** $25xy + 15ax$
c $16ay + 28by$ **d** $32xy - 24ax$
e $42abc - 35bx$ **f** $56ab + 42by$
g $6ab - 9ay + 12a$ **h** $4xy + 6ay - 10cy$
i $21xy - 15x + 12ax$ **j** $15kxy + 25axy$
k $18cx + 12axy + 15x$ **l** $21abc + 56acx$
m $15axy + 18ax - 30xy$ **n** $abc + 5bc - 15bcx$
o $24xy - 16x^2 + 8ax$ **p** $2x^2 + 4x - 6xy$
q $5x - 10xy - 25x^2$ **r** $6ab - 9a^2b + 3abc$

End points

You should be able to ... ... so try these questions

A Calculate missing dimensions

A1 This is the plan of a paved area.

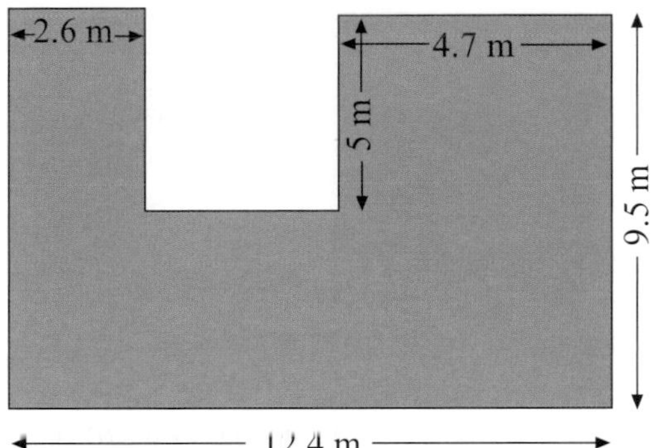

Calculate the total perimeter of the paved area.

A2 This is the plan of a play area.

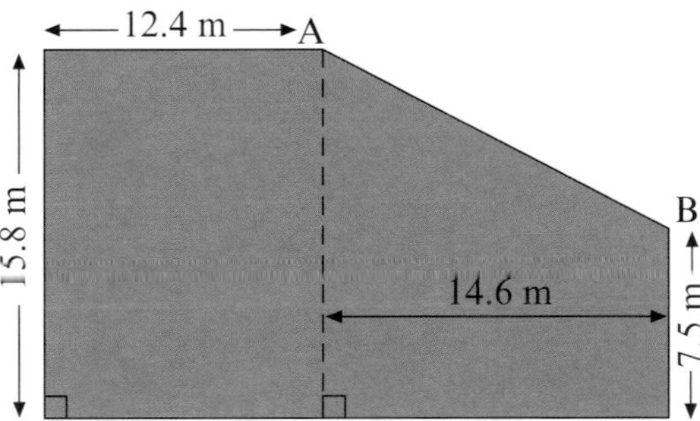

The total perimeter of the play area is 79.4 metres.
Calculate the length of AB.

A3 This is the plan of a circular pond.

The distance around the edge of the pond is 100 m.

a Find the radius of the pond to 1 decimal place.

b Calculate the area of the pond to 3 significant figures.

End points

You should be able to ...	... so try these questions
B Simplify expressions	**B1** Multiply out the brackets from: **a** $4(5f - 4)$ **b** $m(m + n)$ **c** $2p(r - p)$ **d** $ab(a + b)$ **e** $mn(2m + 3n)$ **f** $2xy(3y - 5x)$ **B2** Simplify these expressions. **a** $2(2b - 4) + 6(4 + 3b)$ **b** $4(2b - 4) + 6(5 + 3b)$ **c** $3ab(b - a) + 7ab(b + a)$ **d** $3xy(x - 3) + 7xy(y + 4)$
C Write expressions to solve a problem	**C1** The shape ABCDEF is cut from a square. The square is p centimetres wide. 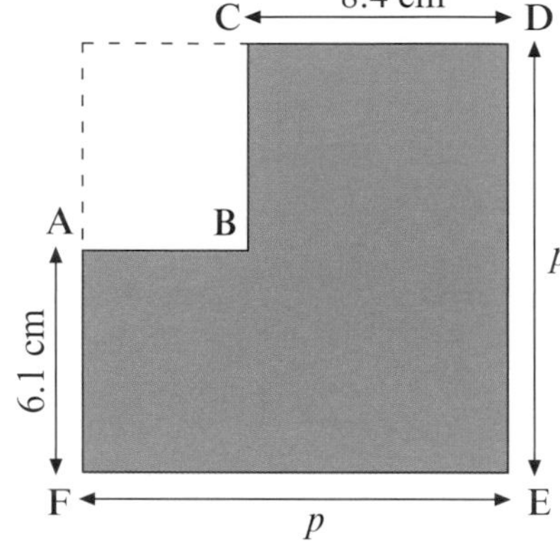 **a** Write an expression for the length of: **i** AB **ii** BC **b** **i** Write an expression for the perimeter of ABCDEF. **ii** What is the value of p if the perimeter is 56 cm? **C2** A rectangle is a centimetres wide. The length is 5 centimetres more than the width. **a** Write an expression for the length of the rectangle in centimetres. **b** Write an expression for the perimeter of the rectangle in centimetres. **c** Show that $a^2 + 5a$ is an expression for the area of the rectangle in square centimetres. **d** What value of a makes the perimeter 56 cm? **e** If $a = 5.1$, what is the area of the rectangle?
D Factorise expressions	**D1** Factorise these fully. **a** $8p + 4q$ **b** $6a - 12b$ **c** $a^2 + ab$ **d** $7m + 3mn$ **e** $8xy + 10y$ **f** $4ab + 6a^2$ **g** $3xy^2 - 5x^2y$ **h** $4pq - 6p^2q$ **i** $2g^2h - 5h^2$ **j** $30xy + 24x^2 - 16xy^2$ **k** $25a^2b - 15b^2a$ **l** $14axy - 21x^2 + 35x^2y$ **m** $100xy + 50ay^2 - 250x^2y$

Some points to remember

- Like terms must have exactly the same letters in.
- When simplifying an expression, add and subtract like terms.
- When factorising an expression check your answer by multiplying out the bracket.

Starting points

You need to know about ...

A Some mathematical terms

Lines AB and CD are **perpendicular** to line XY because they would meet XY at right angles.

XY is also **perpendicular** to AB and CD.

AB and CD are **parallel**.

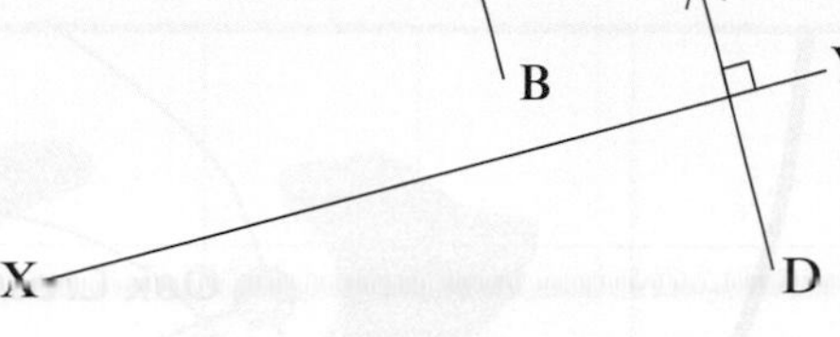

These terms are for parts of a circle.

Arc
Chord
Diameter
Radius
Tangent

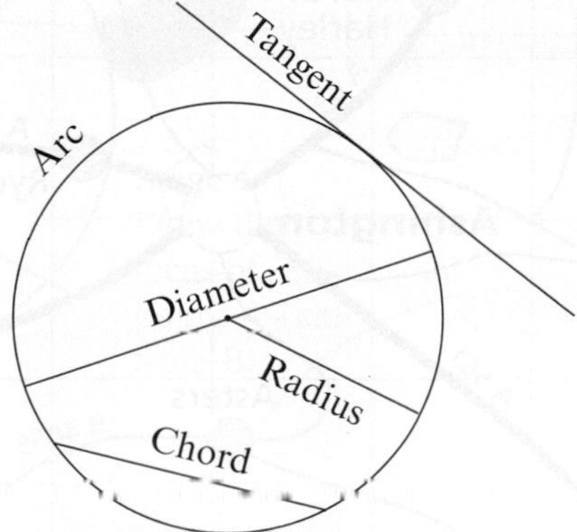

These terms are for different types of triangle.

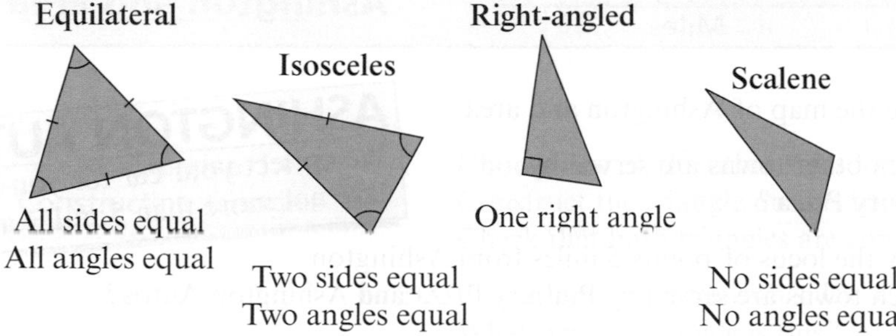

Equilateral: All sides equal, All angles equal

Isosceles: Two sides equal, Two angles equal

Right-angled: One right angle

Scalene: No sides equal, No angles equal

B Congruent triangles

Triangles are said to be **congruent** if they have the same shape or size or if they are reflections of each other.

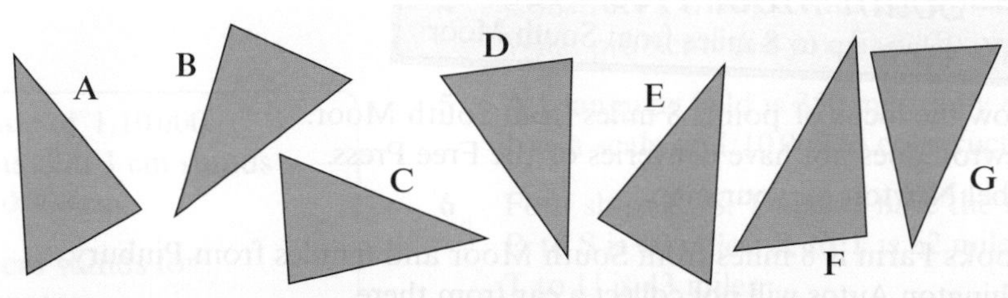

Triangles A and B are **congruent** to each other.

... so try these questions

A1 You will need to measure for some of these questions.

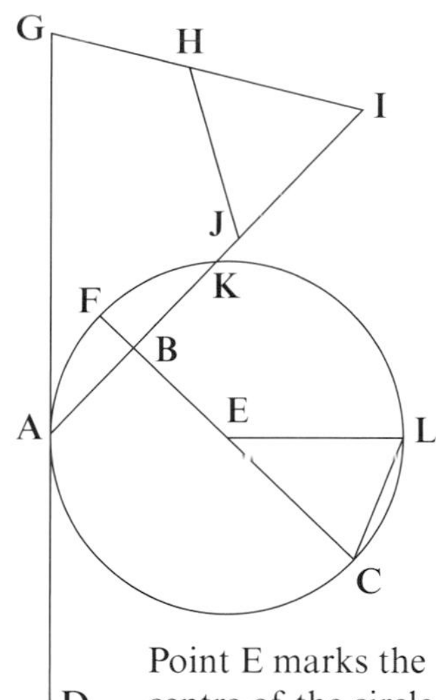

Point E marks the centre of the circle

- **a** Which line is perpendicular to line FC?
- **b** Give a line which is perpendicular to GD.
- **c** Give a line which is a radius of the circle.
- **d** Which triangle is isosceles but not equilateral?
- **e** Which triangle is scalene?
- **f** Which triangle is equilateral?
- **g** Which line is a chord?
- **h** Which line is a tangent?
- **i** Which line is a diameter?

B1 Which of the triangles B to G is not congruent to triangle A?

B2 Draw a different triangle which is congruent to triangle A.

Constructing triangles from other data

ΔABC means Triangle ABC.

∠B means angle B.

You can construct a triangle when you are not given the length of all three sides. You might be given **one side and two angles**.

Example Draw ΔABC where AB = 5 cm, ∠B = 55°, and ∠A = 43°.

Stage 1 Make a rough sketch.
2 Make the side you know (AB) the base and draw it.
3 At A draw an angle of 43° with a protractor.
4 At B draw an angle of 55°.

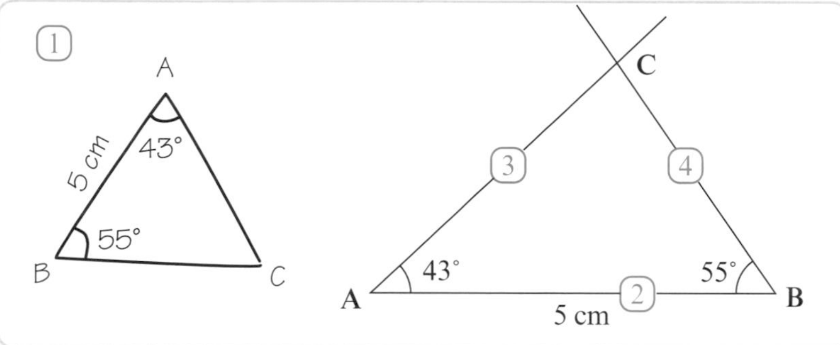

You might be given **two sides and one angle**.

Example Draw ΔRST where RT = 6.5 cm, RS = 4.5 cm and ∠R = 46°.

Stage 1 Make a rough sketch.
2 Make the long side (RT) the base, and draw it.
3 At R draw an angle of 46° and mark S, 4.5 cm from R.
4 Draw the last side TS.

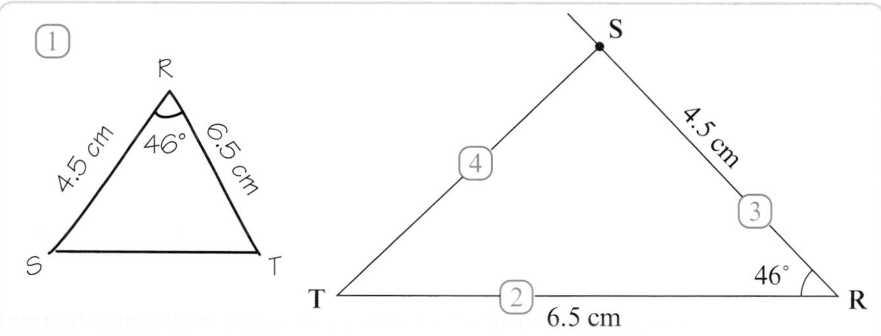

Exercise 11.3
Constructing triangles

1 Construct these triangles.

a ΔDEF, where DF = 6 cm, EF = 5 cm and ∠F = 69°
b ΔGHI, where GI = 7.2 cm, ∠G = 53° and ∠I = 42°
c ΔJKL, where JK = 8 cm, ∠J = 25° and ∠K = 125°
d ΔMNP, where NP = 6.3 cm, MP = 5.2 cm and ∠P = 131°

2 Construct ΔQRS, where RQ = 8.3 cm, ∠R = 35° and ∠S = 68°. You will need to calculate another angle first.

3 Construct these triangles to decide which two look identical.

a ΔIJK, where JK = 5.5 cm, IJ = 8 cm and ∠J = 50°
b ΔLMN, where LM = 8 cm, ∠L = 60° and ∠M = 50°
c ΔPQR, where PR = 6.5 cm, QR = 7.5 cm and ∠R = 70°

Constructing a perpendicular from a point to a line

A line segment is a measurable portion of a straight line. It is finite in length. Straight lines can be infinite.

To construct a perpendicular from point A to the line BC:

- With centre A draw an arc that it cuts the line BC twice [X and Y]

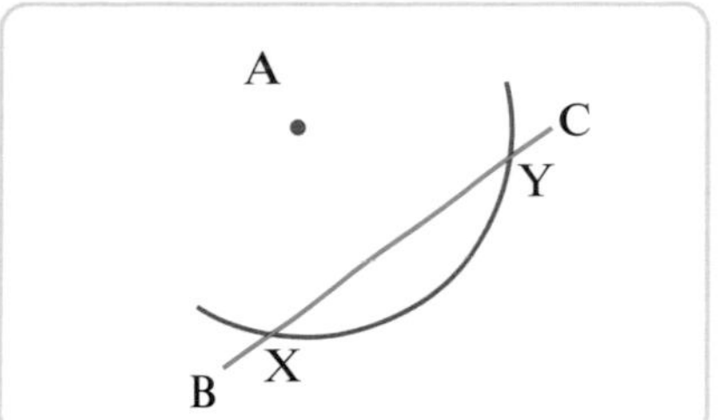

- With centre X open the compass over half the length of the line segment XY and make an arc.
- With the same radius and centre Y make another arc crossing the arc made from X.
- Join the crossing point of the arcs and A, extending the line through to BC.
- The line is perpendicular to BC.

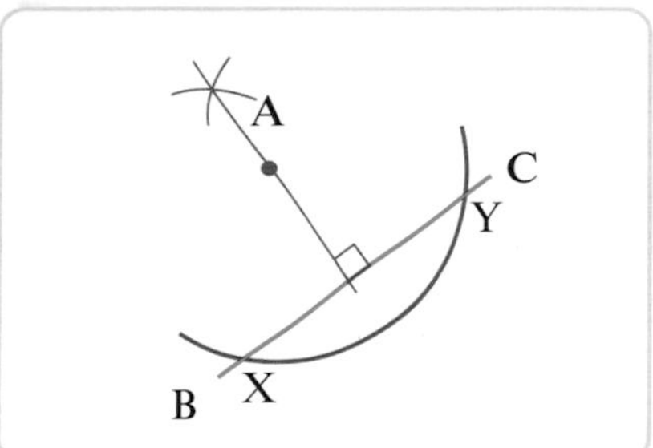

Exercise 11.4
A perpendicular from a point to a line

1 **a** Draw a straight line CD with points A and B either side of the line as in the diagram.

b Construct perpendiculars from A and B to CD.

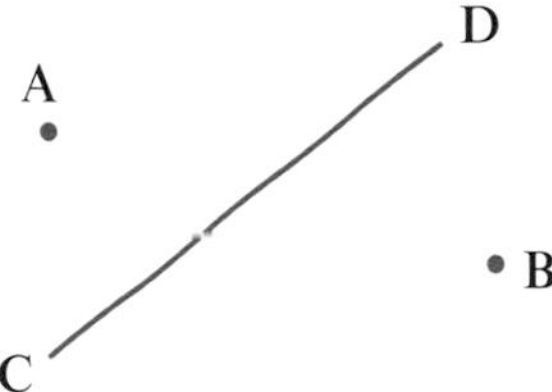

2 **a** Construct this triangle.

b Construct a perpendicular from C to AB.

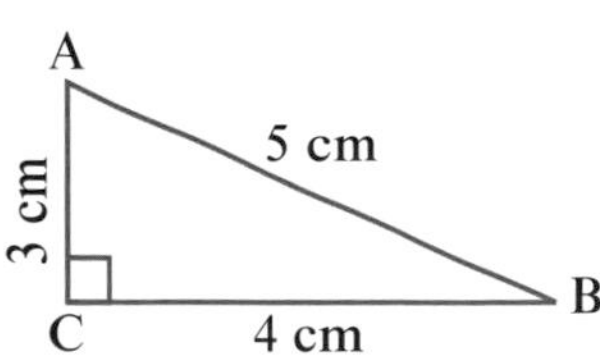

Thinking ahead to ...
perpendicular bisectors

When two lines cross they are said to intersect.

A Draw a straight line and mark two points, A and B, 6 cm apart.

B Draw a circle at A and another with the same radius at B.

C Draw larger circles with equal radii at A and B.
If they intersect, then mark the points of intersection.

D Continue by drawing larger circles and marking the points of intersection.

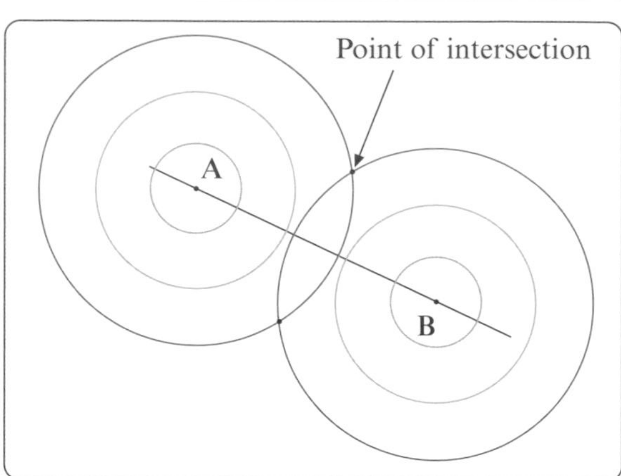

E **a** Draw the locus of all points of intersection of equal circles.
b Describe the link between this locus and the line AB.

Perpendicular bisectors

A line which cuts a straight line exactly in half at right angles is called a **perpendicular bisector**.

To construct the perpendicular bisector of the line EF.

- Draw the line EF
- With centre E draw an arc with a radius greater than half of EF.
- With centre F draw another arc with the same radius.
- Join the two points of intersection.

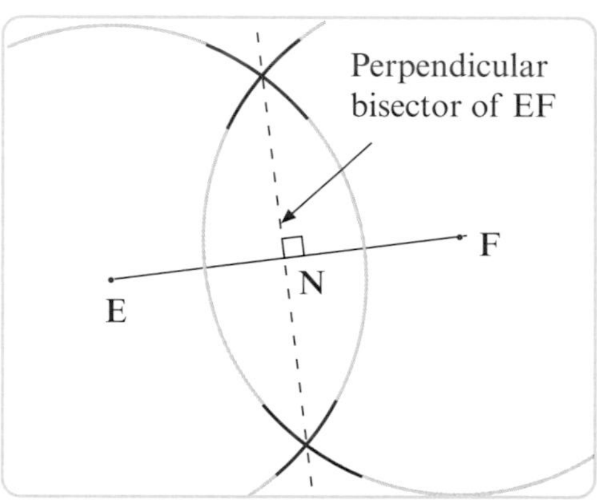

N marks the midpoint of EF.

Exercise 11.5
Perpendicular bisectors

1 **a** Draw a line AB, 6 cm long. Construct the perpendicular bisector of AB.
b Mark any point on the bisector and measure its distance to A and to B.
c What can you say about the distance of any point on the bisector from A and from B?

2 Draw the perpendicular bisectors of lines with these lengths.
Check that both sides are equal in length and that you have right angles.
a 10 cm **b** 7.7 cm **c** 4.6 cm

3 **a** Draw this triangle.
b Construct the perpendicular bisector of each side.
c Where do all three bisectors intersect?
d Does this happen for other triangles?

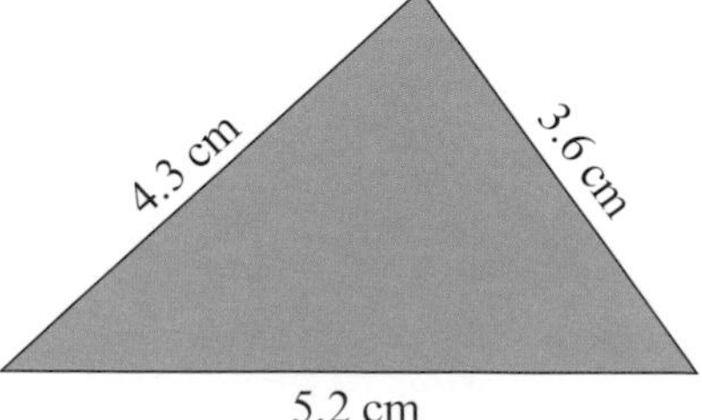

Thinking ahead to ...
bisecting an angle

A pedestrian area is edged by two buildings which meet at an angle of 36°.
The plans say trees must be planted an equal distance from both buildings.

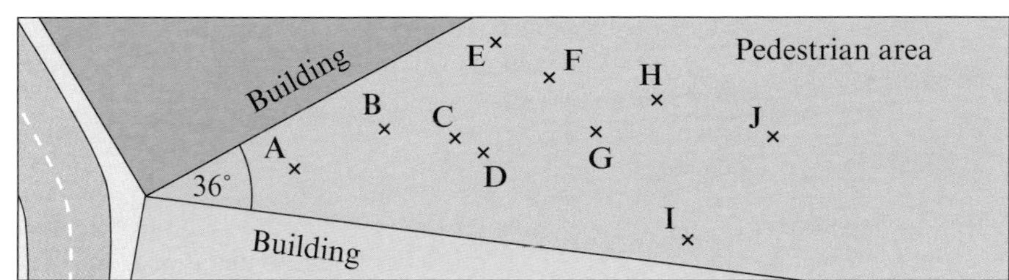

'Equidistant from' means 'the same distance from'.

A Which crosses on the diagram mark where trees could be planted?

B Draw two lines which meet at an angle of 36°.
Mark the locus of all points which are equidistant from both lines.

Bisecting an angle

To bisect an angle means to draw a line which cuts it in two equal parts from its vertex.

It is useful to be able to bisect an angle without measuring it.
You can do this using a pair of compasses.

Stage 1
Draw an angle ABC.

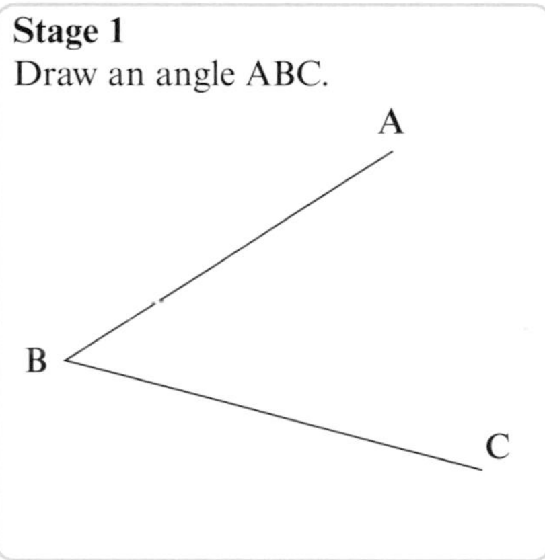

Stage 2
With centre B draw an arc so it cuts AB and BC at D and E.

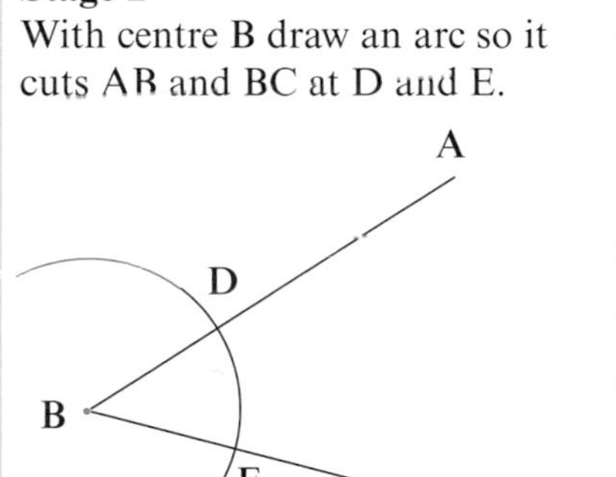

Stage 3
With centre D, draw an arc.
With centre E, draw an arc.

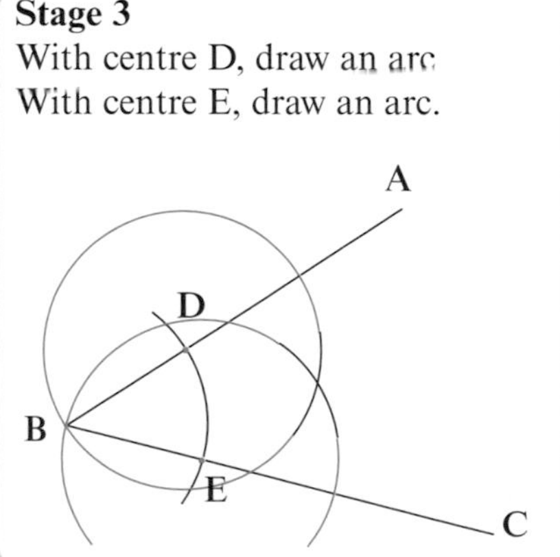

Stage 4
Draw the bisector BF.

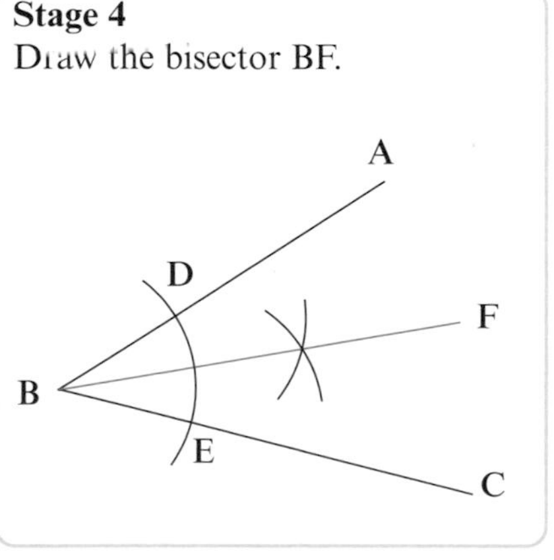

Exercise 11.6
Bisecting an angle

1 Use a protractor to draw each angle then bisect it using compasses.
a 44° **b** 100° **c** 146° **d** 90°

2 Use compasses to draw an equilateral triangle with sides of 8 cm.
What size is each angle?
Bisect one of the angles. What angle have you made?

Meeting conditions

Constructions such as bisecting angles or lines can be used for making scale drawings where conditions have to be met.

Rules that have to be met such as 'the tap must be the same distance from A and B' are known as conditions.

Example

A water tap is to be put in a large garden but it must meet these conditions:
1 It must be the same distance from the two greenhouses, A and B.
2 It must be the same distance from the grape vine wires as from the hedge.

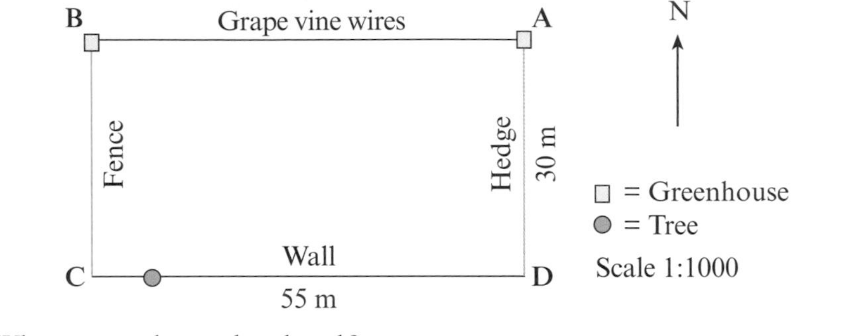

Where must the tap be placed?

To meet condition 1 you draw the perpendicular bisector of BA.
This line is the locus of all points equidistant from B and A.

To meet condition 2 you bisect angle BAD.
This line is the locus of all points equidistant from line BA and line AD.

Where the two loci intersect both conditions are met – so the tap must be at this point.

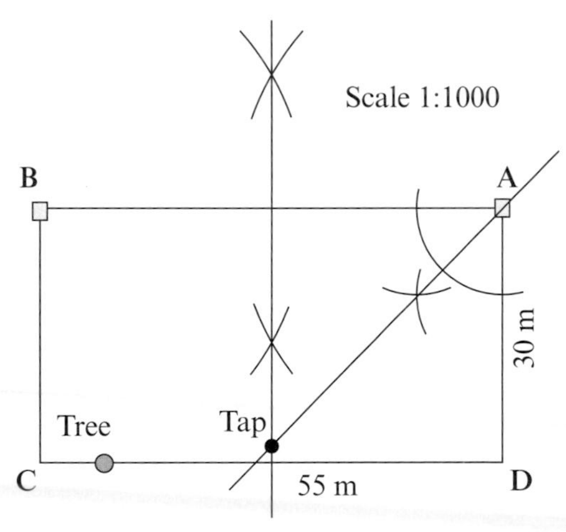

Exercise 11.7
Meeting conditions

1 **a** On the scale diagram above measure the distance in centimetres between the tap and the tree.
b What is this actual distance in the garden?
c Make a scale drawing to show where the tap will be if:
condition 1 stays the same
condition 2 says the tap must be equidistant from the wall and the hedge.

2 An Olympic javelin field has lines which make an angle of 29° to each other.
A thrower aims the javelin so that it flies equidistant from both lines.
The thrower hopes to reach the club record of 88 metres.

This diagram only approximates to how a true javelin field is marked out. The throwing point actually lies on an arc about 2 metres wide which comes at the end of a 36 metre run-up.

a Make a scale drawing of the field for throws up to 100 metres. Use a scale of 1:1000.
b Mark the locus of all points 88 metres from the throwing point.
c Construct the locus of points equidistant from the sidelines.
d Mark where the thrower hopes the javelin will land.

3 Two lighthouses are 3.6 miles apart on a straight coastline.
A ferry sails into port by keeping the same distance from both lighthouses.
A fishing boat sails so that it is always 3 miles from the coast.

a Make a scale drawing of the coast to show the position of the lighthouses. Use a scale of 1 cm to 0.5 miles.
b Show and label the course taken by the fishing boat.
c Construct and label the course taken by the ferry.
d Mark the point where there is the greatest risk of a collision.

Loci and regions

Exercise 11.8
Loci and regions

Goats eat almost every type of plant.
They are often tied by a rope to limit their grazing.

Their rope can be fixed to a ring which can slide along a rail.

For example, a goat is tied in this way to a rail 8 metres long. The rope allows the goat to graze 2 metres from the rail.

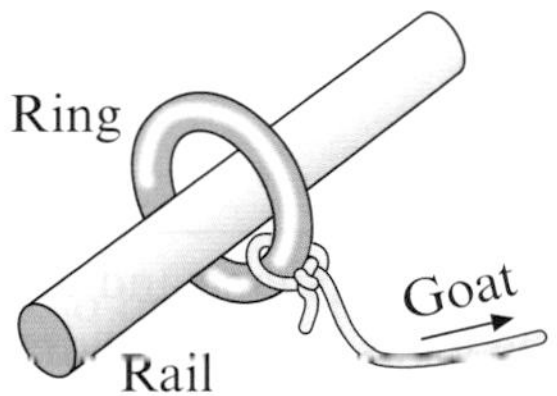

When the ring is halfway along the rail the goat can graze the area inside the circle shown.

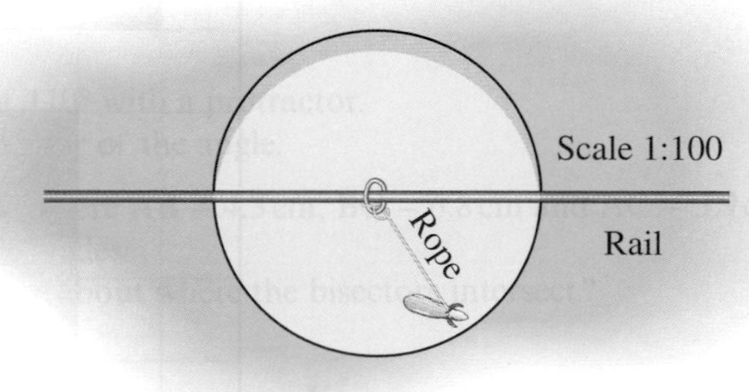

1 Make a scale drawing of the grazing area shown above.

2 On the same diagram draw the grazing areas for other positions of the ring along the rail.
Outline in red the shape of the total area the goat can graze.

Ratio and proportion

- An amount can be shared in a given ratio.

Example Share £140 in the ratio 2 : 1 : 4.

							Total
Ratio	2	:	1	:	4		
Amounts	?	:	?	:	?		£140

- Find the total number of parts:
  2 + 1 + 4 = 7
- Find the amount for one part:
  £140 ÷ 7 = £20
- Multiply each number of parts by the amount for one part.

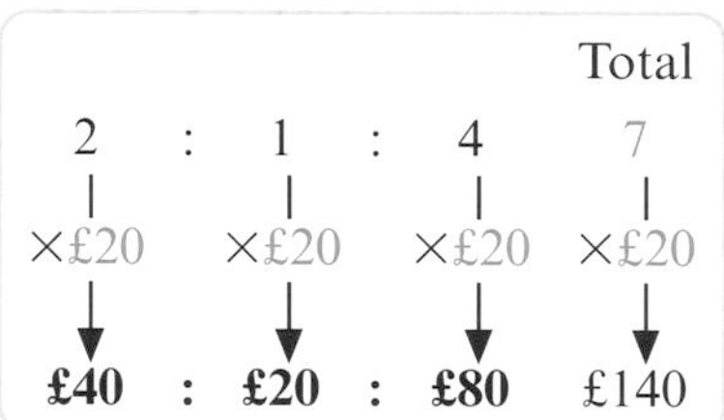

Exercise 12.2
Sharing in a given ratio

'in proportion to' means 'in the same ratio as'

1. Sarah and Denzil share 35 conkers in the ratio 4 : 3.
   Calculate how many conkers each gets.
2. Tim spends his money on sweets, comics and videos in the ratio 1 : 2 : 3.
   Calculate how much he spends on each item from £18.
3. Amy, Ben and Zoe are left £150 by their grandmother.
   The money is shared in proportion to their ages: 5, 3 and 2.
   Calculate how much each child gets.
4. Liz is training for the 100 m hurdles.
   She splits her time between speed work and technique in the ratio 3 : 5.
   Calculate how long:
   a she spends on technique in two hours of training
   b she spends on speed work in four hours of training.

- Some problems need amounts to be kept in proportion.

Example This recipe for Swiss Hot Chocolate serves 2 people.
How much of each ingredient is needed for 5 people?

Swiss Hot Chocolate
- *600 ml milk*
- *140 g drinking chocolate*
- *80 ml whipped cream*

Milk		Chocolate		Cream	Serves
600 ml	:	140 g	:	80 ml	2
?	:	?	:	?	5

- Find the **multiplier:**
  5 ÷ 2 = 2.5
- Multiply each amount by the multiplier.

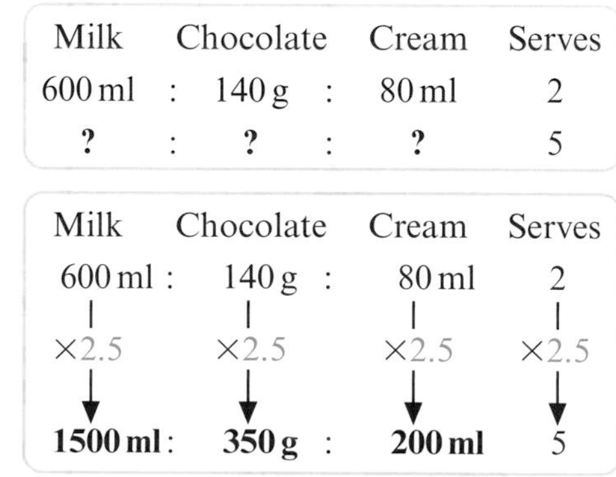

Exercise 12.3
Keeping in proportion

1. This recipe serves 6 people.
   How much of each ingredient do you need for 9 people?

Chilled Chocolate Drink
230 g caster sugar • 300 ml water
50 g cocoa • 1200 ml chilled milk

2 This recipe for chocolate fudge makes 36 pieces.

Chocolate Fudge

- *450 g white sugar*
- *150 ml milk*
- *150 ml water*
- *75 g butter*
- *30 g cocoa*

The multipliers here are less than 1.

a How much cocoa is needed to make 72 pieces?
b How much milk is needed to make 18 pieces?
c Calculate how much of each ingredient you need to make 24 pieces.

3 This recipe makes 24 sweets.

Chocolate Peppermint Creams

- *230 g icing sugar* • *1 egg white*
- *4 drops peppermint essence*
- *100 g plain chocolate*

a Calculate how many drops of peppermint essence are needed to make 42 sweets.
b How many egg whites do you need to make these 42 sweets?

Discuss your answer.

Ratios as fractions

- A ratio that compares two quantities, $p:q$, can be given in the form $\frac{p}{q}$.

Example

The ratio of **width to height** is 4 : 3, i.e.

$$\frac{\text{Width}}{\text{Height}} = \frac{\mathbf{4}}{\mathbf{3}}$$

so the width is $\frac{4}{3}$ of the height.

- In some ratios, you can use the **total** number of parts to give other fractions.

Example The ratio of boys to girls in a class is 1 : 2

Boys :	Girls	Total
1 :	2	3

Ratio of **boys to girls** is 1 : 2

$$\frac{\text{Boys}}{\text{Girls}} = \frac{\mathbf{1}}{\mathbf{2}}$$

Ratio of **boys to total** is 1 : 3

$$\frac{\text{Boys}}{\text{Total}} = \frac{\mathbf{1}}{\mathbf{3}}$$

The number of boys is $\frac{1}{2}$ of the number of girls and $\frac{1}{3}$ of the total.

Exercise 12.4
Ratios as fractions

For Questions **1** and **2** refer to the panel above.

1 Give the ratio of girls to boys in the form:

a Girls : Boys **b** $\frac{\text{Girls}}{\text{Boys}}$

2 Give the ratio of girls to total in the forms $p:q$ and $\frac{p}{q}$.

3 For this photograph, give each of these as a fraction:

a the ratio of width to height
b the ratio of height to width.

4 The ratio of women to men in a fitness class is 1 : 4.
Give the fraction of the class which are:

a women **b** men.

5 The ratio of height to diameter for this tin is 2 : 3.

a What fraction of the diameter is the height?
b Explain why the fraction $\frac{2}{5}$ has no meaning for this tin.

Ratios and enlargement

The ratio is calculated by dividing a length in the **image** by the corresponding length in the **object**.

- The scale factor of an enlargement is the ratio of corresponding lengths.

Example

Calculate the scale factor of this enlargement, and find the height K′L′.

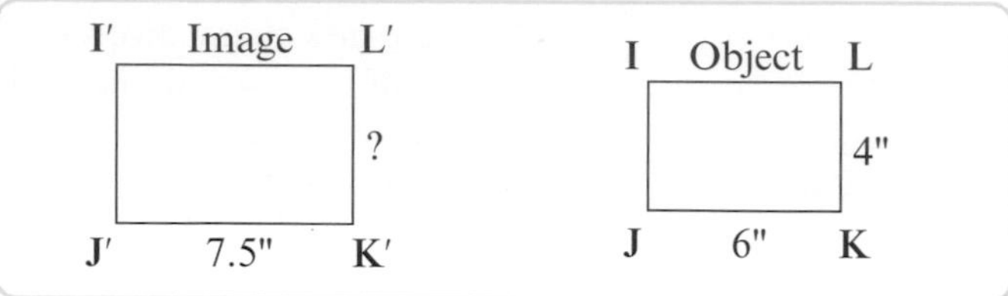

- Divide width J′K′ by the corresponding width JK.

$$7.5:6 = \frac{7.5}{6} = 1.25$$

- Multiply the corresponding height by the scale factor.

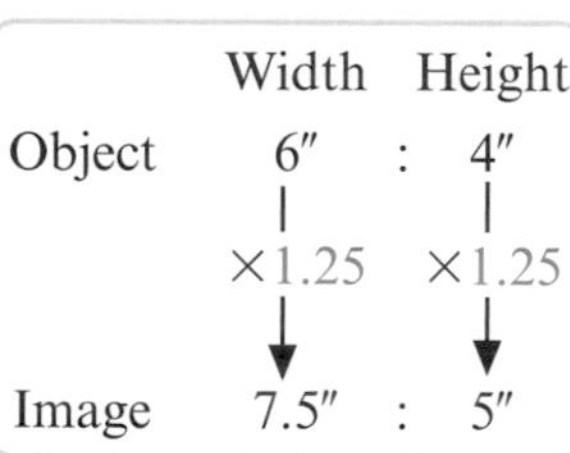

Exercise 12.5
Enlargement

If the image is larger than the object, the scale factor is greater than 1.

If the image is smaller than the object, the scale factor is less than 1.

1

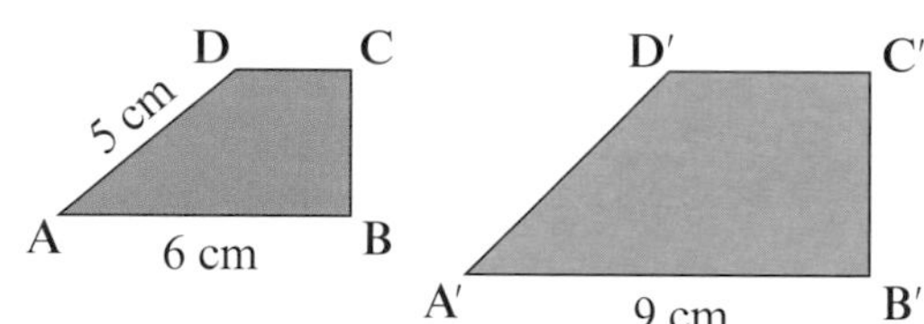

Trapezium A′B′C′D′ is an enlargement of trapezium ABCD.

a Calculate the scale factor of the enlargement.
b Use your scale factor to find the length A′D′.

2

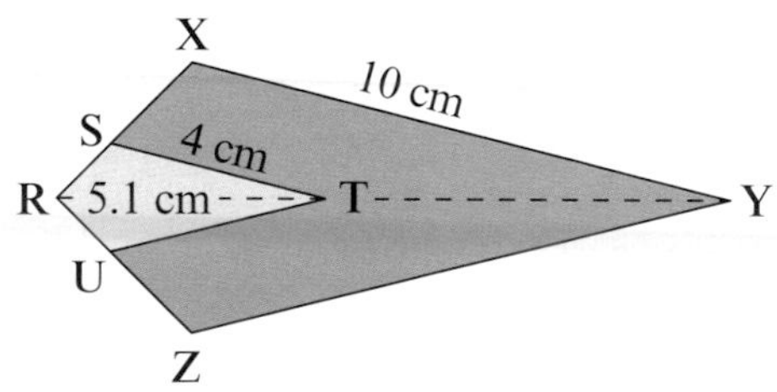

Kite RXYZ is an enlargement of kite RSTU.

a Calculate the scale factor of the enlargement.
b Find the length RY.

3

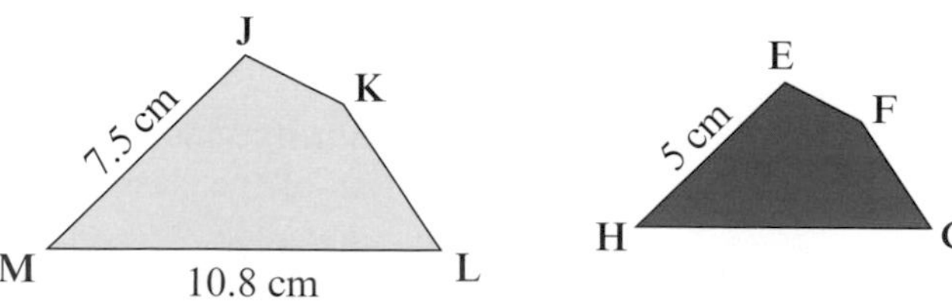

Shape EGFH is an enlargement of shape JKLM.

a Calculate the scale factor of the enlargement.
b Find the length HG.

End points

You should be able to ...	... so try these questions
A Multiply the number of parts in a ratio	**A1** Tropical Fruit Juice is a mix of pineapple and grapefruit in the ratio 5 : 2. Calculate: **a** the amount of pineapple juice to mix with 180 ml of grapefruit juice **b** the total amount of Tropical Fruit Juice produced.
B Share an amount in a given ratio	**B1** Amy, Ben and Zoe are left £400 by their grandfather. The money is shared in proportion to their ages: 10, 8 and 7. Calculate how much each child gets.
C Keep amounts in the same proportion	**C1** This ice cream recipe serves 8 people. **Chocolate Ice Cream** • 150 g caster sugar • 20 g cocoa • 4 eggs • 410 g can evaporated milk • 4 tablespoons boiling water **a** Calculate how much of each of these ingredients you need to serve 20 people: **i** caster sugar **ii** boiling water. **b** How many eggs are needed to serve 6 people?
D Write ratios as fractions	**D1** For this photograph, give the ratio of height to width: **a** in its simplest terms **b** as a fraction  6" 8" **D2** The ratio of girls to boys in a class is 2 : 3. What fraction of the class are girls?

Some points to remember

- Write in headings to help you get the ratio the right way round. For example, a ratio of boys to girls of 1 : 3 can be written as:
  Boys : Girls
  1 : 3
- The scale factor of an enlargement is the ratio of corresponding lengths.

SUMMER

Nine Elms Garden Centre

Watering cans - plastic			
5 litre	8 litre	10 litre	12 litre
99p	£1.29	£1.69	£1.99
Watering cans - traditional			
1 gallon	1.5 gallon	2 gallon	2.5 gallon
£4. 99	£6.49	£7.99	£9.99

Special offers for July

Nine Elms Mulchbags

Large	£1.49
Super	£1.99
Major	£2.99
Professional	£4.99

15% OFF THESE PRICES

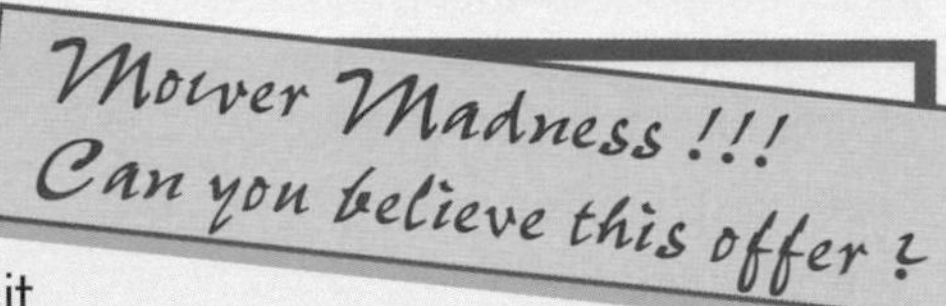

- 20% deposit
- 0% interest
- 6 equal monthly payments

Super Hover	Mowhog	Mastermow
£179.99	£199.99	£249.99

LIQUIDGRO

Add one scoop (15 grams) per gallon

£3.99

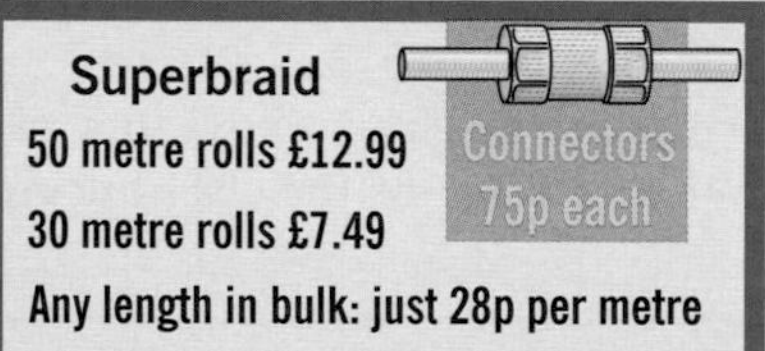

Superbraid
50 metre rolls £12.99
30 metre rolls £7.49
Any length in bulk: just 28p per metre
Connectors
75p each

The easy way to stop weeds

15 metres long
and
1650 mm wide

Only £3.75
per roll
while stocks last

The Super 9 Garden Store – 100% treated timber

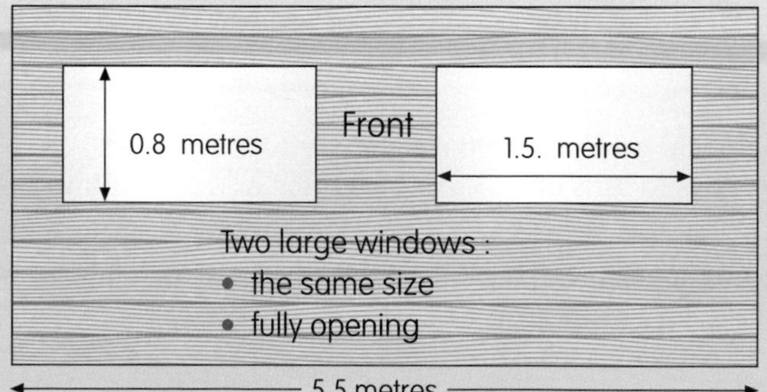

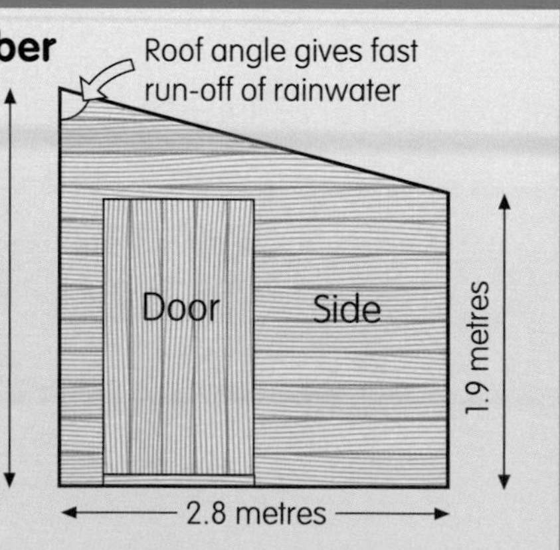

Special Features:

- The back of the store has no windows
- One side has no door or window
- The roof has a 10 cm overhang on all sides

£449.99

1 List the discount price of each size Mulchbag, to the nearest penny.

2 Which plastic watering can do you think is best value for money?
Explain your answer.

3 Which watering can holds most water: the 12 litre or the 2.5 gallon?
Explain your answer.

4 Darren says that a plastic watering can is only about 20% of the cost of a traditional one.
Do you agree? Explain your answer.

5 Which traditional watering can do you think is best value for money?
Explain your answer.

6 A 5 litre plastic watering can weighs 105 grams.
A gallon of water weighs 10 lb.
What do you expect a 5 litre plastic watering can to weigh when it is full?

7 An empty 1.5 gallon traditional can weighs 0.7 lb.
a Give the weight of this can in kilograms.
b Do you expect a full 1.5 gallon traditional can to weigh more, or less, than 7.5 kg?
Explain your answer.

8 **a** To the nearest gram, how much Liquidgro is needed for 10 litres of water?
b To be fairly accurate how many scoops would you put in the 10 litres of water?
Explain your answer.

9 If you measured the Liquidgro accurately:
a How many gallons will the 1 kg pack treat?
b How much spare liquidgro is in the pack?
c What is this spare Liquidgro as:
i a fraction of the pack?
ii a decimal of the pack?
iii a percentage of the pack?

10 The manufacturers of Liquidgro say that a 1 kg pack will treat 300 litres of water.
Is this a fair claim? Explain your answer.

11 Liquidgro costs £285 per tonne to produce.
Packaging costs are 6.8 pence per pack.
Distribution costs are 13.5 pence per pack.

How much profit do the manufacturers make on one tonne of Liquidgro?

12 **a** What is the area of plastic sheet on a roll?
b To the nearest penny, what is the price per square metre?

13 The sheet is rolled on a tube of radius 4.5 cm.
The tube is made from a rectangle of card, with a 1.5 cm overlap allowed for gluing.

What area of card is used for the tube?

14 Which do you think is better value for money: the 50 metre or the 30 metre roll of hose pipe?
Explain your answer.

15 A hockey club needs a hose 180 metres long.
a List three different ways to buy 180 metres of hose in the July special offers.
b What is the cheapest way for the hockey club to buy the hose they need?
Explain your answer.

16 Calculate the total area of glass used for the windows of the Super 9 Garden Store.

17 **a** What shape is the back of the Garden Store?
b What are the dimensions of the back?

18 Calculate the area of one side of the store (including the door).

19 The front, back and both sides of the store have to be sprayed with timber preservative.
Calculate the total area to be sprayed.

20 Preserver is sprayed at 250 ml per m^2.
Will 5 litres of preserver cover one store?
Explain your answer.

21 Calculate the size of the roof angle.

22 **a** What shape is the roof of the store?

One dimension of the roof is 5.7 metres (including overhangs).
b What is the other dimension of the roof?
c Calculate the area of the roof.

The roofing used weighs 4.4 kg per m^2.
d What is the weight of the roof?

23 Labour charges are 40% of the price of the store which takes $12\frac{1}{2}$ hours to make.
a What is the labour charge for a store?
b Calculate the labour charge per hour.

The door takes 45 minutes to make.
c What is the labour charge for a door?

24 For each mower, give the monthly payment.

143

Toujours Paris

Ile de la Cité

This boat-shaped island in the River Seine is where Paris was first inhabited by Celtic tribes over 2000 years ago. It is where Notre Dame is situated. This cathedral is a superb example of French medieval architecture and is particularly known for its wonderful stained glass rose windows.

The width of this South Window is 13 metres.

In the Ile de la Cité you will also find Point Zéro. This is a geometer's mark from which all distances in France are measured.

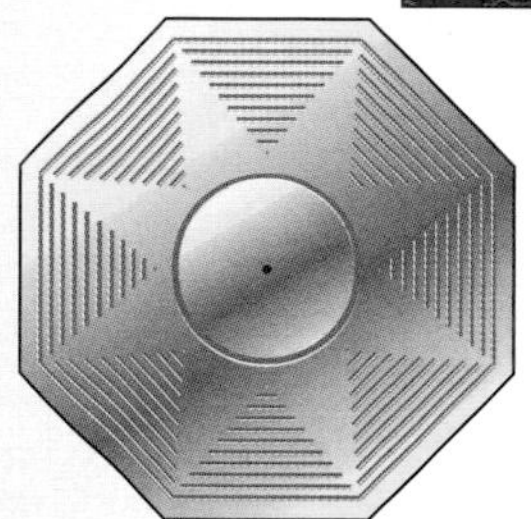

It measures 29 cm across and each side is 12 cm long.

Paris au Quotidien

Arrondissements: Il faut le savoir, Paris est divisé en 20 arrondissements se déroulant en spiraleà partir du 1er (le quartier du Louvre).

Banques: Ouvertes en général du lundi au vendredi de 9h à 16h30, quelques (rares) agences le samedi. Les Caisses d'Epargne ouvrent plus souvent le samedi et ferment le lundi.

Change: On ne peut pas tout prévoir à l'avance; vous pourrez changer vos devises dans les gares, les aéroports, les grandes agences de banque, les points change (ouverts tard le soir), ainsi qu'à notre bureau d'accueil des Champs-Elysées.

Daily Life in Paris

Districts: You should know that Paris is divided into 20 districts numbered in a circular direction, starting with the 1st district (the Louvre area).

Banks: They are generally open from Monday to Friday from 9 am to 4.30 pm, some (rare) branches on Saturday. The savings banks are more often open on Saturday and closed on Monday.

Exchange: You cannot foresee everything; you can therefore change your foreign currency in railway stations, airports, major bank branches, exchange offices (open late in the evening), as well as in the visitors office on the Champs-Elysées.

Eiffel Tower Factfile – 1996

- Built in 1889 for the Universal Exhibition
- Built by Gustave Eiffel (1832–1923)
- Built from pig iron girders
- Total height 320 metres
- The world's tallest building until 1931
- Height to 3rd level is 899 feet
- There are 1652 steps to the third level
- Two and a half million rivets were used
- The tower is 15 cm higher on a hot day.
- Its total weight is 10100 tonnes
- 40 tons of paint are used every 4 years
- On a clear day it is possible to see Chartres Cathedral, 72 km away to the South West.
- The tower is visited by about $5\frac{1}{2}$ million people every year

Admission charge 56 Francs

DAY TRIPS BY EUROSTAR

Waterloo Station (London) to Paris

Celebrate that special occasion in style with a day trip to Paris!

ADULT £

CHILD (4–11 YRS) £

Entry fees in Paris – 1996

Eiffel Tower	56 FF
Louvre	45 FF
Pompidou Centre	35 FF
Picasso Museum	28 FF
Museum of Modern Art	27 FF
Versailles Palace	45 FF
Parc de la Villette	45 FF
Cluny Museum	28 FF

Paris Lucky dip – Superb value !

In the hat are four tickets to the:
Eiffel Tower,
Picasso Museum,
Versailles Palace and
the Museum of Modern Art.
Pick two tickets at random from the four.
Entry fee – only 83 Francs.

Datafile

Population of Paris (in 1982) 2 188 918
In 1996, £1 sterling was equivalent to 7.54 French Francs.
1 metre = 3.281 feet
1 kilometre = 0.62 miles

1 How many lines of symmetry has the South Window of Notre Dame?

2 What order of rotational symmetry has the South Window?

3 What is the circumference of the South Window?

4 Calculate the area of the South Window.

5 For Point Zéro in the Ile de la Cité, the outer polygon is regular.
  a What is the name of this polygon ?
  b Calculate the size of an exterior angle.
  c Calculate the size of an interior angle.

6 Calculate the total area of Point Zéro.

7 How many planes of symmetry do you think the Eiffel Tower has?

8 How many tons of paint will have been used on the tower from when it was built up to the year 2000?

9 How much higher than the third level is the total height of the tower?
Give your answer to the nearest metre.

10 What is the mean height of a tower step in cm?

11 The ratio of steps to the third level, to steps to the first level is about 9 to 2.
About how many steps are there to the first level?

12 Use standard form to give:
  a the weight of the tower in tonnes
  b the number of rivets used.

13 Round the number of steps to the third level to:
  a the nearest ten
  b the nearest hundred
  c the nearest thousand.

14 Which one of the following gives the approximate percentage that the tower grows on a hot day?
  a 5% **b** 0.5% **c** 0.05% **d** 0.005%

15 What is the approximate bearing of:
  a Chartres from Paris
  b Paris from Chartres?

16 In 1996, what was the Eiffel Tower entry fee equivalent to in £ Sterling?

17 **a** Approximately how much money in Francs was made from entry fees to the Eiffel Tower in 1996?
  b What was this equivalent to in £ Sterling?

18 Give the population of Paris in 1982 to:
  a 2 sf **b** 3 sf **c** 4 sf **d** 5 sf.

19 Compare the French and English texts in the extract on Daily Life in Paris.
What is the relative frequency of a vowel in each language? (Vowels are a, e, i, o, and u.)

20 Compare the French and English extracts.
Which language uses the longest words?
Describe how you decided.

21 For the Paris Lucky Dip use the following shorthand:

E – Eiffel Tower
P – Picasso Museum
V – Versailles Palace
M – Museum of Modern Art

  a List each pair of tickets that could be picked at random.
  b What is the probability that a pair of tickets is picked which includes the Versailles Palace?
  c What is the probability of picking P and M?
  d Give the probability of picking a pair of tickets
    i worth more than the entry fee
    ii worth less than the entry fee.
  e Is the seller likely to make a profit or loss on every hundred entries? Give your reasons.

22 These two bills were for Eurostar day trips to Paris.

Enclosed are tickets for:
2 adults and 3 children
Total charge £465

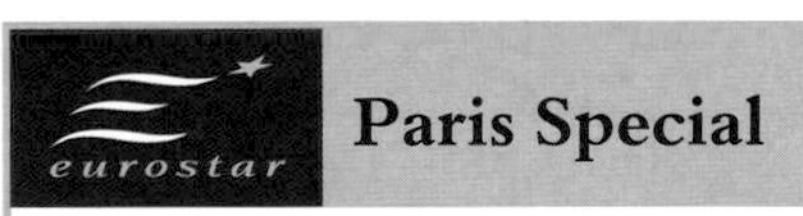

Enclosed are tickets for:
3 adults and 1 child
Total charge £386

Calculate:
  a the total cost for these 5 adults and 4 children
  b the cost of an adult's ticket
  c a child's ticket
  d the total cost for 1 adult and 2 children.

Starting points

You need to know about ...

A Units for measuring distance

- Kilometres (km), metres (m), centimetres (cm) and millimetres (mm) are **metric** units.
- Miles, yards, feet and inches are **imperial** units.

Metric	Imperial
1 km = 1000 m	1 mile = 1760 yards
1 m = 100 cm	1 yard = 3 feet
1 cm = 10 mm	1 foot = 12 inches

Some approximate conversions

Think of: 5 miles as 8 km
$1\frac{3}{4}$ pints as 1 litre
30.5 cm as 1 foot
2.2 lb as 1 kg

B Interpreting graphs

- Graphs that help you change from one unit to another are called **conversion graphs**.

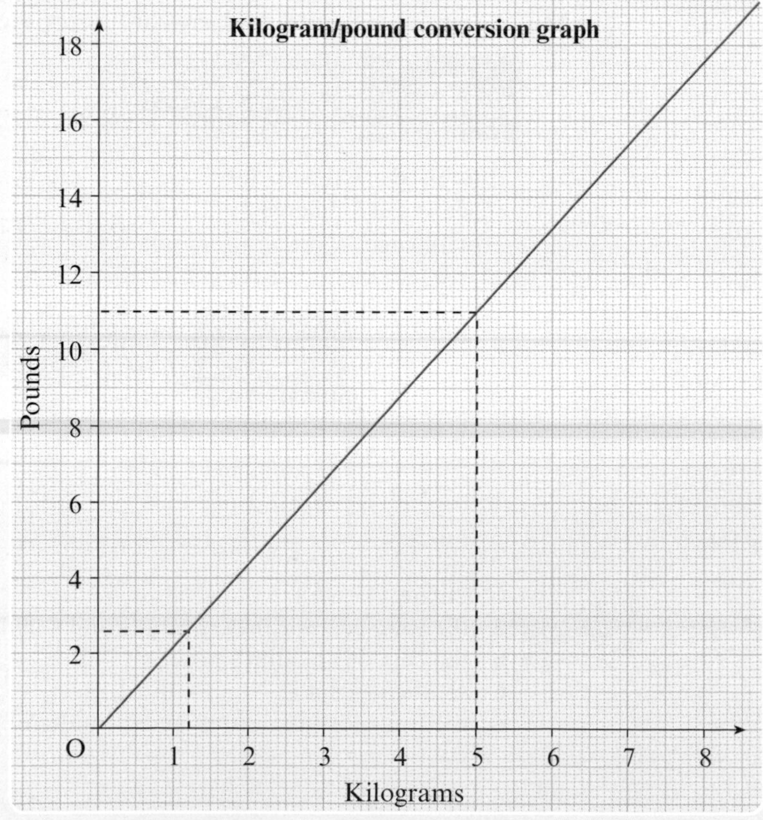

Examples 5 kilograms ≈ 11 pounds
2.6 pounds ≈ 1.2 kilograms

... so try these questions

A1 Write these in metres.
a 5 km **b** 340 cm

A2 Write these in metres. Choose an appropriate degree of accuracy for each answer.
a 3 miles **b** 5 feet
c 2 yards

A3 Write these in centimetres, correct to the nearest cm.
a 40 mm **b** 4 km
c 3.9 m **d** 7 inches

A4 Write these in kilometres, correct to 2 dp.
a 5130 metres **b** 7.1 miles

A5 Write 20 km in miles, correct to 1 dp.

B1 Use the graph to give an estimate of these in pounds, to 1 dp.
a 7 kilograms
b 3.8 kilograms

B2 Use the graph to give an estimate of these in kilograms, to 1 dp.
a 16 pounds **b** 5 pounds
c 7.8 pounds **d** 4.5 pounds

B3 Use the graph to estimate the number of pounds in 1 kilogram, to 1 dp.

B4 1 gallon ≈ 4.55 litres.
a Use this approximation to copy and complete this table up to 8 gallons.

Gallons	1	2	3
Litres	4.55	9.1	

b Make a conversion graph for gallons/litres up to 8 gallons.
c Use your graph to give an estimate of:
i 2.5 gallons in litres
ii 15 litres in gallons.

C Calculating with time

- Times can be written:
  - in 12-hour time using am or pm
  - in 24-hour time.

Examples 6:20 am is 06:20 in 24-hour time.
6:20 pm is 18:20 in 24-hour time.

- Some units for measuring time are hours (h), minutes (min) and seconds (s):

1 day = 24 hours
1 hour = 60 minutes
1 minute = 60 seconds

- Times can be written in different ways.

Examples 250 minutes is 4 hours and 10 minutes.
3 minutes and 12 seconds is 192 seconds.

C1 Write these times using am or pm.
a 09:30 **b** 14:21

C2 Write these in 24-hour time.
a 2:23 am **b** 5:25 pm

C3 How many minutes are in 1 h and 15 min?

C4 Write 400 minutes in hours and minutes.

D Timetables and time intervals

This shows part of a bus timetable from Paignton to Heathrow.

PAIGNTON	0625	0825	0955	1145	1345	1700
TORQUAY	0640	0840	1010	1205	1400	1715
Newton Abbot	0700	0900	1030	1225	1425	1735
EXETER	0730	0930	1100	1300	1500	1805
Taunton	↓	↓	↓	↓	↓	1855
Calcot Coachway	↓	↓	1345	↓	1745	↓
HEATHROW AIRPORT	1105	1310	1440	1635	1840	2145

- Times for each different bus are shown in vertical columns.

Example The 0825 bus from Paignton stops at
0840 in Torquay,
0900 in Newton Abbot, …

- The arrows show that the bus does not stop.

Example The 0825 bus from Paignton does not stop at Taunton or Calcot Coachway.

This distance table shows distances in miles between four cities.

London			
117	Birmingham		
159	**76**	Sheffield	
397	**292**	**248**	Glasgow

- To find the distance between two cities, read down and across.

Example

The distance from London to Sheffield is 159 miles.

D1 What is the latest time you can catch a bus from Exeter to reach Heathrow before 6:00 pm?

D2 In hours and minutes, calculate how long each bus takes to go from Paignton to Heathrow.

D3 Which bus takes the longest time to travel from Torquay to Exeter?

D4 After 3:00 pm, what is the time of the first bus from Torquay to Exeter?

D5 It takes David 8 hours to drive from London to Glasgow. About how long do you think it would take him to drive from London to Sheffield?

D6 Suki drives from Glasgow to London via Birmingham. She leaves at 2.15 pm and arrives in Birmingham at 7.15 pm. She leaves Birmingham at 7.45 pm. About what time do you think she will arrive in London?

Distance, time and constant speed

A **constant speed** is a steady speed.

An object travelling at a constant speed does not slow down or get faster.

At a constant speed of 9 metres per second, how far would a car travel in 20 seconds?

- In 1 second the car travels 9 metres.
- So in 20 seconds the car would travel 9 × 20 = 180 metres.

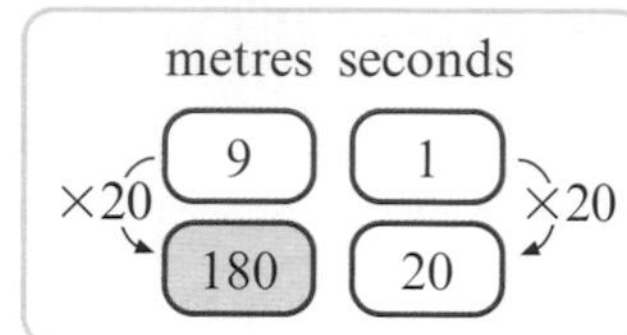

Distance = Speed × Time

At a constant speed, a car travels 8 kilometres in 5 minutes.
How fast is it travelling?

- In 5 minutes the car travels 8 kilometres.
- In 1 minute the car travels 8 ÷ 5 = 1.6 kilometres.
- So the speed of the car is 1.6 kilometres per minute.

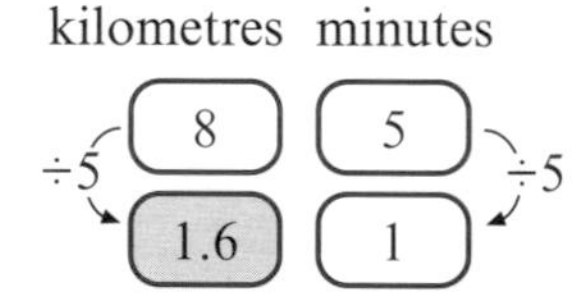

Speed = Distance ÷ Time

To find the formula for speed:

Distance = Speed × Time

Now divide both sides by Time:

$$\frac{\text{Distance}}{\text{Time}} = \text{Speed}$$

At a constant speed of 20 miles per hour, how long would it take for a car to travel 50 miles?

- To travel 20 miles takes 1 hour.
- So to travel 50 miles takes 50 ÷ 20 = 2.5 hours.

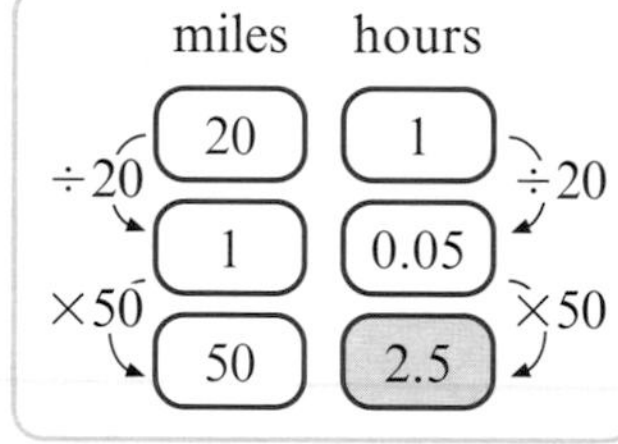

Time = Distance ÷ Speed

To find the formula for time:

Distance = Speed × Time

Now divide both sides by Speed:

$$\frac{\text{Distance}}{\text{Speed}} = \text{Time}$$

Exercise 13.1 Calculating with constant speeds

Accuracy
In this exercise, give distances in metres, to 1 dp, and times to the nearest second.

1 At the Cairngorm Ski Area, you can go up and down the ski slopes by chairlift. The chairs on the White Lady Chairlift travel at a constant speed of 2.55 metres per second.

a How far would you travel on this chairlift in:
i 2 seconds **ii** 10 seconds **iii** 4.5 seconds?

b At this speed, how many metres would a chair travel in:
i 1 minute **ii** 1 hour?

Chairs on the White Lady Chairlift travel 1054 metres to the top of the slope.

c How long does it take a chair to travel to the top of the slope:
i in seconds **ii** in minutes and seconds?

2 An escalator is a set of stairs that moves at constant speed.

a Why do you think the speed of an escalator is constant?

This table shows data on some escalators in London Underground stations.

m/s stands for metres per second.

Station	Normal speed (m/s)	Length (m)	No. of steps
Alperton	0.46	13.7	102
The Angel	0.75	60.0	318
Chancery Lane	0.60	9.1	84
Kentish Town	0.66	44.1	237

Assume that each person stands still on one step of the escalator.

b Calculate how long it takes to travel up each escalator.
c At The Angel, 250 people travel up the escalator.
Each person stands on the step just below the person in front.
If the first person steps onto the escalator at 2.00 pm, when will the last person reach the top of the escalator?

For problems with hours and minutes, it is often easier to work in minutes.

Example

A plane travels 670 miles at a constant speed in 1 hour and 42 minutes.
Calculate its speed in miles per hour.

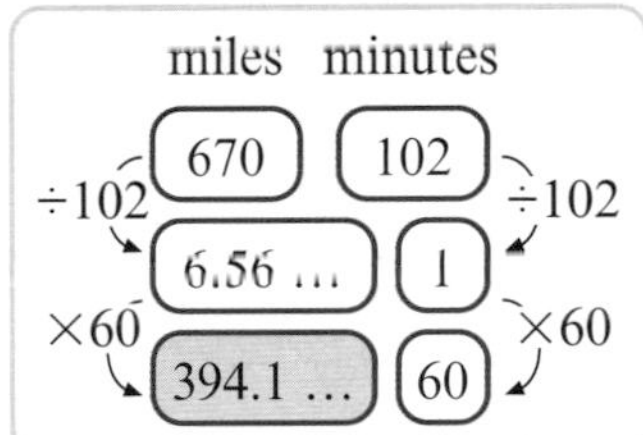

- Time taken is 60 + 42 = 102 minutes.
- Speed = Distance ÷ Time
  = 670 ÷ 102
  = 6.5686 ... miles per minute
- In 60 minutes (1 hour) the plane travels
  6.5686 … × 60
  = 394.1176 ... miles
- So the speed is about 394.1 miles per hour.

Exercise 13.2 Calculating with hours and minutes

Accuracy
In this exercise, give all answers correct to 1 dp unless stated otherwise.

1 A plane travels 1300 miles at a constant speed in 2 hours and 30 minutes. Calculate its speed in miles per hour. Do not use a calculator.

2 A student has tried to solve a problem that involves hours and minutes.

A plane travels 1115 miles at a constant speed in 1 hour and 50 minutes. Calculate its speed in miles per hour.

Speed = distance ÷ time
= 1115 ÷ 1.50
≈ 743.3 miles per hour. ✗

a Explain the mistake you think she has made.
b Solve the student's problem to find the speed of the plane.

3 A Douglas DC-10 jet cruises at a constant speed of 620 miles per hour.

a Find this speed in miles per minute, correct to 2 dp.
b At this speed, how many miles would it travel in 10 minutes?

mph stands for miles per hour.

4 A Boeing 757 jet cruises at a constant speed of 403 mph.

a Find this speed in miles per minute, correct to 2 dp.
b How many minutes would it take to travel 120 miles at this speed?

5 Concorde cruises at a constant speed of 1350 mph.

a At this speed, how long would it take to travel 800 miles?
b How far would it travel in 2 hours and 35 minutes?

Constant speeds and graphs

These diagrams show the positions of three cars on a road at one-second intervals. Each car is travelling at a constant speed.

The scale shows the distance along the road from the bridge in metres

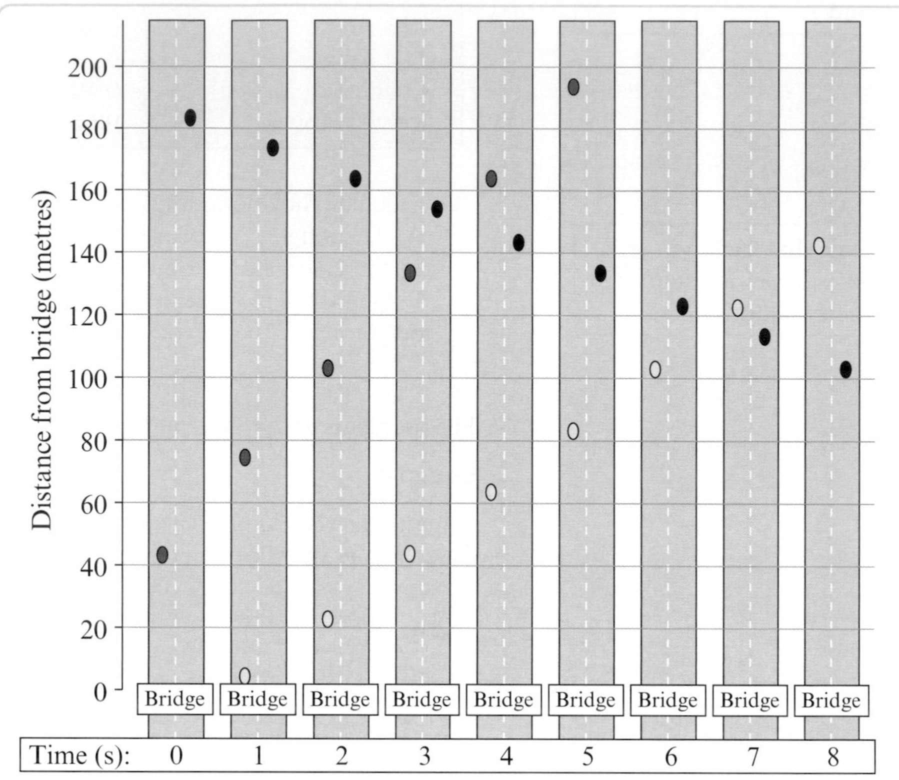

This is part of a **distance–time graph** for the red car.

- The shortest distance is measured from each car to the bridge.
- The red car travels 30 metres every second.

  It has a constant speed of 30 metres per second or 30 m/s.

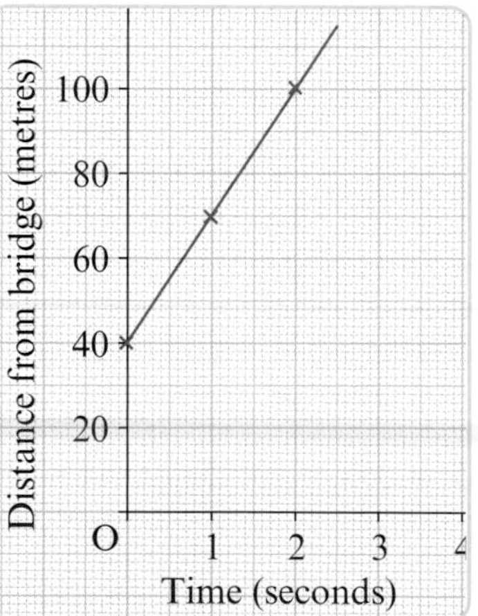

Exercise 13.3
Distance-time graphs

Title your graph 'Distance–time graph for cars'.

For Question **1** refer to the panel above.

1 **a** **i** Draw a set of axes from:
- 0 to 8 seconds on the horizontal axis
- 0 to 200 metres on the vertical axis.

ii Draw the complete distance–time graph for the red car.

b Use your graph to estimate the distance of the red car from the bridge after 4.5 seconds.

c After how many seconds was the red car 120 metres from the bridge?

d How far was the black car from the bridge after 6 seconds?

e Draw the graphs for the yellow car and the black car on your axes.

f i Find the speeds of the yellow car and black car in metres per second.
 ii Which was travelling fastest: the red, yellow or black car?
 iii How can you tell this from your graph?

g i After how many seconds did the red car pass the black car?
 ii Explain how you found your answer.

2 This distance–time graph shows part of the journeys of four cars on a motorway.

The distances are measured from a speed camera.

The colour of each line shows the colour of the car.

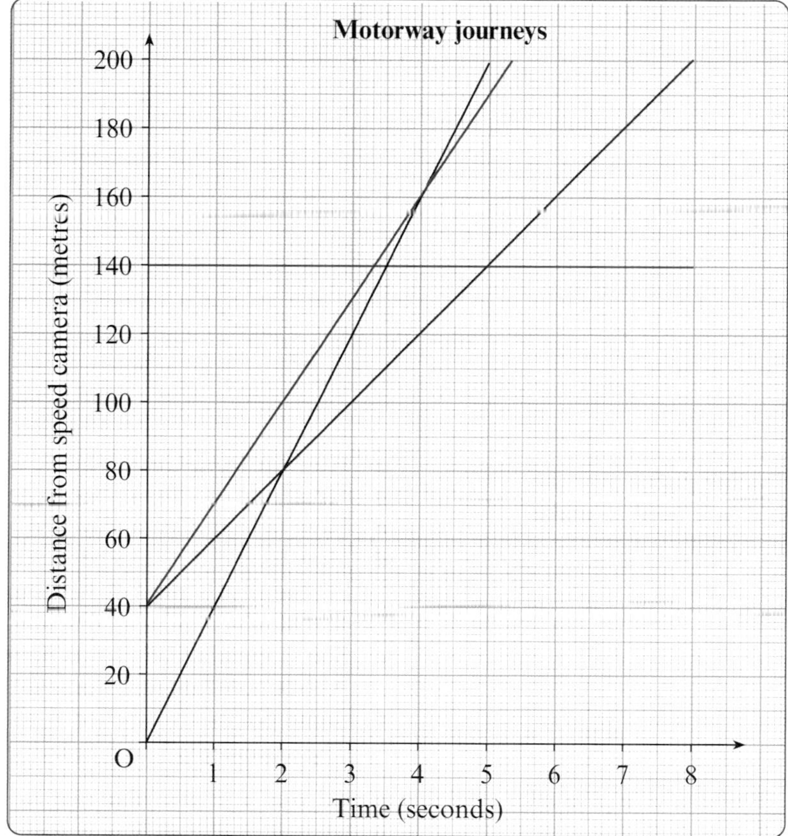

Use tracing paper over the graph if you need to make marks to help you.

a Estimate how far each car was from the speed camera after 2.5 seconds.

b Which car was travelling fastest?

c After how many seconds did:
 i the black car pass the blue car
 ii the blue car pass the red car?

d Find the speed of each car in m/s.

e What do you think happened to the red car?

3 This diagram shows the position of three cars and their speeds in m/s. Each car is travelling in the direction of the arrow at a constant speed.

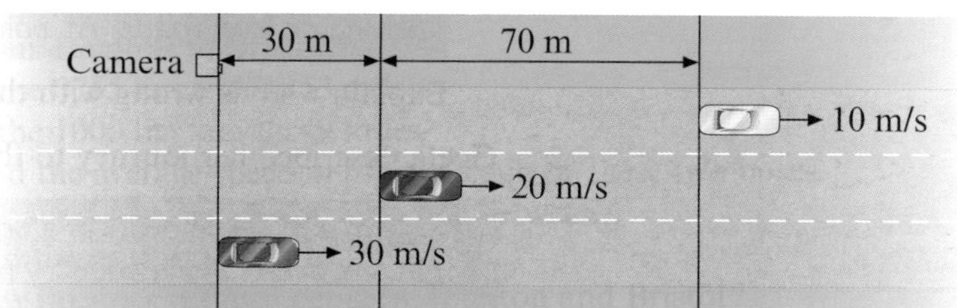

Use a set of axes from:
- 0 to 10 seconds on the horizontal axis
- 0 to 300 metres on the vertical axis.

a Draw a distance–time graph for the next 10 seconds.

b How far is the red car from the camera when it passes the blue car?

Starting points

You need to know about ... | ... so try these questions

A Naming shapes, sides and angles

- The vertices of a shape can be labelled with letters. The sides and angles can be named using these letters.

Example

- This is ΔPQR (triangle PQR).
- The **sides** of PQR are PQ, QR and PR. PQ is also used for the length of the line PQ.
- The **angles** in PQR are: $P\hat{Q}R$ (or ∠ PQR or $\hat{Q}$), $Q\hat{R}P$ (or ∠ QRP or $\hat{R}$), and $R\hat{P}Q$ (or ∠ RPQ or $\hat{P}$).

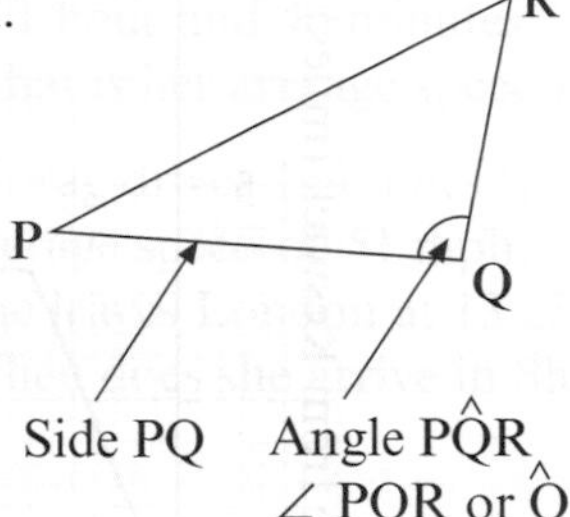

A1

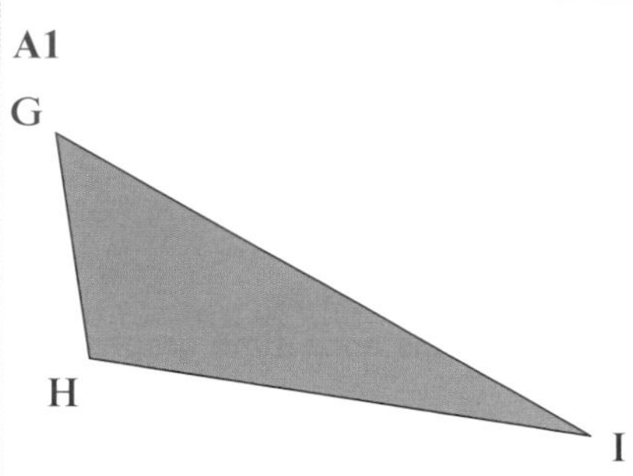

In triangle GHI:

a which side is the longest?
b which angle is the largest?
c which angles are acute?
d which is the shortest side?

B Similar triangles

- When two shapes are similar:
  - the corresponding angles are equal
  - the lengths of the corresponding sides are in the same ratio.

Two triangles are similar if either:

- corresponding angles are equal, or
- the lengths of corresponding sides are in the same ratio.

So to identify similar triangles you only need to check one of these properties.

- This is one way to identify **corresponding sides** in a pair of similar triangles.

In triangles ABC and PQR AC and PR are **corresponding sides** because they are opposite equal angles.

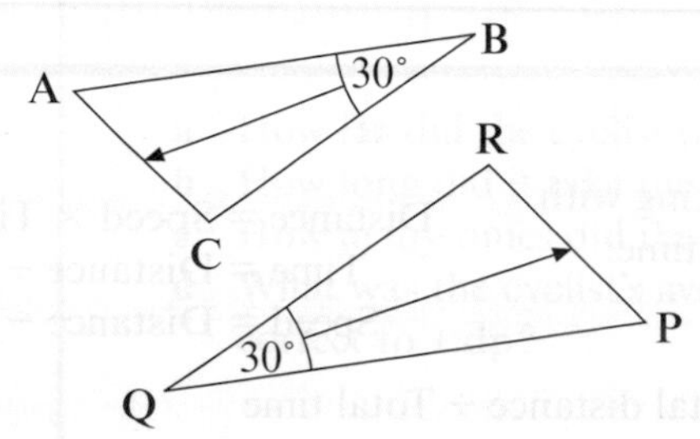

B1

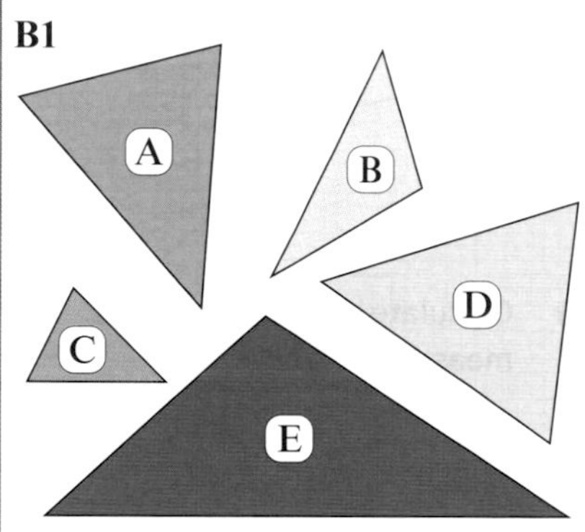

Using triangles A to E find two pairs of similar triangles.

B2 ΔEFG and ΔXYZ are similar.

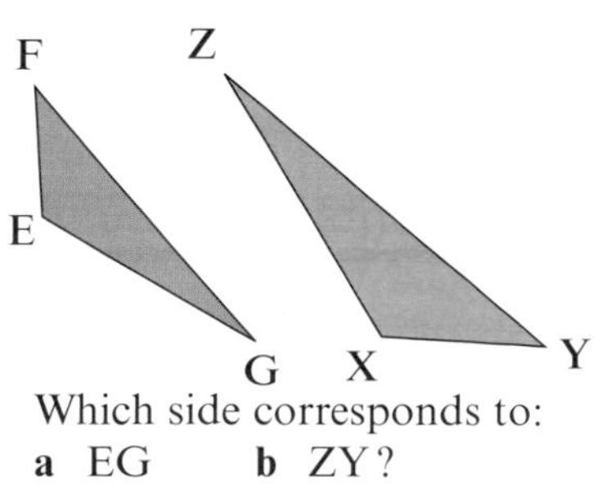

Which side corresponds to:
a EG **b** ZY?

Thinking ahead to ... similar triangles

A Triangles ABC and PQR are similar.

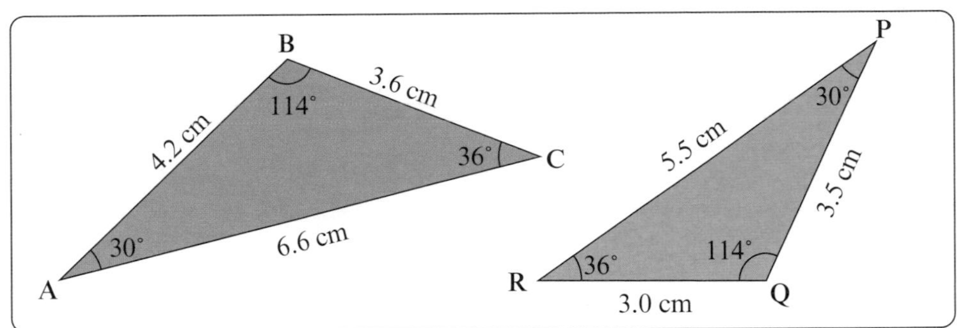

a In triangle ABC, the shortest side is 3.6 cm and the longest side is 6.6 cm.

Check that the ratio $\frac{\text{Shortest side}}{\text{Longest side}}$ equals 0.55 (to 2 dp).

b For triangle PQR calculate the ratio $\frac{\text{Shortest side}}{\text{Longest side}}$ (to 2 dp).

c What do you notice?

B **a** Calculate the ratio $\frac{\text{Longest side}}{\text{Shortest side}}$ (to 2 dp) for triangles ABC and PQR.

b What do you notice?

Similar triangles

Triangles will only be similar if:

- all corresponding angles are equal
- any pair of corresponding ratios are equal

- In any triangle you can compare the length of two sides using a ratio.

Example

In triangle KLM:

$$\frac{KL}{ML} = \frac{1.9}{2.0} = 0.85$$

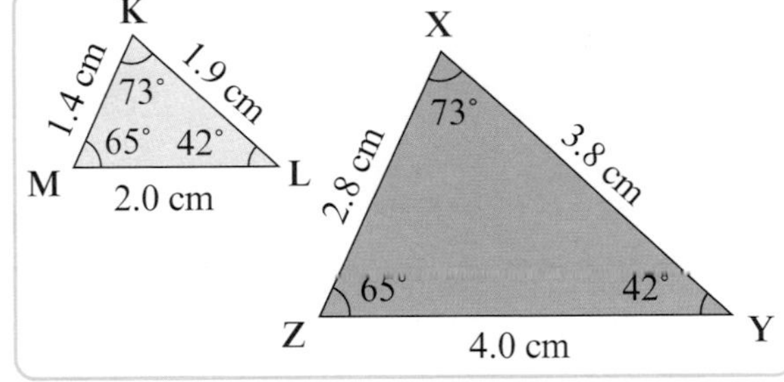

In triangles KLM and XYZ:

KM corresponds to XZ and ML corresponds to ZY.

So $\frac{KM}{ML}$ and $\frac{XZ}{ZY}$ can be called **corresponding ratios**.

$$\frac{KM}{ML} = \frac{1.4}{2.0} = 0.7 \text{ and } \frac{XZ}{ZY} = \frac{2.8}{4.0} = 0.7 \qquad \text{So } \frac{KM}{ML} = \frac{XZ}{ZY}$$

- For two similar triangles, **any** pair of corresponding ratios are equal.

So in triangles KLM and XYZ:

$\frac{ML}{KM}$ and $\frac{ZY}{XZ}$ are corresponding ratios, so $\frac{ML}{KM} = \frac{ZY}{XZ}$

$\frac{KL}{KM}$ and $\frac{XY}{XZ}$ are corresponding ratios, so $\frac{KL}{KM} = \frac{XY}{XZ}$

Exercise 14.1
Ratios of sides in similar triangles

Accuracy
For this exercise round each answer to 2 dp.

1 Triangles ABC, FDE and IHG are similar.

a **i** List the ratios that are equal to $\frac{GH}{GI}$

ii Check by calculating the value of these ratios.

b **i** List the ratios that are equal to $\frac{GI}{GH}$

ii Check by calculating the value of these ratios.

c List other sets of corresponding ratios in triangles ABC, FDE and IHG. Find the value for each set.

It may help if you redraw the triangles in the same orientation.

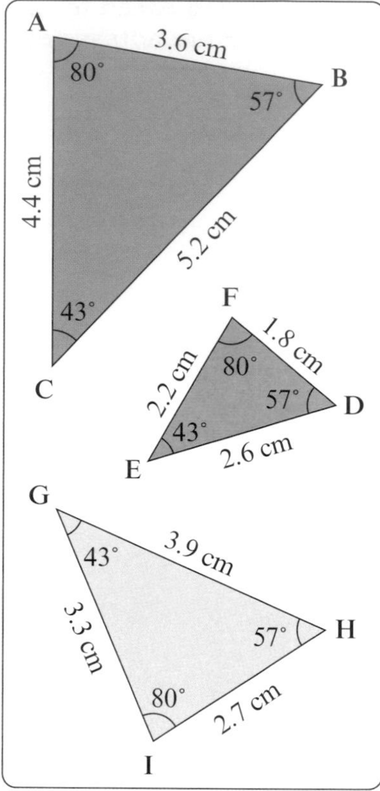

2

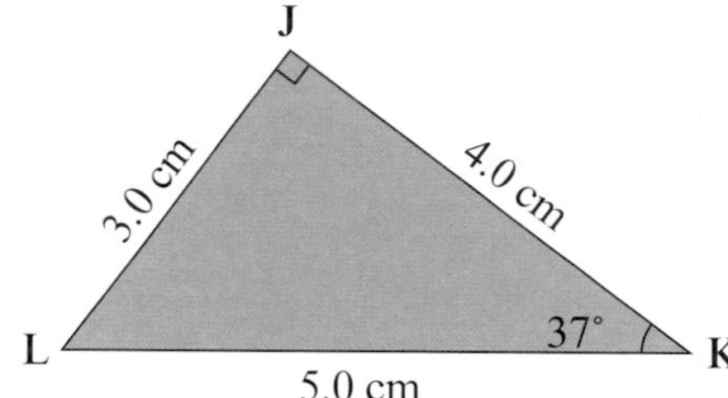

M 3.2 cm N

4.0 cm 2.4 cm

53°

O

a Explain why triangles JKL and NMO are similar.

b Find the value of all the corresponding ratios in triangles JKL and NMO.

3 For two similar triangles, how many pairs of corresponding ratios are there?

Right-angled triangles

- The sides of a right-angled triangle can be labelled like this:

the hypotenuse (the longest side)	hyp
the side opposite x	opp
the side adjacent to (next to) x	adj

The **hypotenuse** is the side opposite the right angle.

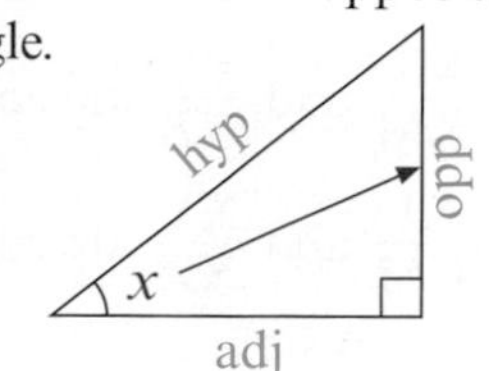

adj x hyp opp

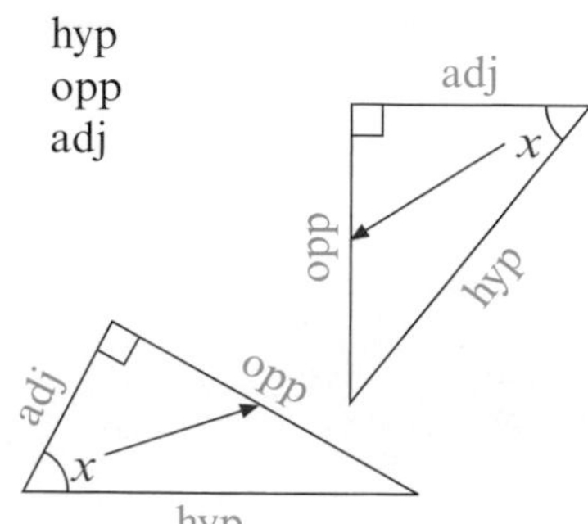

Exercise 14.2
Ratios of sides in right-angled triangles

1 Triangle A is a 30° right-angled triangle.

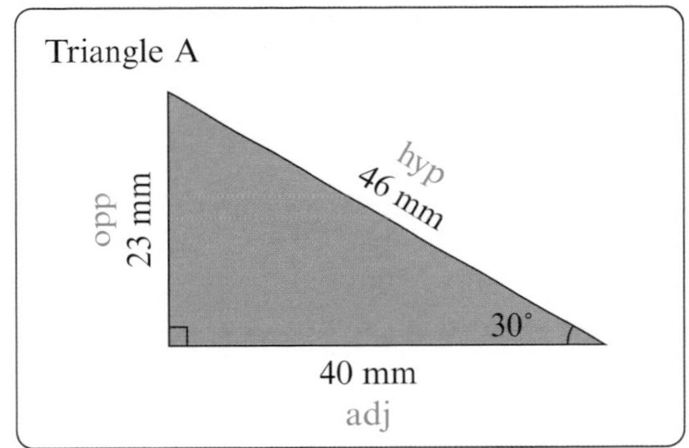

Make a ratio table like this and fill in the values for triangle A.

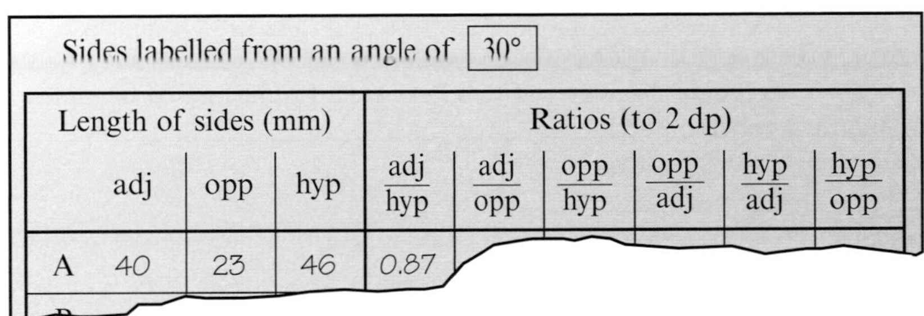

Sides labelled from an angle of 30°

Length of sides (mm)				Ratios (to 2 dp)					
	adj	opp	hyp	adj/hyp	adj/opp	opp/hyp	opp/adj	hyp/adj	hyp/opp
A	40	23	46	0.87					

2 Draw some other 30° right-angled triangles, and for each triangle:

Draw your triangles on squared paper.

- **a** label the sides from the 30° angle as hyp, opp or adj
- **b** measure the length of each side to the nearest millimetre
- **c** calculate the ratios for each pair of sides and fill in your ratio table.

3 Describe anything you notice about your ratios.

4 **a** What do you think will happen to the ratios for other right-angled triangles?
b Test your ideas on other right-angled triangles.

Trigonometric ratios

- The ratios of the sides in a right-angled triangle are called **trigonometric ratios**.

In triangle ABC,

sine:	sin 30°	= 0.50
cosine:	cos 30°	= 0.87
tangent:	tan 30°	= 0.58
		(to 2 dp)

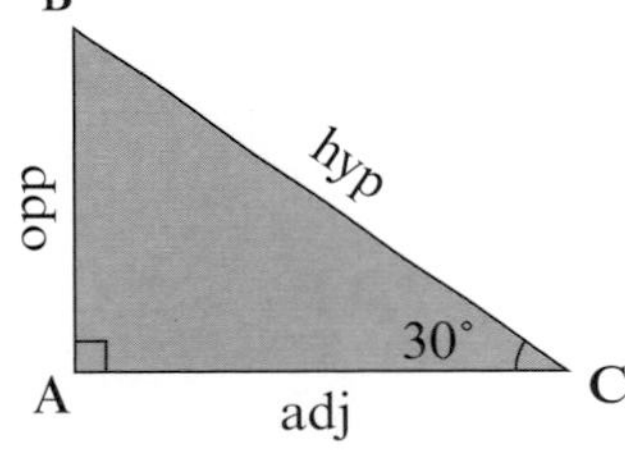

Exercise 14.3
Trigonometric ratios

Your ratios may not match the values of sin 30°, cos 30° and tan 30° exactly.

1 **a** Find these keys on your calculator: 

b Use them to find the values of sin 30°, cos 30° and tan 30°.
c Check the values match those given above.
d Match each value to one of your ratios for the 30° triangles that you have drawn.

2 **a** How can you calculate the sine, cosine and tangent of an angle?
b Check your ideas on the triangles you drew in Exercise 14.2.

Sine, cosine and tangent of an angle

A mnemonic (pronounced *ne-mon-ik*) is a memory aid.

This is a mnemonic used to remember the definitions for the sine, cosine and tangent of an angle:

Skive off homework
Cheat at homework
Telling off after

◆ These are definitions of three trigonometric ratios (or 'trig' ratios):

sine: $\sin x = \frac{\text{opp}}{\text{hyp}}$

cosine: $\cos x = \frac{\text{adj}}{\text{hyp}}$

tangent: $\tan x = \frac{\text{opp}}{\text{adj}}$

These trig ratios can be used to calculate any side or any angle in **right-angled triangles**.

Exercise 14.4
Calculating trigonometric ratios

Accuracy
For this exercise round each answers to 2 dp.

1 These are all right-angled triangles.

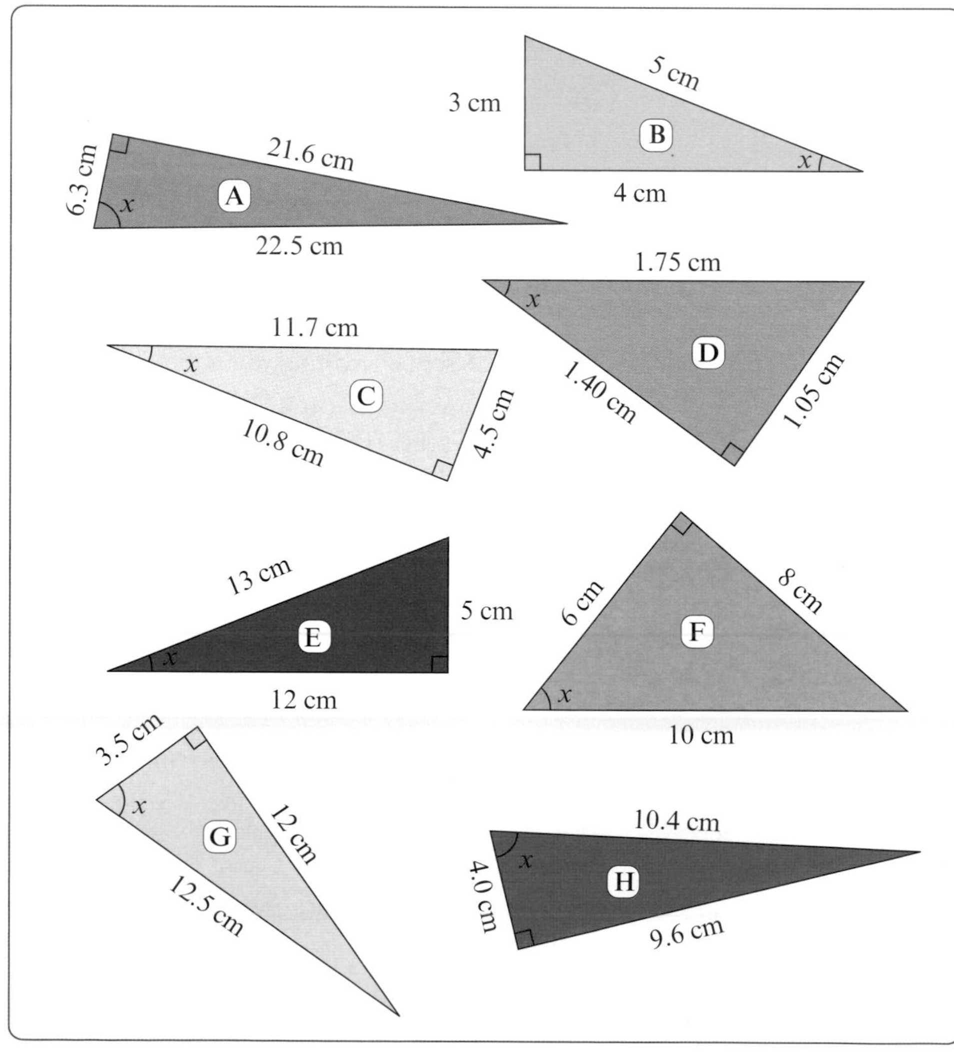

a Sketch each of these triangles and label the sides: opp, adj, hyp.

b For each triangle, find the value of sin x, cos x and tan x.
i Give each value as a fraction
ii Give each value as a decimal.

Finding a side

- Each trig ratio can be written in different ways.

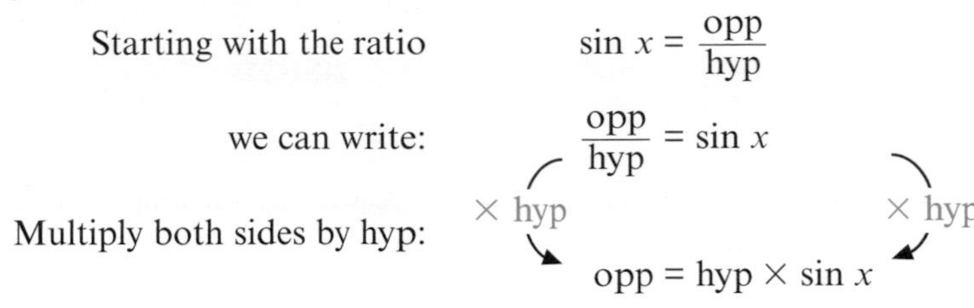

Starting with the ratio $\sin x = \frac{\text{opp}}{\text{hyp}}$

we can write: $\frac{\text{opp}}{\text{hyp}} = \sin x$

Multiply both sides by hyp: $\text{opp} = \text{hyp} \times \sin x$

So $\sin x = \frac{\text{opp}}{\text{hyp}}$ gives: opp = hyp × sin x

In the same way, $\cos x = \frac{\text{adj}}{\text{hyp}}$ gives: adj = hyp × cos x

and $\tan x = \frac{\text{opp}}{\text{adj}}$ gives: opp = adj × tan x

- The length of any side in a right-angled triangle can be calculated using trigonometry.

Example In ΔABC calculate the length of AC.

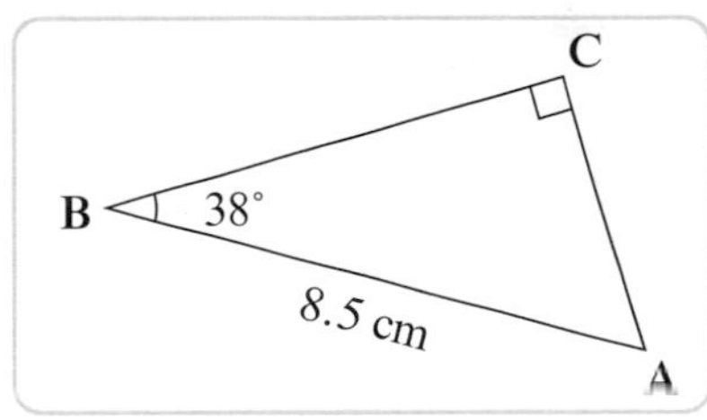

Use the full calculator value for sin 38° and round at the end of the calculation.

On some calculators this set of key presses can be used to calculate 8.5 × sin 38°

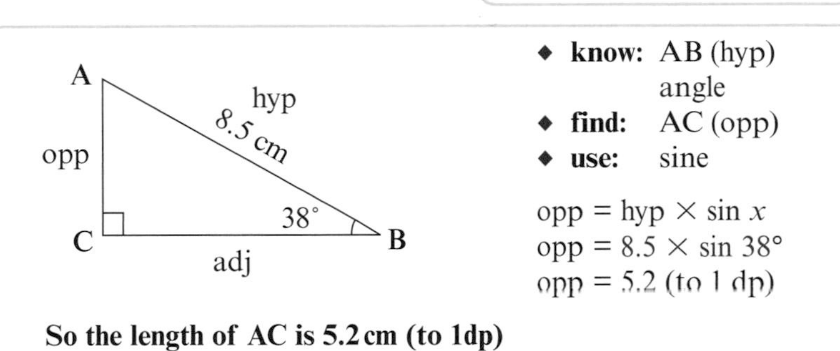

- **know:** AB (hyp) angle
- **find:** AC (opp)
- **use:** sine

opp = hyp × sin x
opp = 8.5 × sin 38°
opp = 5.2 (to 1 dp)

So the length of AC is 5.2 cm (to 1dp)

Exercise 14.5
Finding lengths in right-angled triangles

Accuracy
For this exercise round each answer to 1 dp.

1 List the key presses you use to calculate 8.5 × sin 38° on **your** calculator.

2 Sketch each of these triangles and calculate the length of the blue side.

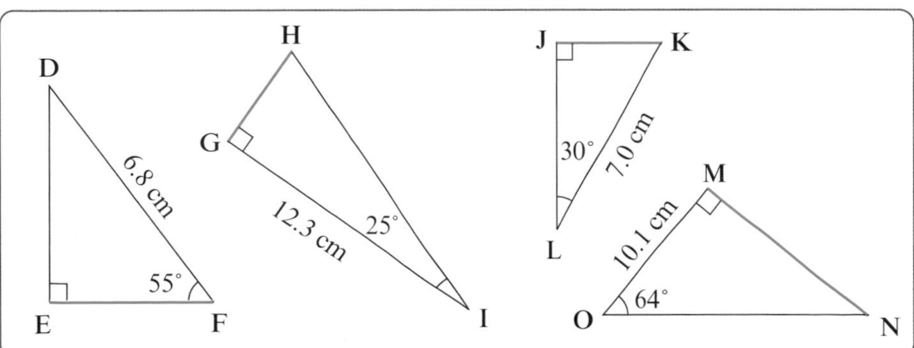

3 This is part of a calculation to find BC in ΔABC.

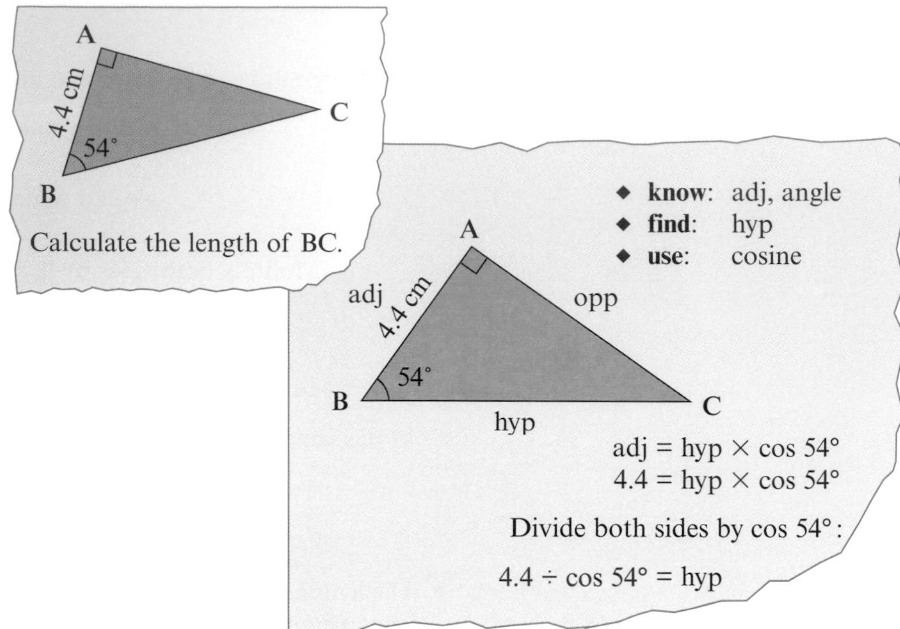

Use the full calculator value for cos 54° and round at the end of the calculation.

On some calculators this set of key presses can be used to calculate 4.4 ÷ cos 54°

a For **your** calculator, list the key presses to calculate 4.4 ÷ cos 54°

b What is the length of BC?

4

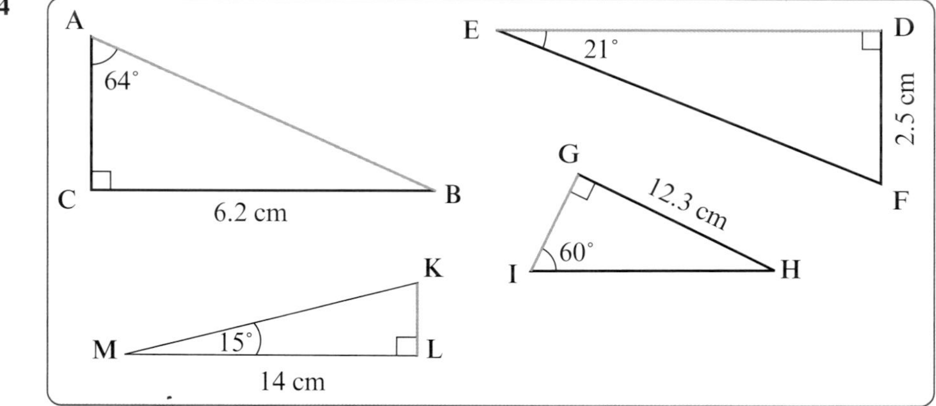

Calculate the length of AB, DE, KL and GI in these triangles.

5

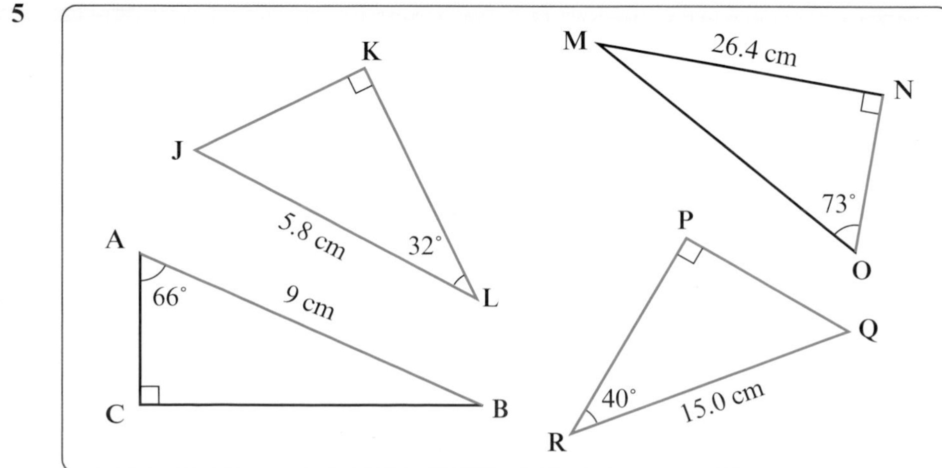

Calculate the length of JK, NO, AC and PR in these triangles.

Finding an angle

These are some pairs of inverse operations:

to add / to subtract
to square / to square-root

On some calculators the inverse button is marked

  or 

- When you know the value of one trigonometric ratio for an angle, you can use the inverse function on a calculator to find the size of the angle.

Example

Find x when $\sin x = 0.375$
On some calculators, these are the key presses.

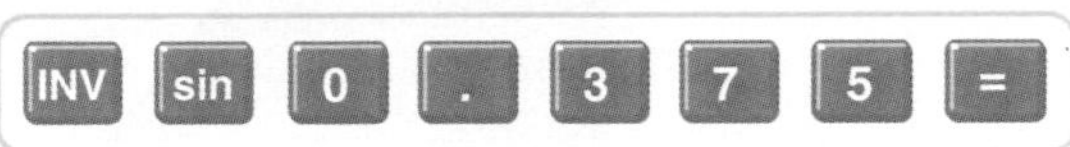

$x = \mathbf{22°}$ (to the nearest degree)

Exercise 14.6
Calculating angles

Accuracy
For this exercise round each angle to the nearest degree.

When you use trigonometric ratios to solve problems it is useful to write down:

- what you know
- what you are trying to find
- what ratio you need to use.

1 For your calculator, list the key presses to find x when $\sin x = 0.375$

2 Find the angle x when:

a $\sin x = 0.3584$ **b** $\tan x = 1.0256$ **c** $\cos x = 0.351$

3 Find the angle x when:

a $\tan x = \frac{5}{13}$ **b** $\cos x = \frac{3.5}{8.16}$ **c** $\tan x = \frac{8.9}{5.4}$

Do not round the value before using the inverse operation.

4 This is part of a calculation to find x in triangle ABC.

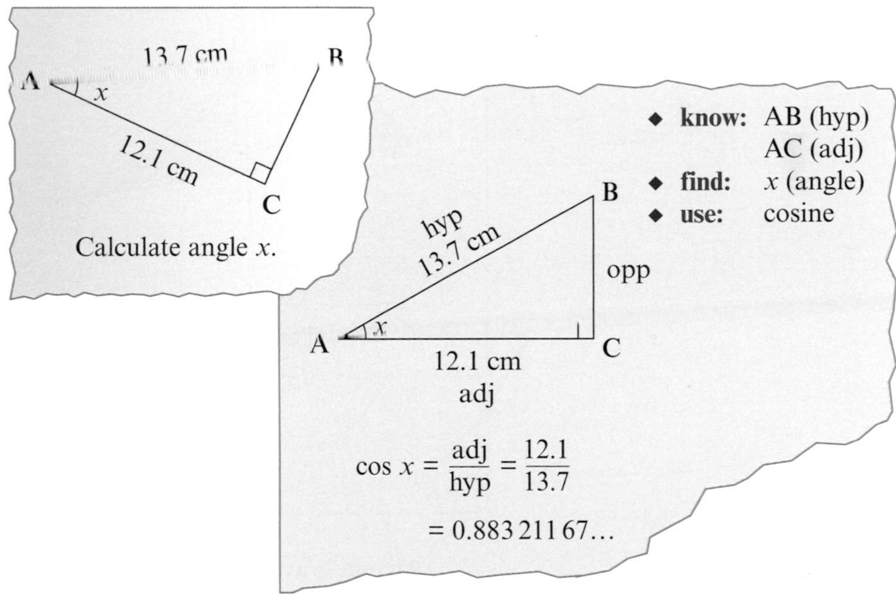

What is the size of angle x?

5

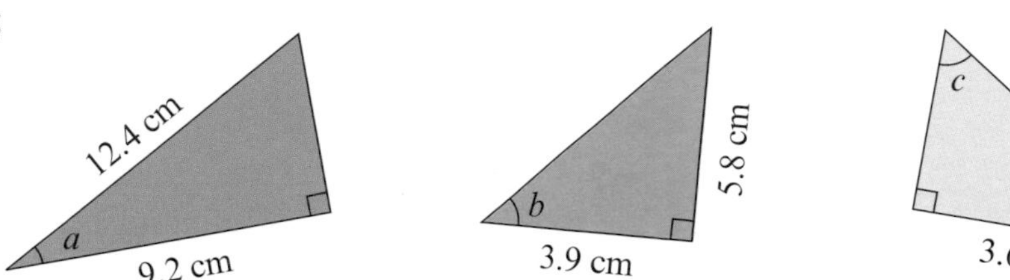

Calculate the size of angles a, b and c.

6 For each triangle, calculate the size of the lettered angle.

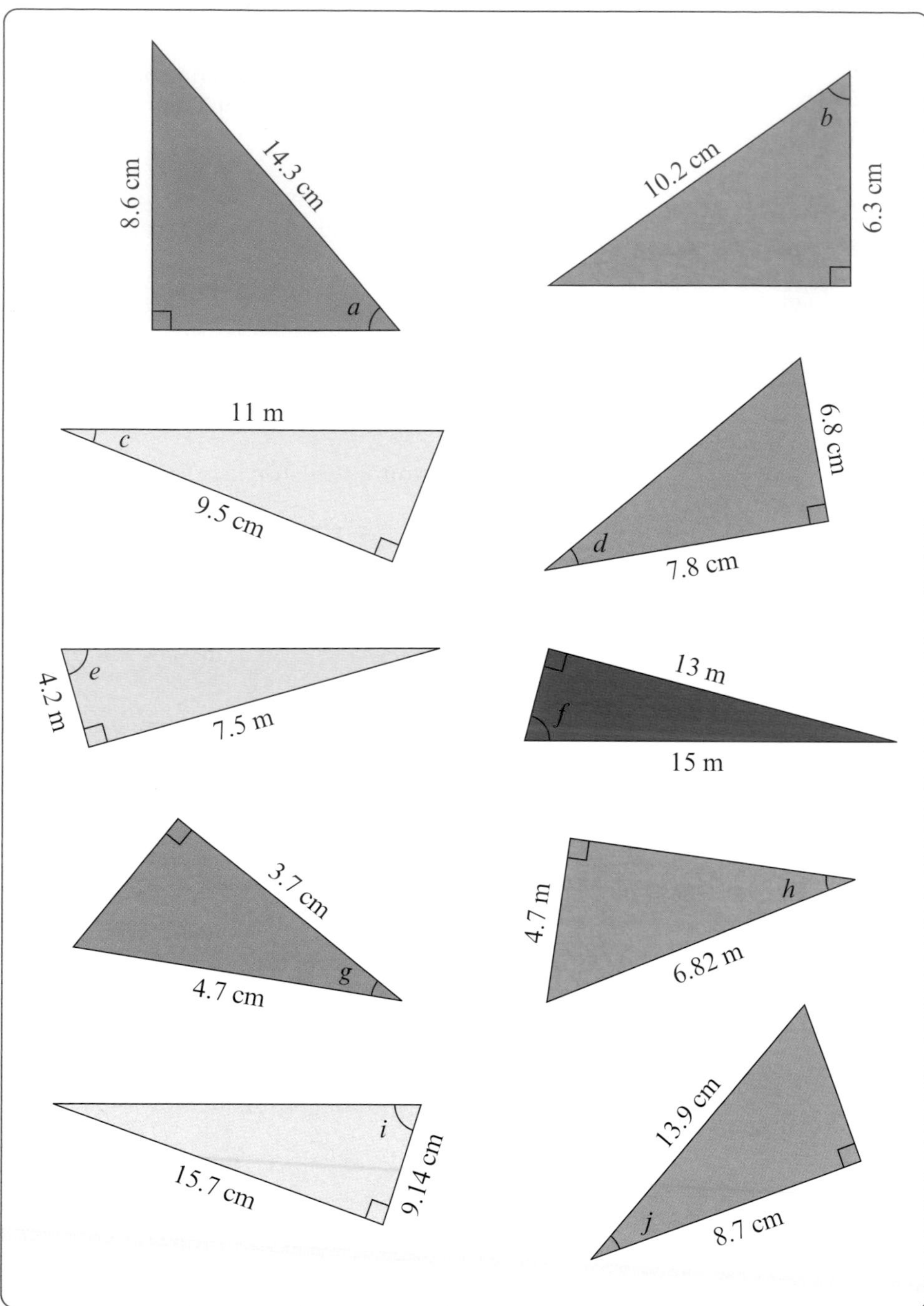

7 Ali and Wayne calculate cos x for a triangle.
These are their answers:

Ali $\cos x = 0.798449612\ldots$

Wayne $\cos x = 1.252427184\ldots$

a Which value for cos x must be incorrect?
b Explain why.

Thinking ahead to ...
finding sides and angles

A In triangle RST:
  a Is RS greater than or less than 5.6 cm?
  b Explain your answer
B Estimate the length of RS.
C Calculate the length of RS to 2 dp.

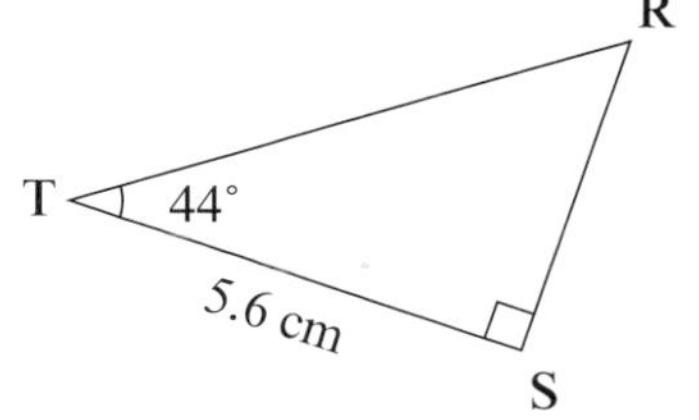

Finding sides and angles

Exercise 14.7
Estimating and calculating sides and angles

> **Accuracy**
> For this exercise round each answer to 3 sf.

1 The lengths and angles marked in blue on these triangles are incorrect.

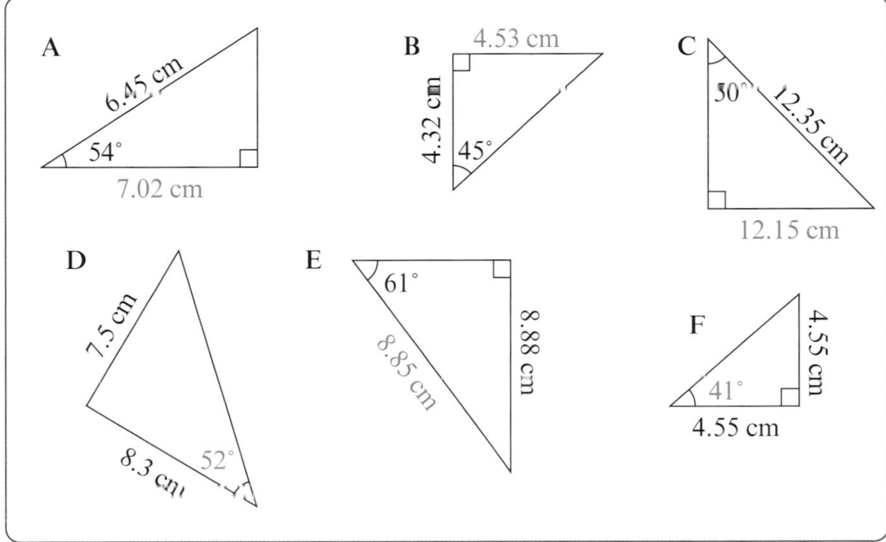

a For each triangle, without calculating, explain why the length or angle marked in blue is wrong.
b Calculate the correct length or angle for each triangle.

2 Sketch each triangle and calculate the length or angle marked with a letter.

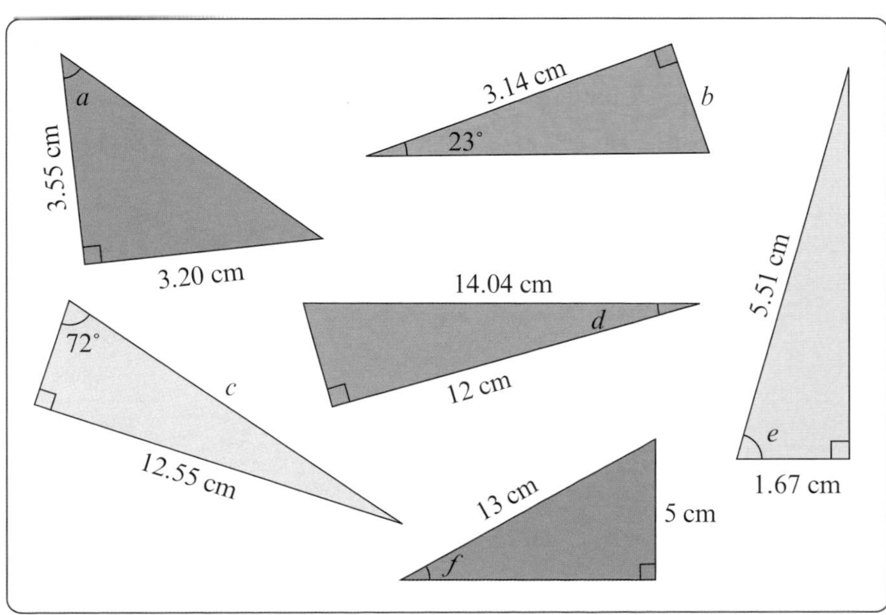

Trigonometry – solving problems

Exercise 14.8
Solving problems

Accuracy
For this exercise:
- round each angle to 2 dp
- round each distance to the nearest cm.

To write a length written in metres to the nearest centimetre you can round it to 2 dp.

Example

5.3167 m
= 5 metres 31.67 cm
= 5 metres 32 cm to the nearest cm
= 5.32 m (to 2 dp)

25.12906 m
= 25.13 m to the nearest cm

1 This ladder is 5 metres long.
The foot of the ladder is 1.25 metres from the wall.

a The ladder is at an angle x to the horizontal. Calculate the angle x.
b The ladder reaches a height of h metres up the wall. Calculate the height h to the nearest centimetre.

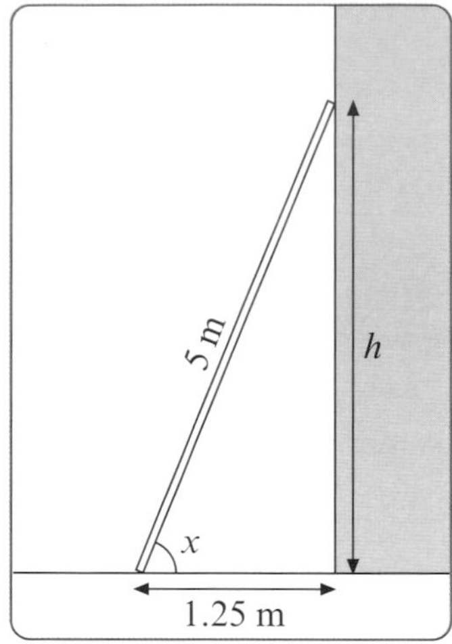

2 The firm Light Ladders recommends that the angle x, between the ladder and the horizontal, should be between 72° and 76°.

The length of this ladder is 4.5 metres.
The distance from the foot of the ladder to the wall is d metres.

a Find the distance d when the angle x is 72°.
b Find d when x is 76°.
c When x is 72° how far up the wall does the ladder reach?

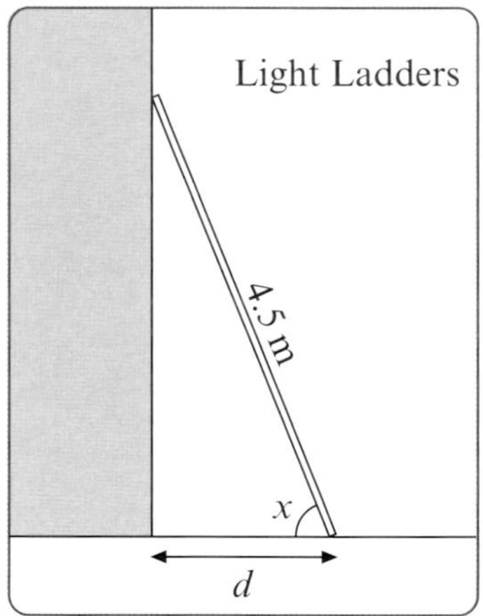

3 This is part of the safety label Light Ladders put on their extending ladder.

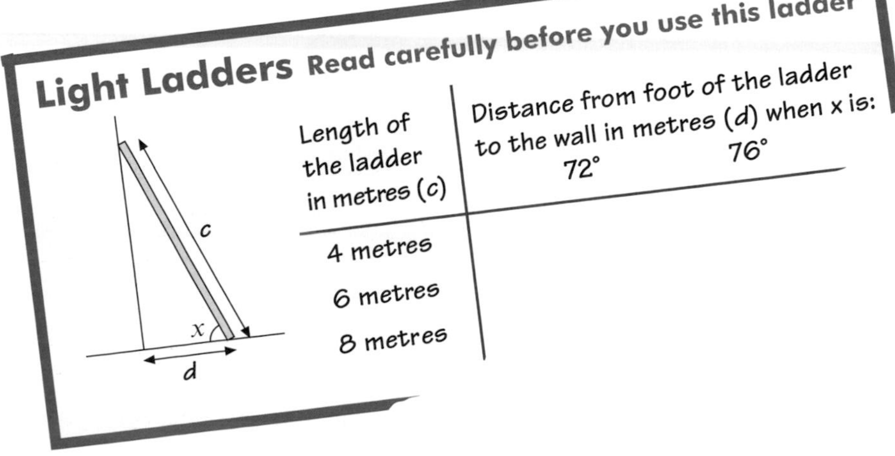

Light Ladders Read carefully before you use this ladder

Length of the ladder in metres (c)	Distance from foot of the ladder to the wall in metres (d) when x is: 72°	76°
4 metres		
6 metres		
8 metres		

Make the safety label for this extending ladder.

4 This diagram shows some steps and a ramp outside a building.

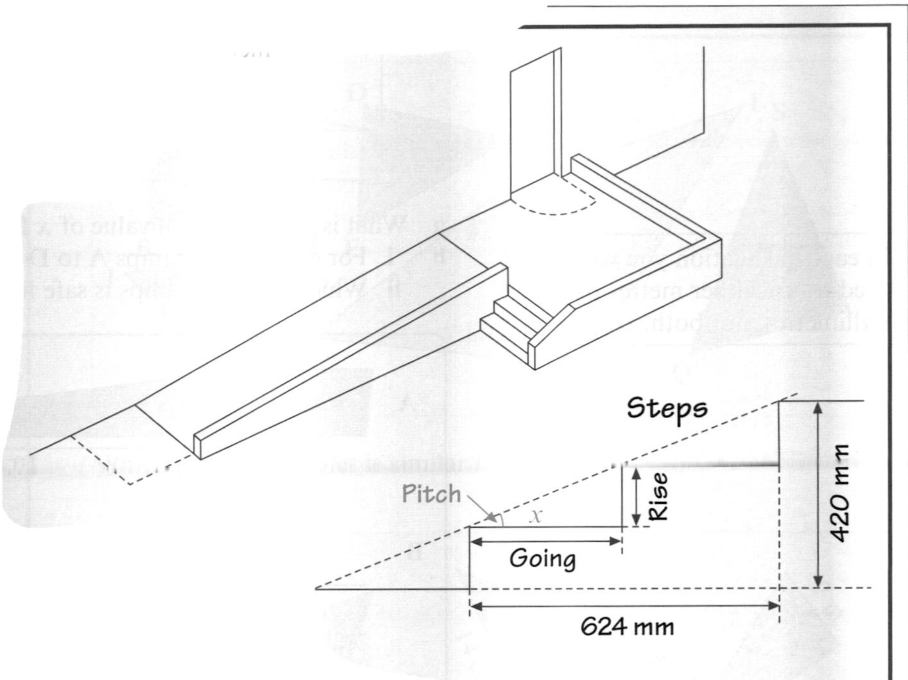

This is a sketch of a cross-section of the steps.
All the steps are the same size.

a For one step what is the length of the going?

b For each step what is the length of the rise?

c Complete this diagram for one step.

d Calculate the angle x for these stairs.

Angle x is known as the pitch of the stairs.

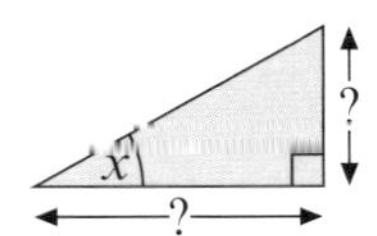

5 This is a cross-section of the ramp.

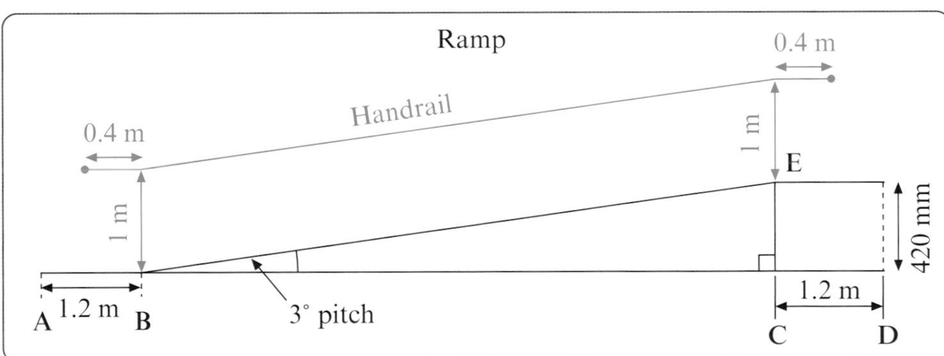

The pitch of the ramp is 3° and the height is 420 millimetres.

a Give the height of the ramp in metres.

b Calculate the length of the ramp, BE, in metres.

c A handrail is fixed to the wall 1 metre above the ramp.
It extends 0.4 metres beyond each end of the ramp.
What is the total length of the handrail?

d Calculate the horizontal distance BC.

e At each end of the ramp there must be a landing at least 1.2 metres long.
Calculate the horizontal distance AD.

In each calculation you will need to use either metres or millimetres, not both.

Starting points

You need to know about ...

A Substituting into simple formulas

Example A formula for the approximate area (A) of a circle with radius r is: $A = 3r^2$. What is A when $r = 10$?

$$A = 3 \times 10^2 = 3 \times 100 = 300$$

So a circle of radius 10 cm has an approximate area of 300 cm^2.

B Flowcharts

Flowcharts can be used to solve problems.

Example I think of a number. I subtract 5 and then multiply by 6. My answer is 42. What number was I thinking of?

- A flowchart for the problem: Number → [– 5] → [× 6] → 42
- A reverse flowchart: 12 ← [+ 5] ← [÷ 6] ← 42

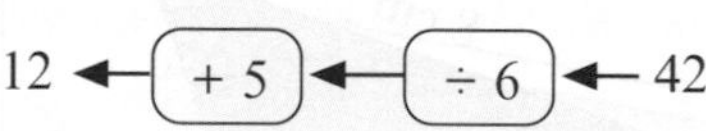

So the number I was thinking of is 12.

C Balancing equations

Balancing can be used to solve problems.

Example I think of a number. I subtract 5 and then multiply by 6. My answer is 42. What number was I thinking of?

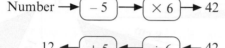

- An equation for the problem: $6(n - 5) = 42$
- Solve the equation by balancing:

$$6(n - 5) = 42$$
$$6n - 30 = 42$$
$+30$ (both sides)
$$6n = 72$$
$\div 6$ (both sides)
$$n = 12$$

So the number I was thinking of is 12.

D Simplifying expressions

- Collect together any like terms.

Example

$$5a^2 + 3a + 10 + 4a - 2$$
$$= 5a^2 + 3a + 4a + 10 - 2$$
$$= 5a^2 + 7a + 8$$

- Complete any multiplications as far as possible.

Examples

- $p \times p \times 5 = 5p^2$
- $3a \times 5b = 15ab$
- $3x \times 5x = 15x^2$

... so try these questions

A1 What is the approximate area of a circle with radius 4 cm?

A2 For each formula, find the value of A when $x = 5$.

a $A = 4x - 1$ **c** $A = 7(x + 2)$

b $A = \frac{x}{2}$ **d** $A = 30 - 2x$

B1 Use flowcharts to solve these number puzzles.

a I think of a number. I multiply by 4, then add 5. My answer is 53. What number was I thinking of?

b I think of a number. I subtract 8, then multiply by 3. My answer is 93. What number was I thinking of?

C1 Solve these equations:

a $4(n - 1) = 76$

b $2n + 5 = 5n - 1$

C2 I think of a number. I multiply by 7, then add 9. My answer is 65. What number was I thinking of?

a Write an equation for this puzzle.

b Solve it to find the number.

D1 Simplify these expressions:

a $4x + 3 + x + 5$

b $2a + 3b + 7a - b$

c $2x^2 + 4x + 3x^2 - 2x$

d $2a + 5a^2 + 4a - 7$

e $t \times t \times 3$

f $5c \times 2d$

g $4k \times 3k$

h $w \times 7w$

Thinking ahead to ... rearranging formulas

A A cook book gives this formula to find the time to cook a piece of lamb.

> Allow 30 minutes per pound and an extra 20 minutes.

a How long would it take to cook 6 pounds of lamb?

b What weight of lamb would be cooked in 1 hour and 50 minutes?

Rearranging formulas

The formula that links cooking time (T) with weight (W) can be written as:

$$T = 30W + 20$$

- The formula shows how to find T when you know W, but it can be rearranged to show how to find W when you know T.

Flowchart method for rearranging formulas

- Draw a flowchart for the formula.

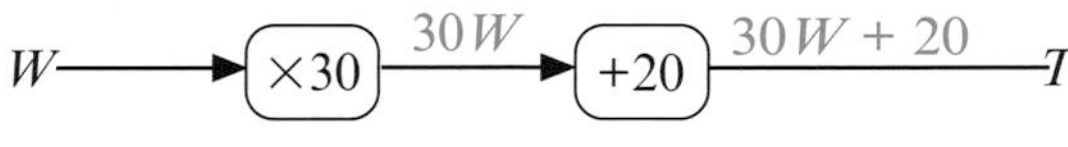

- Reverse the flow chart to rearrange the formula.

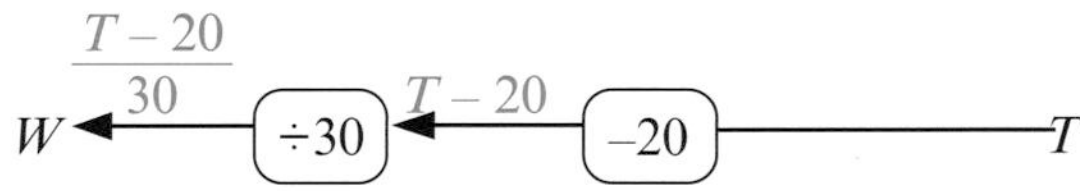

- The rearranged formula is: $W = \frac{T - 20}{30}$

Addition 'undoes' subtraction and vice versa. For example:

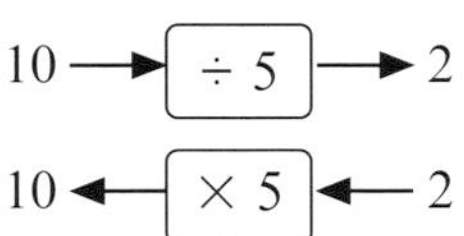

Multiplication 'undoes' division and vice versa. For example:

10 → ÷ 5 → 2

10 ← × 5 ← 2

Balancing method for rearranging formulas

- Add, subtract, multiply or divide **both** sides of the formula by equal amounts.

$$T = 30W + 20$$

-20 on both sides:

$$T - 20 = 30W$$

$\div 30$ on both sides:

$$\frac{T - 20}{30} = W$$

- The rearranged formula is: $W = \frac{T - 20}{30}$

Exercise 15.1 Rearranging linear formulas

1 Make p the subject of each formula:

a $t = p + 1$	b $s = 7p$	c $y = 5p - 2$	d $v = 10 - p$
e $h = 12 - 9p$	f $r = p + q$	g $t = pq$	h $x = 2p + f$
i $k = lp - m$	j $g = \frac{p}{9}$	k $m = \frac{p}{n}$	l $s = \frac{p}{2} + 1$

2 The formula that gives the cost in pence (c) of placing an advertisement in a local paper, where n is the number of words is:

$$c = 15n + 50$$

a Find the cost of a 65-word advert.

b Rearrange the formula so that it begins, $n = \dots$.

c How many words were used in an advert that cost £19.85?

d With £10.00, what is the maximum number of words you could use?

> For the formula you need to write £19.85 in pence.

3 You can estimate the distance between you and a storm using the formula:

$$d = \frac{t}{5}$$

where d is the distance in miles and t is the number of seconds between the lightning and the thunder.

When a formula begins, $t = \ldots$, then t is the subject of the formula.

a Make t the subject of the formula.
b For a storm 1.5 miles away, how many seconds will be between the lightning and the thunder?

4 The formula for the sum (S) of the interior angles of a polygon is:

$$S = 180(n - 2)$$

where n is the number of sides.

The interior angles of this polygon are marked in red.

a What is the sum of the interior angles of a hexagon?
b Make n the subject of the formula.
c The sum of the interior angles of a polygon is 3240°.
How many sides has the polygon?
d Explain why it is not possible to draw a polygon where the sum of the interior angles is 600°.

5 The formula for the number of vitamin pills (n) that Jim has left after d days is:

$$n = 100 - 3d$$

a How many vitamin pills would Jim have left after 14 days?
b How many vitamin pills does he take each day?
c Copy and complete this diagram to make d the subject of the formula.

$n = 100 - 3d$
$+3d$ $n + 3d =$? $+3d$
$-n$ $3d =$? $-n$
? $d =$? ?

d After how many days does Jim have 10 pills left?

6 In the summer, an ice-cream seller uses this formula to estimate the number of ice-creams (n) she will sell in a day at x pence each.

$$n = 800 - 5x$$

a About how many ice-creams will she sell in a day at 50p each?
b Make x the subject of the formula.
c What price will she need to charge to sell 500 ice-creams?
d Explain why she is unlikely to charge £1.70 for an ice-cream.

7 The formula that links distance (d), speed (s) and time (t) is:

$$d = st$$

To make s the subject of the formula, $d = st$:

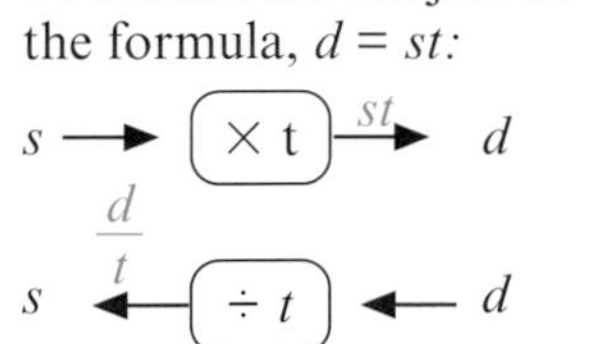

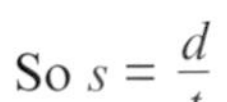

So $s = \frac{d}{t}$

a How far will a car travel in 2 hours at a speed of 47 mph?

With s as the subject, the formula $d = st$ can be written:

$$s = \frac{d}{t}$$

b Calculate the speed of a plane that travels 1800 miles in 3 hours.
c Copy and complete this flow chart for $d = st$:

t → [ ] → d

d Reverse the flow chart and make t the subject of the formula.
e Find the time it takes to travel 125 km at a speed of 50 km/h.

A formula for converting kilometres (k) to miles (m) is: $m = \frac{5}{8}k$

- Making k the subject gives a formula for converting miles to kilometres.

$$m = \frac{5}{8}k$$
$$\times 8 \quad 8m = 5k \quad \times 8$$
$$\div 5 \quad \frac{8m}{5} = k \quad \div 5$$

- So $k = \frac{8m}{5}$

Exercise 15.2
Rearranging formulas with fractions

1 Make w the subject of each formula.

a $k = \frac{3}{5}w$ **b** $\frac{3}{4}w = x$ **c** $\frac{3}{5}w + 1 = k$ **d** $3x = 4 + \frac{1}{2}w$

e $3y - 2 = \frac{1}{3}w + 1$ **f** $\frac{1}{3}w = 2k - 4$ **g** $\frac{3}{5}w + x = 4$ **h** $\frac{w+1}{3} = k$

i $\frac{1}{2}(w - 2) = 3k$ **j** $\frac{2}{3}w = \frac{1}{2}k + 1$ **k** $\frac{3}{4}w + 1 = 1 - k$ **l** $\frac{1}{2}w - 2 = \frac{1}{2}k$

2 A formula for converting kilograms (k) to pounds (p) is:

$$p = \frac{11}{5}k$$

a How many pounds are in 8 kilograms?
b Make k the subject of the formula.
c Convert 3 pounds to kilograms.

3 The formula for the area, (A), of a triangle with base length b and height h is:

$$A = \frac{1}{2}bh$$

a Make h the subject of the formula.
b Find the height of a triangle with area $100\,cm^2$ and base length 2.5 cm.

4 Temperature in °F can be converted to °C using this formula:

$$C = \frac{5(F - 32)}{9}$$

a Change 68 °F to °C.
b Copy and complete this flowchart:

$F \rightarrow [-32] \xrightarrow{F-32} [\;\;] \rightarrow [\;\;] \rightarrow C$

c Reverse the flowchart to give a formula that converts °C into °F.
d Change 24 °C to °F.

5 Make y the subject of each formula:

a $z = \frac{3}{4}y$ **b** $m = \frac{1}{3}xy$ **c** $d = \frac{1}{2}y + 4$ **d** $k = \frac{4}{7}y - h$

$\frac{18d}{5t} = 18d \div 5t$

6 When d metres are travelled in t seconds, a formula for average speed (s) in km/h is:

$$s = \frac{18d}{5t}$$

a In 1992, Linford Christie ran 100 metres in 9.96 seconds. What was his average speed in km/h, correct to 2 dp?
b Make d the subject of the formula.
c Make t the subject of the formula.

End points

You should be able to ... ... so try these questions

A Substitute in formulas

A1 When a stone is thrown straight up with a speed of u m/s, the formula:

$$h = ut - 5t^2$$

gives its approximate height, h metres, after t seconds.

A stone is thrown upwards with a speed of 60 m/s.
What is its height after:

a 2 seconds **b** 5 seconds **c** 10 seconds?

A2 The formula for the volume (V) of a cone is:

$$V = \frac{1}{3}\pi r^2 h,$$

where r is the radius of the base and h is the height.

A cone has a base of radius 2.5 cm and a height of 8 cm.
Calculate its volume in cm^3.

B Rearrange formulas

B1 A car hire company uses this formula to calculate the cost in pounds, C, of hiring a car for n days:

$$C = 30n + 20$$

a Make n the subject of the formula.
b With £470, for how many days could you hire a car?

B2 Asif uses this formula to calculate the cost of running his car for a week:

$$C = 10 + \frac{1}{5}m,$$

where C is the cost in pounds and m is the number of miles he drives.

a Make m the subject of the formula.
b Asif wants the cost of running his car to be £30 or less per week. What is the maximum number of miles he can drive in a week?

B3 Make k the subject of each formula:

a $h = 5k^2$ **b** $m = \dfrac{n + k}{5}$ **c** $A = k^2 + 9$

C Multiply out brackets

C1 Which of these expressions is equivalent to $(t + 5)(t - 3)$?

A $t^2 + 8t + 15$ **B** $t^2 + 8t - 15$ **C** $t^2 + 2t - 15$

C2 For each of these, multiply out the brackets and simplify.

a $(a + 1)(a + 8)$ **b** $(b + 4)^2$ **c** $(5c + 2)(c + 3)$
d $(d + 2)(d - 1)$ **e** $(2e + 5)(3e - 10)$ **f** $(f - 7)(3f - 2)$

D Factorise expressions

D1 Which pair of expressions multiply to give $k^2 + 5k - 14$?

$k + 2$ $k - 7$ $k - 2$ $k + 14$ $k + 7$ $k - 1$

D2 Factorise these expressions.

a $x^2 + 12x + 11$ **b** $x^2 + 9x + 14$ **c** $x^2 + 4x - 5$
d $x^2 + x - 6$ **e** $x^2 - x - 20$ **f** $x^2 - 5x + 6$

Starting points

You need to know about ...

A Finding the median of a frequency distribution

- You can always find the median of ungrouped data by:
  - listing the data in order
  - finding the middle value (or the middle pair of values).

1996 Olympic Games – Pole Vault Final
Best height cleared by each finalist

Height (metres)	5.60	5.70	5.80	5.86	5.92	Total
Frequency	4	3	1	3	3	14

5.60 5.60 5.60 5.60 5.70 5.70 5.70 5.80 5.86 5.86 5.86 5.92 5.92 5.92

The median is halfway between the middle pair, so
median height = **5.75 m**

- For a distribution with a large total frequency, using cumulative frequencies is quicker than listing the data.

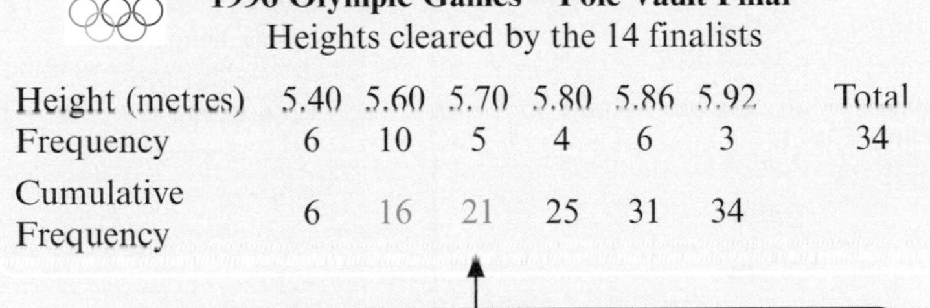

1996 Olympic Games – Pole Vault Final
Heights cleared by the 14 finalists

Height (metres)	5.40	5.60	5.70	5.80	5.86	5.92	Total
Frequency	6	10	5	4	6	3	34
Cumulative Frequency	6	16	21	25	31	34	

The middle pair (the 17th and 18th) are both 5.70, so median height = **5.70 m**

B Calculating the mean of a frequency distribution

- To calculate the mean of ungrouped data:
  - multiply each value by its frequency
  - calculate the total of all the values

1996 Olympic Games
Pole Vault Final
Heights cleared by finalists

Height (metres)	Frequency		Total at each height
5.40	6	5.40×6	32.4
5.60	10	5.60×10	56
5.70	5	5.70×5	28.5
5.80	4	5.80×4	23.2
5.86	6	5.86×6	35.16
5.92	3	5.92×3	17.76
Totals	34		193.02

  - divide the total of all the values by the total frequency.

Mean height $= \frac{193.02}{34} =$ **5.68 m** (to 3 sf).

... so try these questions.

1996 Olympic Games
Men's High Jump Final

Height (metres)	Frequency: Best height cleared	Frequency: All heights cleared
2.15	–	3
2.20	–	9
2.25	4	12
2.29	3	8
2.32	4	6
2.35	1	3
2.37	1	1
2.39	1	1
Totals	14	43

A1 For the best-height distribution, find the median height cleared by listing the data.

A2 For the all-heights distribution, use cumulative frequencies to find the median height cleared.

1996 Olympic Games
Women's High Jump Final

Height (metres)	Frequency: Best height cleared	Frequency: All heights cleared
1.80	–	8
1.85	–	14
1.90	–	13
1.93	5	14
1.96	4	9
1.99	2	5
2.01	1	3
2.03	1	2
2.05	1	1
Totals	14	69

B1 For the best-height distribution, calculate the mean height cleared.

B2 For the all-heights distribution, calculate the mean height cleared.

C Measures of spread

- The range is the simplest measure of how spread out a set of data is.

1996 Olympic Games – Pole Vault Final
Best height cleared by each finalist

Height (metres)	5.60	5.70	5.80	5.86	5.92	Total
Frequency	4	3	1	3	3	14

Range = Highest value – Lowest value
= 5.92 – 5.60
= **0.32 m**

- The interquartile range measures the spread of the middle 50% of the distribution by ignoring the lowest 25% and the highest 25%.

5.60 5.60 5.60 5¦60 5.70 5.70 5.70¦5.80 5.86 5.86 5¦86 5.92 5.92 5.92

Interquartile range = Upper quartile – Lower quartile
= 5.86 – 5.60
= **0.26 m**

D Comparing sets of data

- You can compare sets of data using two types of value: an average, and a measure of spread.

1996 Olympic Games – Pole Vault Finals
Best heights cleared by finalists

	1992	1996
Median	5.625 m	5.75 m
Interquartile range	0.35 m	0.26 m

On average, finalists jumped 12.5 cm higher in 1996 than in 1992. The finalists were more closely matched in 1996 because the best heights cleared were less spread out than in 1992.

E Using grouped data

- Data which is collected in groups, or grouped to make it easier to present, is called a **grouped frequency distribution**.

1996 Olympic Games – Great Britain's Athletics Team

Age	18–22	23–27	28–32	33–37	38–42	Total
Frequency	9	37	27	5	4	82

- Each group of data is a **class**: the size of a class is the **class interval**.

The 18–22 class has a class interval of 5 years.

- You can show grouped data on a **grouped frequency diagram**.

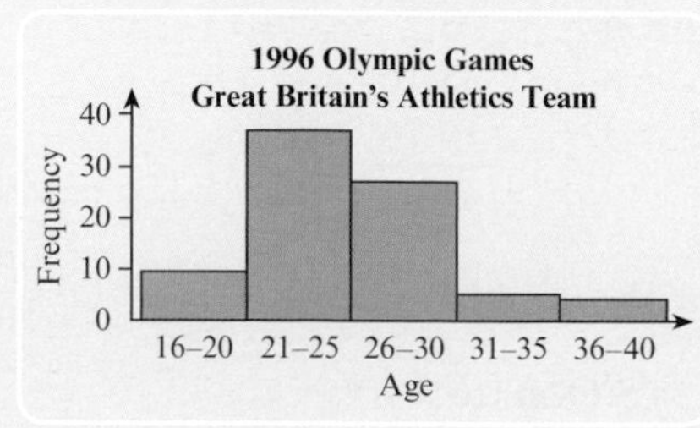

C1 Use the data for the 1996 high jump finals on page 185 to calculate the range of the best heights cleared for:
a men **b** women.

C2 Calculate the interquartile range of the best heights cleared for:
a men **b** women.

1992 Olympic Games
High Jump Finals
Best heights cleared by finalists

Median – men	2.295 m
Mean – women	1.91 m
Interquartile range	
– men	0.06 m
– women	0.085 m

D1 Use your answers to Questions **A1**, **B1**, and **C2** to compare the best heights cleared in 1992 and 1996 by:
a men **b** women.

1996 Olympic Games
Great Britain's Athletics Team

Women Age	Women Frequency	Men Age	Men Frequency
18–23	8	18–21	3
24–29	13	22–25	17
30–35	10	26–29	14
36–41	3	30–33	8
Total	34	34–37	5
		38–41	1
		Total	48

E1 Give the class interval for:
a women **b** men.

E2 Draw a grouped frequency diagram to show the ages of:
a women **b** men.

Histograms and frequency polygons

- A grouped frequency diagram is also called a **histogram**.
- The class with the highest frequency is called the **modal class**.
- A **frequency polygon** is a set of straight lines joining the middle points on the top of each bar of a histogram.

The frequency is the number of golfers.

153 golfers played in the 1st and 2nd rounds. These scores are for the top 77 golfers after the 2nd round. Only these 77 golfers played in the 3rd and 4th rounds.

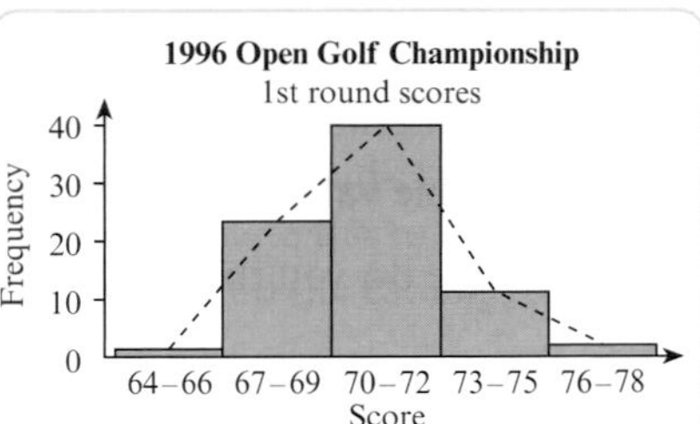

- You can compare frequency distributions on the same diagram by plotting the frequency polygon for each distribution.

The middle scores of each of the classes are 65, 68, 71, 74, 77, 80.

These can be called the **mid-class values**.

The frequencies are plotted against the middle scores.

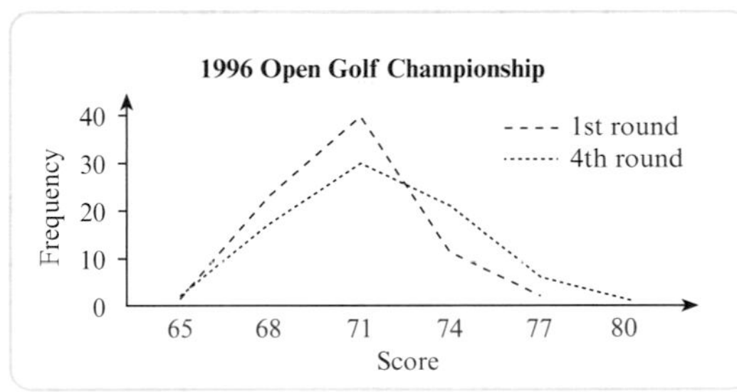

Exercise 16.1
Histograms and frequency polygons

1 Use the histograms to give the modal class for:

a the 1st round scores **b** the 4th round scores

2 Golfers aim for as low a score as possible in a round.

a Do you think the golfers did better in the 1st round or the 4th round?
b Use the frequency polygons to explain your answer.

3 These are the ages of the 153 golfers who played in the 1996 Championship.

Qualifiers are those golfers who played in all 4 rounds. Non-qualifiers only played in the first 2 rounds.

Age	17–27	28–38	39–49	50–60	
Frequency					Total
Qualifiers	18	34	22	3	77
Non-qualifiers	14	52	9	1	76

a List the eleven different ages in the 17–27 class.
b Which age is the mid-class value?
c List the mid-class values for each class in the age data.

4 **a** Copy these axes on to squared paper.
b Label your horizontal axis.
c Draw frequency polygons to compare the ages of the qualifiers and non-qualifiers.

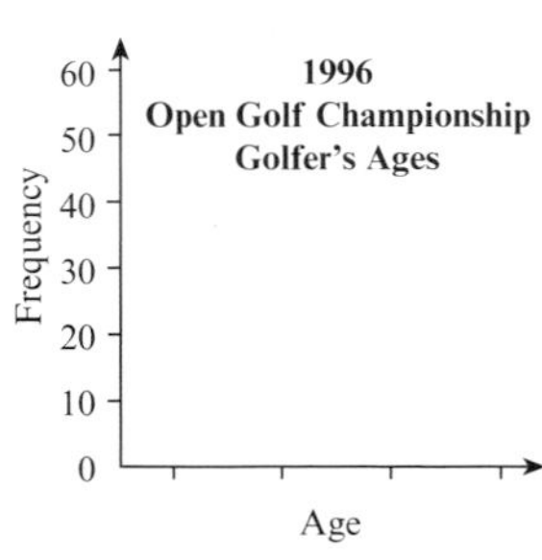

5 Do you think frequency polygons would give a good comparison if there were 97 qualifiers and 56 non-qualifiers? Give reasons for your answer.

Thinking ahead to ... measuring and accuracy

A Which of these numbers could be rounded to 12.7?

12.73 | 12.76 | 12.68 | 12.749 | 12.75

B A number has been rounded to 1 dp to give 9.3.
Give three possible values for the number:

a to 2 dp b to 3 dp.

Measuring and accuracy

- Data which results from measuring, such as heights, distances, weights and times, can be given to different degrees of accuracy.
  For instance, times are often given to:
  the nearest one-tenth (0.1) of a second, or
  the nearest one-hundredth (0.01) of a second, or
  the nearest one-thousandth (0.001) of a second.
- The **limits of accuracy** of a measurement are the values between which the exact measurement must lie.
  For instance, the limits of accuracy of a time measured as 9.83 seconds are:
  9.825 seconds and **9.835** seconds.

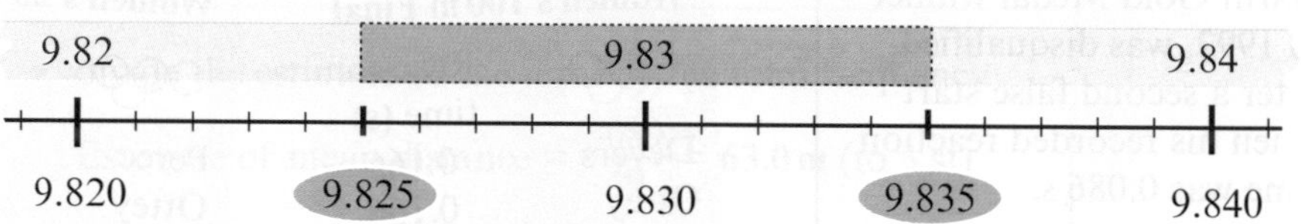

Exercise 16.6
Measuring and accuracy

1 Draw a diagram to show the limits of accuracy of:

a 9.84 seconds b 7.12 metres c 0.82 metres d 0.174 seconds

2 Give the limits of accuracy of these measurements.

a 12.38 seconds b 9.2 metres c 7.3 seconds d 7.30 seconds

- Times and distances in athletics events are not measured to the **nearest** unit.
  All times are rounded up to the **next** one-hundredth of a second.
  For instance, the limits of accuracy of a time given as 9.83 seconds are:
  9.820 seconds and **9.830** seconds.

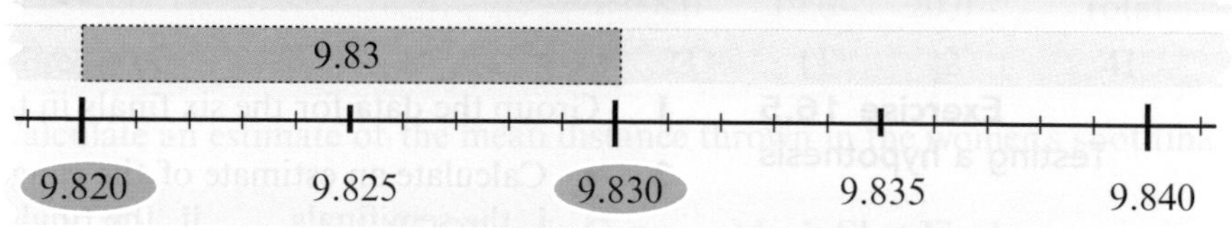

Exercise 16.7
Measuring in athletics

All distances are rounded down to the next centimetre.

1 Gail Devers and Merlene Ottey were given the same time in the final of the women's 100 m.
Give the limits of accuracy of their time.

Women's 100 m Final

	Time (s)
Devers	10.94
Ottey	10.94
Torrence	10.96
Sturrup	11.00

2 Jackie Joyner-Kersee jumped 7.00 metres in the women's long jump final.
Give the limits of accuracy of her distance.

3 Explain why it is appropriate in athletics events to round times up to the next unit, and distances down to the next unit.

Drawing a cumulative frequency curve

1996 Olympic Games – Decathlon (Day 2)
Points scored in each of the 5 events by the top 10 decathletes

Points	400–499	500–599	600–699	700–799	800–899	900–999	1000–1099	Total
Frequency	1	0	8	10	17	10	4	50

- To draw a cumulative frequency curve:
  - construct a cumulative frequency table

Points	<400	<500	<600	<700	<800	<900	<1000	<1100
Cumulative frequency	0	1	1	9	19	36	46	50

If you include a cumulative frequency of 0 in your table then you have a point to start the curve from: (400, 0)

  - plot the cumulative frequencies on a graph
  - join the points with a smooth curve.

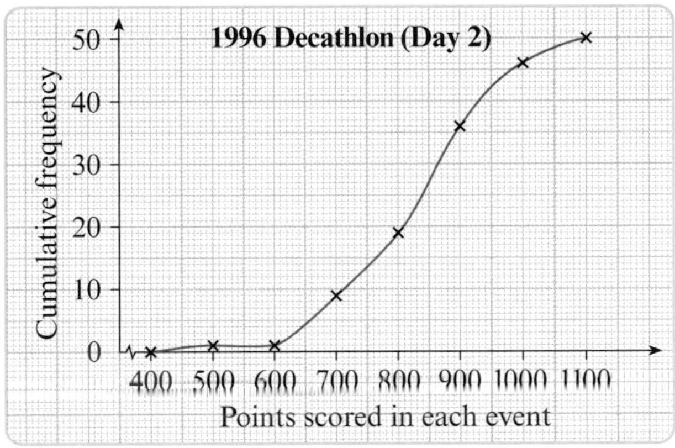

Exercise 16.8
Drawing cumulative frequency curves

1

1996 Olympic Games – Decathlon (Day 1)
Points scored in each of the 5 events by the top 10 decathletes

Points	700–799	800–899	900–999	1000–1099	Total
Frequency	8	23	16	3	50

a Copy and complete the cumulative frequency table below.

Points	<700	<800	<900	<1000	<1100
Cumulative frequency	0				

b Use your table to draw a cumulative frequency curve.

2

Steve Backley (GB) won the javelin silver medal with the very first throw of the final.

1996 Olympic Games – Men's Javelin Final

Distance (metres)	76–	78–	80–	82–	84–	86–	88–	Total
Frequency	2	4	14	13	8	6	1	48

a Copy and complete the cumulative frequency table below.

Distance (metres)	<76	<78	<80	<82	<84	<86	<88	<90
Cumulative frequency	0	2	6					

b Draw a cumulative frequency curve for the men's javelin final.

3 Using your answers to Question **2**, can you think of a way to find the median distance thrown in the final? Explain your method.

Estimating the median and the interquartile range

There were 48 throws in the final, so the **exact** median distance is halfway between the 24th and 25th longest.

It is impossible to find these distances from the table, so you can only **estimate** the median.

- A cumulative frequency table shows which class the median is in.

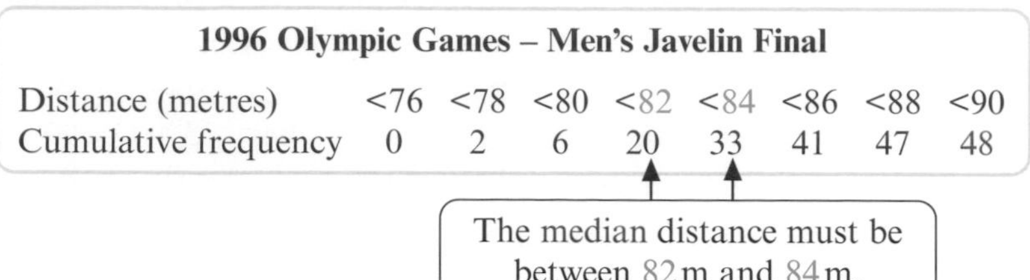

1996 Olympic Games – Men's Javelin Final

Distance (metres)	<76	<78	<80	<82	<84	<86	<88	<90
Cumulative frequency	0	2	6	20	33	41	47	48

The median distance must be between 82 m and 84 m.

As you are only estimating distances within the classes, you can divide the 48 throws into four quarters, using the **12th**, **24th**, and **36th** longest (cumulative frequencies 12, 24, and 36).

- To estimate median and quartiles from a cumulative frequency curve:
  - divide the cumulative frequency into four quarters
  - go across to the curve, then go down and read off each value.

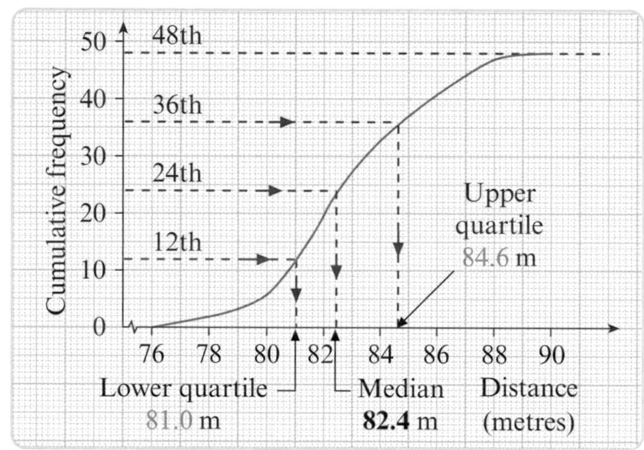

Estimate of median distance = **82.4 m**

Estimate of interquartile range = 84.6 – 81.0
= **3.6 m**

Exercise 16.9
Estimating the median and interquartile range

1

1996 Olympic Games – Men's Hammer Final

Distance (metres)	72–	74–	76–	78–	80–	Total
Frequency	1	11	23	13	4	52

a Copy and complete the cumulative frequency table below.

Distance (metres)	<72	<74	<76	<78	<80	<82
Cumulative frequency	0	1	12			

b Draw a cumulative frequency curve for the men's hammer final.

2 Use your cumulative frequency curve to estimate:
a the median distance thrown **b** the interquartile range.

3 Compare the distances thrown in the men's hammer final and the men's javelin final.

4 **a** From page 193, copy the cumulative frequency table and the cumulative frequency curve for the decathlon (day 2).
b Estimate the median points score.
c Estimate the interquartile range of the scores.

Use cumulative frequencies 12.5, 25, and 37.5 to divide the data into four quarters.

Use cumulative frequencies 12.5, 25, and 37.5 to divide the data into four quarters.

5 Use your answers to Exercise 16.8, Question **1** to estimate:

a the median points score on day 1 of the decathlon
b the interquartile range of the scores.

6 Compare the points scored on days 1 and 2 of the decathlon.

7

1996 Olympic Games – Women's Javelin Final

Distance (metres)	56–	58–	60–	62–	64–	66–	Total
Frequency	7	11	10	10	7	1	46

a Make a cumulative frequency table for the women's javelin final.
b Draw a cumulative frequency curve.

8 **a** Decide how to divide the data into four quarters.
b Estimate the median distance thrown and the interquartile range.

Estimating cumulative frequencies

◆ You can estimate cumulative frequencies from a cumulative frequency curve.

Example Estimate how many throws were:
a less than 81.5 m **b** greater than 85 m.

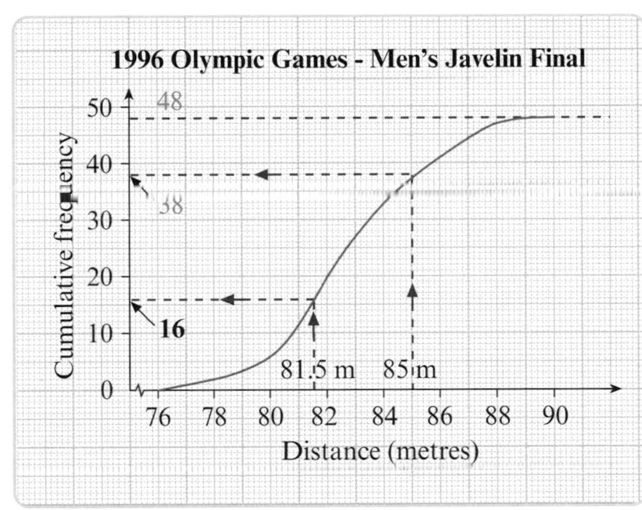

a Estimated number of throws less than 81.5 m = **16**

b Estimated number of throws greater than 85 m = 48 – 38
= **10**

Exercise 16.10
Estimating cumulative frequencies

1 Use your cumulative frequency curve from Exercise 16.9, Question **1** to estimate how many throws were less than 75 m.

2 The winning throw in the 1976 men's hammer final was 77.52 m. Estimate how many throws were greater than this in 1996.

Jonathan Edwards (GB) won the silver medal in the triple jump.

3

1996 Olympic Games – Men's Triple Jump Final

Distance (metres)	15.5–	16.0–	16.5–	17.0–	17.5–	18.0–	Total
Frequency	1	7	14	6	2	1	31

a Draw a cumulative frequency curve for the men's triple jump final.
b Estimate how many jumps were greater than 16.8 m.

Moving averages

A **moving average** is used to smooth out the changes in a set of data that varies over a period of time.

A moving average can often give a better idea of any trend shown in a set of data.

Example

This data gives the umbrella sales per week for a department store.

This set is the raw data gained from sales.

Week	1	2	3	4	5	6	7	8	9	10
Umbrellas sold	4	8	3	7	5	0	16	2	6	10

What can you say about the trend for umbrella sales?

Taking 3-week moving averages:

These are called

3-point moving averages

because they each use these three data values.

Average for weeks 1 to 3 : $(4 + 8 + 3) \div 3 = 5$
2 to 4 : $(8 + 3 + 7) \div 3 = 6$
3 to 5 : $(3 + 7 + 5) \div 3 = 5$
4 to 6 : $(7 + 5 + 0) \div 3 = 4$
5 to 7 : $(5 + 0 + 16) \div 3 = 7$
6 to 8 : $(0 + 16 + 2) \div 3 = 6$
7 to 9 : $(16 + 2 + 6) \div 3 = 8$
8 to 10 : $(2 + 6 + 10) \div 3 = 6$

This set of data is smoother than the raw data set.

So the 3-weekly moving averages of sales are:

5, 6, 5, 4, 7, 6, 8, 6

You can see the effect of moving averages on a graph.

Each moving average is plotted at the end of the group of 3 pieces of data.

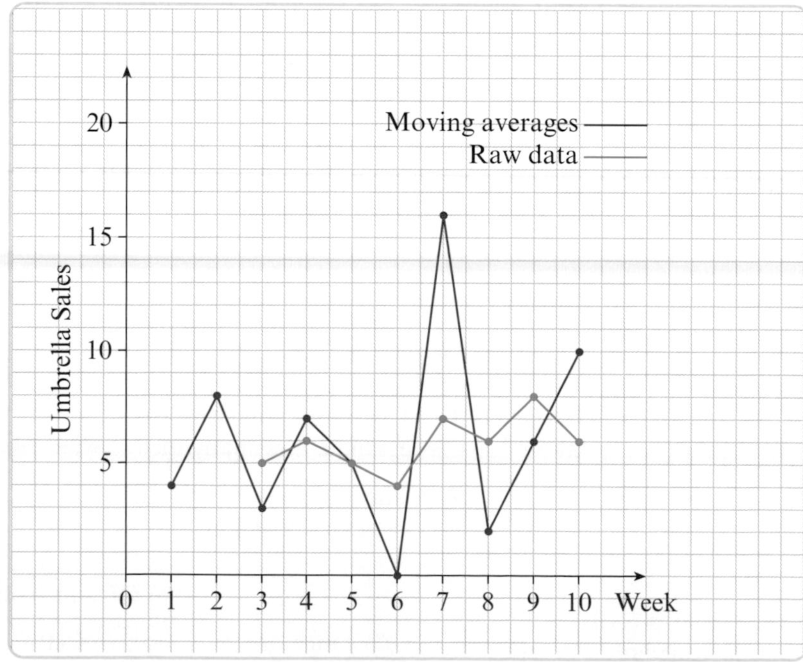

As for the trend, if anything sales of umbrellas seem to be rising.

Exercise 16.11
Moving averages

A moving average can become a decimal value.

1 Calculate the three-point moving averages for each data set.

a 24, 7, 5, 20, 2, 14, 8, 11, 8, 11, 23
b 8, 12, 16, 11, 3, 13, 2, 12, 1, 23, 3, 19
c 11, 14, 2, 20, 8, 5, 5, 35, 2, 15, 4, 11

2 Calculate the five-point moving averages for each data set.

a 18, 20, 27, 34, 16, 7, 52, 30, 24, 12, 8, 6
b 12, 9, 23, 14, 6, 8, 12, 15, 32, 40, 7
c 16, 22, 31, 18, 12, 4, 6, 19, 22, 25
d 14, 24, 0, 7, 17, 6, 23, 37, 2, 14, 18

3 Calculate the four-point moving averages for each data set.

a 12, 23, 14, 16, 7, 4, 18, 24, 35, 16, 7, 14
b 7, 24, 16, 35, 4, 18, 19, 18, 23, 6, 52
c 19, 6, 8, 7, 62, 34, 25, 32, 28, 46, 70
d 56, 14, 22, 34, 6, 7, 19, 0, 42, 57, 62, 31

4 This data gives walking boot sales per week.

Week	1	2	3	4	5	6	7	8	9	10	11	12
Sales (pairs)	5	12	14	1	3	6	10	5	6	15	6	5

a Calculate the four-point moving average for the data.
b On one pair of axes draw graphs to show:
  i the raw data
  ii the four-point moving averages.
c Comment on any trend you can identify in boot sales.

5 This data gives watch sales per week from a catalogue shop.

Week	1	2	3	4	5	6	7	8	9	10	11	12	13	14	15	16	17
Watches sold	23	28	14	16	9	23	8	14	21	19	3	18	19	1	9	13	8

a Calculate the four-point moving average for the data.
b On one pair of axes draw graphs to show:
  i the raw data
  ii the four-point moving averages
c Comment on any trend you can identify in watch sales.

End points

You should be able to ... / ... so try these questions

A Use histograms and frequency polygons

1996 Olympic Games – Reaction Times
Semifinals & Finals – 100 m, 200 m, 100 m/110 m Hurdles

Reaction time (s)		0.12–	0.14–	0.16–	0.18–	0.20–	0.22–	0.24 –	0.26–	Total
Frequency	Men	7	22	23	10	3	2	2	1	70
	Women	5	10	30	17	5	3	1	0	71

A1 Draw a histogram to show the reaction times for:
a men **b** women.

A2 Give the modal class for:
a men **b** women.

B Estimate the median and the interquartile range

B1 **a** Make a cumulative frequency table for the men's reaction times.
b Draw a cumulative frequency curve.
c Use your cumulative frequency curve to estimate:
i the median reaction time **ii** the interquartile range.

C Calculate estimates of the mean and the range

C1 Estimate the range of the reaction times in:
a the women's hurdles **b** the men's hurdles.

C2 Calculate an estimate of the mean reaction time in:
a the women's hurdles **b** the men's hurdles.

D Compare grouped frequency distributions

D1 Compare the reaction times in the women's and men's hurdles.

E Estimate cumulative frequencies

E1

1996 Olympic Games – Women's Triple Jump Final

Distance (metres)	13.5–	14.0–	14.5–	15.0–	Total
Frequency	15	21	10	1	47

a Draw a cumulative frequency curve for the women's triple jump final.
b Use your curve to estimate how many jumps were:
i less than 13.8 m **ii** greater than 14.4 m

F Use moving averages

F1 This data shows numbers of students late for school.

Week	1	2	3	4	5	6	7	8	9	10	11	12	13	14
Lates	12	6	3	18	9	6	6	15	9	12	0	18	18	3

a Calculate the three-point moving average for the data.
b On a pair of axes draw graphs to show:
i the raw data
ii the moving average data.
c Comment on any trend you can identify in the data.

Some points to remember

- You cannot find the modal value of a set of data when it is grouped, but you can identify the modal class (the class with the highest frequency).
- You cannot find **exact** values of the mean, the median, the range and the interquartile range, of grouped data, but you can calculate an **estimate** of each.
- When you estimate the mean of grouped data, usually the smaller the class interval the closer the estimate is to the exact mean.

Starting points

You need to know about ...

A Writing fractions as decimals and percentages

- To write a fraction as a decimal:
  - divide the numerator by the denominator.

$\frac{3}{8} = 3 \div 8 = 0.375 = 37.5\%$ $\quad$ $\frac{5}{6} = 5 \div 6 = 0.83... = 83.\dot{3}\%$

$\frac{7}{4} = 7 \div 4 = 1.75 = 175\%$ $\quad$ $\frac{5}{3} = 5 \div 3 = 1.6... = 166.\dot{6}\%$

B Calculating a percentage of a given amount

- To calculate 50% of an amount simply:
  calculate half the amount by dividing by 2 ($50\% = \frac{1}{2}$)
  To calculate 25% of an amount simply:
  calculate a quarter of the amount by dividing by 4.
  Some other percentages can be calculated in a similar way.

  To calculate 65% of an amount is not as easy.
  This is one way: calculate 1%, and then use it to find 65%.

Example Find 65% of 420 kg:

100%	420
Divide both sides by 100:	
1%	4.20
Multiply both sides by 65:	
65%	273

So 65% of 420 kg is 273 kg.

C Interpreting a calculator display in calculations with money

Interpreting a calculator display correctly is important, particularly when you are dealing with units of money.

For example, when you calculate 24% of £15 the calculator may display the answer as:

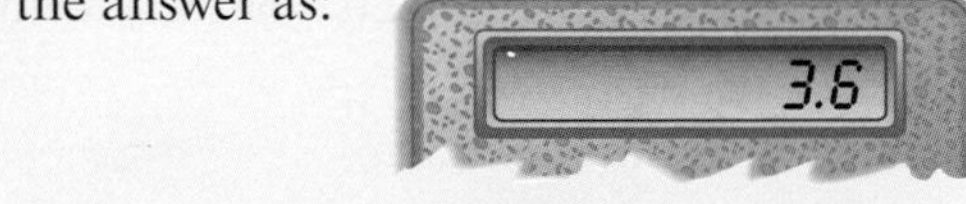

The calculation is in pounds so the answer is: **£3.60**
(£3.60 is 24% of £15)

When you calculate 24% of £1.50 the calculator displays the answer as:

The calculation is in pounds so the answer is: **£0.36**
(£0.36 is 24% of £1.50)

But an amount of money is not usually written as £0.36.
It is more likely that the amount is given as 36 pence.
So 24% of £1.50 is 36 pence.

... so try these questions

A1 Write each of these as a decimal and as a percentage:

a $\frac{3}{4}$ **b** $\frac{5}{8}$ **c** $\frac{8}{5}$ **d** $\frac{5}{4}$

e $\frac{6}{5}$ **f** $\frac{13}{20}$ **g** $\frac{8}{32}$ **h** $\frac{9}{6}$

i $\frac{7}{16}$ **j** $\frac{16}{7}$ **k** $\frac{9}{8}$ **l** $\frac{18}{10}$

B1 Calculate:
a 50% of 650 miles
b 75% of £400
c 10% of 20 marks
d 72% of 3450
e 25% of 40 marks
f 36% of 475 kg
g 58% of £415
h 12% of 3 kg
i 7% of £2.50
j 4.5% of 360 metres
k 38% of £4.50
l 15% of 12 cm
m 25% of £25
n 80% of 3 miles
o 40% of 25 km
p 15% of £3200
q 18% of 615 miles
r 36% of 25 cm
s 60% of 14 metres
t 12% of 3.65 km.

C1 In each of these calculations interpret your calculator display to give the answer in the units asked for.
a Calculate 16% of £85 (answer in pounds).
b Calculate 14% of 4 cm (answer in millimetres).
c Find 44% of 3.5 kg (answer in grams).
d Find 9% of £2.27 (answer in pence).
e What is 78% of 65 mm:
i in millimetres
ii in centimetres?
f What is 12.5% of 72 pence:
i in pence
ii in pounds?
g Give 35% of 4 tonnes in:
i tonnes **ii** kilograms.

Writing one number as a percentage of another

You can also compare the numbers the other way round, e.g.

36 as a percentage of 18

As 36 is twice 18:
36 as a percentage of 18 is 200%

When you compare two numbers, you can think of:
one number as a percentage of the other.

For instance, with the numbers 18 and 36 one comparison is:
18 is half of 36, or 18 is 50% of 36.

So 18 as a percentage of 36 is 50%.

When you compare numbers you might have to approximate one number as a percentage of the other.

Example

With the numbers 17 and 60, 17 is a little more than one quarter of 60.
So 17 as a percentage of 60 is a little more than 25%.

Exercise 17.1
Comparing numbers

1 What is 35 as a percentage of 70?

2 Give 64 as a percentage of 16.

3 Is 22 as a percentage of 28 a little more, or a little less than 75%? Explain your answer.

4 A number p as a percentage of 80 is 10%.
  i What is the number p?
  ii Explain how you calculated your answer.

5 Roughly, what is 15 as a percentage of 32?

6 Two numbers k and j are chosen so that this rule is true:

k as a percentage of j is 75%

  a Jo chooses the value 12 for k. What is the value of j?
  b Rashid chooses the value 28 for j. What is the value of k?
  c List four values you choose for k, and for each give the value of j that makes the rule true.

7 Ian chooses a whole number n and describes it in this way:
n as a percentage of 40 is a little more than 20% but not as much as 25%.
  a What is the number chosen by Ian?
  b Explain how you calculated Ian's number.

8 This table shows the amount Jim spent each month on travel and food.
For each month give the amount spent on travel as a percentage of the total amount Jim spent.

	Amount spent on travel	Amount spent on food	Total amount spent
Oct	£6	£5	£24
Nov	£8	£20	£32
Dec	£12	£15	£60
Jan	£12	£16	£48
Feb	£12	£2	£16
Mar	£2	£5	£20

9 For the amount spent on food as a percentage of the total spent:
  a In which months was this exactly 25%?
  b In which month was it a little more than 30%?
  c What was it in November?

With the two numbers 18 and 36:
to find 18 as a percentage of 36, think of the comparison in this way:

- 18 is half of 36 or $18 \div 36 = 0.5$
- 18 is 50% of 36 or $0.5 \times 100 = 50\%$

This can be a single calculation:

$$(18 \div 36) \times 100 = 50\%$$

So 18 as a percentage of 36 is 50%

This method can be used to give any number as a percentage of another.
With any two numbers p and t:

You can **calculate p as a percentage of t in this way**: $(p \div t) \times 100$

Example

Jenny and Bruce walked from John O'Groats to Land's End for charity.
John O'Groats to Land's End is 868 miles.
By the end of day 4 they had walked 113 miles.
What percentage of the total distance is this?

To calculate 113 as a percentage of 868

$$(113 \div 868) \times 100 = 13.018 \ldots$$

By the end of day 4, they had travelled 13.0% (1 dp) of the total distance.

Exercise 17.2
Calculating one number as a percentage of another

1 Calculate, giving your answer correct to 2 dp:

a 16 as a percentage of 36 **b** 52 as a percentage of 80
c 14 as a percentage of 48 **d** 85 as a percentage of 184
e 12 as a percentage of 15 **f** 1.5 as a percentage of 12
g 0.75 as a percentage of 20 **h** 16 as a percentage of 12
i 35 as a percentage of 21 **j** 15 as a percentage of 8.

2 A driving school uses this as part of their advertising:

70% of our students pass their test first time !

In July 1997, they had 55 people who took their test for the first time, and 38 of them passed the test.

a What percentage of those people who took their test for the first time in July 1997, passed the test?
b How accurate is the advertising for the driving school?
Explain your answer.

3 A sports club was sent a bill for repairs to its video camera.
The bill was for a total of £65.70, and only £9.20 of this was for parts.

What is the charge for parts as a percentage of the total bill?
Give your answer correct to the nearest whole number.

4 An old stadium had seating for 23 500 spectators.
The stadium was rebuilt, with seating in the new stadium for 40 000.

a Give the seating of the new stadium as a percentage of the old seating.
Give your answer correct to the nearest whole number.
b Is your answer about what you expected?
Give reasons for your answer.

**Thinking ahead to ...
percentage changes**

Often percentages are used to describe a change in an amount.
For example, in supermarkets special offers can be shown as:

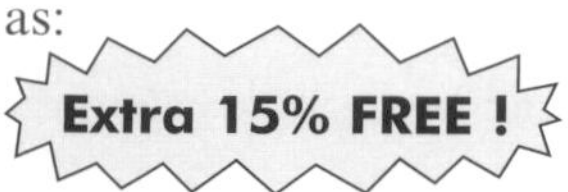

For the customer:

A How much extra is free?
Is it more or less than a quarter of the amount for the normal price?

B What fraction would you use to describe the extra free amount?

C How many ml of the product are in the special offer pack?

Percentage changes

When you increase an amount by a percentage, you will end up with **more than** the **100%** you started with.

For example:
with an increase of 25% there is the 100% you start with, plus the 25% increase.

100% + 25% = 125%

As a percentage of the start value, the end value is 125%.

When a value increases by a percentage, the final value can be calculated in different ways. Two methods are shown here.

Example

Toothpaste is sold in tubes containing 150 ml.
In a special offer, 12% extra toothpaste is put in the tube at the same price.
How much toothpaste is in the special offer tube?

Method 1
- Calculate 12% of 150 ml.
- Add the extra amount to the 150 ml.

So 150 ÷ 100 gives 1%
150 ÷ 100 × 12 gives 12%
So 12% of 150 is 18

The special offer tube contains **168 ml** of toothpaste
(168 = 150 + 18)

Method 2
Think of the toothpaste in this way:

100% of the contents is 150 ml

The special offer tube has 12% extra, so it must contain:

112% of 150 ml

112 % as a fraction is $\frac{112}{100}$

112 % as a decimal is 1.12

Calculate 112% of 150:
150 × 1.12 = 168

The special offer tube contains **168 ml** of toothpaste.

**Exercise 17.3
Increasing a value by a certain percentage**

1 In a special offer, the 440 grams of coffee in a jar is to be increased by 10%.
  a What percentage of the 440 grams of coffee are in the special offer jar?
  b Calculate the amount of coffee in the special offer jar.

2 **a** Increase 350 kg by 20% **b** Increase 447 km by 15%
  c Increase 4.50 metres by 22% **d** Increase £35 000 by 8%

3 In 1992 in the USA there were a total of 143 081 443 registered cars.
By the year 2010 it is estimated that this total will increase by 44%.
Estimate the number of registered cars in the USA in 2010.

4 A crisp manufacturer sells 25 gram bags of crisps. They decide to increase the weight of crisps in a bag by 4%. Give the new weight of crisps per bag.

When you decrease an amount by a percentage, you will end up with **less than** the **100%** you started with.

Example

With a decrease of 25% there is the 100% you start with, minus the 25% decrease.

100% – 25% = 75%

As a percentage of the start value, the end value is 75%.

You can also decrease an amount by a percentage using methods 1 and 2.

For example:

A fast food store decided to decrease the weight of packaging for their regular meals, which weighed 40 grams, by 18%.
Calculate the weight of the new packaging.

Method 1

- Calculate 18% of 40 grams.
- **Take** this weight from the 40 grams.

So $40 \div 100$ gives 1%

$40 \div 100 \times 18$ gives 18%

So 18% of 40 is 7.2

The new regular meal packaging weighs **32.8 grams**

(32.8 = 40 – 7.2)

Method 2

Think of the packaging in this way:

100% of the contents weighs 40 g

The new packaging weighs 18% less, so it must weigh:

82% of 40 grams

82 % as a fraction is $\frac{82}{100}$

82 % as a decimal is 0.82

Calculate 82% of 40:

$40 \times 0.82 = 32.8$

The new regular meal packaging weighs **32.8 grams**.

Exercise 17.4
Decreasing a value by a certain percentage

1 **a** Decrease 380 kg by 30%. **b** Decrease 416 km by 25%.
c Decrease 22 metres by 5%. **d** Decrease £338 000 by 15%.
e Decrease £25.50 by 28%. **f** Decrease 28 600 tonnes by 42%.
g Decrease 5 300 000 by 11%. **h** Decrease 1.6 cm by 25%.

2 Anya grows strawberries for supermarkets.
Last year she used a total of 1560 kilograms of fertilizer.
Next year, she wants to decrease the amount of fertilizer used by 6%.

How much fertilizer would you expect to be used next year?
Give your answer correct to the nearest kilogram.

3 The car ferry *Vista* was built to carry a maximum of 1210 cars.
New safety rules mean that the number of cars must be reduced by 7%.

a What is the maximum number of cars for the *Vista* with the new rules?
b Explain the degree of accuracy you used, and why.

4 Before a bypass was built, an estimated 41 000 cars a day passed through the town of Ashington.
The bypass was supposed to reduce the cars in Ashington by 35%.

a Estimate how many cars passed through Ashington, per day, after the bypass had been built.
b How many fewer cars per day is this?

5 Last year the fishing boat *Emma K* landed 9210 kg of shell fish.
This year they expect shell fish landings to be reduced by 17%.

How much shell fish does the *Emma K* expect to land this year?

6 In a sale items were reduced by 20%. Before the sale a kettle cost £24.50.

a What was the sale price of the kettle?
b How much was saved by someone who bought the kettle in the sale?

Looking ahead to ... percentages and VAT

VAT is short for:
Value-added tax.

VAT is a tax added to the price of goods or services. It was introduced on:
1 April 1973 at a standard rate of 10%.

On 18 June 1979 the standard rate was increased to 15%.

On 1 April 1994 the standard rate was increased to 17.5%.

Shopkeepers, traders, and customers have had to calculate VAT since 1973.

When it was first introduced at 10%, one quick method was:
"Divide by 10 and add it on."

When VAT was increased to 15%, a quick method was:
"Divide by 10, half the answer, and add both amounts on."

A What quick method can you think of to calculate VAT at 17.5%?

B In 1995 Ria fitted a stair carpet and charged £200 + VAT at 17.5%. What was the total charge to fit the carpet?

C Peter bought a cycle tyre and was charged £12.50 + VAT at 17.5%.
i What did he pay in total for the tyre?
ii Explain any rounding you did when calculating the total.

Percentages and VAT

To calculate the total price of goods or services including VAT is the same as: increasing the cost price by the percentage VAT.

Example A garden shed is advertised for: **£114.99** + VAT at 17.5%
Calculate the total charge for the shed.

117% as a decimal is 1.17.
118% as a decimal is 1.18.

117.5% is half-way between 117% and 118%.

As a decimal, 117.5% must be half-way between 1.170 and 1.180

which is 1.175.

Think of the total charge (cost price + VAT) for the shed in this way:

100% of the cost price + **17.5%** of the price (VAT)

The total charge for the shed is: **117.5%** of its cost price

117.5% as a decimal is 1.175

Calculate 117.5% of £114.99:
$114.99 \times 1.175 = 135.113 \ldots$

The total charge for the shed (including VAT) is:

£135.11 (to the nearest penny)

These examples are with VAT at 17.5%.

In all questions that involve VAT you must use the standard rate of VAT at the time.

If you are unsure, ask for the rate of VAT.

Or: you can calculate 17.5% of £114.99 and add this to £114.99

To calculate 17.5% of £114.99
$114.99 \times 0.175 = 20.123 \ldots$

The total charge (including VAT) is: £114.99 + £20.123 ...

£135.11 (to the nearest penny)

Exercise 17.5
Calculating prices that include VAT

1 The cost of these items is given without VAT (ex VAT).
Calculate the charge, including VAT, for each item.

a camera £16	**b** trainers £44.25	**c** bike £185
d toaster £19.40	**e** pen 72 pence	**f** TV £368.42
g fridge £262	**h** mower £24.55	**i** CD £9.35

2 Jenny was told that repairs to her car would be £245.
When she paid she found that the £245 did not include VAT.

a How much did she pay, including VAT?
b How much VAT was added to her bill?

Percentages and interest

From the *Oxford Mathematics Study Dictionary*.

Interest The interest is the amount of extra money paid in return for having the use of someone else's money.

When you borrow, or save money with a Bank, Building Society, The Post Office, a Credit Union, or a finance company, interest is *charged* or *paid*.

Interest is: *charged* on money you borrow and *paid* on money you save.

per annum is a term that means each year.

Interest is: *charged* or *paid* pa (per annum) and *charged* or *paid* at a fixed rate e.g. 6% pa.

You borrow £350 for one year at 9%. How much do you pay back in total?

At the end of a year you will pay back:

100% of the £350 + 9% interest
You will pay back a total of
109% of £350

Calculate 109% of £350:
$350 \times 1.09 = 381.5$

You will pay back a total of **£381.50**

You save £120 for one year at 4%. How much in total will you have?

At the end of a year you will have:

100% of your £120 + 4% interest
You will have a total of
104% of your £120

Calculate 104% of £120:
$120 \times 1.04 = 124.8$

You will have a total of **£124.80**

Exercise 17.6
Calculating interest paid for one year

1 To buy a TV, Ewan borrows £340 for one year at a rate of interest of 14% pa. How much, in total, does Ewan pay back for his loan?

2 To buy a new bike for £275, Jess decides to use £120 of her savings and to borrow the rest of the money for a year with an interest rate of 17% pa.

a How much money will Jess have to borrow to buy the bike?
b How much will she pay back in total for her loan?
c In total, how much will she have paid for the bike?

3 Marie won £2500 in a competition.
She decided to save the money, for a year, at an interest rate of 4.5% pa.

a At the end of the year, with interest, how much will Marie have in total?
b How much interest was she paid?

4 A hockey club was given a grant of £84 250. They did not spend the money straight away, but decided to save it for a year at 6.8% pa interest.
By waiting a year, how much was the grant now worth in total?

Simple interest

Simple interest is a type of interest not often used these days.

- Simple interest is fixed to the sum of money you actually borrow or save.
- When you borrow a sum of money:
simple interest is *charged* for each year on the actual sum borrowed.
- When you save a sum of money:
simple interest is *paid* for each year on the actual money saved.

This is known as the: simple interest formula.

The formula can be given in different forms.

If you use the formula in a different form, make sure you know how to use it to calculate the interest.

The amount of interest added to a loan, or added to a sum of money that is saved can be calculated using this formula:

$$I = P \times R \times T$$

In words, the formula is:

Interest = ***P***rincipal × ***R***ate × ***T***ime

Principal is the amount of money you borrow or save.
Rate is the interest rate pa **as a decimal.**
Time is for how long, usually in years.

Example

Calculate the interest charged on a loan of £750, for 4 years at 7% pa.
Use the formula

$I = P \times R \times T$ with $P = 750$, $R = 0.07$ ($7\% = \frac{7}{100} = 0.07$), $T = 4$

$I = 750 \times 0.07 \times 4$
$I = 210$
The interest charged on this loan is £210.

If the £750 had been saved for 4 years at 7% pa:
The interest paid on the savings would have been £210.

Exercise 17.7
Calculating simple interest

1 Calculate the simple interest charged, or paid for each of these:

a £450 borrowed for 6 years with interest at 12% pa
b £280 saved for 8 years with a rate of interest of 3% pa
c £1500 saved for 15 years at 6% pa. interest
d £6000 borrowed for 10 years at an interest rate of 17% pa
e £25 borrowed for 20 years with interest at 18% pa.

2 To buy new kit a band borrows £3500 over 10 years at 17% interest pa.

a Calculate the amount of interest paid on the loan.
b At the end of the loan, how much in total will have been paid for the kit?

3 Scot forgot about the £150 he had saved at an interest rate of 4.5% pa.
He found it had been earning interest for twelve years.
How much were these savings worth in total at the end of twelve years?

4 A new bridge will cost an estimated £44 million, and take six years to build.
If, at the start, all £44 million is borrowed at a simple interest rate of 8.5%, estimate the total cost of the bridge.

Compound interest

Like simple interest, compound interest is either charged or paid, but not just on the original sum borrowed or saved. For example, £100 saved for 2 years at 4% can be thought of in this way:

- At the end of year 1, the total is £104 (£100 + £4 interest)
- At the end of year 2, the total is £108.16 (£104 + £4.16 interest)

In short, compound interest includes interest on interest already *paid* or *charged.*

Compound interest is used by banks, building societies, and shops.

Calculating compound interest this way is time consuming as the number of years increases. There is a formula you can use.

If money is invested it is saved.

The compound interest formula is:

$$T = P\left(1 + \frac{r}{100}\right)^n$$ where T is the total of the amount plus interest
r is the rate of interest
n is the number of years
P is the amount an invested or borrowed

Example

Calculate the value of an investment of £2500 for 6 years at a compound rate of 4%.

Using the formula $T = P\left(1 + \frac{r}{100}\right)^n$ $\quad r = 4$ and $n = 6$

$$T = 2500\left(1 + \frac{4}{100}\right)^6$$

$$T = 2500(1.04)^6$$

$$T = 2500 \times 1.2653.......$$

$$T = 3163.297.......$$

So after 6 years the investment is worth £3163.30 (nearest penny)

By working year-by-year, you can calculate compound interest and see that it builds to a greater total than simple interest over the same time span.

Example Calculate the interest on £480 saved for 3 years at 7%

Round the answers to each calculation to the nearest penny.

- Interest paid at the end of year 1 is: £33.60 (£480 × 0.07)
  Interest for year 2 will be calculated on £513.60 (£480 + £33.60)
- Interest paid at the end of year 2 is: £35.95 (£513.60 × 0.07)
  Interest for year 3 will be calculated on £549.55 (£513.60 + £35.95)
- Interest paid at the end of year 3 is: £38.47 (£549.55 × 0.07)

The total interest paid is: £33.60 + £35.90 + £38.47 = **£107.97**

Exercise 17.8
Calculating compound interest

1 Calculate the total interest paid on these savings at compound interest:

a £350 for 3 years at 9% pa **b** £1400 for 4 years at 3% pa
c £3600 for 5 years at 6% pa **d** £12 250 for 3 years at 4% pa
e £4050 for 2 years at 12% pa **f** £35 250 for 2 years at 12% pa

2 Shelly won £5000, and put it in a savings scheme for five years.
The savings scheme pays interest at 8% pa compound.

a Calculate the total interest paid on this saving.
b At the end of five years what was the total in Shelly's saving scheme?
c **i** What would the total be if the only rounding was done at the end?
ii What do you think the banks do with the rounding problem?

For questions **3** to **5** use the formula.

Give your answers correct to the nearest penny

3 Calculate each of these for compound interest.

a £1500 is invested for 8 years at 3%. Find the value of the investment at the end of the term.

b Calculate the value of an investment of £4200 over 12 years at 5%.

c £6200 is invested for 8 years at 6%. What is the final value of the investment?

d Find the final value of an investment of £7800 for 5 years at 6%

4 To buy a car Marie takes out a loan of £4500 at 7% compound interest for 4 years.

Find the total amount Marie will pay back on her loan.

5 A club borrows £21500 at 3% compund interest for a period of 12 years. Calculate the amount the club will actually pay back for this loan.

Reverse percentages

When you divide by 117.5, you are calculating 1%.

You then multiply by 100% to calculate the 100%.

This is for VAT at 17.5%

Check on the current rate of VAT to find the number to divide by to calculate 1%.

Calculating the original value of something, before an increase or decrease took place, is called 'calculating a reverse percentage'.

Example The total price of a bike (including VAT at 17.5%) is £146.85
Calculate the cost price of the bike without VAT.

$$\begin{aligned} \text{Total price of bike} &= 100\% \text{ of cost price} + 17.5\% \text{ of cost price} \\ 146.85 &= 117.5\% \text{ of cost price} \\ 146.85 \div 117.5 \times 100 &= 100\% \text{ of cost price} \\ 124.978... &= \text{Cost price} \end{aligned}$$

The price of the bike without VAT is £124.98 (to the nearest penny)

Exercise 17.9
Calculating reverse percentages

1 These prices include VAT, calculate each price without VAT.
Give your answers correct to the nearest penny.

a CD player £135.50	**b** camera £34.99	**c** ring £74.99
d trainers £65.80	**e** TV £186.75	**f** phone £14.49
g calculator £49.99	**h** PC £799.98	**i** tent £98.99

2 Ella bought a pair of boots for £45 in a sale that made an offer '20% off!'.

a What was the non-sale price of the boots?

b How much did Ella save buying the boots in the sale?

3 When Mike sold his bike for £35, he said he made a profit of 35%.
To the nearest pound, how much did Mike pay for the bike?

4 A 600 gram box of cereal is said to hold '35% more than the regular box'.
How many grams of cereal are in the regular box?

End points

You should be able to ...	... so try these questions
A Write one number as a percentage of another	**A1** What is 12 as a percentage of 48? **A2** Give 60 as a percentage of 20. **A3** Correct to 2 dp, what is 18 as a percentage of 64? **A4** In a 456-page book, 35 of the pages have pictures on them. Give the number of pages with pictures as a percentage of the total number of pages in the book, correct to 1 dp.
B Increase a value by a certain percentage	**B1** Increase 480 km by 16%. **B2** A CJ regular size cola is 380 ml. CJ decide to increase the size of their regular cola by 12%. To the nearest ml, give the size of the new regular cola.
C Decrease a value by a certain percentage	**C1** Decrease £55.80 by 6%. Give your answer to the nearest penny. **C2** In 1992, there were a total of 42 154 breakdowns on a motorway section. In 1993 the total number of breakdowns fell by an estimated 7%. Estimate the number of breakdowns in 1993 on this motorway section.
D Work with VAT	**D1** What is the standard rate of VAT today? **D2** The cost of these items is given without VAT (ex. VAT) **a** a camping stove £34.75 **b** a body-board £158.40 Calculate the cost, to the nearest penny, of each item including VAT.
E Calculate reverse percentages	**E1** The price of a TV is £259.99 including VAT. Calculate the cost of the TV 'ex.VAT' (without VAT).
F Calculate simple interest	**F1** Calculate the interest paid on £480 saved for six years, at a simple interest rate of 4%. **F2** £500 was borrowed at a simple interest rate of 9% over 4 years. **a** Calculate the total interest paid on this loan. **b** How much was paid back on the loan in total?
G Calculate compound interest	**G1** Calculate the total of these savings at compound interest: **a** £680 for 3 years at 6% **b** £2575 for 4 years at 3% **c** £3600 for 8 years at 4% **d** £7150 for 12 years at 2% **e** £1500 for 20 years at 6% **f** £318 for 4 years at 1%

Some points to remember

- When you work with a calculator it is easier if you think of, and use, percentages as decimals.
- Check that your answer to a calculation is of the right size, and to a sensible degree of accuracy.
- When you are calculating interest, paid or charged, make sure you know whether it is at a rate of simple interest, or compound interest.

Identifying transformations

Exercise 18.1
Identifying single transformations

The Moors used tessellations of tiles to decorate walls and floors. These are patterns used at the Alhambra Palace in Spain.

> The Moors are Muslims of mixed Berber and Arab descent who live in North West Africa.

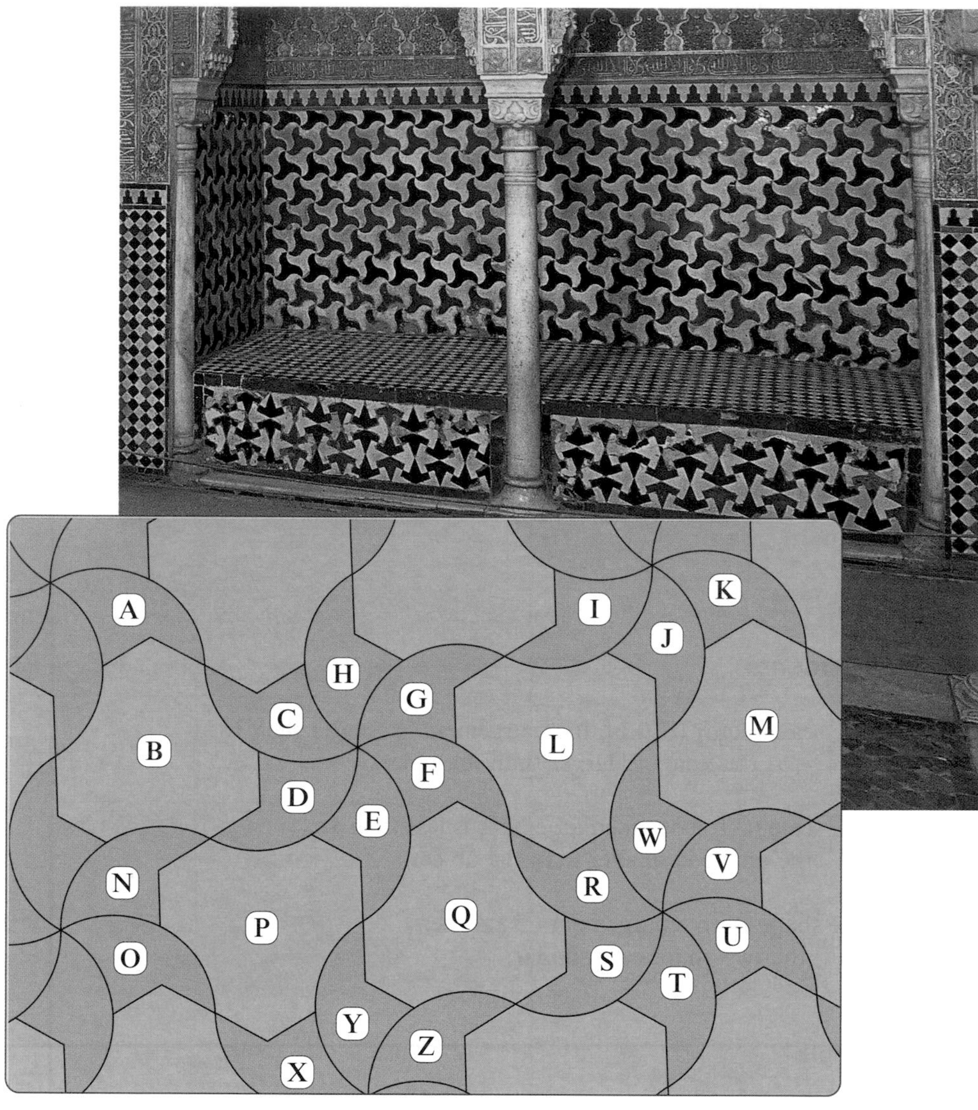

1 This is a tessellation made from tiles of two different shapes.

- **a** **i** How many lines of symmetry has tile A?
  ii How many lines of symmetry has tile B?
- **b** What is the order of rotational symmetry for:
  i tile A **ii** tile B?

> When a shape is transformed the object is said to map on to the image.

2 Each complete tile on the tessellation is labelled.

- **a** What type of transformation will map tile C on to tile H?
- **b** Give two different types of transformation that will map tile B on to tile M.
- **c** Tile C maps on to tile U by a rotation of 180° clockwise. Which other tiles are also an image of tile C after this rotation?
- **d** Which tiles are an image of tile G after a translation?
- **e** List all the tiles that are an image of tile D after a rotation:
  i of 60° clockwise **ii** of 60° anticlockwise.

Single transformations

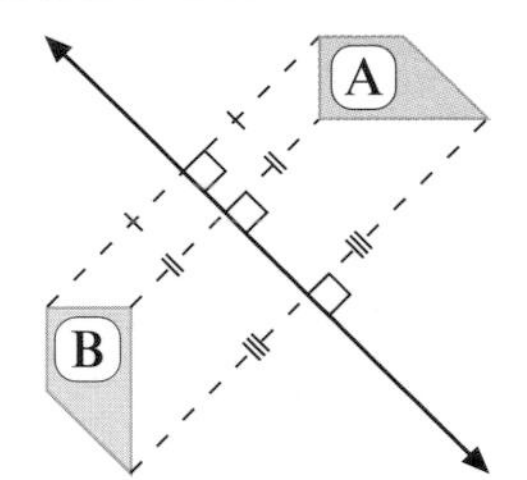

When an object is reflected in a mirror line:

- lines joining corresponding points on the object and image are perpendicular to the mirror line
- the object and image are the same distance from the mirror line.

- To describe a **reflection** fully you need to give the mirror line.

Example

A reflection can map triangle 1 on to triangles 3, 4 or 5.

- A reflection in the line GC maps triangle 1 on to triangle 3.
- A reflection in the line HD maps triangle 1 on to triangle 4.
- A reflection in the line HB maps triangle 1 on to triangle 5.

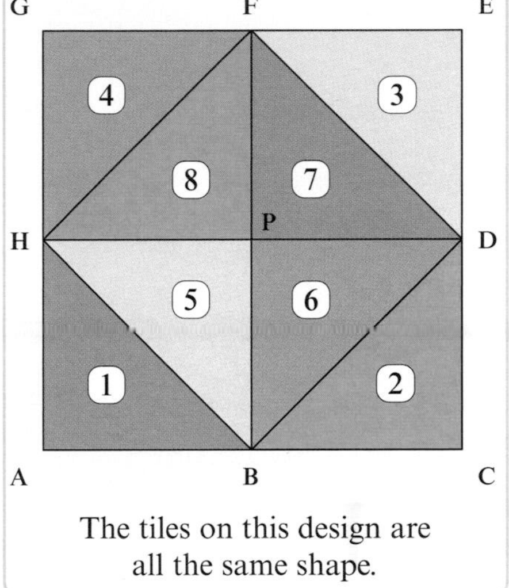

The tiles on this design are all the same shape.

- To describe a **rotation** fully you need to give:
  - the angle and direction of rotation
  - the centre of rotation.

Example

A rotation of 90° anticlockwise ($^{+}90°$) can map triangle 1 on to triangle 2 or 8.

- A rotation of 90° anticlockwise about the point P maps triangle 1 on to triangle 2.
- A rotation of 90° anticlockwise about the point H maps triangle 1 on to triangle 8.

Exercise 18.2 Describing transformations

A rotation of 90° anticlockwise ($^{+}90°$) and a rotation of 270° clockwise ($^{-}270°$) about the same point both map an object on to the same image.

A rotation of 180° anticlockwise ($^{+}180°$) and a rotation of 180° clockwise ($^{-}180°$) about the same point both map an object on to the same image.

1 Use the triangles in square ACEG above to answer these questions.

a Using a rotation of 90° clockwise ($^{-}90°$) about the point P, what is the image of
  i triangle 5 **ii** triangle 2 **iii** triangle 4?

b A rotation maps triangle 2 on to triangle 5. What is:
  i the centre of rotation
  ii the angle of rotation?

c Describe fully a rotation that maps:
  i triangle 7 on to triangle 2
  ii triangle 3 on to triangle 1.

d What is the mirror line for a reflection that maps:
  i triangle 2 on to triangle 6
  ii triangle 2 on to triangle 3?

e Which triangle, after a rotation of $^{+}90°$ about the point P, is the image of triangle 6?

f Which triangle, after a reflection in the line FB, is the image of triangle 3?

g What transformation maps triangle 7 on to triangle 4?

h Describe two transformations that map triangle 4 on to triangle 2.

i What transformation can map triangle 8 on to itself?

Exercise 18.3
Describing transformations

This pattern uses tiles of two different shapes.

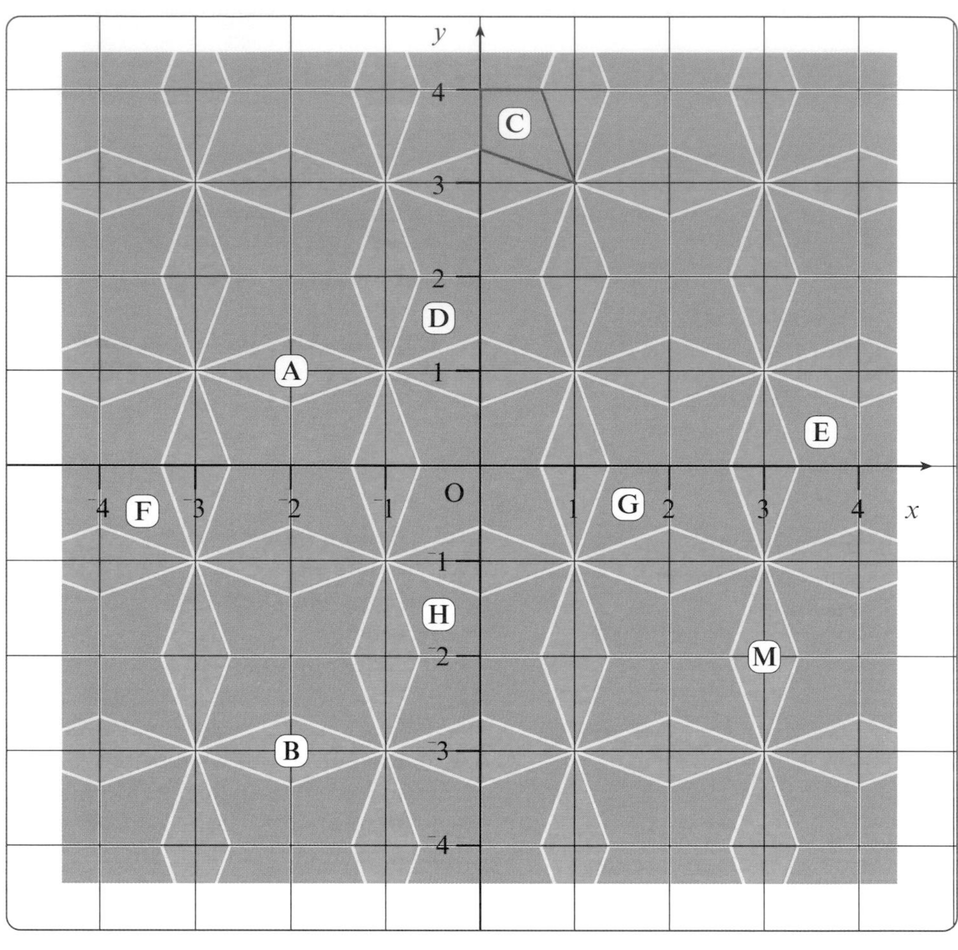

1 You can use vectors and coordinates to describe transformations on a grid.

a What vector translates tile A on to tile B?

b Tile A is mapped on to tile B by a rotation of 180°. What are the coordinates of the centre of rotation?

c What is the equation of the mirror line for the reflection which maps tile A on to tile B?

You could trace the object and try different centres of rotation.

2 Six of the kites are labelled, C to H.

a i Match kites C to H in three pairs so that:

in each pair one kite is a reflection of the other.

ii In a table like this show the equation of the mirror line for each pair.

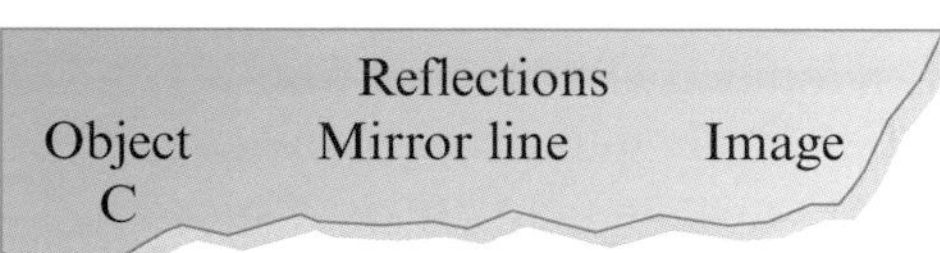

	Reflections	
Object	Mirror line	Image
C		

b i Sort kites C to H into three pairs so that:

in each pair one kite is a translation of the other.

ii Make a table to show the vector for each translation.

Use this diagram for questions **3** to **8**.

3

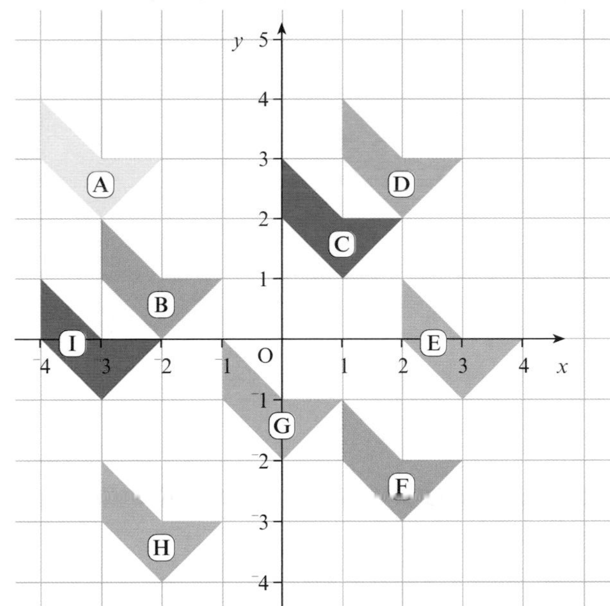

Describe each translation with a vector.

a	A to B	**b**	A to C	**c**	B to D
d	C to B	**e**	B to F	**f**	H to E
g	E to A	**h**	F to A	**i**	D to I

4 Shape B is translated by the vector $\begin{pmatrix} 0 \\ ^-4 \end{pmatrix}$.

Which shape does B map on to?

5 Shape C is mapped on to another shape by a translation described by $\begin{pmatrix} ^-3 \\ ^-5 \end{pmatrix}$.

Which shape is C mapped on to?

6 Shape F is mapped on to another shape by two translations. The translations are $\begin{pmatrix} 1 \\ 2 \end{pmatrix}$ and $\begin{pmatrix} ^-5 \\ 1 \end{pmatrix}$.

a Which shape is F mapped on to?
b Give the vector that describes this mapping of F as a single translation.

7 **a** List the vectors that describe this set of translations:

D → E → F → H → B → C

b Describe the translation of D to C as a single vector.

8 Shape I is mapped on to shape E.
Give the vector that describes this translation.

Exercise 18.4
Transformations

1 **a** Draw triangle A on axes with $^-5 \leq x \leq {}^+5$ and $^-5 \leq y \leq {}^+5$.

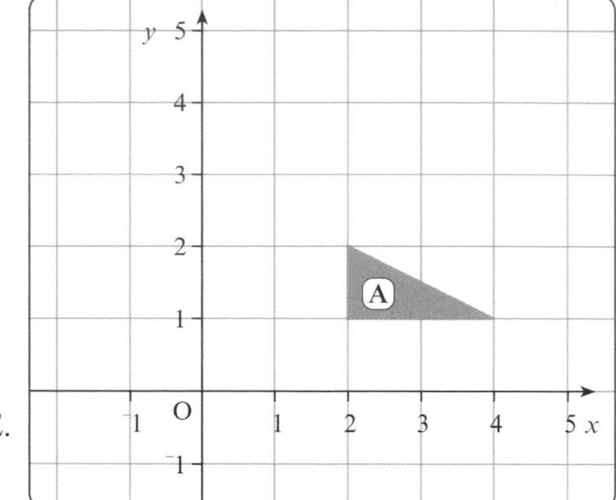

b These transformations map triangle A on to B, C, D and E.

Object	Transformation	Image
A	Reflection in the line $y = x$	B
A	Reflection in the line $x = 1$	C
A	Rotation of 180° about the origin	D
A	Rotation of ⁻90° clockwise about the point (1, 0)	E

Draw and label the images B, C, D and E .

c Describe a transformation that maps B on to C.

d Describe a transformation that maps:
i D on to E **ii** E on to C.

2 Draw axes with $^-5 \leq x \leq 5$ and $^-5 \leq y \leq 5$.

a Plot the shape with vertices at (1, 2), (1, 4), (2, 3), (2, 2). Label the shape A.

b Reflect A in the line $x = 2$. Label the image B.

c Rotate A 90° anticlockwise about the point (0, 2). Label the image C.

d Translate A by $\begin{pmatrix} ^-2 \\ ^-3 \end{pmatrix}$. Llabel the image D.

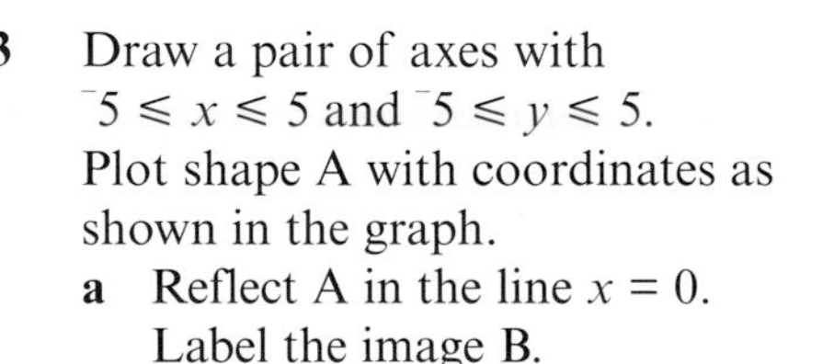

3 Draw a pair of axes with $^-5 \leq x \leq 5$ and $^-5 \leq y \leq 5$.
Plot shape A with coordinates as shown in the graph.

a Reflect A in the line $x = 0$.
Label the image B.

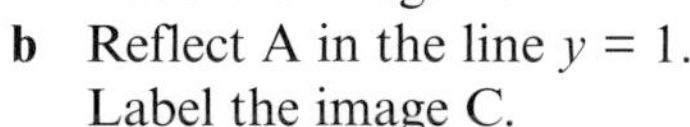

b Reflect A in the line $y = 1$.
Label the image C.

c Rotate A 90° clockwise about the point (2, 2). Label the image D.

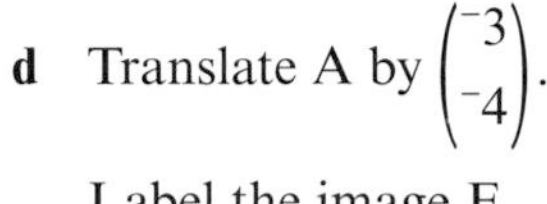

d Translate A by $\begin{pmatrix} ^-3 \\ ^-4 \end{pmatrix}$.

Label the image E.

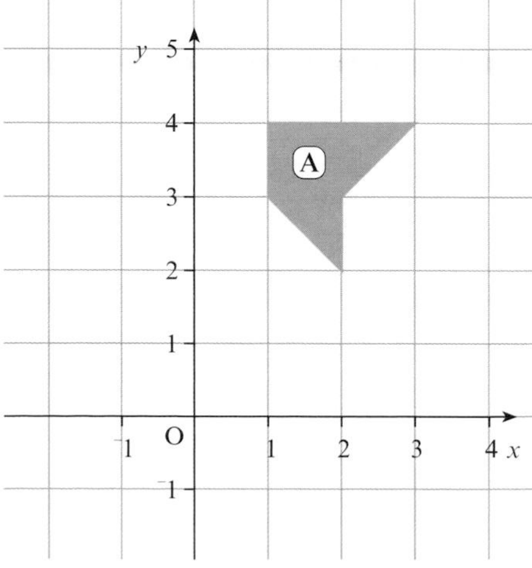

4 Draw a pair of axes with $^-5 \leqslant x \leqslant 5$ and $^-5 \leqslant y \leqslant 5$.
Plot shape A with coordinates as shown in the graph.

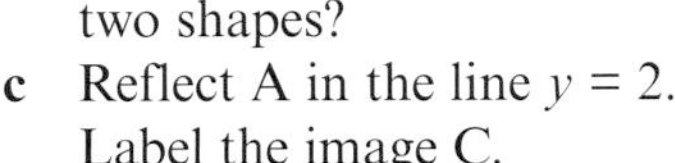

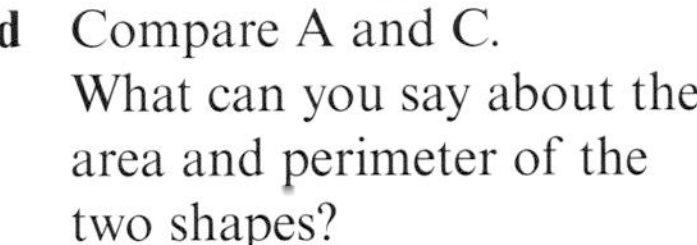

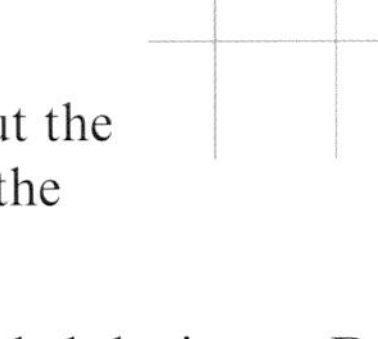

a Rotate A 90° anticlockwise about (0, 2).
Label the image B.

b Compare A and B.
What can you say about the area and perimeter of the two shapes?

c Reflect A in the line $y = 2$.
Label the image C.

d Compare A and C.
What can you say about the area and perimeter of the two shapes?

e Translate A by $\begin{pmatrix} ^-2 \\ ^-5 \end{pmatrix}$. Label the image D.

f Compare A and D.
What can you say about the area and perimeter of the two shapes?

5 A rectangle 3.5 cm wide and 8 cm long is transformed by a rotation of 90° anticlockwise about the point (2, 3).
The image of the rectangle is labelled K.
Find the area and perimeter of rectangle K.

6 Triangle W is transformed by a reflection in the line $y = {}^-2$.
The image of W is labelled R.
Find the area and perimeter of R.

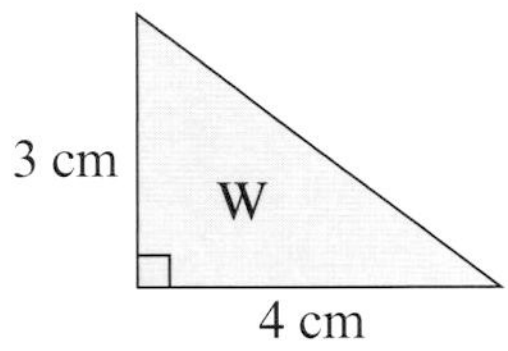

For the axes use:
$^-5 \leqslant x \leqslant 5$ and $^-8 \leqslant y \leqslant 8$.

7 **a** Find the area and perimeter of shape N.

b Draw shape N on a pair of axes.

c Rotate N 90° clockwise about (0, 0).
Label the image L.

d Give the area and perimeter of L.

e Reflect N in the line $y = {}^-1$.
Label the image K.

f Translate N by $\begin{pmatrix} ^-4 \\ ^-6 \end{pmatrix}$. Label the image J.

g What can you say about the area and perimeter of N, L K and J?

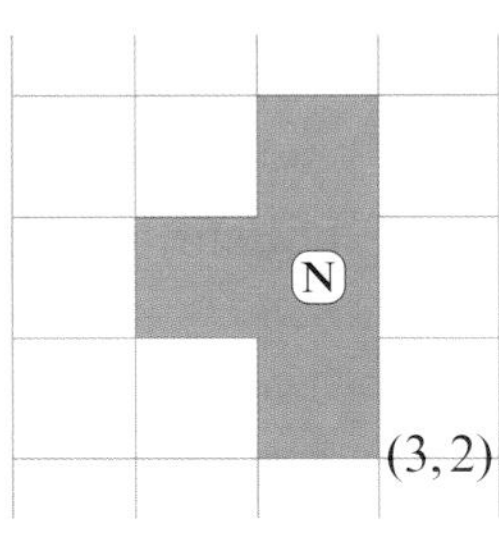

A Use single transformations

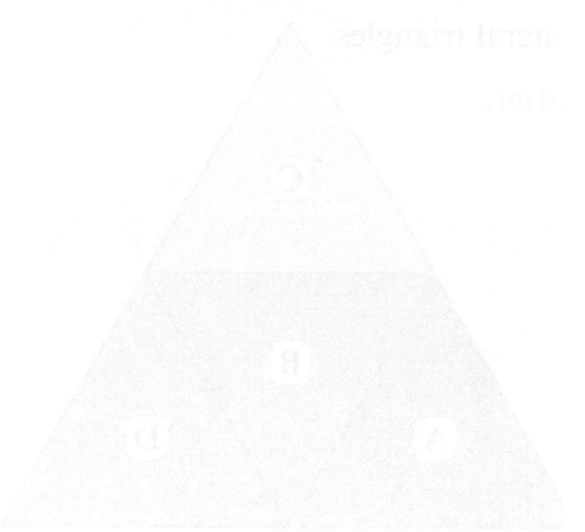

A3 Draw the a pair of axes with:
$^{-}5 \leqslant x \leqslant 5$
$^{-}8 \leqslant y \leqslant 8$
Copy shape A.

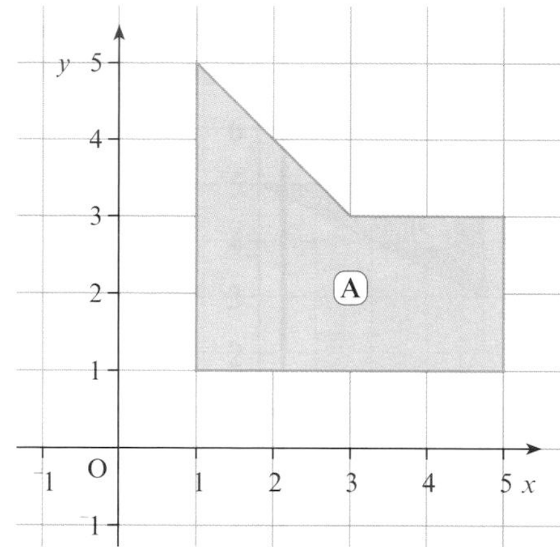

a Enlarge shape A with a SF $\frac{1}{2}$ centre (1, 1).
Label the image B.

b Find the area of shape A.

c What is the area of shape B?

d What fraction of the area of shape A is shape B?

e What can you say about the edge lengths in shape A and shape B?

f Reflect A in the line $y = {}^{-}1$.
Label the image C.

g Compare shapes A and C.
What is the same about the shapes?
What is different about the shapes?

h Translate A by $\begin{pmatrix} ^{-}6 \\ ^{-}7 \end{pmatrix}$, and label the image D.

i Compare shapes A, B, C and D.
What is the same about each of the shapes?

j Shape A was enlarged by SF $\frac{1}{3}$ and the image was labelled E.
What fraction of the area of shape A is shape E?

k What can you say about the edge lengths in Shape A and shape E?

B Use combined transformations

B1 These transformations map the hexagon E, from Question **A2**, on to J, K and L.

Object	First transformation	Second transformation	Image
E	Reflect in $y = ^-1$	Reflect in $x = 0$	K
E	Reflect in $y = ^-x$	Reflect in $y = 0$	L
E	Rotate 180° about (0, 0)	Rotate 180° about (0, $^-2$)	M

a On a new diagram, on axes with $^-6 \leqslant x \leqslant 6$ and $^-6 \leqslant y \leqslant 6$, draw and label E and the images K, L and M.

b What single transformation maps E on to:
i K **ii** L **iii** M?

c These transformations map K on to L, M and E.
Describe the second transformation fully.

	Object	Image	First transformation	Second transformation
i	K	L	Rotate $^-90°$ about ($^-1$, $^-3$)	
ii	K	M	Reflect in $x = 0$	
iii	K	E	Translate $\begin{pmatrix} 2 \\ 2 \end{pmatrix}$	

d What single transformation maps K on to:
i L **ii** M **iii** E.

Some points to remember

- Transformations on a grid

Transformation	Describe by giving:	Object and image are:
Enlargement	the coordinates of the centre the scale factor	similar
Reflection	the equation of the mirror line	congruent
Rotation	the coordinates of the centre the angle of rotation the direction of rotation (+ anticlockwise, − clockwise)	congruent
Translation	the vector $\begin{pmatrix} -\leftrightarrow+ \\ +\updownarrow- \end{pmatrix}$	congruent

Southampton Evening Chronicle *Monday 15 April 1912*

TRAGEDY AT SEA

IT IS WITH great regret that we bring you the news that last night at 10:40 pm the 'unsinkable liner' the Titanic hit an iceberg on her way to New York. The Titanic later sank at 2:20 am with the loss of many lives. It was the liner's maiden voyage and on board were 331 first class passengers, 273 second class 712 third class and a full crew – only 32.2 % of those on board survived. Each first-class passenger had paid £870 for the privilege of making the voyage in this luxury floating palace. To reassure the passengers the orchestra was still playing as the liner was going down and many passengers were so sure the ship could not sink that they refused to board the lifeboats.

The captain had been given repeated warnings of icebergs ahead but chose to steam on at 22.5 knots. It was calm weather with good visibility but the lookouts had not been issued with binoculars. The iceberg is thought to have had a height of about 100 feet showing above the water and a weight of about 500 000 tons. The sea water temperature was only 28° Fahrenheit and this took its toll on those jumping overboard. It is thought that the capacity of the lifeboats was insufficient for the number of people on board.

Survivors were picked up by the liner Carpathia which had heard the SOS when it was 58 miles away. The Carpathia steamed at a staggering 17.5 knots to reach the sinking Titanic. The ship's engineer said this was 25% faster than her usual speed.

HOW FAIR WAS THE RESCUE?

Reports coming in give the final casualty figures from the Titanic. There was not enough lifeboat space for all on board because the Titanic was considered to be the first unsinkable ship. The owners White Star admit that lifeboats could only hold 33% of the full capacity of the liner and 53% of those on board on that fateful night. Breaking the survival figure down by class we find 203 first-class, 118 second-class and 178 third-class passengers were rescued. Nearly a quarter of all crew were saved. Analysis of these figures is taking place to see if all people on board had an equal chance of being rescued.

Strange BUT *true*

Fourteen years before the disaster, and before the Titanic had been built, a story was published which described the sinking of an enormous ship called the Titan after it had hit an iceberg on its maiden voyage.
The comparisons between ships is even more amazing.

	Titan (Fiction 1898)	*Titanic* (True 1912)
Flag	British	British
Month of sailing	April	April
Displacement (tons)	70 000	66 000
Propellers	3	3
Max. speed	24 knots	24 knots
Length	800 feet	882 feet
Watertight bulkheads	19	15
No. of lifeboats	24	20
No. on board (inc crew)	2000	2208
What happened?	Starboard hull split by iceberg	Starboard hull split by iceberg
Full capacity	3000	

Distances at sea are measured in nautical miles and ships' speeds are given in knots.
1 nautical mile is 1852 metres.
1 knot is a speed of 1 nautical mile/hour.

1 In 24-hour time give the time that:
- **a** the Titanic hit an iceberg
- **b** the Titanic sank.

2 How long did it take the Titanic to sink after hitting the iceberg?

3 **a** Calculate the number of crew on board the Titanic on her maiden voyage.
b Calculate approximately how many people died.

4 If all those who survived were in lifeboats, what was the mean number of people per boat?

5 How much money in total was taken in fares for the first class passengers?

6 In 1912 the price of a small house was about £200. In 1996 the same house would cost about £68 000. If fares on a cruise liner increased in the same ratio, what would have been the first class ticket price in 1996?

7 About $\frac{7}{8}$ of an iceberg's height is below water level. Estimate the total height of the iceberg that the Titanic hit.

8 **a** What was the Carpathia's usual speed?
b At 17.5 knots, how long would it take the Carpathia to steam the 58 nautical miles to the Titanic?
c The Carpathia received the SOS message at 12:30 am. Approximately how long after the Titanic sank did she arrive at the scene?

9 You can convert a temperature from degrees Celsius (°C) to degrees Fahrenheit (°F) with this formula:

$$F = \frac{9}{5}C + 32$$

a Make C the subject of the formula.
b Calculate a water temperature of 28 °F in degrees Celsius.

10 **a** How many people could the lifeboats have held in total?
b Use your answer to part **a** to calculate an approximate value for the full capacity of the Titanic.

11 Calculate the relative frequency of survival for:
- **a** first-class passengers
- **b** second-class passengers
- **c** third-class passengers.

12 Calculate the total percentage of the passengers aboard who were rescued.
(About 25% of the crew were rescued.)

13 What is the displacement of the Titanic to 1 sf?

14 The richest person on the boat was Colonel J.J. Astor who was thought to be worth £30 million. Write this number in standard form.

The drive shafts of the Titanic were 201 feet long and fell $3\frac{1}{2}$ feet over their length.

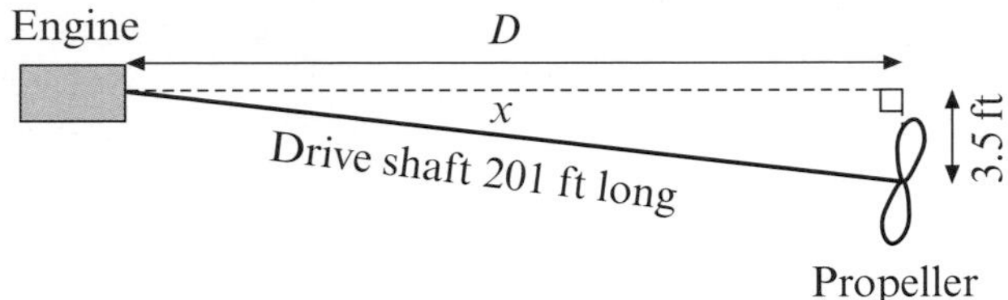

15 Calculate the angle x that a driveshaft made with the horizontal to the nearest degree.

16 Calculate the horizontal distance D to 2 dp.

The Titanic had cranes, known as derricks, for lifting the cargo on to the ship.
This diagram shows a derrick in one position.
The tower and part of the cable are vertical.

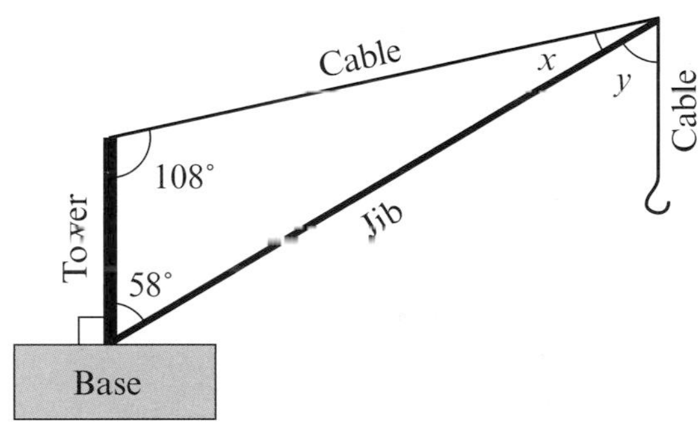

17 What is the angle y? Give your reasons.

18 Calculate the value of angle x. Give your reasons.

At the enquiry after the sinking, White Star, the owners, said that in the previous ten years they had carried 2 179 594 passengers with the loss of only 2 lives.

19 Give the number of passengers carried to:
a 1 sf **b** 3 sf
c 4 sf **d** 5 sf

20 In the ten years between 1981 and 1990 about 1.2×10^7 people from the UK crossed the Atlantic by plane. With the same death rate as White Star gave, roughly how many people would have died in that time?

On 31 May 1682 there was a cloudburst in Oxford which gave 24 inches of water in less than a quarter of an hour.

A slight shower!

Banananananananananananananananananana

On 30 April 1988 in Selinsgrove Pennsylvania USA a banana split was made which was 7.32 km long.

Jo's always on hand

In 1900 Johann Hurlinger of Austria walked 871 miles from Vienna to Paris on his hands. His average speed was 1.58 mph and he walked for 10 hours each day.

OH! WHAT A LITTLE ONE-WHEEL

In March 1994 in Las Vegas Peter Rosendahl of Sweden rode a unicycle 20 centimetre high for a distance of 3.6 metres. The wheel diameter was only 2.5 cm.

Up the pole!

Mellissa Sanders lived in a hut at the top of a pole in Indianapolis USA for two years starting on 10 October 1986. Her hut measured 1.8 metres wide by 2.1 metres deep.

Stacks of cards!

In 1995 Brian Berg of Spirit Lake Iowa USA built a tower of playing cards with 83 stories. The tower was 4.88 metres high. In 1978 James Warnock of Canada held the previous record with 60 stories.

Number types

On 14 October 1993 Mikhail Shestov set a record when he had typed the numbers 1 to 795 on a PC by the time 5 minutes was up. He had made no errors.

Unique cycle on unicycle

Takayuki Koike of Japan road 100 miles on a unicycle in a record time of 6 hours, 44 minutes and 21 seconds on 9 August 1987.

Tall stories

The tallest man in the world was Robert Wadlow from Alton Illinois in the USA who was 8 feet 11.1 inches tall. The tallest man in Scotland was Angus Macaskill from the Western Isles who was 7 feet 9 inches tall. The highest mountain in the USA is Mount McKinley at 20 320 feet and the highest one in Scotland is Ben Nevis at 4408 feet.

Can beans be hasbeens?

Baked beans were first introduced into the UK in 1928. By 1992 they were selling at the rate of 55.8 million cans per year.

Rail Trick

The Katoomba Scenic Railway in New South Wales in Australia is the steepest railway in the world. Its gradient is 1 in 0.8 but it is only 310 metres long. The ride takes about 1 minute 40 seconds and carries about 420000 passengers a year.

Weight watchers

The circumference of the Earth at the equator is 40075 km and its mass is 5880 000 000 000 000 000 000 tons.

Piece on earth

A jigsaw with 1500 wooden pieces was made for the photograph on the cover of the BEEB magazine. The jigsaw was assembled in 1985 by students from schools in Canterbury. It measured 22.31 metres by 13.79 metres.

The swift hare and the XJ tortoise

In 1992 the Jaguar XJ220 set the land speed record for a road car of 217 miles per hour. The spine-tailed swift has been recorded as flying at 220 miles per hour. Will this mean that Brands Hatch is converted for spine-tailed swift racing?

Can can or cannot

A square based pyramid tower of 4900 cans was built by 5 adults and 5 children at Dunhurst School, Petersfield on 30 May 1994 in a time of 25 minutes 54 seconds.

Barmy salami

A salami is usually about 9cm in diameter and about 35cm long, but at Flekkefjord in Norway in July 1992 a giant salami was made which had a circumference of 63.4 cm and was 20.95 metres long.

1 Calculate the height of James Warnock's tower of cards in 1978.

2 Give the weight of the Earth in standard form.

3 Give the circumference of the Earth to:
a 4 significant figures
b 3 significant figures.

4 Calculate the diameter of the Earth at the equator.

5 For Peter Rosendahl's mini unicycle give:
a the height of the bike in metres
b the distance he travelled in millimetres
c the circumference of the wheel in centimetres.

6 How many days was it between when Mellissa Sanders came down from her pole hut and Mikhail Shestov set his number typing record?

7 Convert the height of water that fell on Oxford in less than a quarter of an hour to metres.

8 What was the area of the Canterbury jigsaw?

9 If the Canterbury jigsaw had been out in the Oxford rain, what volume of water would have landed on it?

10 What was the average speed of Takayuki Koike's unicycle ride:
a in miles per hour **b** in kilometres per hour?

11 How long would it take Takayuki Koike to unicycle along the length of the Selinsgrove banana split if he always unicycles at the same average speed?

12 How many of the giant banana splits would fit end to end round the equator? Give your answer in standard form to a suitable degree of accuracy.

13 A salami is shaped roughly like a cylinder. Use this to calculate the approximate volume of a normal salami.

14 **a** Calculate the diameter of the Flekkefjord salami in centimetres.
b Calculate the volume of the Flekkefjord salami in cm^3.
c The density of salami is about 1.01 g/cm^3. Calculate the mass (weight) of the giant salami in kg.

15 If the capacity of Mellissa Sanders's hut on a pole was 6.62 $metres^3$, what was the height of her hut in metres? Give your answer to 3 sf.

16 Give the ratio of the length of the giant banana split to the length of the giant salami in the form $1:n$, to the nearest whole number.

17 **a** What was Robert Wadlow's height in:
i inches (to the nearest inch)
ii centimetres (to 1 dp)
iii metres (to 2 dp)?
b What is the ratio tallest man : highest mountain in the form $1:n$ for:
i the USA **ii** Scotland?

18 For the numbers 1 to 12, fifteen digits are used.
a How many digits are in the numbers 1 to 100?
b How many digits had Mikhail Shestov typed in by the time five minutes was up?
c What was Mikhail Shestov's typing speed in digits per second (to nearest whole number)?

19 For how many days was Johann Hurlinger walking on his hands when he travelled between Paris and Vienna?

20 **a** At what angle does the Katoomba Scenic Railway climb?
b How many metres does the railway rise over its entire length?
c What is the average speed of the train:
i to the nearest metre per second
ii in km per hour?

21 The spine-tailed swift and the Jaguar XJ220 race together at their maximum speeds over a course of 240 miles. How long will the swift have to wait for the Jaguar at the finish line?

22 A baked bean can has a diameter of 7.4 cm and a height of 10.5 cm.
a What is the volume of one can of beans in cm^3?
b What was the total volume of the cans in the Petersfield tower in m^3?

23 The pyramid of cans in Petersfield had one can on the top layer and each can in the stack rested on four cans below.
a How many cans were in:
i the second layer down from the top
ii the third layer down?
b There were 24 layers in the tower. How many cans were on the bottom layer?

24 Imagine that as baked bean cans are bought in the UK they are emptied then lined up end to end round the equator. After roughly how long would they encircle the Earth?

25 Land's End to John O'Groats is 886 miles.
a How many giant salamis long is this?
b If the giant salami was rolled along this distance how many rotations would it make?

26 How many times does a banana split go into a basin plant? Hint: think of a bananagram!

Starting points

You need to know about ... / ... so try these questions

A Relative frequency

Relative frequency is a way of estimating a probability.
It can still be used in equally likely situations but is the **only** way when outcomes are not equally likely. It is found by experiment, by survey, or by looking at data already collected.

Example This data shows the colour of cars passing a factory gate one morning.

Colour	Frequency
Red	68
Black	14
Yellow	2
Green	34
Blue	52
Grey	35
Other	23
TOTAL	228

The relative frequency of red cars is $\frac{68}{228}$ = 0.30 (to 2 dp).

This probability is an estimate and can be used to predict the number of red cars passing the gate on a different morning.

A1 What is the relative frequency of blue cars in the survey? Give your answer to 2 dp.

A2 What is the relative frequency of cars which are:

a not blue
b either yellow or green
c neither green nor red
d either black, grey or blue?

B Tree diagrams

A tree diagram has branches showing different events.
This is a tree diagram for spinning a coin then rolling a 1 to 6 dice.
The only square numbers on the dice are 1 and 4.

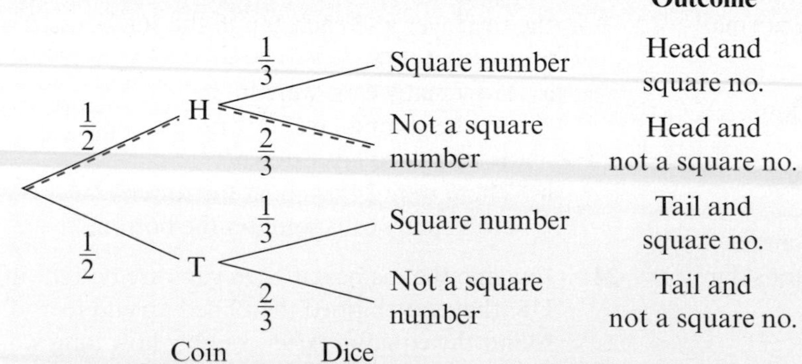

To find the probability of a particular outcome you can multiply the probabilities on the branches that lead to it.

Example To find the probability of a head and a non-square number you multiply the probabilities along the dotted branches.

Probability of a head and non-square no. is $\frac{1}{2} \times \frac{2}{3} = \frac{2}{6}$

B1 What is the probability of a head and a square number?

B2 What is the probability of a tail and a non-square number?

B3 A red 1 to 6 dice and a blue 1 to 6 dice are rolled.

a Draw a tree diagram to show the probability of a triangle number on the red dice and an even or odd number on the blue one.

b What is the probability of getting a non-triangle number on the red with an even number on the blue dice?

Thinking ahead to ...
experimental probability

Theoretical probability is based on the analysis of equally likely outcomes.

It is usually given as a fraction.

From the number of items you can calculate the theoretical probability of picking any one item at random when each item is equally likely.

Example In this bag there are known to be 3 red cubes, 5 blue cubes and 1 yellow cube.

The probability of picking a particular cube is equally likely as picking any other cube.

So the probability of picking a red cube at random is $\frac{3}{9}$ (or $\frac{1}{3}$).

A From this bag what is the probability of picking:

a a blue cube **b** a yellow cube
c a cube which is not blue **d** a cube which is not blue and not yellow?

Probability based on experiment

Relative frequency is usually given as a decimal or a percentage.

You might not know the number, or colour, of cubes in a bag.

The probability of a particular colour can be estimated by doing an experiment to find the relative frequency.

In this experiment a cube is picked at random, its colour recorded, and the cube replaced.

These results are from an experiment:

Colour	Frequency
Red	53
Blue	15
Yellow	67
Total	135

The relative frequency of picking red is $\frac{53}{135}$ = 0.39 (to 2 sf).

This means that it is likely that about 39% of the bag's contents are red.

Exercise 19.1
Relative frequency

1 From the bag above what is the relative frequency of picking:

a a blue cube **b** a yellow cube?

2 What is the relative frequency of not picking blue?

3 Suppose that you know that the total number of cubes in the bag is 20. Estimate the number of:

a red cubes **b** blue cubes **c** yellow cubes.

4 In another experiment, red, green and blue cubes were in a bag. Cubes were picked out at random, examined and replaced. The relative frequency of red was 0.62 and of blue it was 0.21

a What was the relative frequency of green?
b Why is it impossible to know how many cubes were in the bag?

Finding relative frequency

You will need three Multilink cubes and special dice for Exercise 19.2.

- Traffic police need to know the probability of an accident at different road junctions, so that they can suggest changes.
- A railway company needs to know how likely trains are to arrive on time, so it can change its timetable.
- A fairground attendant needs to know your chances of getting a ping-pong ball into a glass jar, so he can make a profit.

In these cases **relative frequency** and not **theoretical probability** is used. Relative frequency can be found either from data that is recorded or by doing an experiment (**experimental probability**).

Exercise 19.2
Relative frequency experiments

1 When a Multilink L-shape is dropped it can come to rest in three positions.

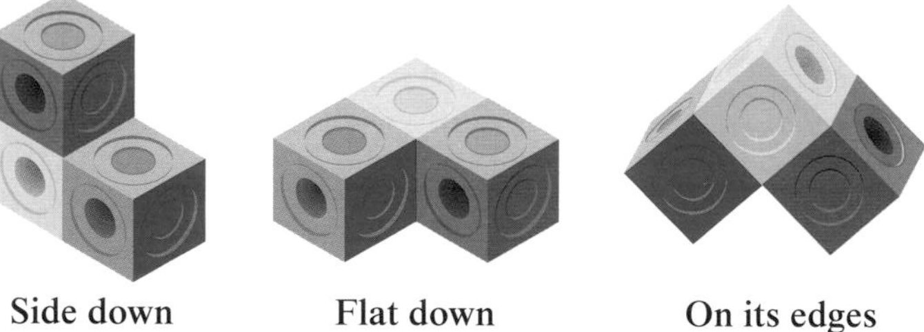

Side down Flat down On its edges

a Make a frequency table like this to record how the shape lands.

Position	Tally	Frequency
Side down		
Flat down		
On its edge		

b Drop an L-shape fifty times and record the outcomes in your table.
c Calculate the relative frequency of the shape landing:
i side down **ii** flat down **iii** on its edge.
d What should the answers to parts **i**, **ii** and **iii** add up to?
e If your L-shape is dropped 231 times how many times would you expect it to land side down?

For Questions **2** to **5** you need three dice with faces marked like this:

Dice A has faces 1, 1, 5, 5, 5, 5
Dice B has faces 3, 3, 3, 4, 4, 4
Dice C has faces 2, 2, 2, 2, 6, 6

2 **a** Roll dice A and B together thirty times and record which dice wins each time.
b What is the relative frequency of dice A winning?

3 **a** Roll dice B and C together thirty times and record which one wins.
b What is the relative frequency of B winning?

Some experiments in probability do not work out quite as you would expect them to.

4 **a** If dice A and C were rolled together, which dice do you think would win most often?
b Do an experiment and record your results.
c What is the relative frequency of A winning?
d Was the result of your experiment what you expected?

5 Roll all three dice together and find the relative frequency that each wins.

Using a relative frequency tree diagram

This diagram shows the relative frequencies of cars leaving a roundabout when they have approached it in the direction of a red arrow.

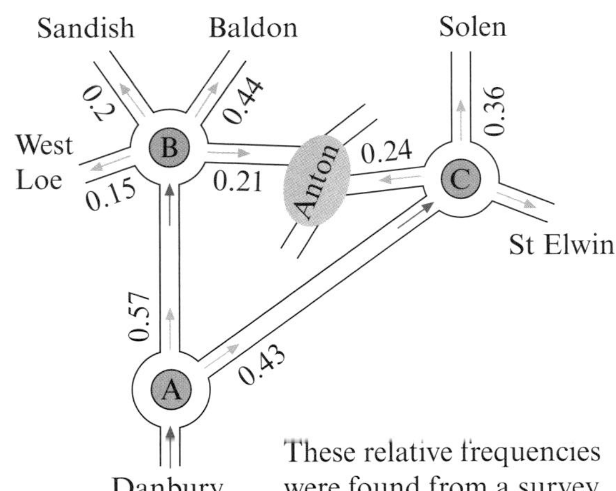

These relative frequencies were found from a survey.

Assume motorists always take the shortest route and that they do not travel on the same road twice.

From this data you can estimate the probability that a car coming from Danbury will take the road to Baldon.

A tree diagram can help to organise the data.

Example Calculate an estimate of the probability that a car coming from Danbury will take the road to Baldon.

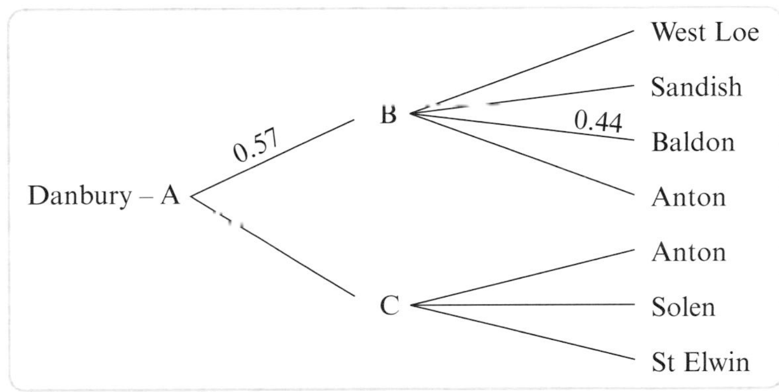

You can multiply the probabilities when you move along the branches of a tree diagram.

An estimate of the probability is $0.57 \times 0.44 = 0.25$ (to 2 dp)

So about 1 car in 4 coming from Danbury takes the Baldon road.

Exercise 19.3
Relative frequency and tree diagrams

1 Calculate an estimate of the probability that a car approaching roundabout C will take the road to St Elwin.
Explain how you worked this out.

2 Copy the tree diagram and fill in all the probabilities.

3 Estimate the probability that a car from Danbury will take:

a the Sandish road **b** the West Loe road
c the Solen road **d** the St Elwin road.

4 Estimate the probability that a car will go from Danbury:

a to Anton via roundabout B
b to Anton via roundabout C
c to Anton by either route.

5 On Wednesday 342 cars approached roundabout A from Danbury.
Estimate the number of these cars that took the road to:

a roundabout B **b** roundabout C **c** Baldon
d Solen **e** Sandish **f** Anton.

Starting points

You need to know about ...

A Using the inequality signs > and <

- These signs can be used with inequalities:

  > stands for **is greater than ...**
  < stands for **is less than ...**

  $x > 5$ means x can have any value greater than 5 but **not 5** itself.
  Here there are an infinite number of values for x.
  This includes non-integer values, such as 15.23 or $8\frac{3}{4}$

- Inequality signs can also be used to order numbers.

  For example: $^{-}23 < {}^{-}4.23 < 1.5 < 37 < 100$
  This can also be written as $100 > 37 > 1.5 > {}^{-}4.23 > {}^{-}23$
  (Note. $^{-}23$ is less than $^{-}4.23$, but 23 is greater than 4.23)

- A range of values can be shown as an inequality.

  $^{-}4 < p < 2$ means p has any value greater than $^{-}4$ but less than 2.
  The numbers $^{-}4$ and 2 are **not** included.
  $^{-}4 < p < 2$ can also be written as $2 > p > {}^{-}4$

- Sometimes you are only interested in the integer values.

  The integer values of x described by the inequality $^{-}3 < x < 5$ are $^{-}2$, $^{-}1$, 0, 1, 2, 3 and 4.

B Sketching linear graphs

- Equations of straight line graphs can be expressed in the form

  $y = mx + c$ where m is the **gradient** of the graph and c is the **y-intercept**

  For example:

  $y = \frac{1}{2}x + 1$ $\qquad$ $y = 8 - 2x$

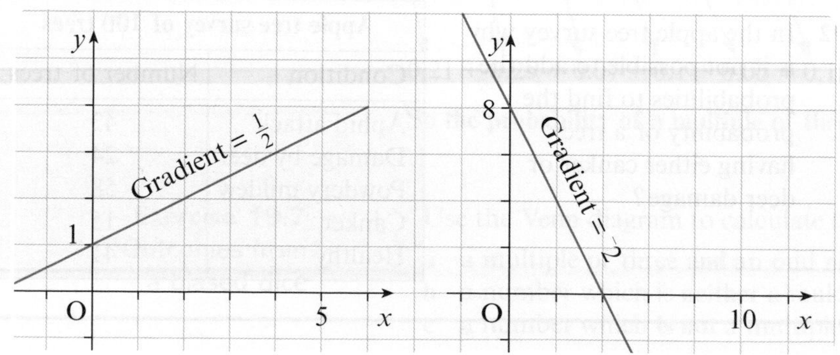

Gradient is $\frac{1}{2}$
y- intercept is 1

Gradient is $^{-}2$
y-intercept is 8

... so try these questions

A1 Replace the ▢ with either > or < to make each of these correct.

a 34 ▢ 12 $\quad$ **b** 2 ▢ 15
c $^{-}12$ ▢ $^{-}56$ $\quad$ **d** 5 ▢ $^{-}2$
e $^{-}2$ ▢ 0 $\quad$ **f** $^{-}17$ ▢ $^{-}5$

A2 Order this set of numbers:
9, 56, $^{-}24$, 5, $^{-}4$, 28, $^{-}30$
a using the sign >
b using the sign <

A3 Which of these numbers is not a possible value for d, where $d < 5$?

12, $^{-}35$, 3.217, $^{-}2$, 5, 4.99, 6.2

A4 What are the integer values of t, where:
a $^{-}6 < t < 3$
b $^{-}1 > t > {}^{-}5$?

A5 Explain why there are no integer values for x where: $^{-}2 < x < {}^{-}1$?

B1 Sketch the graphs of:
a $y = 2x - 2$
b $y = 12 - x$
c $y = 3x$
d $y = {}^{-}4$

Inequalities

- Often a quantity must be kept within a range of values.
  For instance, the speed on a motorway must stay between 30 mph and 70 mph.

> $>$ means **is greater than**
>
> $<$ means **is less than**
>
> $\geqslant$ means **is greater than or equal to**
>
> $\leqslant$ means **is less than or equal to**

The legal speed can be shown in shorthand like this:

$\mathbf{30 < S < 70}$ where S is the speed in mph.

This is called an **inequality**.

This inequality can also be written as $\mathbf{70 > S > 30}$

Since you can also drive at speeds equal to 30 mph and 70 mph this is more accurately shown as:

$\mathbf{30 \leqslant S \leqslant 70}$ or $\mathbf{70 \geqslant S \geqslant 30}$

So all speeds **on or between** 30 mph and 70 mph are legal.

- Sometimes numbers at either end of a range are not included.

Example A firm charges £3.00 each if you buy one switch, £2.50 each if you buy between 1 and 10 switches, and £1.75 each if you buy 10 or more.

For the £2.50 switches this can be written as:

$\mathbf{1 < N < 10}$ where N is the number of switches (1 and 10 are not included).

Exercise 20.1 Writing ranges as inequalities

1 For tomatoes a greenhouse temperature from 45 °F to 80 °F is recommended.
Let T stand for the temperature in °F.
Write the recommended temperature range as an inequality.

2 Write these page ranges as inequalities.

> You sometimes need to choose letters to stand for the variables, such as page number or age.

a Pages 17 to 52 of a book are in colour.
b Pages after 10 and before 64 have photographs.
c The contents finish on page 4, followed by the features pages then the index which starts on page 164. Give the range for the features pages.

3 These are labels for two different drugs.

FERMATOL
For safety reasons this drug must not be used by people older than 65 or younger than 5.

Hypaticain
For safety reasons this drug must not used by the over 65s or by children 5 years and under.

a Write the safe age range for Fermatol as an inequality.
b Write the safe age range for Hypaticain as an inequality.
c What is the difference between the safe age range for each drug?

4 A freezer cabinet must be kept between $^{-}25$ °C and $^{-}18$ °C.
Write this temperature range as an inequality.

> Remember that $^{-}18$ is greater than $^{-}25$.

3 These extracts are from notices or notes.

(A) *Rona Kennedy's exercise bike programme*
You should cycle a distance of between 10 and 20 miles or for at least 15 minutes before breakfast each day.

(B) **PEUGEOT 405 SERVICE NOTE**
Your first service should take place within 6000 miles or 6 months, whichever is the sooner.

(C) *Sungrow Vegetarian Restaurant*
Our party rates apply to groups of between 8 and 20 people. Meal prices range from £6 to £12 a head.

(D) *JENNY'S 15TH BIRTHDAY*
Bike with 14 or more gears. Must cost less than £400.

(E) **Offkit flea spray**
Use only on cats older than 12 weeks and heavier than 2 kilograms.

(F) **Kansas Car Hire**
Prices range from £36 to £62 per day. Maximum five people per vehicle.

(G) **Mike's Supplies**
Dave, we need a scaffold tower next week. It needs to be 15 foot or more tall but must not cost more than £50 per day. Thanks, Mike

(H) **Sandford Superstore**
We are looking for keen sales staff between the ages of 25 and 35 with at least four years experience of vegetarian beef sales.

It is often difficult to tell if the signs $>$ and $<$, or $\geqslant$ and $\leqslant$ fit a situation.

For example, does 'prices up to £40' include a price of exactly £40? It should not, but probably does.

Assume for these extracts that they all mean $\geqslant$ and $\leqslant$.

a For extracts A, B and C
- **i** decide what axes you need and sketch a graph
- **ii** shade the region which meets all the conditions
- **iii** label the other regions on each graph.

b For extracts D to H sketch a graph and shade the matching region. Do not label the other regions.

4 Sketch a graph which **could** describe suitable applicants for the job of a Victorian chimney sweep's assistant. You will need to decide on what axes to use and the inequalities which apply.

In the early 19th Century young boys were sent up chimneys with a hand brush to clean them by unscrupulous chimney sweeps.

The practice was made illegal by parliament in 1833.

Shading regions on graphs

- An inequality such as $\boldsymbol{x > 4}$ **or** $\boldsymbol{^{-}5 < y < 2}$ can be shown as a region on a graph.

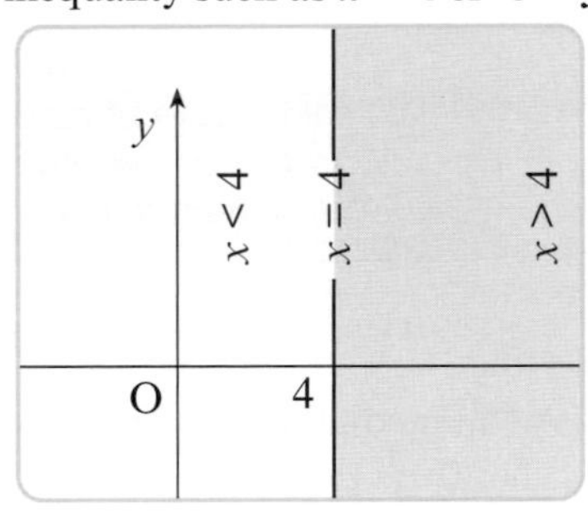

Region shaded ☐ is $x > 4$

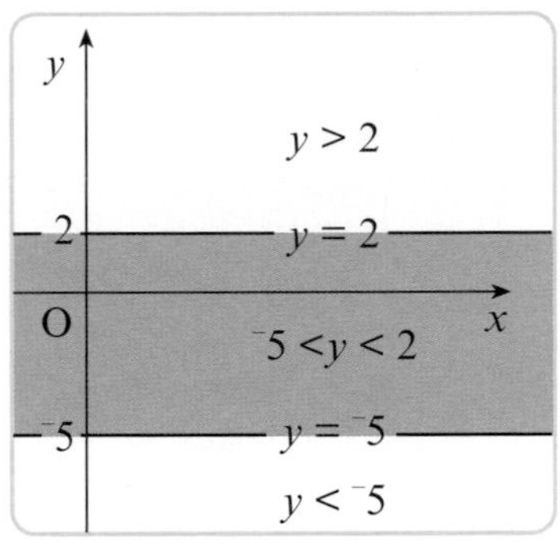

Region shaded ☐ is $^{-}5 < y < 2$

- The conditions satisfied by both inequalities can also be shown on one graph.

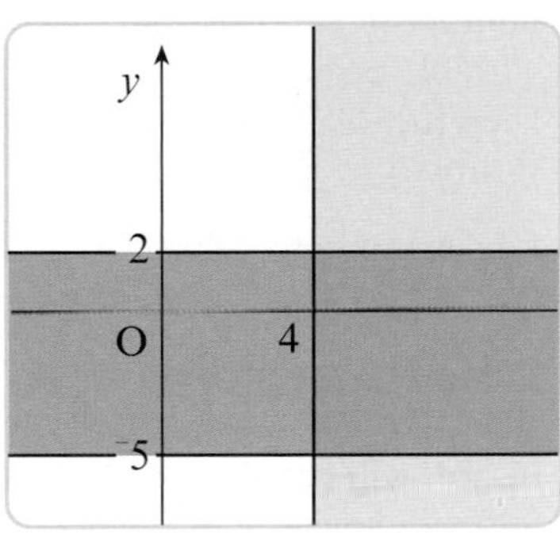

Example

Show the region where $x > 4$ and $^{-}5 < y < 2$.

In this case the shaded region is where the region for $x > 4$ overlaps with the region for $^{-}5 < y < 2$.

This is the part shaded 

Exercise 20.4
Sketching regions

1 Sketch graphs to show these regions:

a $x < 5$ **b** $y > {}^{-}3$ **c** $3 < y < 7$
d $0 < x < 4$ **e** $x < {}^{-}2$ **f** $^{-}4 < y < {}^{-}1$

2 Write an inequality for each shaded region on these graphs.

a

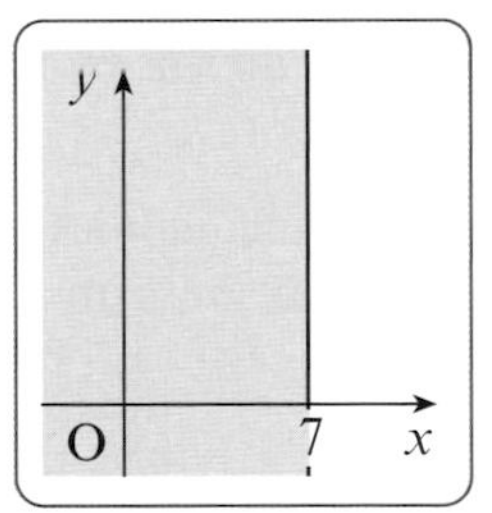

b

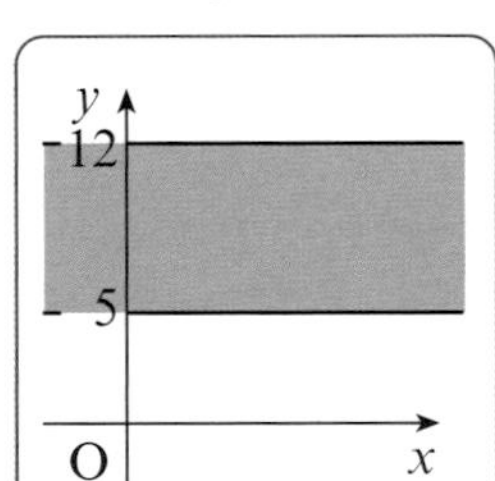

c

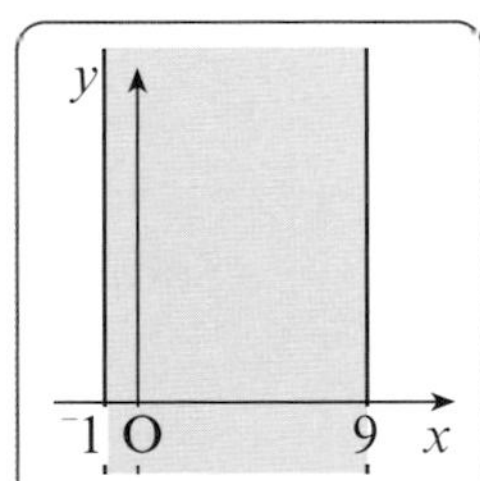

3 Sketch graphs to show the region where

a $y > 5$ and $x > 3$ **b** $y < {}^{-}1$ and $x > 4$
c $^{-}1 < y < 5$ and $x > 2$ **d** $y > 3$ and $2 < x < 8$
e $0 < x < 5$ and $1 < y < 6$ **f** $^{-}2 < y < 6$ and $^{-}5 < x < 0$

4 Give the inequalities which describe the shaded region on this sketch graph.

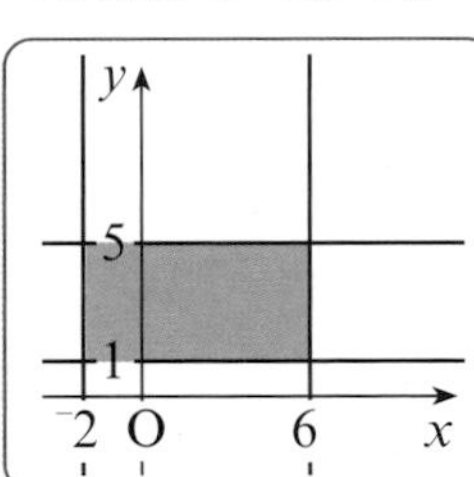

Solving inequalities

- An inequality can have an unknown on both sides – for instance:

> ***Danfield Nurseries***
> We sold 5 trays of roses and 3 single rose plants on Monday.
> On Tuesday we sold 3 trays and 13 single rose plants.
> We sold more rose plants on Monday than on Tuesday.

How many rose plants are on a tray?

As an inequality this can be written as $\mathbf{5x + 3 > 3x + 13}$ where x is the number of rose plants on a tray.

- To solve an inequality you can treat it like an equation.

When you rearrange an inequality try to keep the number in front of the variable positive. (i.e. try to keep the coefficient of x positive).

For example to solve
$3x - 8 \leqslant 5x - 2$

Subtract $3x$ from both sides. Do not subtract $5x$ or you will get $^-2x - 8 \leqslant {}^-2$ and will have the problem of dividing through by $^-2$ and changing the sign.

Example 1

Solve this inequality	$5x + 3 > 3x + 13$
Subtract $3x$ from both sides	$2x + 3 > 13$
Subtract 3 from both sides	$2x > 10$
Divide both sides by 2	$x > 5$

So the number of rose plants on a tray must be greater than 5, i.e. at least 6.

Example 2

Solve this inequality	$3x - 8 \leqslant 5x - 2$
Subtract $3x$ from both sides	$^-8 \leqslant 2x - 2$
Add 2 to both sides	$^-6 \leqslant 2x$
Divide both sides by 2	$^-3 \leqslant x$
So	$x \geqslant {}^-3$

So x can have any values greater than or equal to $^-3$.

The only difference from solving an equation is that if you multiply or divide both sides of an inequality by a negative number the inequality signs will reverse. You can avoid having to do this by keeping the coefficient of x positive.

Exercise 20.5
Solving inequalities

1 Solve each of these inequalities to find the possible values of x. Show each of your answers on a number line.

a $2x > 8$	**b** $5x \leqslant 20$	**c** $9x \leqslant 27$
d $7x \geqslant {}^-28$	**e** $5x \geqslant 12$	**f** $12x \geqslant 108$
g $20x \leqslant 10$	**h** $20 > 4x$	**i** $25x \leqslant 400$
j $150 \leqslant 10x$	**k** $7x \geqslant 49$	**l** $16x \leqslant 64$

2 Solve each of these inequalities. Show each answer on a number line.

a $3y + 6 < 27$	**b** $13 \geqslant s + 5$	**c** $12 > k - 17$
d $5p - 3 \geqslant 27$	**e** $10 < 2b + 3$	**f** $14 < 2x + 8$
g $6a + 50 \leqslant 2$	**h** $5x - 2 > 38$	**i** $6x - 8 < 22$
j $2t + 12 \leqslant 5t$	**k** $4a - 6 \geqslant 22$	**l** $5c - 15 \geqslant 25$
m $13 - 4d > 33$	**n** $3a + 7 \leqslant {}^-32$	**o** $6x + 4 > 40$

3 Solve each of these inequalities. Show each answer on a number line.

a $2a - 5 \geqslant a + 6$	**b** $7k + 4 < 2k - 6$
c $3x + 7 > x - 11$	**d** $4n - 9 \geqslant 2n - 2$

End points

You should be able to … … so try these questions

A Find integer values which satisfy an inequality

A1 For each of these, what integer values of g satisfy the inequality? Show each answer on a number line.

a $^{-}2 < g < 0$ **b** $^{-}5 \leqslant g < 4$
c $7 \geqslant g \geqslant 4$ **d** $^{-}56 \geqslant g > {}^{-}60$
e $^{-}77 < g \leqslant {}^{-}72$ **f** $^{-}3 < g < 4$

B Draw a region to illustrate an inequality

B1 Sketch a graph for Bargain Exhausts.
a Shade in the region that matches these conditions.
b Label the other regions.

B2 Sketch a graph to show the region where $^{-}1 < x < 3$.

B3 Sketch and shade the region on a graph to show where $^{-}1 < x < 2$ **and** $y > 3$.

B4 Write inequalities for the shaded regions in graphs **a** and **b**.

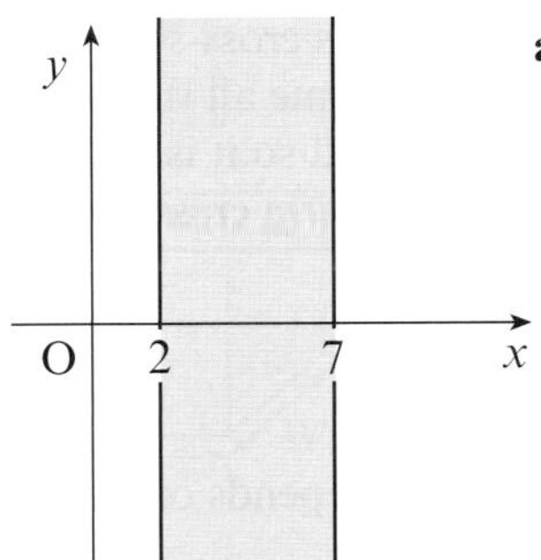

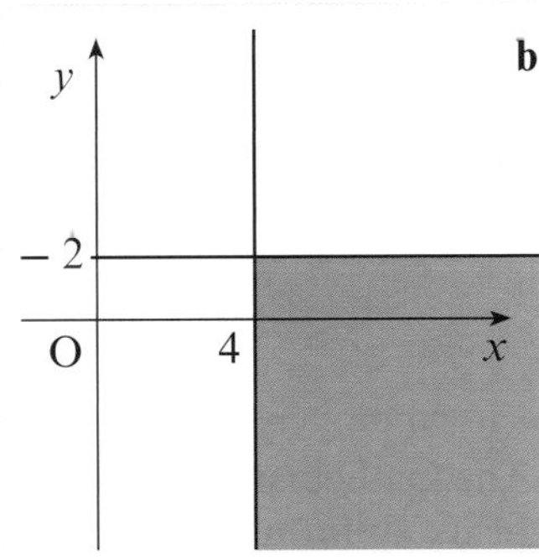

B5 Sketch the graph of $y = 2x - 3$
Shade in the region of your graph where $y < 2x - 3$.

C Solve an inequality to find the range of values for x

C1 Solve these inequalities. Show each answer on a number line.

a $5x \leqslant 30$ **b** $12 < 4f$
c $4a + 5 \geqslant 53$ **d** $5k + 28 > 3$
e $7a - 6 \geqslant 36$ **f** $7d - 10 > 5 - 3d$

Some points to remember

- Always check carefully to see if $\geqslant$ rather than $>$, or $\leqslant$ rather than $<$, is used.
- When you solve an inequality you can treat it like an equation except that:
  - if you multiply or divide both sides by a negative value you must reverse the signs.

2D Views

Showing different views

- A 3D object can be looked at from different directions. Each one may give a different 2D view of the same object.

Example

Sketch views of this object from the directions A, B and C.

The view from A is called the **plan**.

From B the view is called the **front elevation**.

From C the view is called the **side elevation**.

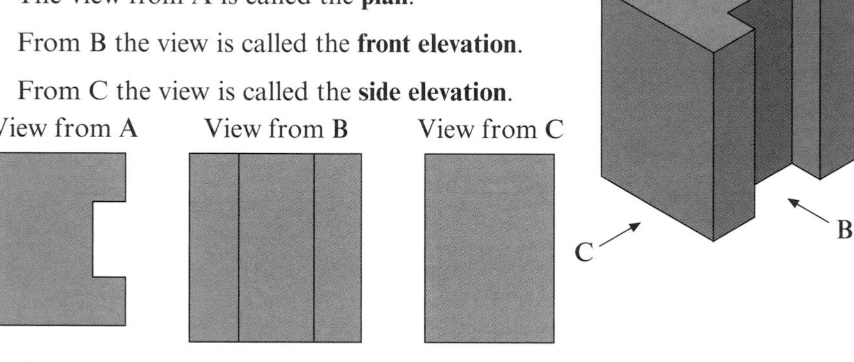

Exercise 21.1
Identifying views

1 Match each view to a direction A, B or C.

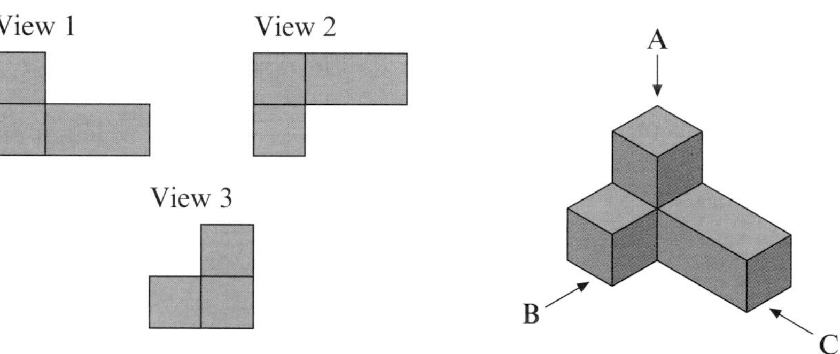

2 Which of these could be views of this church ?

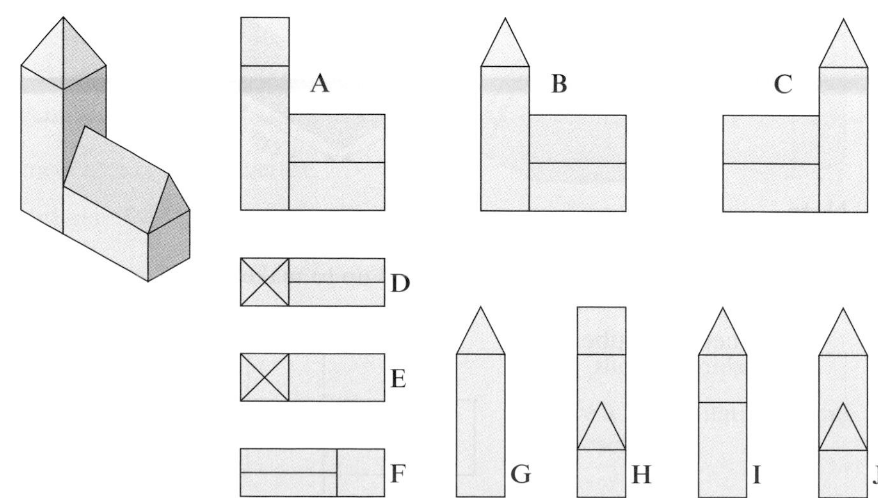

3 Object 1 is made from cubes.
On squared paper, sketch views of the object from the direction of each arrow.

Object 1

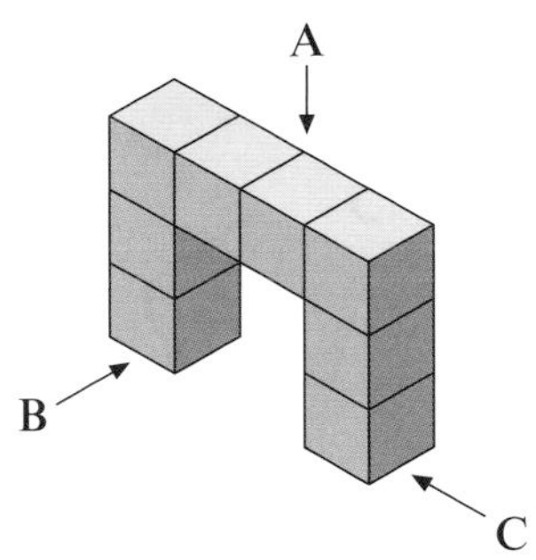

4 **a** How many cubes make up object 2 ?
b On squared paper, sketch the three views of this object.

Object 2

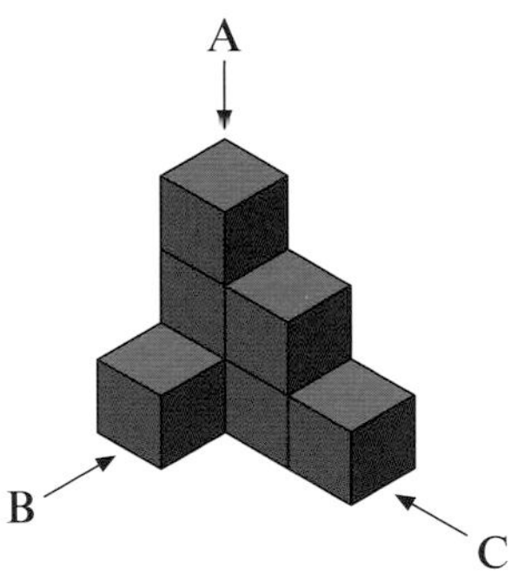

5 **a** If object 3 were made of cubes of this size

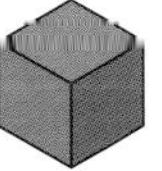

how many would there be ?
b On squared paper, sketch the three views of this object.

Object 3

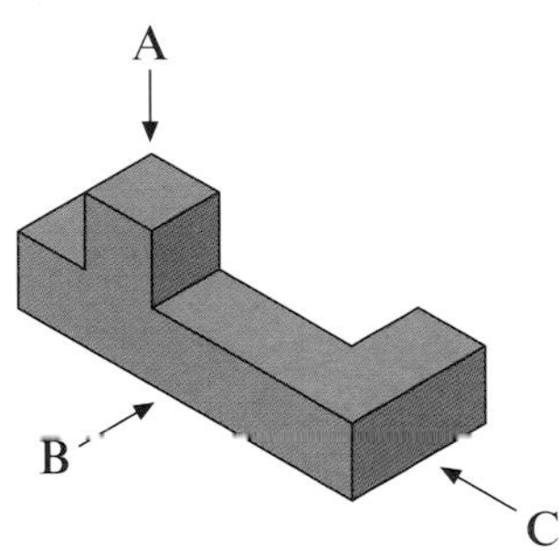

6 Sketch a view of object 4 from each direction A, B and C.

Object 4

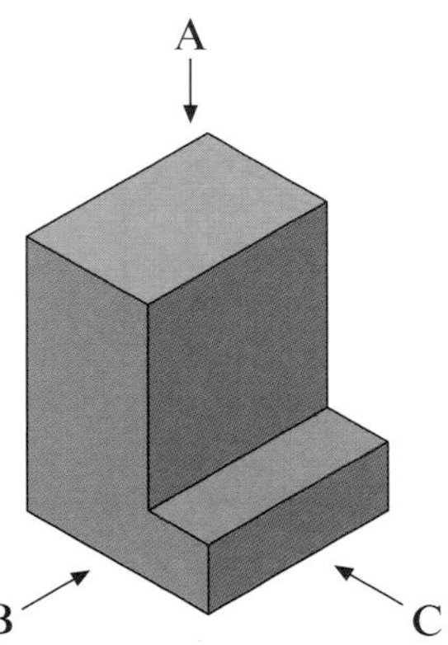

Volume of a prism

The depth of a prism is sometimes called the length or height.

Volume of a prism = Area of a uniform cross-section × Depth

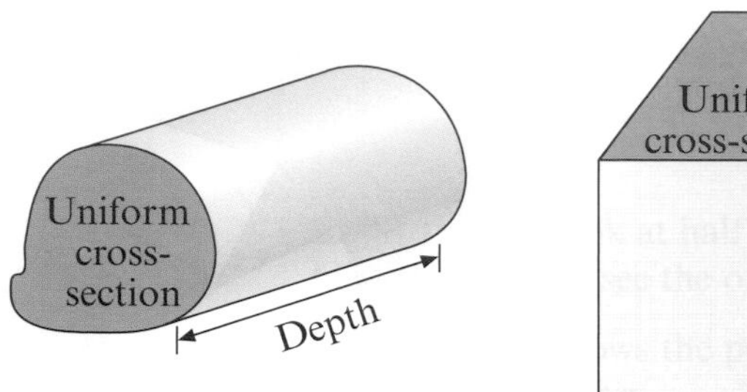

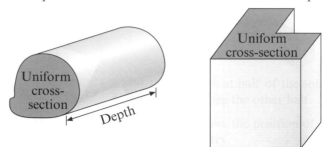

Exercise 21.3
Volume of a prism

1 Calculate the volume of each prism.

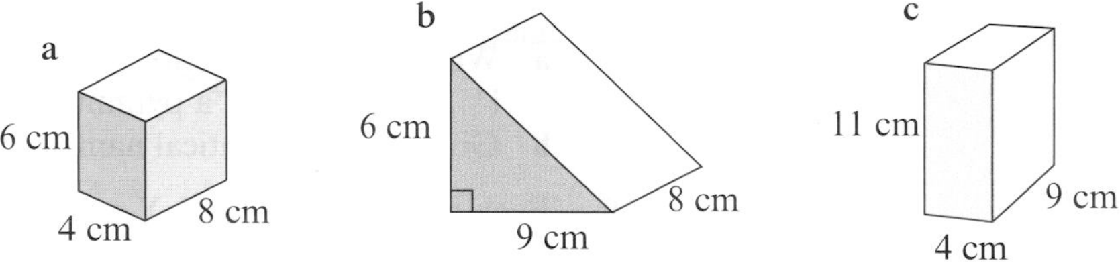

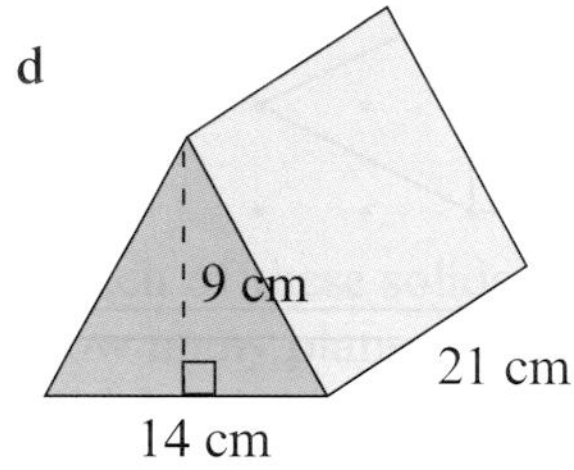

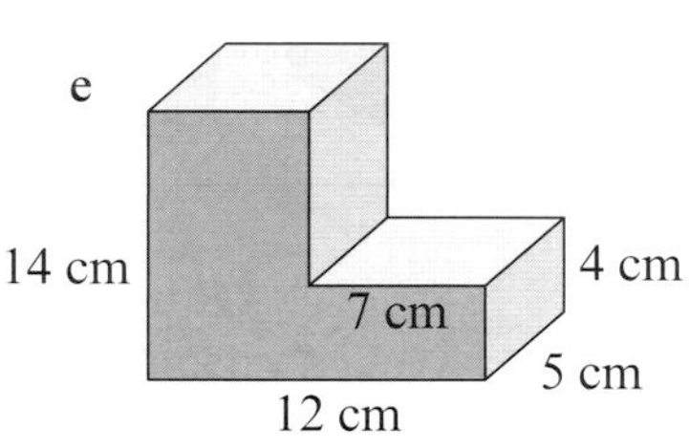

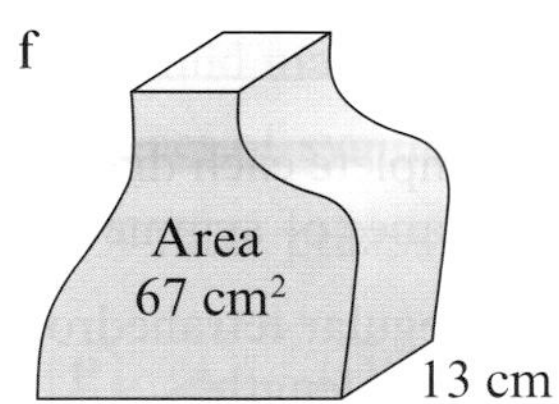

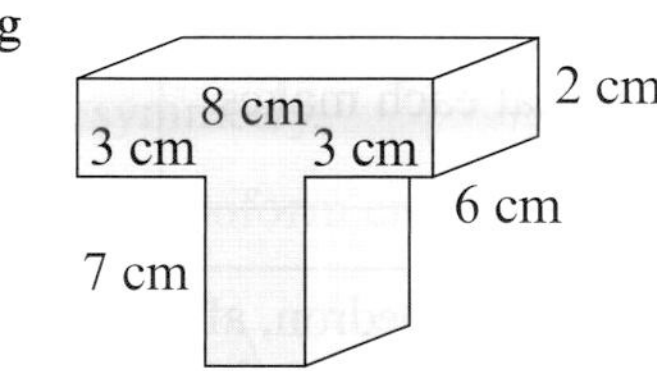

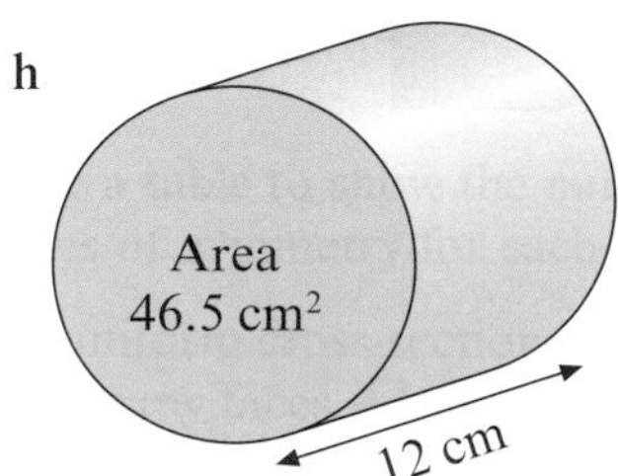

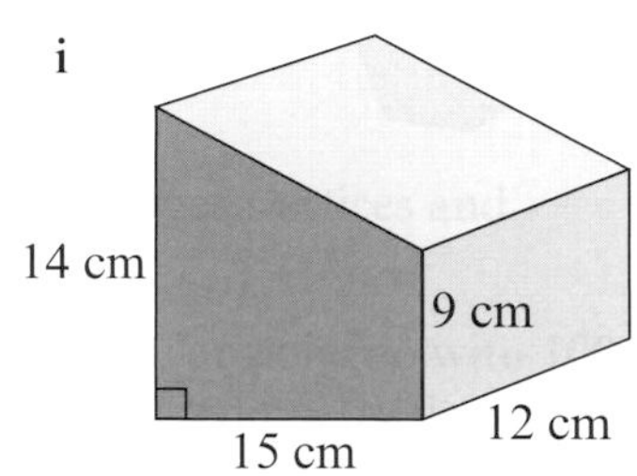

2 This toy is made from a clear plastic cuboid filled with blue liquid. The depth of liquid is 3.5 cm.

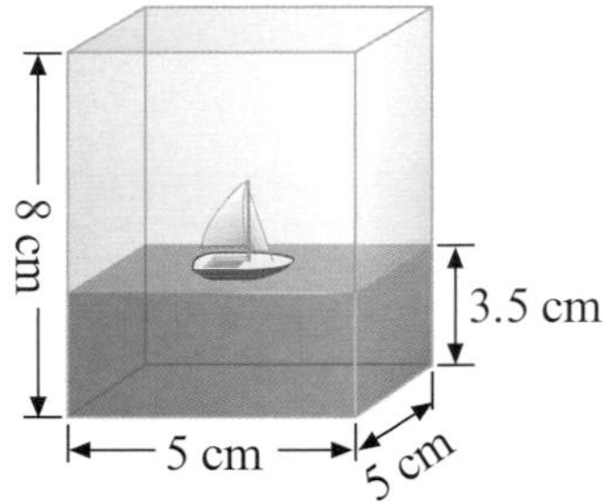

a Calculate the volume of the cuboid.
b Calculate the volume of blue liquid in the cuboid.

3 A box for drawing pins is to be designed in the shape of a cuboid. It is to have a volume of $64\,\text{cm}^3$. This sketch shows one possible box.

Give two other possible sets of dimensions for the box.

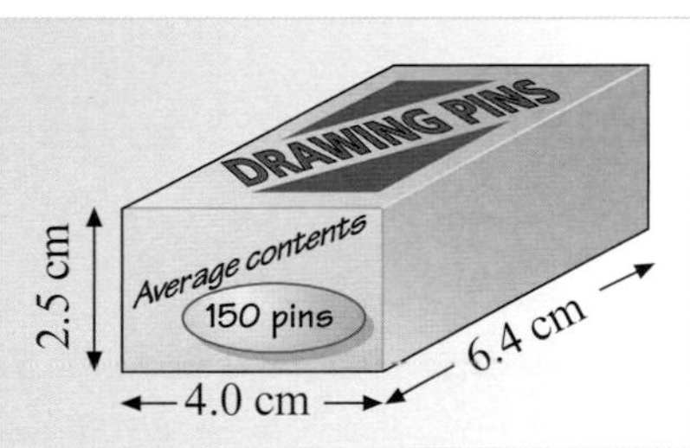

4 A box in the shape of a cuboid is to have a volume of $100\,\text{cm}^3$.

a Give two different sets of dimensions the box might have.
b The manufacturer decides that the box should have a cross-section that is a square.

What might be the dimensions of the box in this case? Explain your answer using a diagram.

5 The diagram shows a triangular prism that is 12 cm long.

The prism has a volume of $756\ \text{cm}^3$.

a Calculate the area of the shaded cross-section.
b Make a sketch of the prism giving possible values for a and b.

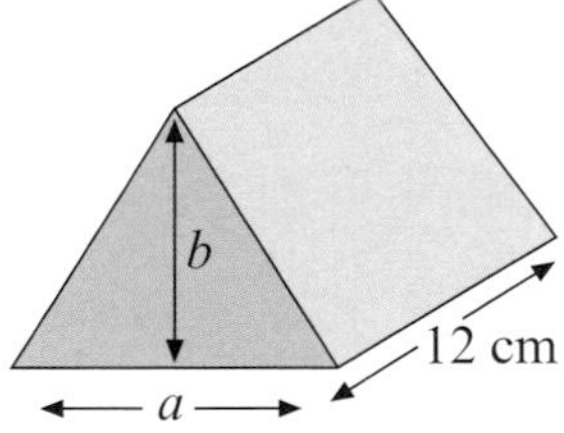

6 A cylinder has a volume of $345\ \text{cm}^3$ and has a height of 15 cm.

a Calculate the area of the cross section of the cylinder.
b Write an expression in π for the radius of the cylinder.

Exercise 21.4
Maximising volume of cylinders

1 This collecting box is in the shape of a cylinder. The diameter of the base is 9.3 cm and its capacity is $1120\,\text{cm}^3$.

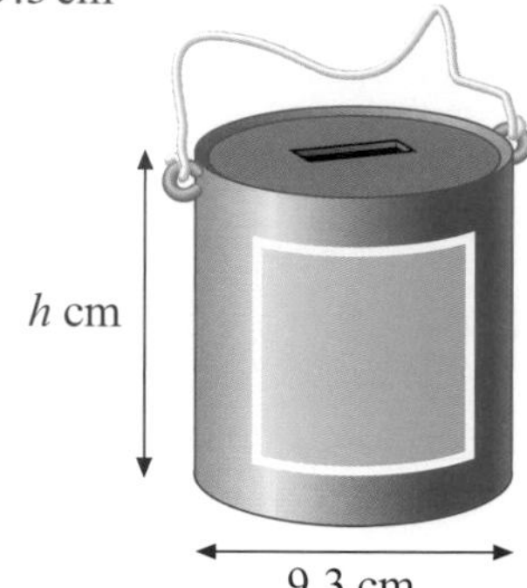

Calculate h correct to 1 dp.

2 A designer in a plastics company is sent this letter.

World Nature Fund – fighting for the environment

18 Westbury Road
LONDON
W2 4ZP

Peterfield Plastics
Unit 381
Cupar Trading Estate
SOUTH KILBRIDE
Lanarkshire

22 August 1996

Dear Ms Barnes,

Our charity has been given a large number of plastic sheets. Each sheet measures 500 mm by 200 mm.

We would like to use these sheets of plastic to make collecting boxes. Our collectors have found that a cylinder is the easiest shape to handle. Of course, we would like the boxes to have the maximum possible volume.

Could you please provide us with your proposed design as soon as possible, including the dimensions of the collecting box.

I look forward to your ideas.

Yours sincerely,

Vita A Green.

a Design a suitable collecting box for the charity.
b What is the diameter and height of your box?
c Write a short report to explain how you decided on these dimensions.
d For your design, what percentage of each sheet of plastic is not used?

The sum of two numbers is found by adding:
for example,
the sum of 2 and 5 is 7.

3 a Sketch two cylinders where the sum of the radius and height is 12 cm.
b Find the volume of each of your cylinders.
c When the sum of the radius and height is 12 cm, what do you think is the maximum possible volume?

4 Choose a different value for the sum of the radius and height of a cylinder and investigate the maximum possible volume.

Exercise 21.5
Volume problems

A new park is planned.
It is to have a children's play area and an open-air swimming pool.

1 This is the design for a sandpit to be dug in the children's play area.
It is in the shape of a cylinder with radius 2.6 m and depth 0.8 m.

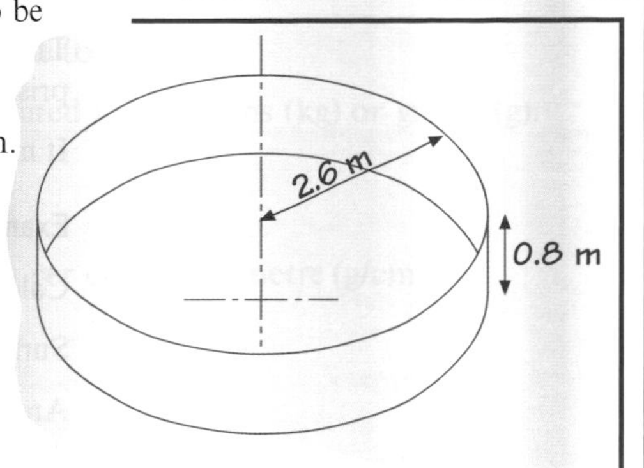

a Calculate the volume of the sandpit in m^3.
b The sandpit is to be filled with sand to a depth of 0.6 m.
Calculate the volume of sand needed.
c The weight of 1 m^3 of the sand is about 1.2 tonnes.
Find the weight of sand needed for the sandpit.

2 This is a plan of the space to be used for swings.

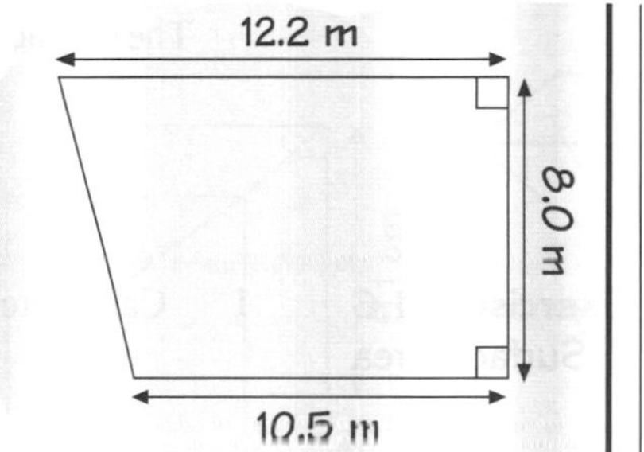

The space is to be dug to a depth of 0.3 m and filled with bark chippings.
a Calculate the volume of bark chippings needed.
b Forestry Products sell bark chippings in bags.
Each bag costs £5.12 and contains 0.07 m^3 of chippings.
i How many bags of chippings should be bought?
ii What is the cost of these bags of chippings?

3 The diagram shows the uniform cross-section of the open-air swimming pool.

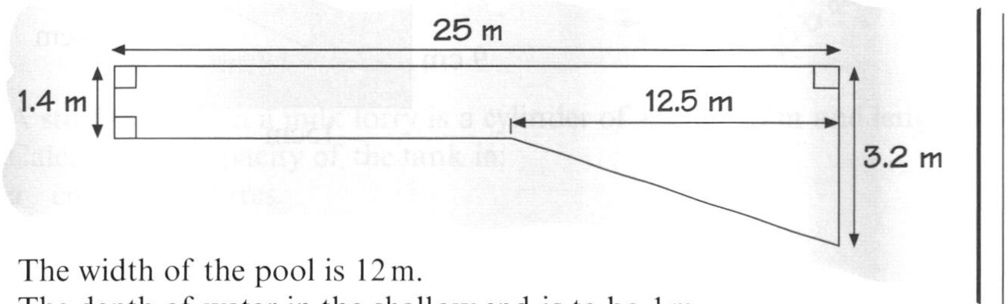

The width of the pool is 12 m.
The depth of water in the shallow end is to be 1 m.
a Calculate the area of the cross-section.
b What is the capacity of the swimming pool?
c Calculate the volume of water in the pool.
d The amount of chlorine added to the water in this pool is 1 cubic centimetre (cm^3) per cubic metre (m^3) of water.
How much chlorine will be added to the water in this pool?

Solving quadratic equations by factorising

A quadratic equation is:

- an equation with one variable (only one letter, e.g. n)
- an equation where the highest power of the variable is 2 (e.g. n^2)

$n^2 + 3n - 4 = 0$ $\quad$ $3(a^2 + 1) = 0$ $\quad$ $2v^2 + 3v - 40 = 0$

are all examples of quadratic equations.

> One way to think of a value that will satisfy an equation, is to think of a value for the variable that makes the equation true.
>
> The value of p that satisfies the equation
> $3p = 12$
> is 4
>
> The solution to the equation
> is $p = 4$
> and
> $p = 4$ satisfies $3p = 12$

An equation is solved when you find a value, or values for the variable that satisfy the equation.

> One way to solve a quadratic equation is to factorise it.
>
> **Example** To solve $n^2 + 3n - 4 = 0$
>
> $n^2 + 3n - 4$ factorises to give: $(n + 4)(n - 1)$
>
> So $\quad (n + 4)(n - 1) = 0$
>
> That means: $\quad$ either $(n + 4) = 0$ ... (as $0 \times (n - 1) = 0$)
> $\quad$ or $(n - 1) = 0$ ... (as $(n - 4) \times 0 = 0$)
>
> For $(n + 4) = 0$ $\quad$ n must have a value of $^{-}4$
> For $(n - 1) = 0$ $\quad$ n must have a value of $^{+}1$
>
> So $\quad n^2 + 3n - 4 = 0$ has two values for n that satisfy it.
> The values of n are: $n = {}^{-}4$ or $n = {}^{+}1$

The quadratic equation $n^2 + 3n - 4 = 0$ has two solutions $n = {}^{-}4$ or $n = {}^{+}1$.
All quadratic equations have no more than two solutions.

Exercise 22.1
Solving quadratic equations by factorising

1 Factorise and solve these quadratic equations.

a $n^2 + 13n - 14 = 0$	**b** $n^2 - 5n - 14 = 0$	**c** $n^2 - 13n - 14 = 0$
d $b^2 + 2b - 15 = 0$	**e** $b^2 + 14b - 15 = 0$	**f** $b^2 - 2b - 15 = 0$
g $a^2 + 7a - 30 = 0$	**h** $a^2 + 13a + 30 = 0$	**i** $a^2 + a - 30 = 0$
j $x^2 + 4x - 5 = 0$	**k** $x^2 + 6x + 5 = 0$	**l** $x^2 - 6x + 5 = 0$
m $y^2 - 8y + 15 = 0$	**n** $y^2 - 16y + 15 = 0$	**o** $y^2 - 14y - 15 = 0$
p $k^2 + 2k - 63 = 0$	**q** $k^2 - 16k + 63 = 0$	**r** $k^2 + 62k - 63 = 0$

2 Match each quadratic equation with a pair of factors from List A, and solve the equation.

a $d^2 + 3d - 10 = 0$
b $d^2 - 5d - 6 = 0$
c $d^2 + 3d - 40 = 0$
d $d^2 - 8d - 33 = 0$
e $d^2 - 14d + 33 = 0$

List A	
$(d + 8)(d - 5)$	$(d - 11)(d + 3)$
$(d - 2)(d + 5)$	$(d - 11)(d - 3)$
$(d + 4)(d - 10)$	$(d + 1)(d - 6)$

> Factorising can be used to solve a quadratic equation when one side is equal to 0.
>
> So, you might need to rearrange an equation before you factorise.
>
> To solve $x^2 + 4x = 5$
> rearrange to $x^2 + 4x - 5 = 0$
> Now factorise and find the solution or solutions.

3 **a** Factorise $x^2 - 8x + 16$.
b Explain why $x = {}^{+}4$ is the only solution to $x^2 - 8x + 16 = 0$.

4 For each of these equations:

- rearrange
- factorise
- find two solutions

a $c^2 + 5c = 6$	**b** $y^2 = 7 - 6y$	**c** $p^2 + 9 = 6p$
d $x^2 = 3x + 10$	**e** $w^2 = 4w - 4$	**f** $b^2 + 14 = 15b$
g $k^2 - 3k = 70$	**h** $v^2 + 6v = {}^{-}9$	**i** $t^2 = 3t + 28$
j $y^2 - 7y = 18$	**k** $g^2 - 8g = {}^{-}16$	**l** $a^2 = 21 - 4a$
m $p^2 - 4 = 3p$	**n** $w^2 = 2w + 24$	**o** $u^2 + 10 = {}^{-}7u$

Solving quadratic equations by trial and improvement

The trial-and-improvement method can be used to find a solution to a quadratic equation.

- You can use this method if you find an equation difficult to factorise.

Example

Find a solution to this equation $x^2 - 3x = 8$

	$x^2 - 3x = 8$	
Try $x = 4$	$4 \times 4 - 3 \times 4 = 4 \neq 8$	($\neq$ means *does not* equal)
Try $x = 5$	$25 - 15 = 10 \neq 8$	
Try $x = 4.5$	$20.25 - 13.5 = 6.75 \neq 8$	
Try $x = 4.6$	$21.16 - 13.8 = 7.36 \neq 8$	
Try $x = 4.7$	$22.09 - 14.1 = 7.99$	
Try $x = 4.71$	$22.18 - 14.13 = 8.05 \neq 8$	(too big)

As an exact value for x is often not possible by trial and improvement you have to accept an approximate answer.

So for $x^2 - 3x = 8$ $\quad x = 4.7$ is a solution (correct to 1 dp)

You can, of course, try any value you like for x.
You must remember that trial and improvement is not like guessing one value, and then another, and so on.

Be systematic with the values you try. It might take a little longer than a lucky guess, but in the end you will find a solution.

Exercise 22.2
Solving quadratic equations by trial and improvement

1 Find a value for x from these quadratic equations, by trial and improvement.

a $x^2 + 2x = 5$ **b** $x^2 - 2x = 6$ **c** $x^2 + 3x = 5$
d $x^2 + 4x = 2$ **e** $x^2 - 3x = 1$ **f** $x^2 + 5x = 3$
g $x^2 - x = 3$ **h** $x^2 + 6x = 2$ **i** $x^2 + 4x = 8$

2 Find by trial and improvement a value for x that satisfies the equation

$x^2 + 4x - 7 = 0$

3 Find a value for x that satisfies the equation

$x^2 - 5x = 2$

Give your answer correct to 1 dp.

4 Find a solution to each equation by trial and improvemement.

a $2x^2 - x = 5$ **b** $3x^2 + x = 7$ **c** $2x^2 - 3x = 4$

5 Explain why trial and improvement is not a good method to find a value for x that satisfies the equation:

$x^2 + 4x - 5 = 0$

8 **a** Copy and complete this table of values for the graph of $y = x^2 + 2x - 3$.

x	$^-4$	$^-3$	$^-2$	$^-1$	0	1	2	3
x^2	16	☐	4	1	☐	1	☐	9
$+2x$	−8	−6	☐	−2	0	2	4	☐
−3	−3	−3	−3	−3	☐	−3	−3	−3
y	5	0	$^-3$	$^-4$	$^-3$	☐	5	☐

b Explain any symmetry you can find in your table of values.

c Draw axes with:
values of x from $^-4$ to $^+3$, and values of y from $^-5$ to $^+25$.

d **i** Plot the points from your table of values.
ii Draw and label the graph of $y = x^2 + 2x - 3$.

9 **a** Copy and complete this table of values for the graph of $y = x^2 - 2x - 3$ for values of x from $^-4$ to $^+3$.

> The value of $-2x$ is calculated by : $^-2 \times$ (the value of x)
>
> So $^-2 \times {}^-4 = {}^+8$
> $^-2 \times {}^-3 = {}^+6$
> so on.

x	$^-4$	$^-3$	$^-2$	$^-1$	0	1
x^2	16	9	4	1	0	1
$-2x$	+8	+6	+4	+2	0	−2
−3	−3	−3				
y	21	12				

b On the same axes as Question **8**, draw the graph of $y = x^2 - 2x - 3$.

c Compare your graphs of $y = x^2 + 2x - 3$ and $y = x^2 - 2x - 3$
i How are your graphs the same?
ii How are your graphs different?

10 **a** Draw up a table of values for the graph of $y = x^2 + 3x - 4$ with values of x from $^-4$ to $^+4$.

b Draw a pair of axes with:
values of x from $^-4$ to $^+4$, and values of y from $^-10$ to $^+25$.

c On your axes, draw a graph of $y = x^2 - 3x - 4$.

d Give the coordinates of the points where your graph crosses the x-axis.

11 **a** Draw up a table of values for the graph of $y = x^2 - 3x - 4$ with values of x from $^-4$ to $^+4$.

b Draw a graph of $y = x^2 - 3x - 4$.

c Give the coordinates of the points where your graph crosses the x-axis.

12 **a** Draw up a table of values for the graph of $y = x^2 - 3x - 10$ with values of x from $^-5$ to $^+5$.

b From your table of values predict where the graph of $y = x^2 - 3x - 10$ will cross the x-axis.
Explain your prediction.

c Draw a pair of axes with:
values of x from $^-5$ to $^+5$, and values of y from $^-15$ to $^+30$.

d Draw the graph of $y = x^2 - 3x - 10$.

13 **a** Draw up a table of values for the graph of $y = x^2 + 2x - 8$, with values of x from $^-5$ to $^+5$.

b Draw the graph of $y = x^2 + 2x - 8$.

c Give the coordinates where $y = x^2 + 2x - 8$ crosses the x-axis.

Solving quadratic equations graphically

This is the graph of $y = x^2 + 5x - 6$, for values of x from $^-8$ to $^+6$.

$y = 0$ is the equation of the x-axis.

We can use the graph to solve the equation $x^2 + 5x - 6 = 0$.

The graph of $y = x^2 + 5x - 6$ crosses the graph of $y = 0$ where:

$x = {}^-6$ and where $x = 1$

At $x = {}^-6$, and $x = 1$ the two equations must be equal.
So, we can write:

$$x^2 + 5x - 6 = 0$$

From the graphs, the values of x that satisfy the equation are:
$x = {}^-6$ or $x = 1$

You can be sure of these solutions, as the values of x from the graph are both integer values.

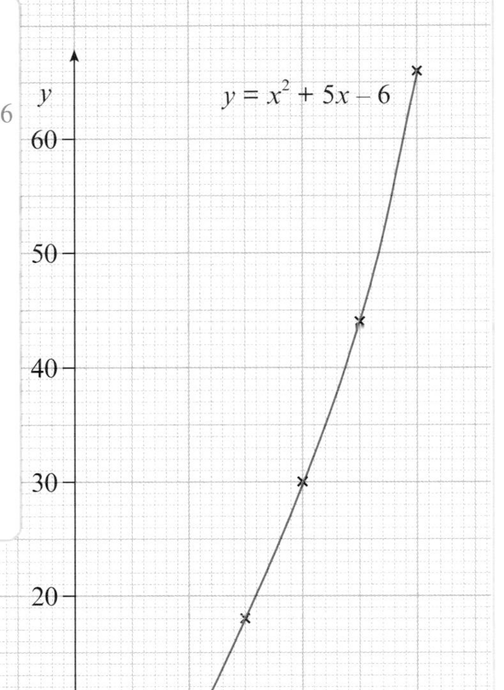

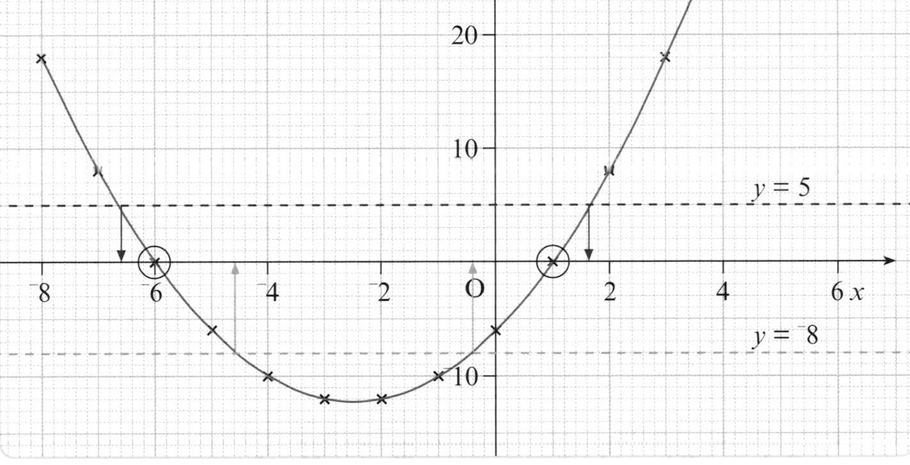

These values of x, from the graph are not integer values. The values are not exact, they are a good approximation. They answer the question, from the graph.

For more accurate solutions, use the values from the graph as a starting point for the trial-and-improvement method.

The graph of $y = x^2 + 5x - 6$ can be used to solve many more equations.

Example From the graph solve $x^2 + 5x - 6 = 5$.
The graph of $y = x^2 + 5x - 6$ crosses the graph of $y = 5$
so we can say: $x^2 + 5x - 6 = 5$
The two equations are equal where $x \approx {}^-6.6$ and where $x \approx 1.6$

The solutions for $x^2 + 5x - 6 = 5$ are: $x \approx {}^-6.6$ and $x \approx 1.6$

Example From the graph solve $x^2 + 5x - 6 = {}^-8$.
The graph of $y = x^2 + 5x - 6$ crosses the graph of $y = {}^-8$
so we can say: $x^2 + 5x - 6 = {}^-8$

The solutions for $x^2 + 5x - 6 = {}^-8$ are: $x \approx {}^-4.6$ and $x \approx {}^-0.4$

C Types of question

- You can use different types of questions on a questionnaire:
  - multi-choice questions

> **7** What do you sleep on?
> *Please tick one box only* ☐ Back ☐ Front ☐ Side

  - multi-choice questions with a scale

> **8** How heavy a sleeper are you? HEAVY ——— LIGHT
> *Please tick one box only*
>
> | Very | Fairly | Average | Fairly | Very |
> |---|---|---|---|---|
> | ☐ | ☐ | ☐ | ☐ | ☐ |

  - branching questions

> **9** Do you suffer from regular sleepless nights?
> *Please tick one box only* ☐ Yes ☐ No
> *If your answer is* NO *then go to Question 12*

  - questions with more than one answer.

> **12** What helps you get a good night's sleep?
> *Put a 1 in the box for the most helpful, a 2 for the next most helpful, and so on*
> ☐ A hot drink just before bed
> ☐ Eating just before bed
> ☐ Exercise during the evening
> ☐ Relaxation breathing
> ☐ Other *(please state)* ______

C1 Design a questionnaire about sleep which includes different types of question.

D Using a scatter diagram

- You can use a scatter diagram to investigate if there is a link between two sets of data.

Question – does the time taken to get to sleep depend on the amount of light in the room?

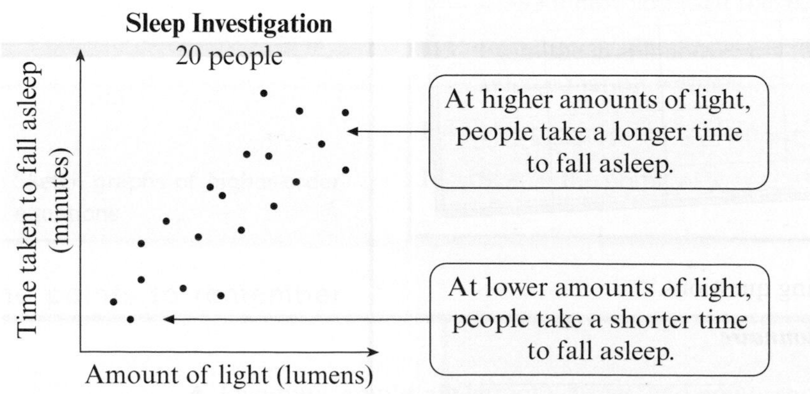

The time taken to fall asleep does depend on the amount of light: as the amount of light increases, the time taken to fall asleep also increases.

D1

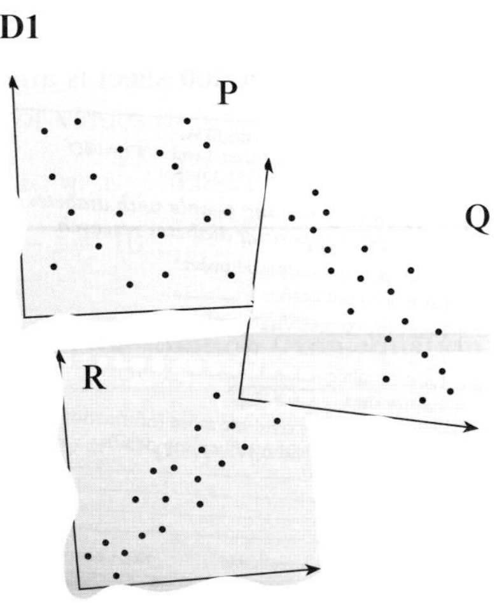

Match each scatter diagram to each pair of sets of data.

a	Time taken to get to sleep	against	Time since drank coffee
b	Amount of sleep	against	Room temperature
c	Amount of sleep	against	Height of bed off floor

Designing and criticising questions

- When you create a questionnaire, your questions must be carefully designed to:
  - make them easy to answer
  - make sure the answers give you the data you need.

Poor question		Improved question
How much sleep did you get last night? ☐ Less than average ☐ About average ☐ More than average	This question is **not clear**: the words used need to be more exact. (Different people are likely to have different ideas of what is meant by 'average'.)	How much sleep did you get last night? ☐ Less than 8 hours ☐ About 8 hours ☐ More than 8 hours
Do you agree that we need at least 8 hours sleep each night?	This is a **leading** question: it leads people into giving a certain answer. (The question seems to expect the answer 'Yes'.)	Do you think we need at least 8 hours sleep each night? ☐ Yes ☐ No ☐ Not sure
What do you sleep on?	This question is **ambiguous**: it could have more than one meaning. (The question is meant to be about sleeping position, but could be answered 'a bed'!)	What do you sleep on? ☐ Back ☐ Front ☐ Side

Exercise 23.1
Designing and criticising questions

1 **a** Explain why this question is not clear.
b Write an improved question.

> When do you usually go to bed?
> ☐ Early ☐ Late

2 **a** Explain why this is a leading question.
b Write an improved question.

> You get a worse night's sleep on a soft bed, don't you?

3 **a** Explain why this question is ambiguous.
b Write an improved question.

> Where do you sleep best?

4

> ***Leisure Centre Survey***
> 1 Do you agree that the town needs a new leisure centre? ☐ Yes ☐ No
> 2 Would you be a frequent user of the centre? ☐ Yes ☐ No
> 3 Would you use the courts? ☐ Yes ☐ No
> 4 How much would you be prepared to pay to use the pool? ☐ Less than £1.50 ☐ More than £2.50

This questionnaire has been written to survey local people about a new leisure centre.

a Criticise each of the questions.
b Write an improved question for each one.

- A survey asks people to give an opinion about something, or asks about facts which are easy to remember.

Example

TV Survey

1 What is your favourite TV channel? ☐ BBC1 ☐ BBC2 ☐ ITV ☐ Channel 4 ☐ Channel 5

2 Did you watch TV last night? ☐ Yes ☐ No

- The data needed for some investigations can only be collected:
  - over a period of time

    How much time do you spend in a week watching each TV channel?

  - by designing an experiment.

    People take longer to get to sleep the more light there is in the room.

The data collection sheet used for these types of investigation can be called an **observation sheet**.

Exercise 23.2
Experiments

To design your experiment:
- decide what data you need
- decide how to collect it
- design an observation sheet.

1 Design an observation sheet to collect data on how much time people spend in a week watching each TV channel.

2 Design an experiment to test this hypothesis.

Body Matters
Your waist is roughly two times the distance around your neck.

3 **a** Carry out your experiment.
b Analyse the data you collect.

4 Do you think the hypothesis is true or false? Explain why.

5 Design an experiment to answer this question.

Body Matters
How many times do people blink in a day?

6 **a** Carry out your experiment.
b Analyse the data you collect.
c Interpret your results to answer the question.

7 Design an experiment to test this hypothesis.

Body Matters
Taller people do not have as good a sense of balance as shorter people.

8 **a** Carry out your experiment.
b Analyse the data you collect.

9 Do you think the hypothesis is true or false? Explain why.

Correlation

- You can describe the link between two sets of data using the term **correlation**.

Sleep Experiment 1
Does the length of time you take to fall asleep depend on how light the room is?

These results show **positive** correlation: an increase in one set of data tends to be matched by an increase in the other set.

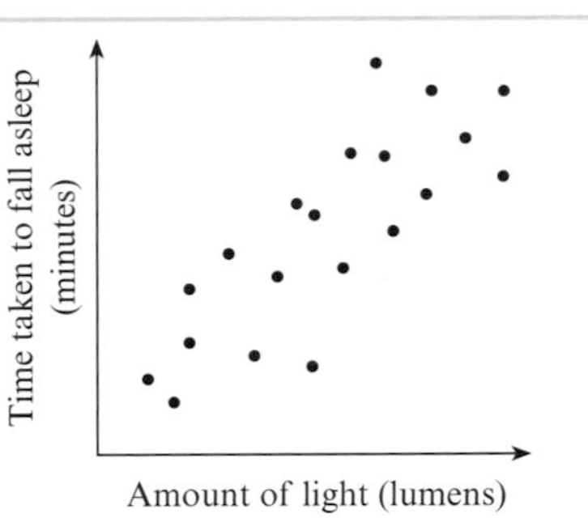

The result of experiment 2 may not be what you expect. It happens because sleep is part of your daily rhythm of sleeping and waking.

Going to bed late means that you will soon reach your time for waking, and vice versa.

A daily rhythm, like this sleep/wake example, is called a **circadian rhythm**.

Sleep Experiment 2
Does the length of time you sleep depend on the length of time since you last slept?

These results show **negative** correlation: an increase in one set of data tends to be matched by an decrease in the other set.

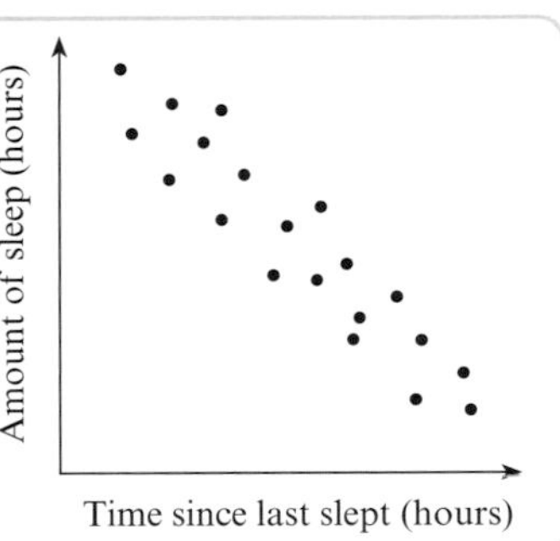

Exercise 23.3 Correlation

1

Mean semi-detached house prices in towns near London – 2nd Quarter 1996

Distance from London (miles)	44	66	41	75	86	47	68	36	62	77	57	53
Mean house price (£000's)	93	78	98	72	63	97	71	104	86	64	78	88

a Draw these axes: horizontal 0 to 90, vertical 50 to 130.
b Plot the house price data on your diagram.
c Is the correlation positive or negative?

A negative number of dioptres shows short-sightedness; a positive number of dioptres shows long-sightedness.

2

Eye Tests for 10 people

Pressure in eye (mmHg)	12.1	11.7	15.2	19.1	11.2	18.9	15.9	17.3	13.0	16.6
Refractive power of lens (dioptres)	3.6	¯3.9	5.1	10.4	¯6.4	3.0	¯6.9	6.5	¯8.4	0.8

a Draw these axes: horizontal 10 to 20, vertical ¯12 to 12
b Plot the eye test data on your diagram.
c Is the correlation positive or negative?

This is called drawing a line **by eye** or **by inspection**.

3 For each of your scatter diagrams, draw a line through the middle of the plots, like this:

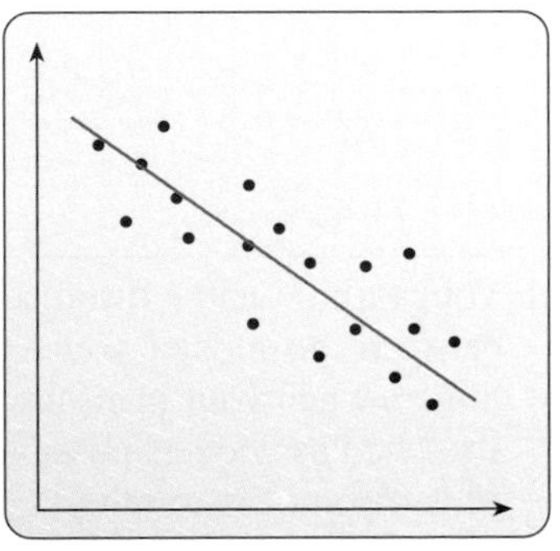

4 Which scatter diagram did you find it easier to draw the line on? Explain why.

Using a line of best fit

- A line drawn through the middle of the plots on a scatter diagram is called a **line of best fit**. The stronger the correlation, the easier it is to draw this line.

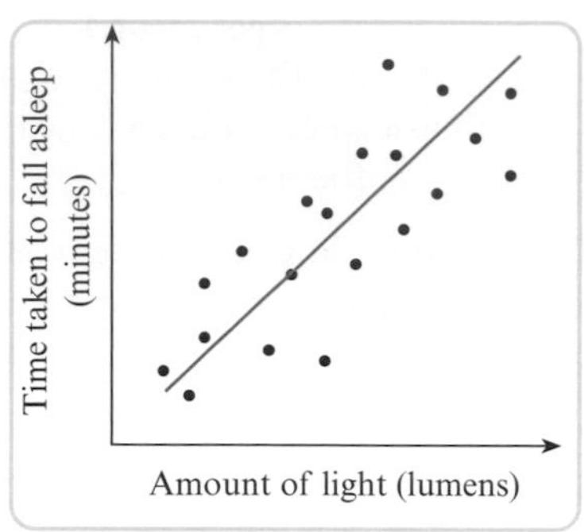

This is **moderate** positive correlation because the plots are well scattered around the line of best fit.

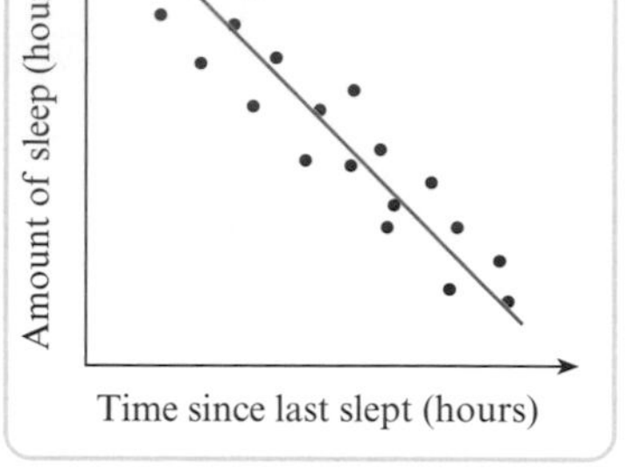

This is **strong** negative correlation because the plots are quite close to the line of best fit.

- When it is not possible to draw a line of best fit, there is no link between the two sets of data: there is **no correlation**.

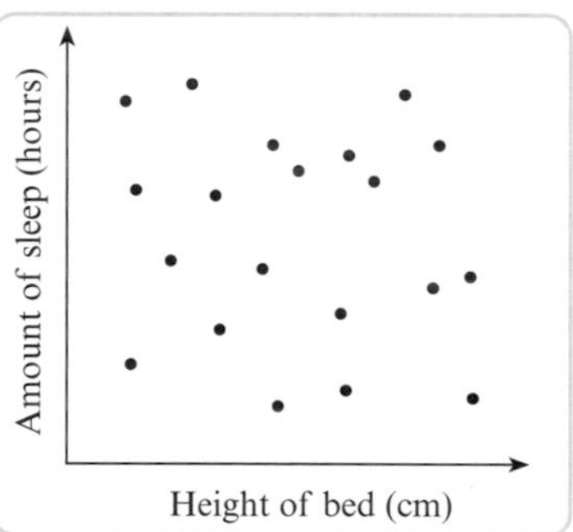

Exercise 23.4
Describing correlation

1 Use this scatter diagram to describe the correlation between income and percentage of income given to charity.

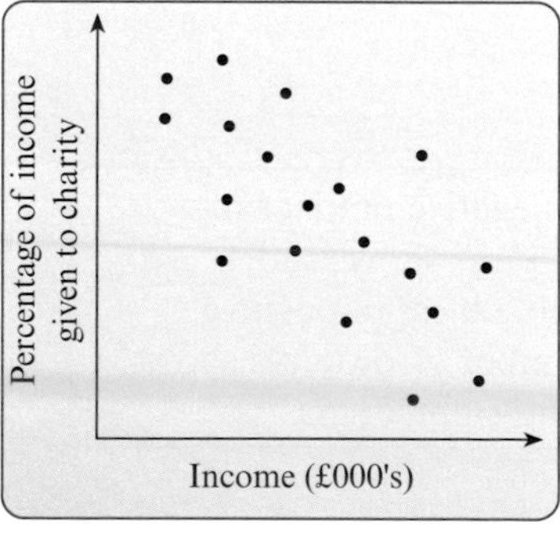

2 Use your scatter diagram from Exercise 23.3 Question **1** to describe the correlation between distance of town from London and mean house price.

3 Use your scatter diagram from Exercise 23.3 Question **2** to describe the correlation between pressure in eye and refractive power of lens.

4 Design an experiment to answer this question:
'Is there any correlation between your fathom and your height?'

> Your fathom is the distance between the ends of your fingers when your arms are stretched as wide as possible. This distance is roughly six feet for an adult.

5 **a** Carry out your experiment.
b Plot the data you collect on a scatter diagram.
c Draw a line of best fit.
d Use your scatter diagram to describe any correlation.

Estimating values from a line of best fit

In part **a** you are estimating within the range of data. This is called **interpolation**. In part **b** you are estimating outside the range of data. This is called **extrapolation**, and is less reliable

- It is possible to estimate values from a line of best fit.

Example

a Estimate the height of a person with head circumference 56 cm.
b Estimate the head circumference of a person 195 cm tall.

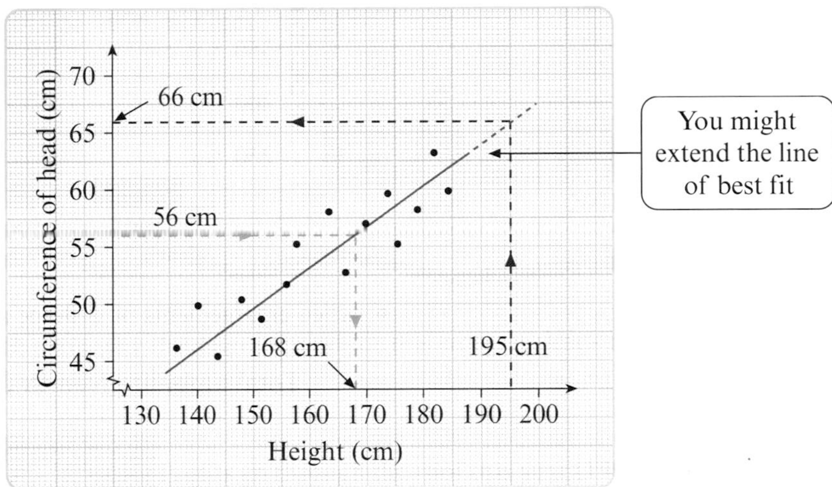

You might extend the line of best fit

a Estimated height of person with head circumference 56 cm = **168 cm**
b Estimated head circumference of person 195 cm tall = **66 cm**

Exercise 23.5
Estimating values

Use your scatter diagram from Exercise 23.3 Question **1** for Questions **1** to **3**.

1 Estimate the distance from London of a town with a mean house price of:
a £90 000 **b** £65 000

2 Estimate the mean house price for a town:
a 70 miles from London **b** 55 miles from London

3 Extend your line of best fit to estimate:
a the distance from London of a town with a mean house price of £120 000
b the mean house price for a town 25 miles from London.

4

Natural Births – Length of Pregnancy & Weight of Baby

Length of pregnancy (days)	271	287	283	274	271	279	263	276	283	270
Weight of baby (kg)	2.5	4.2	3.8	3.3	4.5	3.4	2.9	4.1	4.3	3.5

This data has been collected to test the hypothesis:
'A longer pregnancy leads to a heavier baby.'
Plot this data on a scatter diagram.

5 **a** Draw a line of best fit on your scatter diagram.
b Describe the correlation between length of pregnancy and weight of baby.
c Use your line of best fit to estimate:
i the length of pregnancy for a baby that weighs 3.5 kg
ii the weight of a baby with a length of pregnancy of 280 days.

6 Do you think it would make sense to extend this line of best fit? Why?

- In any **right-angled triangle** you can use Pythagoras' rule and trigonometry.
  - To find **an angle** when you know **two sides** → use **trigonometry**
  - To find **a side** when you know an **angle** and **a side** → use **trigonometry**
  - To find **a side** when you know **two sides** → use **Pythagoras' rule**

Exercise 24.3
Areas and perimeters

Accuracy
For this exercise round your answers to 3 sf.

1 This is part of a calculation to find the area of ΔABC.

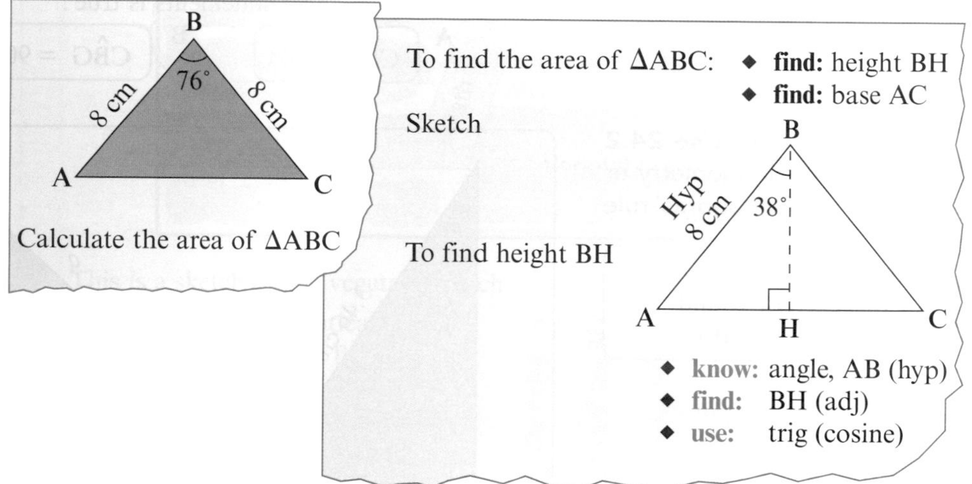

a Calculate the height BH.
b Calculate the length of AC.
c Calculate the area of ΔABC.
d What is the perimeter of ΔABC?

2 The quadrilateral OABC is made from two right-angled triangles. AB and BC are 4 cm long, and OB is 8 cm.

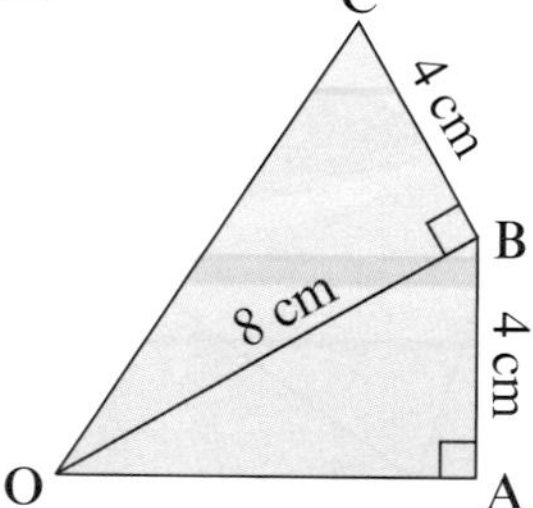

a In ΔOAB calculate:
  i the angle $A\hat{O}B$ **ii** OA
b In ΔOBC calculate:
  i the angle $C\hat{O}B$ **ii** OC
c What is the perimeter of OABC?
d Calculate the area of OABC.

3 If this pattern of triangles is continued the points ABCDE … make a spiral.

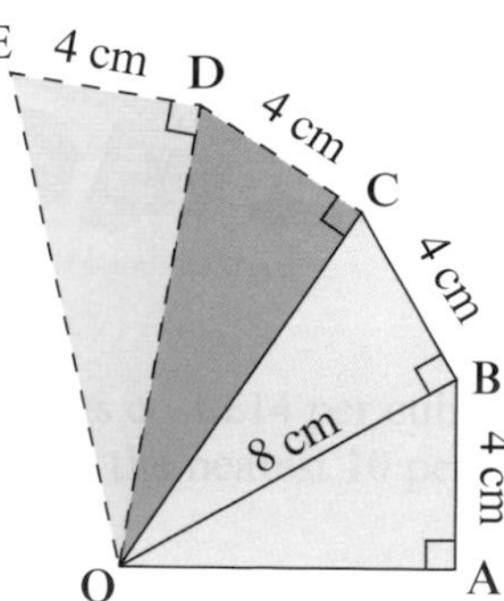

a Calculate the lengths:
  i OD **ii** OE
b Calculate the angles:
  i $D\hat{O}C$ **ii** $E\hat{O}D$
c As the pattern continues:
  i what happens to the length of lines from O?
  ii what happens at O to the angles in the triangles?

Angles of elevation and depression

- From a point A the **angle of elevation** of a point B is the angle between the horizontal and the line of sight from A to B.

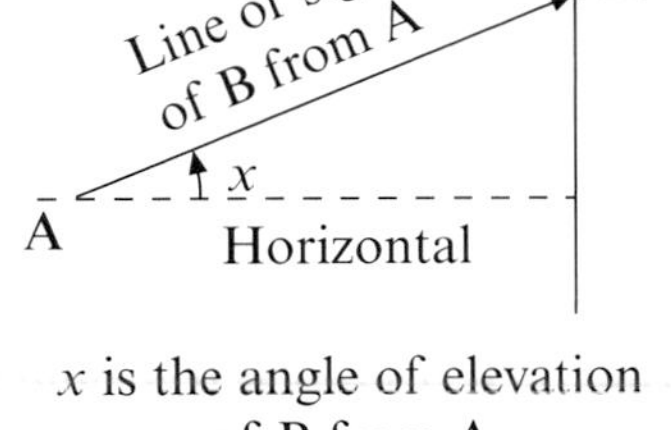

x is the angle of elevation of B from A

- From a point A the **angle of depression** of a point C is the angle between the horizontal and the line of sight from A to C.

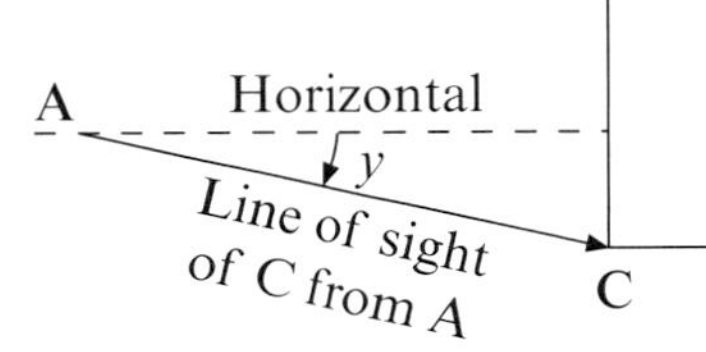

y is the angle of depression of C from A

Exercise 24.4
Angles of elevation and depression

> **Accuracy**
> For this exercise round your answers to the nearest whole number.

1 Jan measures the distance and the angle of elevation or depression to points A and B from T.

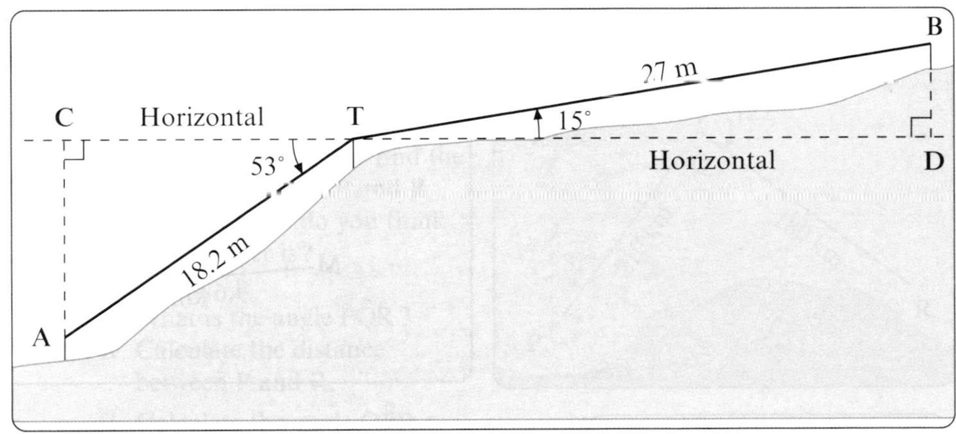

a The angle of depression from T to A is 53° and the distance from T to A is 18.2 metres. The point C is vertically above A.
 i Calculate the vertical height AC.
 ii Calculate the horizontal distance CT.

b The angle of elevation from T to B is 15° and the distance from T to B is 27 metres. The point D is vertically below B.
 i Calculate the vertical height BD.
 ii Calculate the horizontal distance DT.

c On this sketch AE shows the vertical height between A and B, and EB the horizontal distance between A and B.
Calculate:
 i the vertical height AE
 ii the horizontal distance EB
 iii the distance AB.

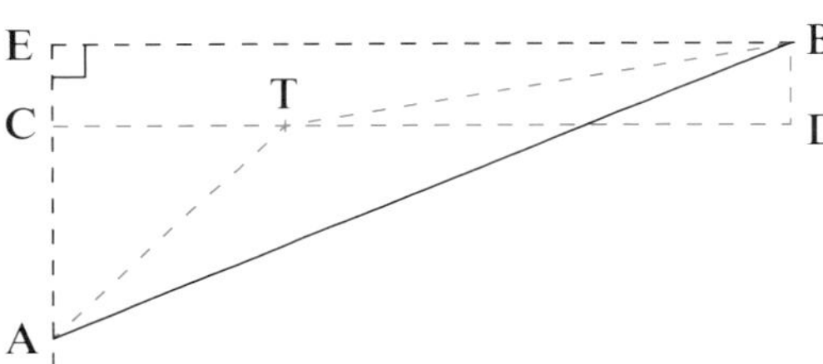

Solving equations

1 Simplify each of these expressions.

a $7k + 8k - 9k$ **b** $2l + 3m + 4l - 2m$
c $4n - 3p + n - 4p$ **d** $3q + 5q - 4q + 6r$
e $5s - 7t - 4s + 7t$ **f** $6 + 5u + 3 + 7v$
g $8w - 7x - 2w + 9x$ **h** $10 - 5y + 1 + 2y$
i $3z + 5 + 6 - 8z - 9 + 10z$

2 Some expressions are arranged in a square.

$4g + h$	$3g - h$	$5g$
$5g - h$	$4g$	$3g + h$
$3g$	$5g + 2h$	$4g - h$

a Find the value of each expression in the square when $g = 7$ and $h = 3$.
b Draw the square with these values.
c Is it a magic square?
d For each row, column and diagonal, find the total of the three expressions.
e Describe how you could change one expression to make this square into a magic square.

3 Solve these equations.

a $7x + 1 = 3x + 21$ **b** $3x - 2 = x + 7$
c $5(x + 1) = 24$ **d** $4(x + 3) + 9 = 13$
e $3x - 2 = 2(x + 1)$ **f** $4x + 5 = 2x - 1$
g $6x - 8 = 10 - 3x$ **h** $13 - 4x = 2x + 4$
i $11 - x = 3(x - 1)$ **j** $2x + 11 = 4 - 5x$
k $21 - 3x = 15 - 2x$ **l** $3x - 10 = 10 - x$
m $4x + 23 = 2(4 - x)$ **n** $4(x - 1) = 2(x + 7)$
o $3(5 - 2x) = 7(x + 4)$ **p** $5(10 - x) = 4(3x + 4)$

4

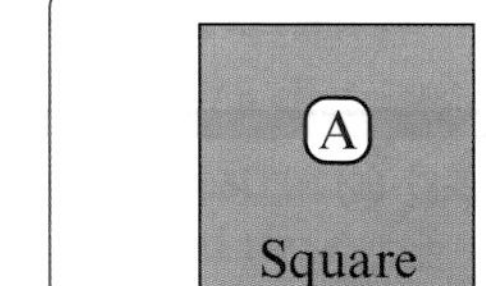

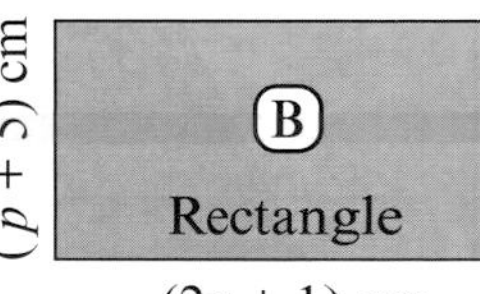

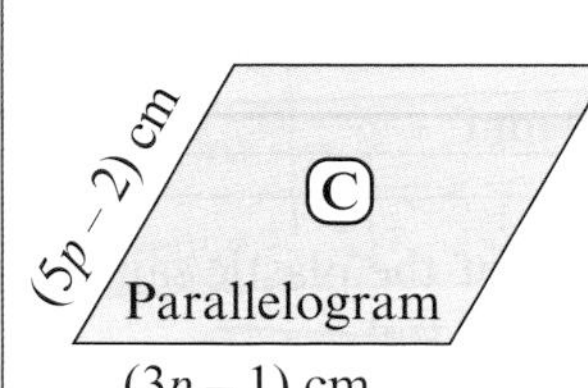

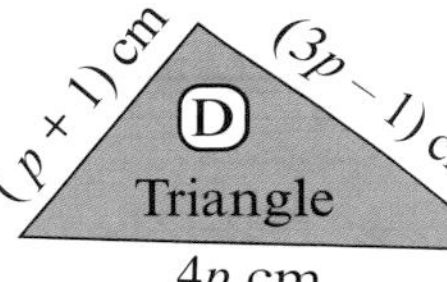

a Find the perimeter of each shape when $p = 7$.
b For each shape, write an expression for the perimeter in terms of p.
c Find a value of p that gives shape D a perimeter of 24 cm.
d Which value of p gives shape C a perimeter of 34 cm?
e Find a value of p that gives shape B a perimeter of 21 cm.
f Find a value of p so that the perimeters of shapes A and B are equal.
g Find a value of p so that the perimeters of shapes A and C are equal.
h Find a value of p so that the perimeters of shapes B and C are equal.
i Explain why the perimeters of shapes A and D are equal for any value of p.

5 For each puzzle A and B, write an equation and solve it to find the number.

A I think of a number, double it and subtract 3. I get the same answer if I subtract my number from 12. What is my number?

B I think of a number, add 1 and multiply by 2. I get the same answer if I subtract 5 and multiply by 3. What is my number?

6 For each pair of equations:

a On one set of axes, draw a graph for each equation.
b Use your graphs to find values for x and y that fit both equations.

A $y = x - 6$, $x + y = 8$ **B** $y = 2x$, $y = 4x - 3$
C $y = 2x + 9$, $y = 3 - x$ **D** $y = 6x - 10$, $y - x = 4$

7 For each pair of equations, use algebra to find the values of x and y.

A $4x + y = 9$, $2x + y = 6$ **B** $3x + y = 37$, $2x + 5y = 29$
C $3x + 5y = 20$, $2x + 3y = 14$ **D** $5x + 4y = 5$, $2x + 6y = 13$
E $3x + y = 18$, $11x - y = 10$ **F** $2x - 3y = 3$, $6x + y = 29$
G $3x - y = 16$, $x + 2y = 3$ **H** $3x - 2y = 1$, $2x + 5y = 7$
I $5x - 2y = 10$, $4x - 2y = 7$ **J** $2x - y = 7$, $9x - 3y = 24$

Estimation and approximation

1 Round each number to the degree of accuracy given:

- **a** 5.674 (2 dp)
- **b** 12.652 (1 dp)
- **c** 2143 (nearest ten)
- **d** 534.687 (2 dp)
- **e** 34.648 (nearest whole number)
- **f** 2639.2 (nearest thousand)
- **g** 13 468.284 52 (2 dp)
- **h** 0.0666 (2 dp)
- **i** 63.68 (nearest ten)
- **j** 59.999 (1 dp)
- **k** 8502 (nearest thousand)
- **l** 68.499 99 (nearest whole number)
- **m** 56.289 64 (3 dp)

2 Round each of these to the number of significant figures given.

- **a** 56.83 (3 sf)
- **b** 16 389 (3 sf)
- **c** 2.456 (2 sf)
- **d** 45.923 (1 sf)
- **e** 15.777 77 (4 sf)
- **f** 725 184 (2 sf)
- **g** 94.56 (1 sf)
- **h** 93 747 656 (5 sf)
- **i** 564.23 (4 sf)
- **j** 196.5 (1 sf)
- **k** 6.7849 (2 sf)
- **l** 15.682 (4 sf)
- **m** 673 492.35 (4 sf)
- **n** 0.035 62 (2 sf)
- **o** 0.027 95 (3 sf)
- **p** 3.0004 (3 sf)

3 By approximating each number to 1 sf work out approximate answers to each of these.

- **a** 84.3×452.53
- **b** 4.876×37.71
- **c** 5683.2×9.372
- **d** 458.12×518
- **e** $734.6 \div 2.316$
- **f** $56.8243 \div 7.8931$
- **g** $41\,952 + 77\,442$
- **h** $34.6296 + 87.3$
- **i** $6834 - 1939.453$
- **j** $45.95 \times 2.943\,56$
- **k** 74.68×4.87
- **l** 7468×0.487
- **m** 7.468×48.7

4 Work out approximate answers then calculate each of these exactly.

- **a** 56.6×21.5
- **b** 4924×3.7
- **c** 246.3×9.67
- **d** 54.27×21.6
- **e** 17.63×1839
- **f** 3.45×0.054
- **g** 0.42×264
- **h** 38.6×25.002

5 For each of these rectangles, estimate its area. Explain how you calculated your estimate.

a

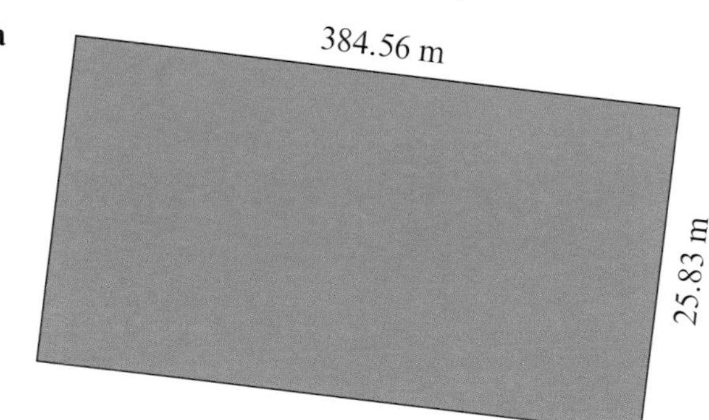

b

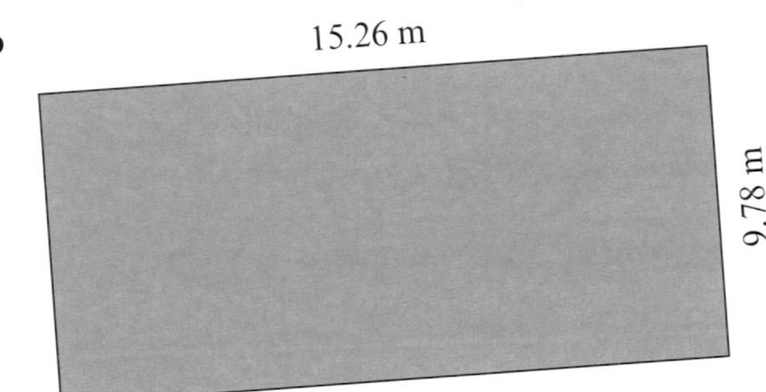

c

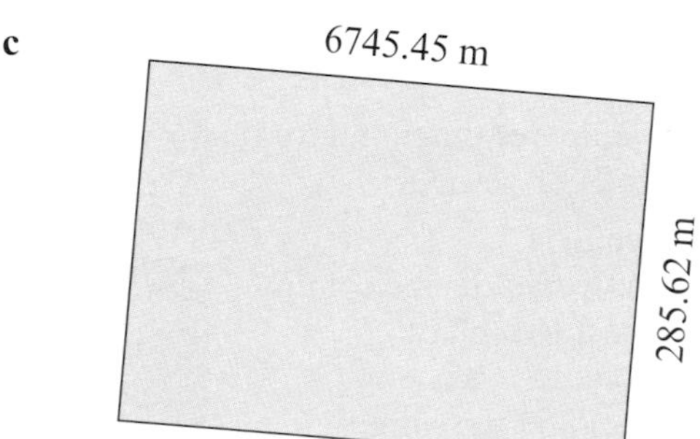

Probability A

1 With a 1 to 6 dice what is the probability that you score:

a 5 b an even number
c a multiple of 3 d a prime number?

2 For this wheel, what is the probability that the pointer stops on:

a B b A
c either A or D
d C e not C
f N g A or B

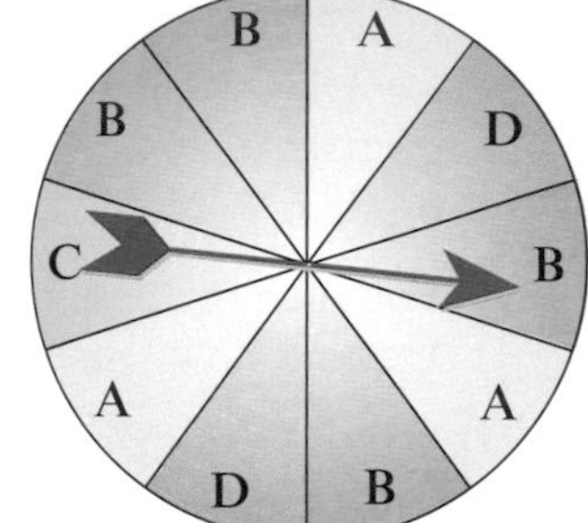

3 For the wheel give an event which has a probability of $\frac{3}{5}$.

4

Draw a sample space diagram to show the outcomes when rolling a 1 to 4 dice and a 1 to 8 dice together.

5 From your sample space diagram, give the probability of getting:

a at least one 3
b at least one 6
c a total of 7 by adding the scores
d a 5 and a 2
e a 3 and a 4
f two prime numbers
g two non-prime numbers
h a total which is less than 6
i a total which is greater than 6
j a total which is a multiple of 3
k two numbers the same
l two numbers which are different.

6 A cube dice has three red faces and three blue faces. Draw a tree diagram to show three rolls of the dice.

7 From your tree diagram give the probability in three rolls of getting:

a three reds
b exactly two of one colour
c at least two blues
d no blues
e all colours the same
f no colours the same.

8 This tree diagram is for two spinners. The outcomes are not all equally likely.

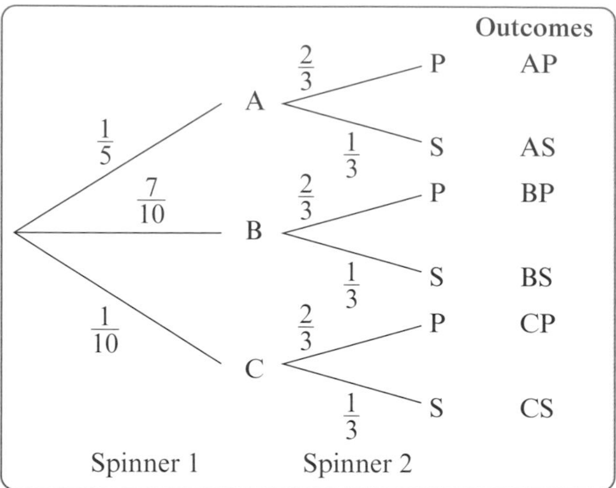

By multiplying probabilities give the probability of:

a B and P b C and S
c A and P d B and S
e A and S f C and P

9 Four CDs P to S are stacked in random order.

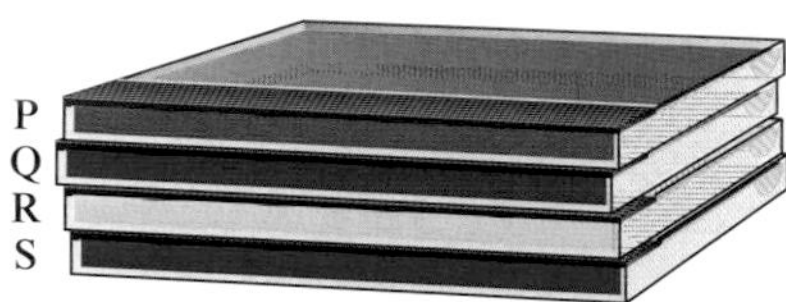

a List all the different arrangements that are possible.
b Give the probability that the CD at the bottom is red.
c What is the probability that the top and bottom CDs are blue?
d What is the probability that there is a blue CD at the top?
e Give the probability that the two blue CDs are next to each other.

10 Five mugs are hanging from hooks. A pair of mugs are chosen at random.

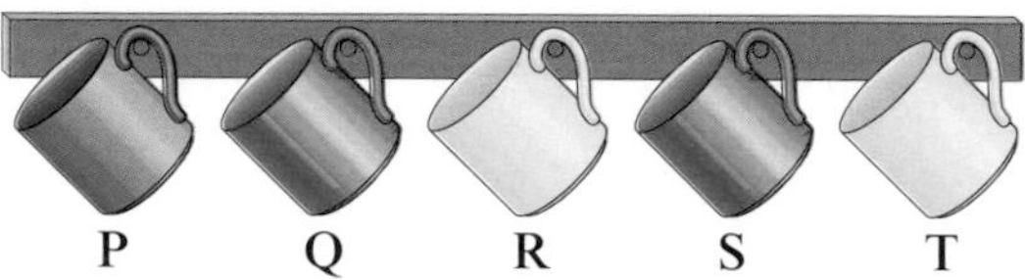

a List every pair it is possible to pick.
b What is the probability of picking:
  i a pair of the same colour
  ii a pair where only one mug is red
  iii a pair of yellow mugs?

Using algebra A

1 This shape is cut from a rectangle.

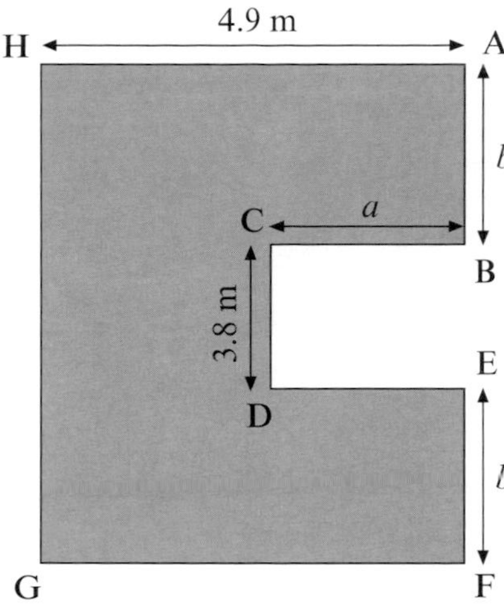

The length of AB is b metres and BC is a metres.

a Write an expression for the length of GH.
b Write an expression for the perimeter of this shape in terms of a and b.
c What is the perimeter if $a = 1.8$ and $b = 4.1$?

2 The dimensions of this trapezium are in centimetres.

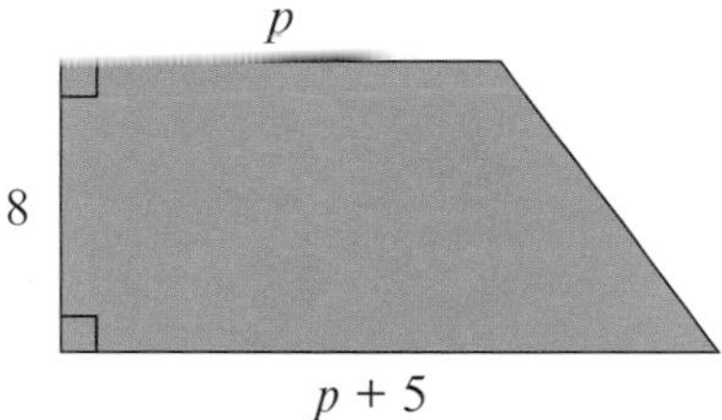

a Write an expression for the area of the trapezium.
b What is the area when $p = 6.4$?
c What value of p gives an area of $54\,\text{cm}^2$?

3 Find three pairs of equivalent expressions.

A	$2(2a + 4)$	B	$4a + 4$	C	$2a + 6$
D	$4(a + 2)$	E	$2a + 8$	F	$2(a + 2)$
G	$2a + 5$	H	$2(a + 3)$	I	$4(a + 1)$

4 Multiply out these brackets.

a $3(a + 6)$ **b** $5(6 + z)$ **c** $3(n - 4)$
d $3(2n + 4)$ **e** $x(x + 2)$ **f** $4p(p - 3)$
g $3b(3 + b)$ **h** $a(2a - 8)$ **i** $4p(3p - 4)$

5 The length of a rectangle is 8 cm greater than its width.

a If the rectangle is w centimetres wide, write an expression for the length of the rectangle.
b Write an expression for the perimeter of the rectangle in terms of w.
c What value of w gives a perimeter of 72 cm?

6 The length of a rectangle is three times its width. The perimeter is 80 cm. What is the area?

7 Multiply these terms.

a $8p \times 6q$ **b** $9y \times y$ **c** $mn \times mn$
d $pq \times 4p$ **e** $8m \times mn$ **f** $9b \times 2a^2$
g $a^3 \times g^3$ **h** $4p \times 5p^2$ **i** $7a^3 \times 4b$

8 Multiply out these.

a $6(4a - 3b)$ **b** $p(n - p)$ **c** $s(s + t)$
d $6m(n - m)$ **e** $5x(x + y)$ **f** $u(4u + 3)$
g $p(5q + 3r)$ **h** $3n(4m + 7n)$ **i** $9a(3b + a)$

9 Which of these expressions is equivalent to $4a(2ab + 3a) + a(5b - a) + 4b(2a^2 + 5a)$?

A $35a^2b + 11a^2$

B $12a^2b + 14ab + 7a^2$

C $41ab + 10a$

D $16a^2b + 11a^2 + 25ab$

10 Simplify these.

a $8n + 4m - 6n + 3m$ **b** $12a^2 - a + 9a^2$
c $b^3 - b^3 + 2b$ **d** $8x - 4xy - 2y + 6xy$
e $p + 8q - 5q + 3p$ **f** $8p^2 + 3pq + 5q^2 - pq$
g $4m^2 + mn - 6n + nm$ **h** $x + 2xy + x - 4xy$

11 Multiply out these brackets and simplify.

a $6(7a - 2b) + 7(2a + b)$
b $9(3x + y) + 4(y - 2x)$
c $x(6x - 2) + 2x(5x + 3)$
d $8x(2y + 3x) + x(6x - 3y)$

12 Write an expression for the width of rectangles A and B.

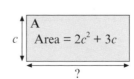

A
c
Area = $2c^2 + 3c$
?

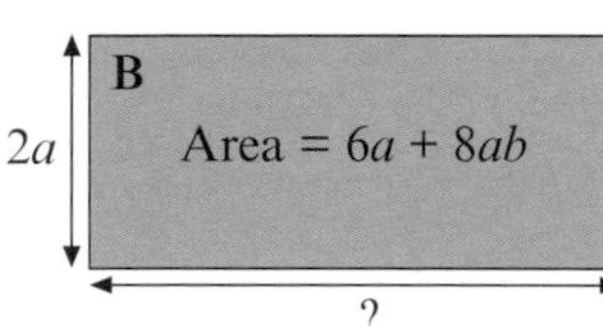

13 Factorise these fully.

a $7p + 3pq$ **b** $m + 3mn$ **c** $8pq + 4q$
d $3y - 12xy$ **e** $2b^2 + 10b$ **f** $18ab + 24bc$
g $x^2y - 2xy$ **h** $9a^2b + 6ab^2$ **i** $12m^2n - 9mn$

Constructions and Loci

1 Construct these triangles accurately. Show all your construction lines.

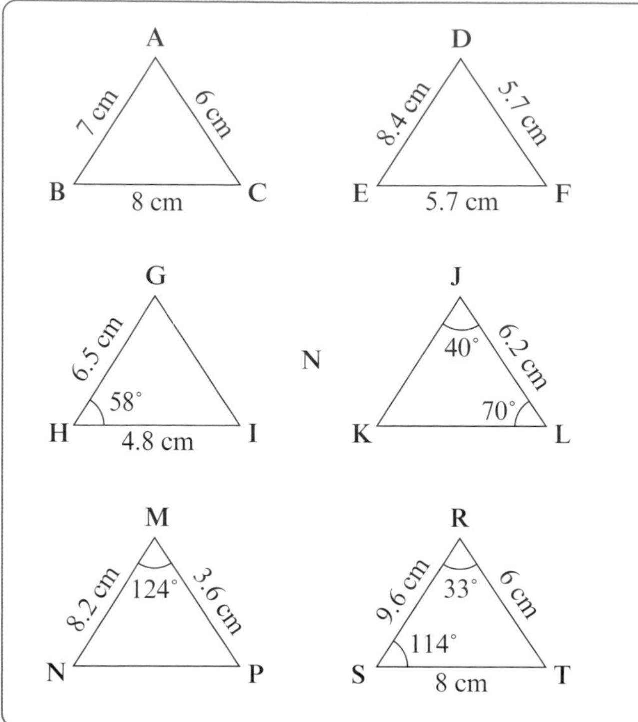

2 Copy each line full size and contruct a perpendicular bisector of it.

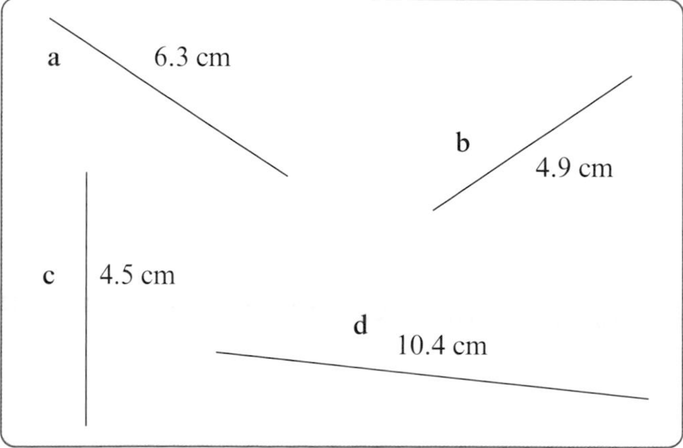

3 Draw each angle accurately and construct its bisector.

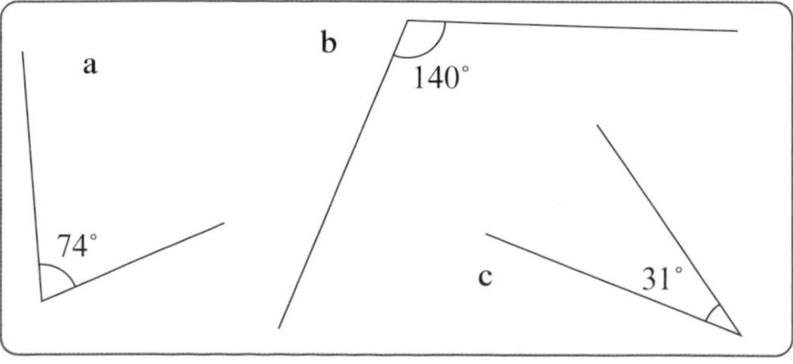

4 Draw each of the following shapes full size and show the locus of points 2 cm from each one.

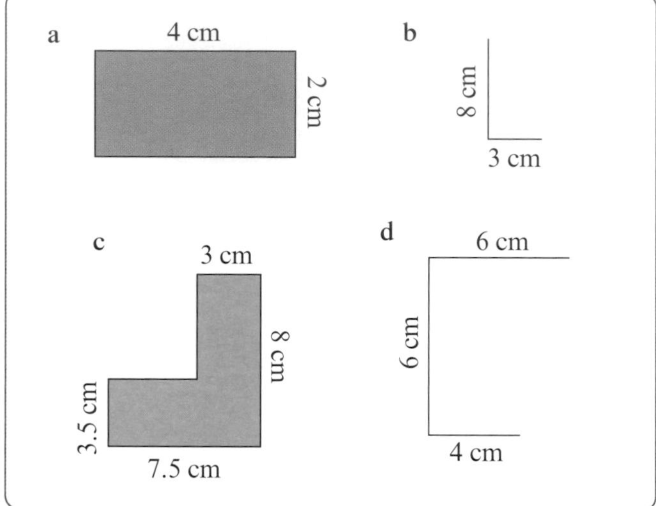

5 Explain how you would construct a perpendicular from a point to a line, with compasses and a ruler. Use a diagram in your answer.

6 A field for the village fete is shaped as a triangle with these dimensions.

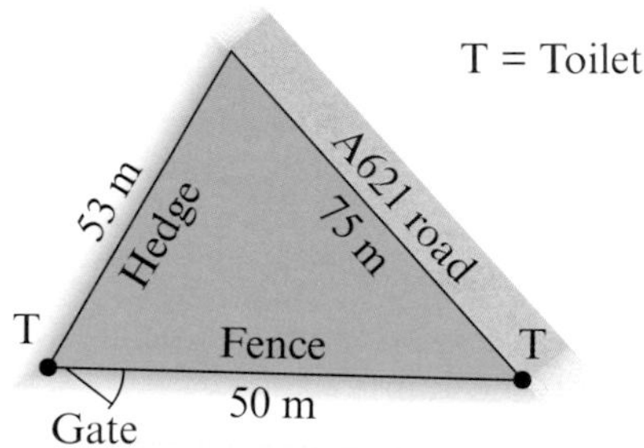

Colonel Briggs Shilton has set out these conditions for the position of the drinks tent.

Condition 1 – The tent must be equidistant from the hedge and the fence.
Condition 2 – The distance from each toilet to the tent must be equal.

a Construct a scale drawing of the field to a scale of 1 to 1000.
b Find by construction the position of the tent.
c Why is this position not a sensible one?
d Give some different conditions which you think puts the tent in a better position.

Ratio

1 Green dyes are a mix of blue and yellow dyes.

Copy the table below and calculate the missing amounts.

Dye	Blue	:	Yellow
Grass	1	:	3
Lime	3	:	7
Pea	5	:	9

Dye	Blue	Yellow	Total
Grass	120 ml		
Lime	450 ml		
Pea		360 ml	
Lime		175 ml	
Pea	600 ml		
Grass		510 ml	
Lime	645 ml		
Pea		495 ml	
Grass		720 ml	
Lime			1200 ml
Grass			860 ml
Pea			700 ml
Grass			1280 ml
Lime			890 ml
Pea			1050 ml

2 Share each of these amounts in the given ratio.

a	£84 2:5	**b**	£248 5:3	**c**	195 g 9:4
d	£120 2:1:3	**e**	200 g 3:4:1	**f**	360 ml 2:3:4
g	420 g 2:3:2	**h**	£495 1:1:7	**i**	99 cm 4:5:2
j	2 m 5:2:1	**k**	1.4 kg 1:2:4	**l**	2.7 cm 1:6:2

3 This recipe makes 25 biscuits.

Chocolate Biscuits
50 g icing sugar • 225 g margarine
100 g plain chocolate • 225 g flour

Calculate how much of each ingredient is needed to make:

a 45 biscuits **b** 10 biscuits

4

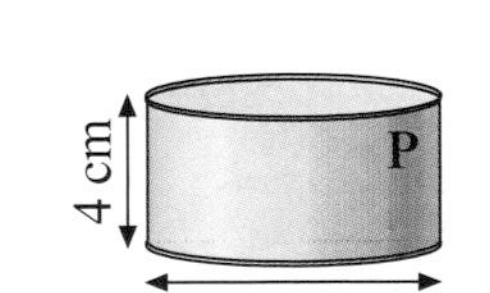

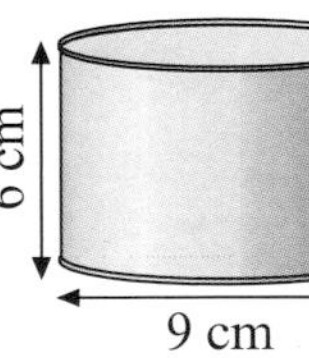

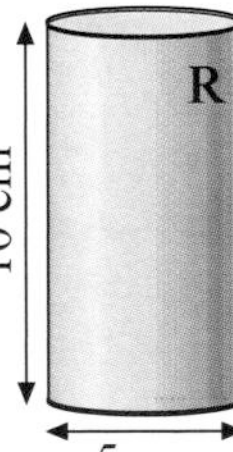

For each tin, give the ratio of height to diameter in its simplest terms.

5 (A 2 : 5) (B 3 : 1) (C 4 : 3) (D 1 : 3) (E 3 : 5)

These are the ratios of men to women in five self-defence classes.
What fraction of each class are men?

6

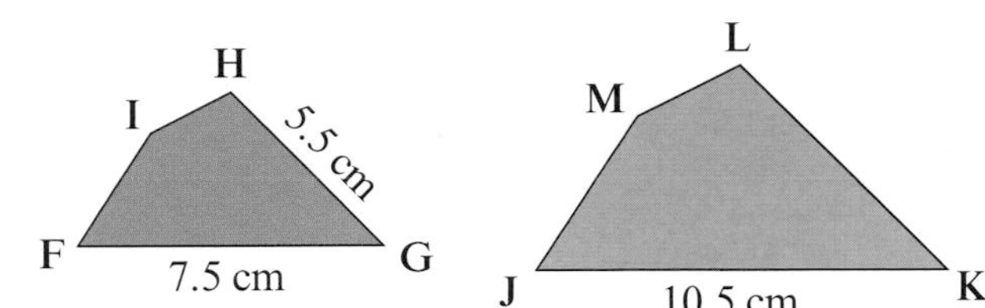

Shape JKLM is an enlargement of shape FGHI.

a Calculate the scale factor of the enlargement.
b Find the length LK.

7

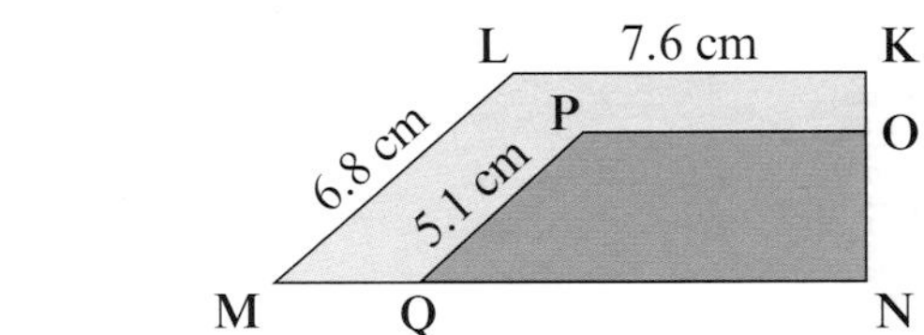

NOPQ is an enlargement of NKLM.

a Calculate the scale factor of the enlargement.
b Find the length PO.

8 A lottery win of £395 775 is shared by 4 winners in the ratio 2:3:4:6.

a Find the amount each winner received.
b Explain how you can check your answer to part **a**.

9 The length and width of a rectangle are in the ratio 3:5.
The perimeter of the rectangle is 56 cm.

a Find the length of the rectangle.
b Find the width of the rectangle.
c Calculate the area of the rectangle.

10 A paint colour called Zulu is mixed in this ratio:
125 ml of red : 275 ml of blue : 100 ml of yellow.

a Give this ratio in its simplest form.
b A mix uses 225 ml of yellow to make the colour called Zulu.
How much of each of the other colours is used?
c 1750 ml of Zulu paint is needed.
Give the amounts of each colour used for the mix.

Distance, speed and time

1 A chairlift travels at a constant speed of 2.6 metres per second.
What distance does one of the chairs travel in:

a 2 seconds **b** 5 seconds
c 9.3 seconds **d** 1 minute
e 3 minutes and 38 seconds?

2 At a constant speed, a plane takes 2 hours to travel 1208 miles.
Find the plane's speed in miles per hour.

3 An escalator is 32.8 metres long and travels at a speed of 0.75 m/s.
How long does it take to travel to the top of this escalator?

4 It takes 20 seconds to go up an escalator that is 14 m long. What is its speed in m/s?

5 Write these speeds in metres per second, to 1 dp.

a 129 metres per minute
b 2 kilometres per second
c 432 metres per hour
d 20 kilometres per hour

6 Write these speeds in kilometres per hour.

a 2 kilometres per minute
b 40 000 metres per hour
c 65 100 metres per hour
d 6 metres per second

7 Write these speeds in miles per minute, to 2dp where appropriate

a 60 mph **b** 30 mph
c 90 mph **d** 45 mph
e 25 mph **f** 42 mph

8 Write these speeds in miles per hour.

a 0.6 miles per minute
b 1.2 miles per minute
c 0.45 miles per minute
d 0.1 miles per second

9 How far does a plane travel in 3 h 15 min at a constant speed of 960 kilometres per hour?

10 How far can a car travel at a constant speed of 51 mph in:

a 1 h 30 min **b** 45 min
c 15 min **d** 1h 20 min
e 2 h 40 min **f** 3 h 45 min?

11 Calculate the time taken in hours and minutes, to the nearest minute, to travel 60 km at a speed of:

a 120 km/h **b** 30 km/h
c 75 km/h **d** 9 km/h
e 55 km/h **f** 49 km/h
g 50 km/h **h** 150 km/h.

12 This sketch graph shows two different vehicles starting from Bristol. One is a motorcycle the other a truck.

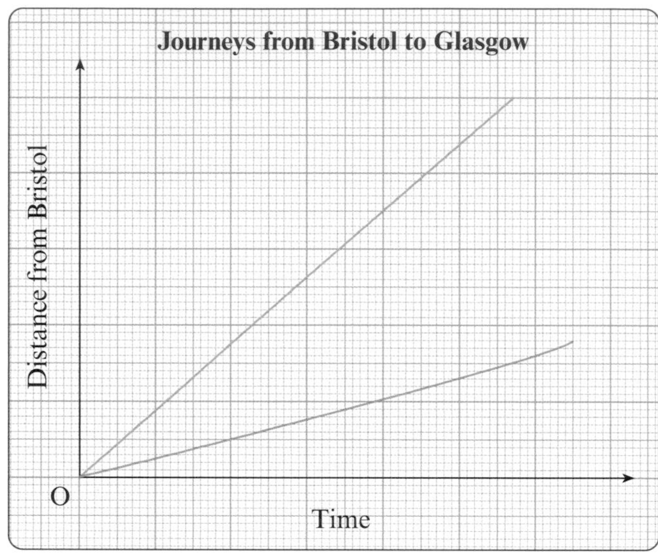

Which line do you think shows the truck?
Explain your answer.

13 Calculate the average speed (in mph or km/h to 2 dp) of a car which travels:

a 90 km in 2 h
b 115 km in 3 h
c 40 miles in 50 min
d 50 km in 35 min
e 70 miles in 1 h 20 min
f 100 km in 1 h 12 min
g 400 miles in 7 h 30 min
h 500 metres in 1 min.

14 Mandy left Glasgow at 10:00 am and cycled 28 miles at a speed of 10 mph.
Amin left Glasgow at 10:20 am and travelled along the same route at a speed of 12 mph.

a Show both their journeys on one graph.
b About what time did Amin pass Mandy?

15 Andy left Taunton at 3:00 pm and took 45 minutes to drive 50 miles to Bristol. He stayed there for 2 h 30 min. He then returned along the same route and arrived home at 7:05 pm.

a Draw a graph to show Andy's complete journey.
b At what time did he begin his return journey to Taunton?
c Calculate his average speed for the journey to Bristol giving your answer to 1dp.
d What was his average speed on the return journey?

Trigonometry

1 Find angle x to the nearest degree when:

a $\sin x = \frac{5}{6}$ **b** $\cos x = \frac{3}{8}$

c $\tan x = \frac{9}{5}$ **d** $\sin x = \frac{11}{15}$

e $\tan x = \frac{3}{13}$ **f** $\cos x = \frac{5}{8}$

2 In each of these triangles calculate the length marked with a letter, to the nearest millimetre.

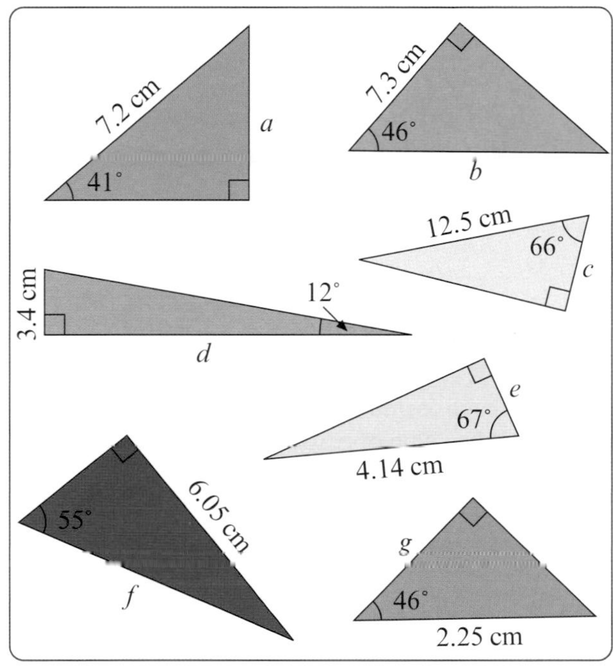

3 In each of these triangles calculate the angle x to the nearest degree.

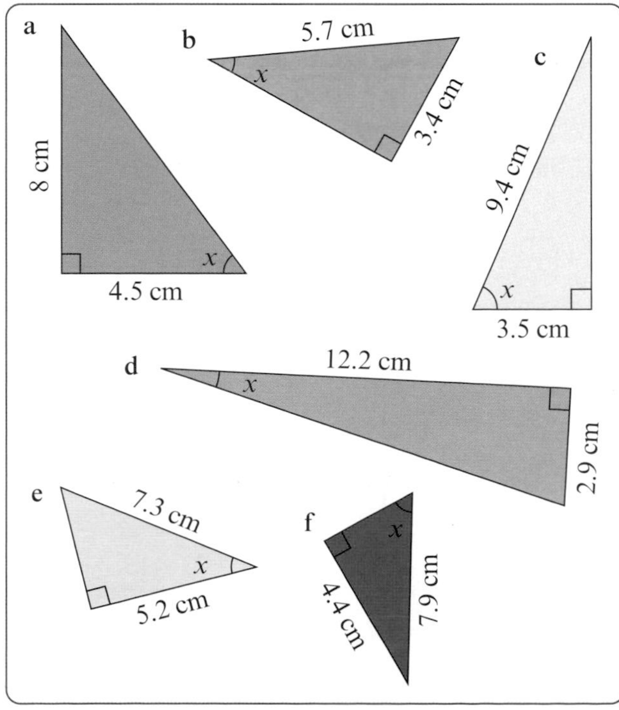

4 In this question give each answer correct to 3 sf.

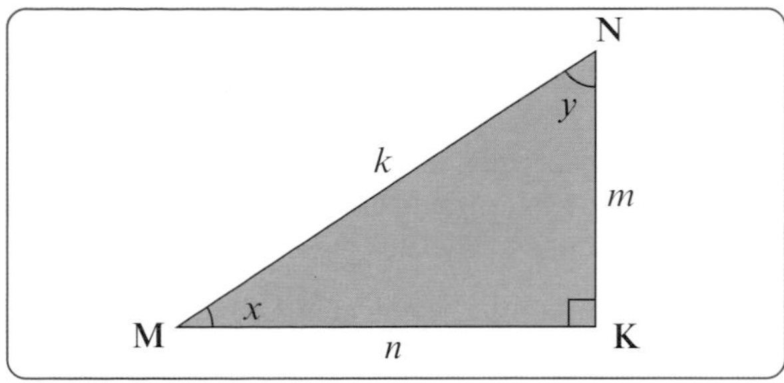

In ΔNKM:

a calculate x when:
- **i** $n = 4.60$ cm and $k = 10.60$ cm
- **ii** $m = 8.40$ cm and $n = 4.62$ cm
- **iii** $k = 5.65$ cm and $m = 3.00$ cm

b calculate y when:
- **i** $n = 6.62$ cm and $k = 10.60$ cm
- **ii** $m = 15.00$ cm and $n = 10.50$ cm

c calculate m when:
- **i** $x = 55°$ and $k = 3.58$ cm
- **ii** $y = 16°$ and $n = 41.50$ cm
- **ii** $x = 38°$ and $k = 4.05$ cm

d calculate k when:
- **i** $y = 25°$ and $m = 19.00$ cm
- **ii** $x = 61$ and $n = 2.40$ cm

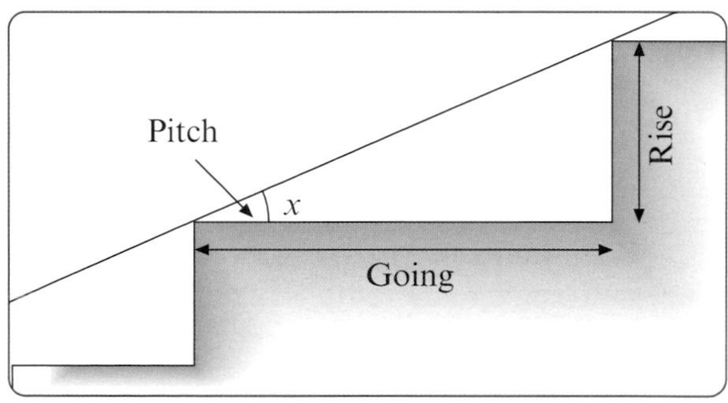

5 Calculate the pitch x for each of these stairs, correct to 2 dp.

Stair	Rise (mm)	Going (mm)
A	160	310
B	173	294
C	153	319

6 For safety the pitch x of these stairs must be between 32° and 22°.

a For a going of 303 mm what is the greatest rise, to the nearest mm?

b For a rise of 175 mm what is the greatest going, to the nearest mm?

Using algebra B

1 A formula for making tea in a pot for p people is:

$t = p + 1$

where t is the number of spoons of tea.

a Find the number of spoons of tea needed for 5 people.
b Rearrange the formula to make p the subject.
c Calculate p when $t = 3$.

2 A formula for the perimeter of a square (P) of edge length x is:

$P = 4x$

a Make x the subject of the formula.
b Find the edge length of a square that has a perimeter of 41 cm.

3 A cook book gives this formula for the time in minutes (T) to cook a turkey that weighs W pounds:

$T = 20W + 20$

a How long would it take to cook a turkey that weighs 12.5 pounds?
b Make W the subject of the formula.
c What is the weight of the largest turkey that is cooked in 2 hours?

4 A formula for converting litres (l) to pints (p) is:

$p = \frac{7}{4}l$

a Use the formula to convert 5 litres to pints.
b Make l the subject of the formula.
c Convert 8.5 pints to litres, correct to 2 dp.

5 The formula for the area of a rectangle (A) with length l and width w is:

$A = lw$

a Make w the subject of the formula.
b Find the width of a rectangle 10 cm in length with an area of 62 cm^2.

6 The formula for the perimeter of a rectangle (P) with length l and width w is:

$P = 2l + 2w$

a Make w the subject of the formula.
b Find the width of a rectangle 10 cm in length with a perimeter of 28.4 cm.

7 Make k the subject of these formulas:

a $V = 4k - 25$ **b** $y = 3(k - 1)$
c $j = \frac{k}{7}$ **d** $w = \frac{3}{4}k$
e $2v + k = 5$ **f** $k - b = 7$
g $d = \frac{2}{5}k + 1$ **h** $h = \frac{k + 3}{8}$
i $p = w + k$ **j** $q = 20 - 5k$
k $v = 13 - \frac{1}{3}k$ **l** $v = s + tk$
m $h = \frac{k + a}{b}$ **n** $h = 2gk$
o $2y + 5k = 3$ **p** $3k - 2h = 9$

8 Make p the subject of these formulas:

a $A = p^2$ **b** $v = 5p^2$
c $g = p^2 + 1$ **d** $q = 3p^2 - 5$
e $j = \frac{1}{4}p^2$ **f** $F = \frac{2}{3}p^2$
g $p^2 + z = 2$ **h** $K = 2\pi p^2$
i $A = 2p^2 + 1$ **j** $H = 3p^2 - 8$

9 A $y = (x + 2)(x + 4)$ B $y = (x + 3)(x - 1)$
C $y = 2x^2 + 3x - 5$ D $y = x^2 - x - 2$

For each formula, find the value of y when:

a $x = 6$ **b** $x = 3$ **c** $x = 1.5$

10 For each formula in x, y and z, find the value of x when $y = 5$ and $z = 2$.

a $x = 3yz + y$ **b** $x = 4z(y - 3)$
c $x = \sqrt{y^2 + 12z}$ **d** $x = 2y^2 - 3z^2$
e $x = (5z + y)^2$ **f** $x = 2z^3 - y$
g $x = \frac{y + 3z}{2y}$ **h** $x = \frac{1}{y} + \frac{1}{z}$

11 For each of these multiply out the brackets and simplify:

a $(n + 2)(n + 5)$ **b** $(f + 1)(f + 10)$
c $(w + 2)(w + 6)$ **d** $(2n + 1)(n + 3)$
e $(4d + 7)(d + 3)$ **f** $(2y + 9)(2y + 5)$
g $(x - 3)(x + 6)$ **h** $(v + 3)(v - 2)$
i $(b + 3)(b - 5)$ **j** $(m - 7)(m - 1)$
k $(h - 1)(h - 5)$ **l** $(t - 3)(t - 7)$
m $(2p + 1)(3p - 1)$ **n** $(5x + 6)(3x - 2)$
o $(6y + 3)(2y - 5)$ **p** $(3g - 7)(4g - 5)$

12 $x + 1$ $x + 2$ $x + 3$ $x + 6$ $x + 4$ $x + 12$

Which pair of expressions multiply to give:

a $x^2 + 7x + 12$ **b** $x^2 + 8x + 12$
c $x^2 + 7x + 6$ **d** $x^2 + 5x + 6$
e $x^2 + 13x + 12$ **f** $x^2 + 6x + 8$?

13 Factorise:

a $x^2 + 12x + 11$ **b** $x^2 + 8x + 15$
c $x^2 + 4x + 4$ **d** $x^2 + 9x + 20$
e $x^2 + 3x - 4$ **f** $x^2 + 2x - 15$
g $x^2 + 4x - 12$ **h** $x^2 - x - 6$
i $x^2 - 11x + 10$ **j** $x^2 - 2x - 8$
k $x^2 - 3x + 2$ **l** $x^2 - 4x + 4$

Grouped data

Table A

1996 Olympic Games Reaction Times Sprint Hurdles (Semi-Finals & Finals)

Reaction time (s)	Frequency Men	Frequency Women
0.120–	1	1
0.130–	4	2
0.140–	1	2
0.150–	4	2
0.160–	5	5
0.170	4	7
0.180–	2	5
0.190–	3	0
Totals	24	24

Use Table A for Questions 1 to 12

1 Draw a histogram to show the reaction times for:

a men **b** women.

2 Give the modal class for:

a men **b** women.

3 Draw frequency polygons to compare the reaction times of men and women.

4 Calculate an estimate of the range of the times for:

a men **b** women.

5 Calculate an estimate of the mean reaction time for:

a men **b** women.

6 For these two distributions, the totals of the reaction times are just as useful for comparison as the means. Explain why.

7 Draw a cumulative frequency curve for:

a men **b** women.

8 Estimate how many reactions times were less than 0.135 s for:

a men **b** women.

9 Estimate how many reactions times were greater than 0.175 s for:

a men **b** women.

10 Estimate the median reaction time for:

a men **b** women.

11 Calculate an estimate of the interquartile range for:

a men **b** women.

12 Use your answers to Questions **4**, **5**, **10**, and **11** to compare the two distributions.

Table B

1996 Olympic Games GB Athletics Team

Age	Frequency Track	Frequency Field
18–22	7	2
23–27	27	10
28–32	20	7
33–37	3	2
38–42	2	2
Totals	59	23

Use Table B for Questions 13 to 15

13 Draw a histogram to show the distribution of:

a track athletes **b** field athletes.

14 Explain why drawing frequency polygons on the same diagram would not give a good comparison of the ages of track athletes and field athletes.

15 Calculate an estimate of the mean age of:

a track athletes **b** field athletes.

C **1996 Olympic Games – Men's 20 km walk**

Time (min)	80–	84–	88–	92–	96–	100–	Total
Frequency	18	21	9	2	1	1	52

D **1996 Olympic Games – Men's 50 km walk**

Time (min)	220–	230–	240–	250–	260–	Total
Frequency	8	13	9	4	2	36

E **1996 Olympic Games – Women's 10 km walk**

Time (min)	41–	42–	43–	44–	45–	46–	47–	48–	Total
Frequency	1	4	7	3	12	5	4	2	38

16 For each distribution in table C, D and E:

a draw a histogram
b calculate an estimate of the range of the times
c calculate an estimate of the mean time
d draw a cumulative frequency curve
e estimate the median time
f calculate an estimate of the interquartile range.

17 This data shows sales of burgers at a motorway service area.

Week	1	2	3	4	5	6	7	8	9	10	11	12
Burger sales	44	32	28	32	40	16	24	20	28	24	36	16

a Calculate the 4-point moving averages for the data.
b On a pair of axes draw graphs of the raw data and the moving average data.
c Comment on any trend you can identify in burger sales.

Working with percentages

1 Calculate, giving answers correct to 2 dp where appropriate:

a 38 as a percentage of 60
b £14 as a percentage of £55
c 68 as a percentage of 24
d 1550 as a percentage of 2500
e 3500 km as a percentage of 75 000 km
f 12.5 kg as a percentage of 40 kg
g £125.50 as a percentage of £150
h 132 miles as a percentage of 868 miles
i £9.38 as a percentage of £56.45
j 15 650 km as a percentage of 1.5 million km.

2 Give answers to each of these correct to 2 dp where appropriate.

a Increase 25 kg by 18%
b Increase 1400 km by 65%
c Increase 3560 miles by 4%
d Increase 1350 yards by 35%
e Increase £25 645 by 6%
f Increase 137 500 tonnes by 5.5%
g Increase 0.7 cm by 50%
h Increase 42 mm by 12.5%
i Increase £35.99 by 7%
j Increase 365 ml by 15%.

3 Give your answers to these correct to 2 dp where appropriate.

a Decrease 485 ml by 12%
b Decrease £45.75 by 20%
c Decrease 65 mm by 65%
d Decrease 0.8 cm by 8%
e Decrease 15 875 tonnes by 75%
f Decrease £15 944 by 34%
g Decrease 1760 yards by 28%
h Decrease 5682 miles by 56%
i Decrease 3600 km by 35.5%
j Decrease 48 kg by 48%.

4 The price of each item is given ex. VAT.
Calculate the price including VAT at today's standard rate.

a crash helmet £185.85
b cycle tyre £11.69
c fishing rod £44.86
d steam iron £21.75
e microwave oven £268.55
f VCR £159.99
g personal CD player £135.38
h CD £9.24
i phone £49.99
j multi-media PC £1499.

5 A printer is advertised for £132 + VAT.
Calculate the total price of the printer.

6 Callum sees the same model TV advertised by two shops in this way:

TV World	£199.99 inc. VAT
Price busters	£169.99 ex. VAT

a From which shop would you advise Callum to buy the TV?
b Give a reason for your answer to part **a**.

7 Calculate the simple interest charged or paid on each of these:

a £675 borrowed for 5 years at 12% pa
b £12 400 borrowed for 2 years at 17% pa
c £170 saved for 4 years at 3% pa
d £65 saved for 9 years at 4% pa
e £1500 borrowed for 3 years at 18% pa
f £4600 borrowed for 4 years at 12.5% pa
g £25 saved for 3 years at 5% pa
h £150 saved for 6 years at 4.5% pa
i £200 borrowed for 2 years at 22.9% pa
j £175 borrowed for 3 years at 18% pa.

8 Calculate the compound interest charged or paid on each of these (use the formula):

a £500 saved for 3 years at 6% pa
b £1400 borrowed for 2 years at 19% pa
c £250 saved for 2 years at 7.5% pa
d £105 saved for 4 years at 3%
e £12 500 borrowed for 3 years at 16%
f £32 saved for 3 years at 4%
g £750 borrowed for 3 years at 21%
h £3675 borrowed for 2 years at 17%
i £125 saved for 3 years at 8%
j £500 saved for 3 years at 4.5%.

9 These prices include VAT.
Calculate each price ex. VAT.

a freezer £299.99
b CD £12.99
c camera £44.99
d phone £9.99
e climbing boots £75
f tent £89.95
g kettle £26.99
h PC £129.99
i calculator £18.99
j TV £139.99

10 Jo bought a ski jacket in a 15% off sale for £85.45.

a What was the pre-sale price of the jacket?
b How much did she save in the sale?

11 In 1995 a ferry company made a profit of £3 600 000.
This was 14% more than the profit for 1994.

Calculate the profit made in 1994.

Transformations

1 Draw the pentagon A on axes with: $^-4 \leq x \leq 10$ and $^-6 \leq y \leq 8$.

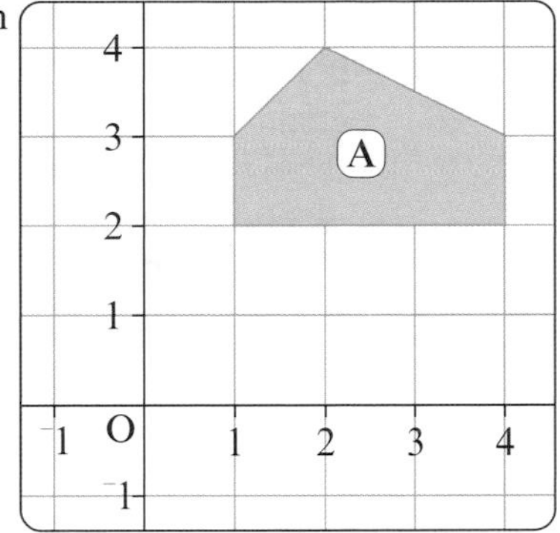

2 Draw the image of A after:

a an enlargement SF 2 with centre (6, 5)

b an enlargement SF $\frac{1}{2}$ with centre (7, $^-4$).

3 These transformations map A on to B, C, D, E and F.

Object	Transformation	Image
A	Rotate 90° anticlockwise about (0, 0)	B
A	Rotate 90° clockwise about (1, 1)	C
A	Reflect in $x = 5$	D
A	Rotate 90° anticlockwise about (6, 2)	E
A	Reflect in $y = {}^-x$	F

a On a new diagram, on axes with: $^-6 \leq x \leq 10$ and $^-6 \leq y \leq 6$ draw and label the images B, C, D, E and F.

b Describe fully the transformation that maps:
 i C on to A ii D on to A.

c For each of these mappings which pentagon is the image of B?

	Object	Transformation	Image
i	B	Rotate 180° about (0, 1)	
ii	B	Reflect in $y = 0$	
iii	B	Translate $\begin{pmatrix} 8 \\ ^-4 \end{pmatrix}$	

d Describe fully the transformation that maps C on to E.

4

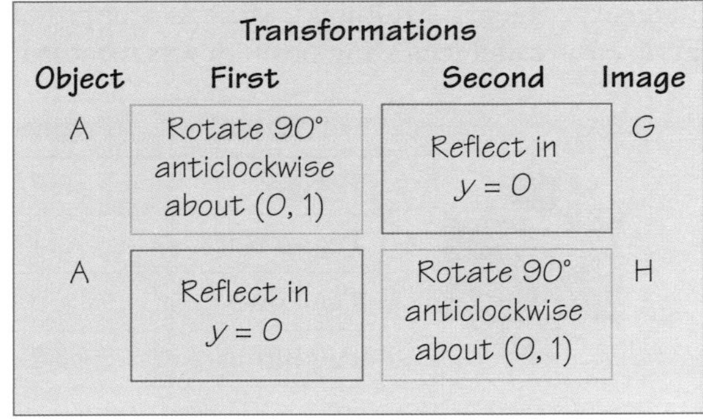

Object	Transformations: First	Second	Image
A	Rotate 90° anticlockwise about (0, 1)	Reflect in $y = 0$	G
A	Reflect in $y = 0$	Rotate 90° anticlockwise about (0, 1)	H

a On a new diagram draw the images G and H.

b Compare G and H and comment on any differences.

c In this table each pair of transformations maps G on to H. Describe the second transformation fully.

	Object	Image	First transformation	Second transforma
i	G	H	Translate $\begin{pmatrix} 6 \\ 0 \end{pmatrix}$	
ii	G	H	Rotate 90° clockwise about (1, $^-2$)	
iii	G	H	Rotate 180° about (0, $^-1$)	

d What single transformation maps G on to H?

5 These transformations map A on to J, K and L.

Object	Transformations: First	Second	Image
A	Enlarge SF 2 with centre (0, 1)	Reflect in $y = 1$	J
A	Reflect in $x = 1$	Enlarge SF 2 with centre (5, 0)	K
A	Enlarge SF 2 with centre (5, 0)	Reflect in $x = 4$	L

a On a new diagram, on axes with: $^-10 \leq x \leq 12$ and $^-10 \leq y \leq 10$ draw the images J, K and L

b What single transformation maps
 i J on to K ii L on to K?

Probability B

In an experiment a Multilink cube was dropped on to a hard surface and its resting position was recorded.

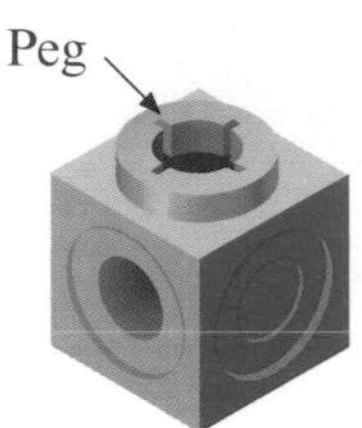

Position	Frequency
Peg up	7
Peg to side	51
Peg down	5
Peg tilting	2

1 Give the relative frequency to 2 dp of the cube coming to rest:

a peg down
b peg to side
c peg tilting
d peg up.

2 Why should all the unrounded answers to Questions **1a** to **1d** add up to 1?

3 What is the relative frequency of a cube resting:

a without the peg down
b with either a peg up or a peg down
c with neither a peg to the side nor a peg up?

4 From the results do you think that each face of the cube is equally likely to be on the top. Explain your answer.

5 The same Multilink cube is dropped 150 times on the same surface. Estimate the number of times it will come to rest:

a peg down
b peg to the side
c peg tilting.

6 Do an experiment yourself to find the relative frequency of different positions a Multilink cube can come to rest.

a Compare your relative frequencies with the results of the experiment above.
b How could you improve the accuracy of your experiment to find the relative frequencies?

7 The probability that Mike is late for work on a Monday is 0.4 and on a Tuesday it is 0.2

a Draw a tree diagram to show the outcomes and probabilities.
b Estimate the probability that Mike is late on Monday and Tuesday.
c Estimate the probability that Mike is late on neither day.
d Estimate the probability that he is late at least once in the two days.

8 Would you use theoretical probability or relative frequency to find the probability that:

a a person in the UK is right- or left-handed
b the next volcanic eruption occurs in May
c the next person you meet in school is female
d the next wine gum in the tube is black
e a toothpaste tube will leak before it is finished
f if you visited a foreign country at random you would be expected to drive on the left
g your toast lands jam side down on the floor if you drop it
h it rains when you have no waterproofs?

A biased spinner has seven sides. On the sides are the numbers 7, 15, 24, 25, 29, 30, 36.

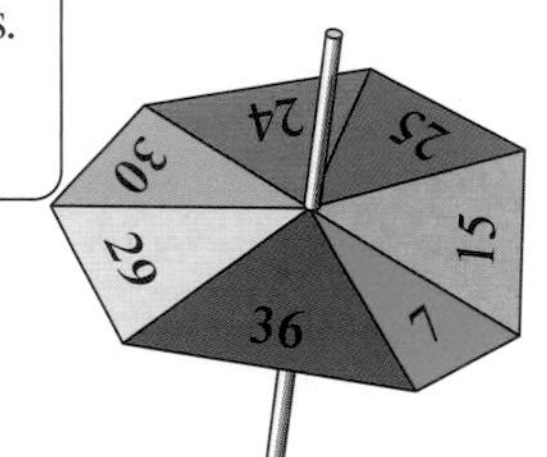

The probability of the spinner landing on each number is given in this table.

Number	7	15	24	25	29	30	36
Probability	0.04	0.15	0.19	0.14	0.14	0.13	0.21

9 **a** What is the probability of the spinner landing on a number less than 20?
b Why can you add the probabilities in this case?

10 **a** Complete the Venn diagram to show multiples of 3 and multiples of 5 for the spinner.

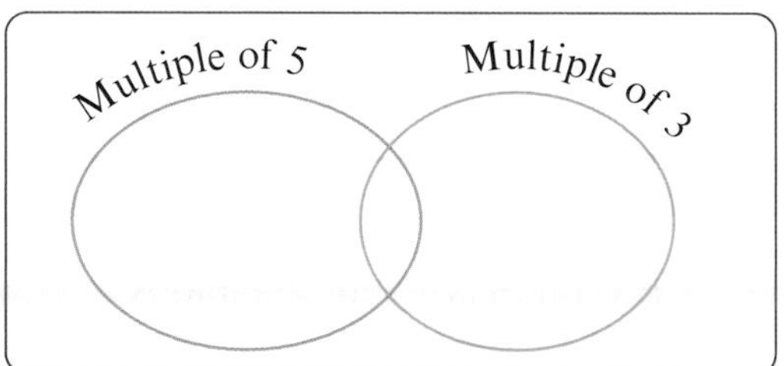

b Why can you not add probabilities in this case?

11 Calculate the probability of getting in one spin:

a a multiple of 5 and a multiple of 3
b either a multiple of 3 or a multiple of 5
c neither a multiple of 3 nor a multiple of 5
d a multiple of 3 but not a multiple of 5.

12 Find the probability of getting either a square number or an even number.

Inequalities

1 Copy each of these and insert > or < in place of the box to make it correct.

a 9 □ $^-4$ **b** $^-8$ □ $^-3$
c 4 □ $^-2$ **d** $^-2$ □ 2
e $^-5.6$ □ $^-4.23$ **f** $^-8$ □ $^-11$

2 Which of these numbers is not a possible value for t, where $t \leqslant {}^-7$?
$^-8$, 4, 71, $^-3.2$, $^-7.003$, $^-28.6$, $^-7$, $^-0.33$

3 What integer values for k satisfy these inequalities? Show each answer on a number line.

a $9 > k \geqslant 4$ **b** $^-3 \leqslant k \leqslant {}^-1$
c $^-5 \leqslant k < 1$ **d** $^-4 \geqslant k > {}^-7$
e $^-56 < k < {}^-57$ **f** $24 \leqslant k \leqslant 24$
g $^-14 \geqslant k > {}^-16$ **h** $6 > k > 0$

4 Write two other inequalities in h which describe the same integer values as $^-2 < h < 2$.

5 Explain why these two inequalities are different types of inequality from each other.

- This chair is suitable for people with weights given by $5 \leqslant w \leqslant 15$, where w is their weight in stones.

- The waiters in a restaurant are given by $5 \leqslant w \leqslant 15$ where w is the number of waiters.

6 This diagram shows the inequality $^-2 < x \leqslant 3$.

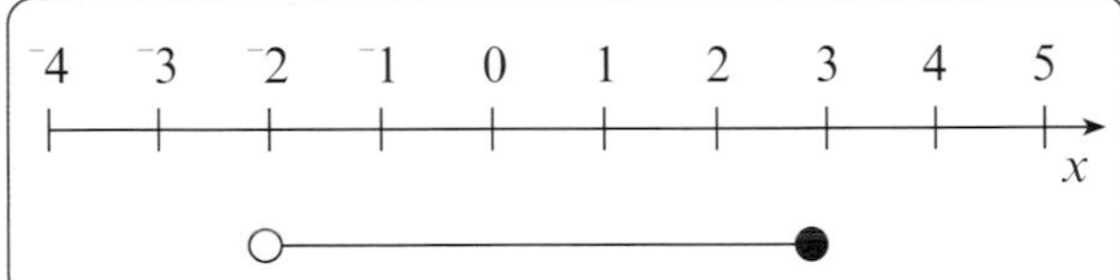

Draw similar diagrams to show these inequalities:

a $7 \leqslant x \leqslant 10$ **b** $^-4 \leqslant x < 1$
c $^-5 < x < {}^-2$ **d** $0 < x \leqslant 4$

7 Solve the following inequalities.

a $3x \geqslant 27$ **b** $6t < {}^-42$
c $42 \geqslant 4p$ **d** $k^2 \leqslant 121$
e $1 + 3w < 7$ **f** $5q + 7 > 53.5$
g $19 \leqslant 3t - 5$ **h** $7 - 2h \geqslant 3$
i $4(3f - 2) < 28$ **j** $c^2 - 5 < 76$
k $36 \leqslant 3(2g + 3)$ **l** $3x + 5 > x - 1$
m $3j - 2 < 2j + 17$ **n** $5d + 20 \geqslant 6 - 2d$
o $24 - 8v > 6 - 4v$ **p** $2(x+3) \leqslant x - 7$
q $5 - 3s > 5 + 3s$ **r** $9u + 6 \leqslant 7u$

8 The conditions on x are given by the inequality $^-4 \leqslant x < 3$.
What is:

a the smallest possible value of x^2
b the largest possible value of x^2?

9 Sketch graphs to show the regions which satisfy the following conditions.

Ⓐ ***Anita Southgate Floral Blinds***
Blinds fit windows between 1 metre and 2 metres wide and up to 1 metre 20 cm high.

Ⓑ **ANTOK SUPPLY SERVICES**
Drivers should be between 21 and 55 with a clean driving licence. We need drivers who have had experience of working with at least four previous companies.

Ⓒ ***Clarkson Hi Fi Sale***
Each rack holds up to 45 compact discs. We have different models with prices from £12.

Ⓓ **Opus 4123R Fax Machine**
Will take fax rolls up to 214 millimetres wide and up to 50 metres long.

Ⓔ **REGENT CAR SALES**
Our new models will do up to 50 miles per gallon at speeds between 50 and 60 miles per hour.

10 Sketch graphs and shade the regions which satisfy the following conditions.

a $x > 4$ **b** $^-3 < y < 4$
c $y < 6$ **d** $4 > x > 1$
e $x < 4$ and $y > 6$
f $^-1 < x < 2$ and $y < {}^-2$
g $0 < x < 4$ and $0 > y > {}^-4$
h $y > x - 2$
i $x + y < 6$
j $y < 2x + 2$

11 What two inequalities define the shaded region on this sketch?

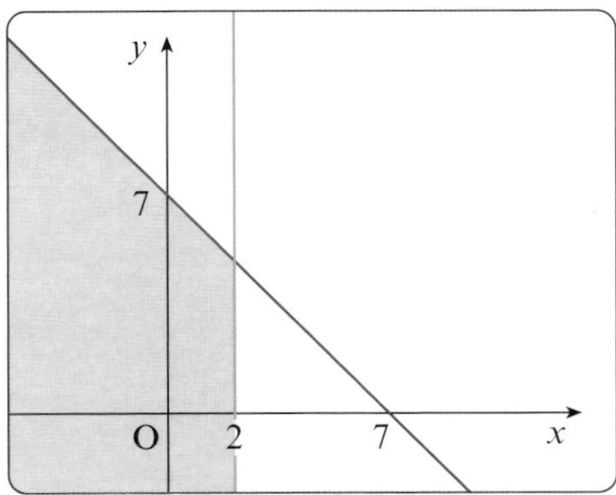

Working in 3D

1 Each of these solids is made from three cubes.

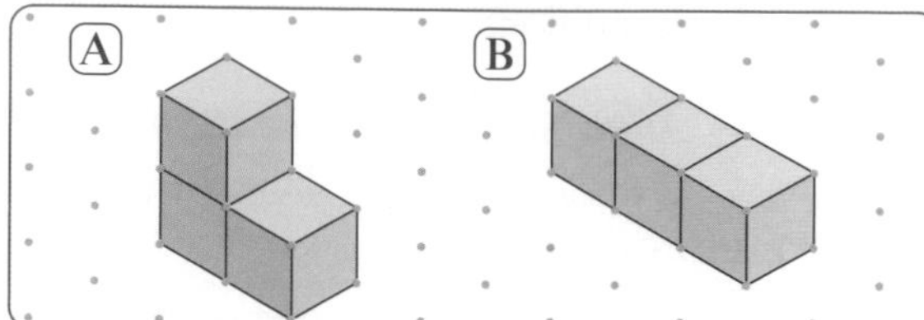

How many planes of symmetry has each solid?

2 a Draw a solid made from five cubes with 1 plane of symmetry.
b Draw a solid made from six cubes with 3 planes of symmetry.

3 How many planes of symmetry has:
a a cuboid with no square faces
b a cube?

4

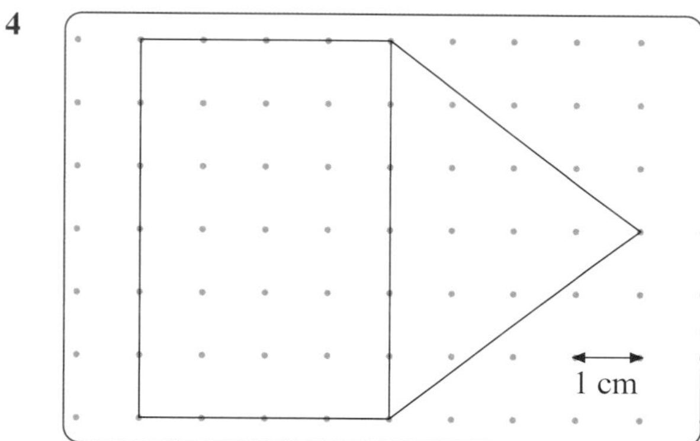

a Copy and complete this diagram to make the net of a prism.
b What is the volume of the box that can be made from the net?
c What is the surface area of the box that can be made from the net?

5 Solids X and Y are prisms.

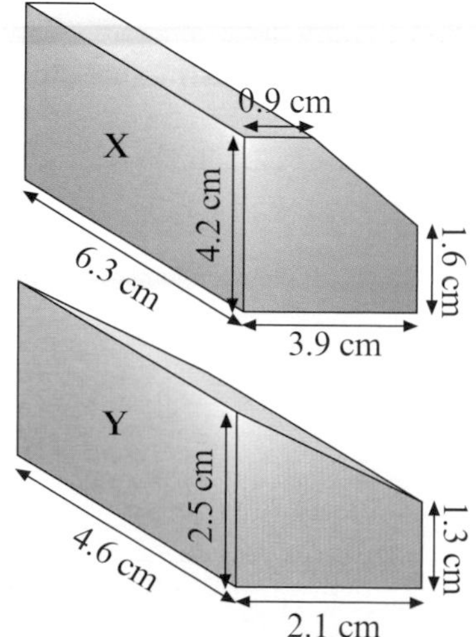

Calculate the volume of each solid.

6 Each prism P and Q has a volume of $1000\,\text{cm}^3$.

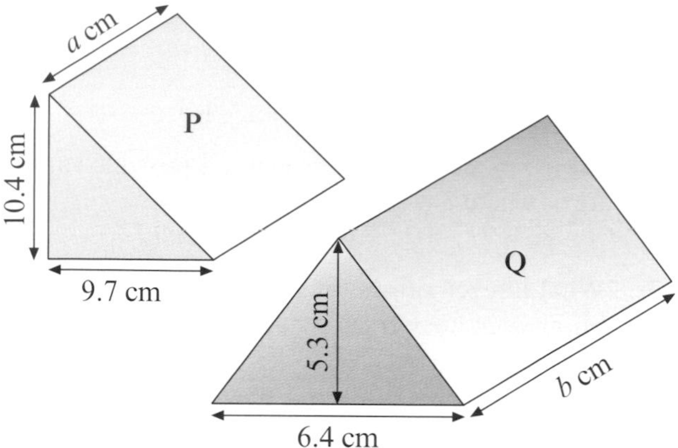

Find the values of a and b correct to 1 dp.

7 Each of these containers is in the shape of a cylinder.

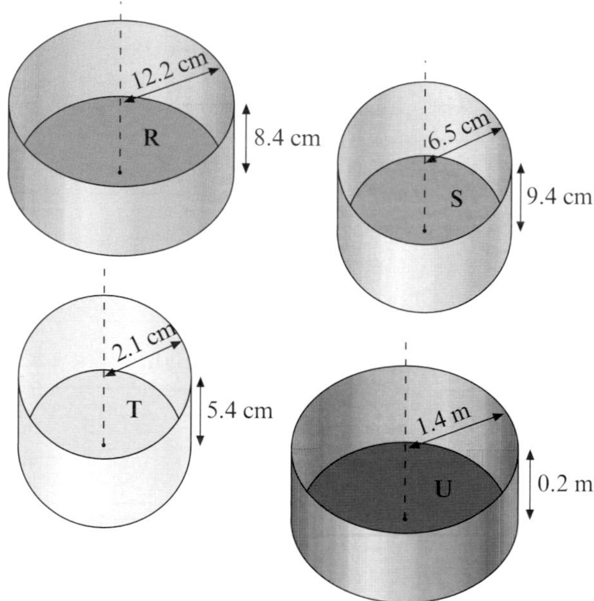

Find the volume of each container, to 1 dp:
a in ml b in litres.

8 In the following expressions, r, h and l each represent a length.
Decide if each expression represents:
a length, area, volume or none of these.
Give reasons for each answer.

a $4h$ b h^2
c rh d rhl
e $2(h + l)$ f pr^2
g $hl + r$ h $\frac{1}{3}rh^2$
i $\sqrt{h^2 + l^2}$ j $prh - 3l^2$
k $r^3 + h^2 + rl$ l $h^3 + rl^2$
m $5(rh + hl + rl)$ n $prh(h + l)$
o $2h(r - l)$ p $\frac{4}{3}p(r + h)$

IN FOCUS 22

Quadratics

1 Factorise and solve these quadratic equations:

a	$c^2 + 5c + 6 = 0$	**b**	$g^2 + g - 20 = 0$
c	$h^2 - 6h - 16 = 0$	**d**	$t^2 + 3t - 54 = 0$
e	$p^2 - 4p - 45 = 0$	**f**	$k^2 + 3k - 40 = 0$
g	$x^2 - 7x + 10 = 0$	**h**	$v^2 + 9v - 36 = 0$
i	$d^2 + 8d + 15 = 0$	**j**	$h^2 - 9h + 18 = 0$
k	$n^2 + 6n - 72 = 0$	**l**	$y^2 - 7y - 30 = 0$
m	$a^2 + 5a - 36 = 0$	**n**	$x^2 + 8x - 48 = 0$
o	$u^2 - 12u + 27 = 0$	**p**	$c^2 + 18c + 77 = 0$
q	$g^2 - g - 72 = 0$	**r**	$y^2 + y - 42 = 0$
s	$k^2 + 15k + 56 = 0$	**t**	$x^2 - 7x - 60 = 0$

2 Find a value for x by trial and improvement. (Give your answer correct to 1 dp.)

a	$x^2 - x = 8$	**b**	$x^3 + x = 5$
c	$x^2 - 2x = 4$	**d**	$x^3 - x = 4$
e	$2x^2 - x = 5$	**f**	$2x^3 - x = 6$

3 **a** Draw up a table of values for the graph of:

$$y = x^2 - 4x$$

with values of x from $^{-}2$ to $^{+}5$.

b From your table of values, draw and label the graph.

4 On a pair of axes **sketch**, and label these graphs:

a $y = x^2$ **b** $y = x^2 + 2$ **c** $y = x^2 - 3$

5 **a** Draw up a table of values for the graph of:

$$y = x^2 - 5x + 4$$

with values of x from 0 to $^{+}5$.

b From your table of values, draw and label the graph.

6 **a** Draw up a table of values for the graph of:

$$y = x^2 + 2x - 8$$

with values of x from $^{-}5$ to $^{+}3$.

b From your table of values draw and label the graph.

c Use your graph to solve the equation:

$$x^2 + 2x - 8 = 0$$

Explain your answers.

7 **a** With values of x from $^{-}7$ to $^{+}4$, draw up a table of values for the graph of:

$$y = x^2 + 3x - 18$$

b Use your graph to solve the equaton:

$$x^2 + 3x - 18 = 0$$

c Explain how you can check your answer.

8 **a** Draw the graph of $y = x^2 - 6x + 5$ for values of x from $^{-}1$ to $^{+}6$.

b Use your graph to solve the equation:

$$x^2 - 6x + 5 = 0$$

c On the same axes draw a graph of:

$$y = 2$$

d Use your graphs to solve the equation:

$$x^2 - 6x + 5 = 2$$

Give reasons for your answer.

9 **a** Draw the graph of $y = x^2 + 4x - 21$ for values of x from $^{-}8$ to $^{+}4$.

b Use your graph to solve the equation:

$$x^2 + 4x - 21 = 0$$

c On the same axes draw a graph of $y = {}^{-}5$.

d Use your graphs to solve the equation:

$$x^2 + 4x - 21 = {}^{-}5$$

e **i** Explain how you would solve the equation:

$$x^2 + 4x - 21 = 4$$

ii Solve the equation: $x^2 + 4x - 21 = 4$

10 Show how each of these equations can be made by rearranging $x^2 + 8x - 20 = 0$.

a $x^2 - 20 = {}^{-}8x$ **b** $20 = x^2 + 8x$
c $x^2 = 20 - 8x$

11 Show how each of these equations can be made by rearranging $x^2 - 7x = 60$.

a $x^2 - 60 = 7x$ **b** $x^2 = 7x + 60$
c $x^2 - 7x - 60 = 0$

12 Rearrange the equation $x^2 - 55 = 6x$ to make three other quadratic equations.

13 Make three other quadratic equations by rearranging $60 - 11x = x^2$.

14 With values of x from $^{-}4$ to $^{+}3$:

a Draw the graph of $y = x^2 + 3$.
b On the same axes draw a graph of $y = 2x + 4$.
c Use your graphs to solve the equation:

$$x^2 + 3 = 2x + 4$$

d Either by drawing another graph or using these graphs solve the equation:

$$x^2 - 2x - 1 = 0$$

15 Sketch the graph of $y = x^3 + 1$.

16 For values of x from $^{-}3$ to $^{+}3$ draw a graph of $y = x^3 + 2x$.

17 Use trial and improvement to solve the equation:

$$x^5 = 20$$

(Give your answer correct to 3 sf.)

Processing data

1

> ***Town Centre Survey***
>
> 1 Do you come into the town centre often? ☐ Yes ☐ No
>
> 2 Do you agree that the town centre should be pedestrianised? ☐ Yes ☐ No
>
> 3 What do you think about buses? ______________

a Criticise each of these questions.
b Write an improved question for each one.

2

> ***Body Matters***
> Right-handed people are more likely to have a stronger left eye than right eye.

Design an experiment to test this hypothesis.

3 **a** Carry out your experiment.
b Analyse the data you collect.

4 **a** Do you think the hypothesis is true or false?
b Explain why.

5

> ***Body Matters***
> Do right-handed people fold their arms differently to left-handed people?

Design an experiment to answer this question.

6 **a** Carry out your experiment.
b Analyse the data you collect.
c Interpret your results to answer the question.

7

Motorbikes – Size of Engine & Price

Engine size (cc)	250	900	600	125	650	900	500	750
Price (£)	4500	8100	6700	2400	5100	6700	3400	9200

a Draw axes: horizontal 0 to 1200
vertical 0 to 12 000
b Plot the motorbike data on your diagram.
c Draw a line of best fit.
d Describe any correlation between size of engine and price.

8 Estimate the price of a motorbike with engine size:
a 800 cc **b** 450 cc

9 Estimate the engine size of a motorbike costing:
a £4000 **b** £6500

10 Extend the line of best fit on your scatter diagram to estimate:
a the price of a motorbike with a 1000 cc engine
b the engine size of a motorbike costing £10 000.

11

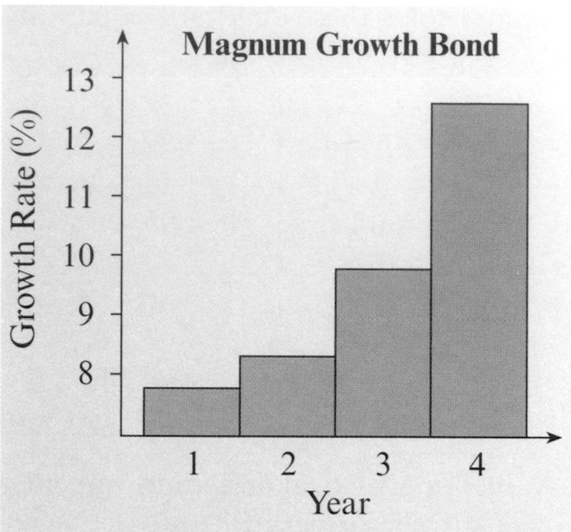

Explain why this diagram is misleading.

12

Explain why this diagram is misleading.

13 The graph was used under the headline 'Fantastic improvement in safety'.

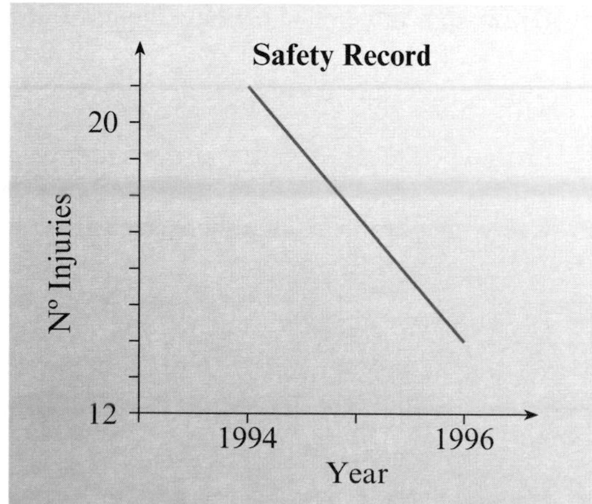

a What data does the graph actually show?
b Explain what was done to the graph to support the headline.
c Draw a graph of the data showing how you think it should look.
d What can you say about the number of injuries in 1995?

Trigonometry and Pythagoras' rule

1

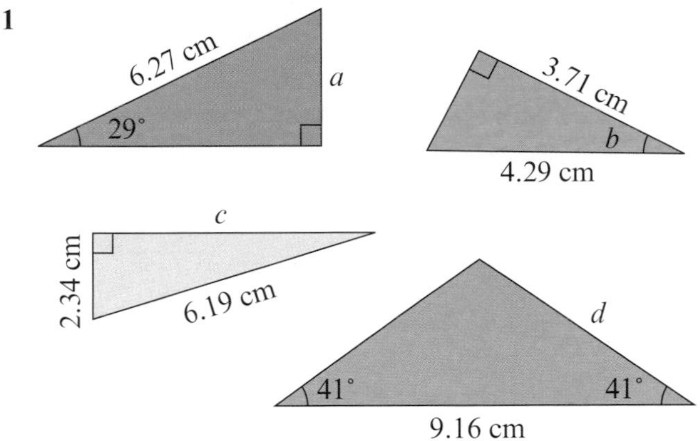

For each triangle calculate the side or angle marked with a letter, correct to 3 sf.

2

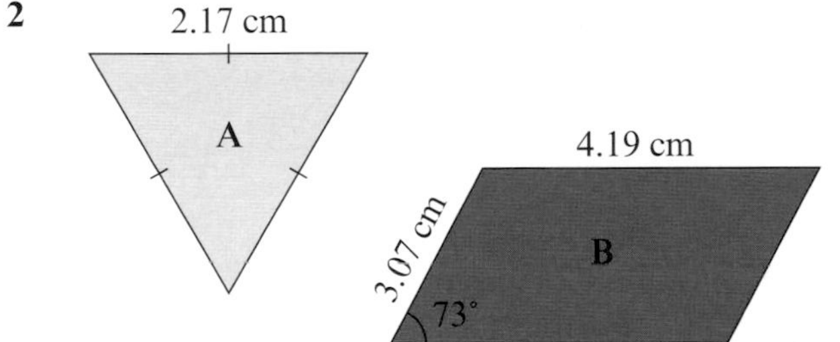

Calculate the area and perimeter of shapes A and B, correct to 2 dp.

3 This diagram shows the top of a building T and a person at G.
The horizontal distance GB is 5.19 m.
The point E is 1.82 m above the ground.
The angle of elevation from E to T is 39°.

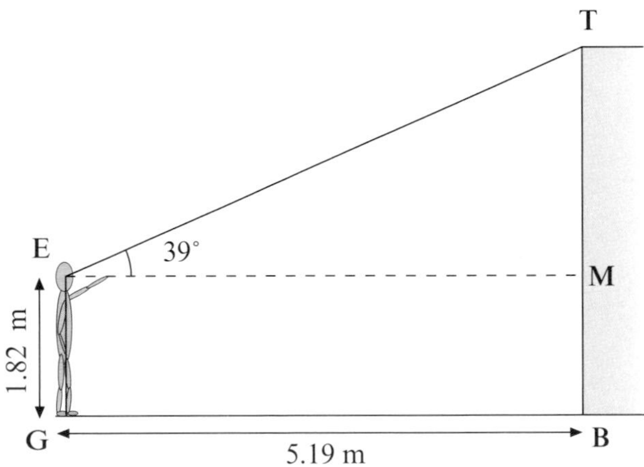

a Calculate, to the nearest 0.01 m:
  i the vertical height MT
  ii the total height TB.
b Calculate the angle of depression from E to B to the nearest degree.

4 This diagram shows the position of three markers A, B and C on a map.
N is a point due north of A and due west of B.

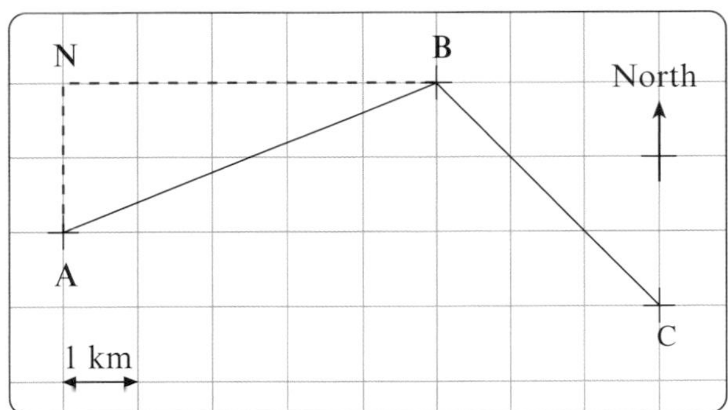

a How far is A:
  i west of B   ii south of B?
b Give the bearing of :
  i C from B   ii B from A   iii A from C
c In Δ ABN, calculate:
  i AB in km to 2 dp
  ii $N\hat{B}A$ to the nearest degree.
d Calculate the distance BC in km to 2 dp.
e Calculate the distance between A and C.

5 Double-decker buses are tested for stability for angles up to 28° from the horizontal.

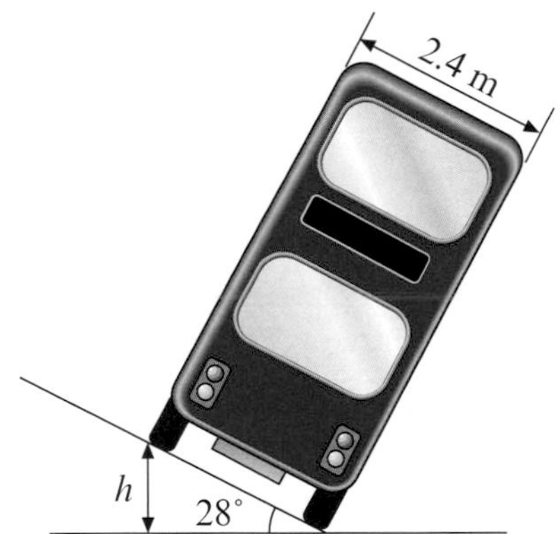

a Calculate the height h in metres, for a double-decker when it is tipped to the maximum test angle, correct to 1 dp.
b For a single-decker bus the test angle is smaller. A single decker bus 2.8 metres wide is tested and the height h is 1.31 metres.
Find the test angle, correct to 1 dp.

Circle properties

1 Draw any circle, and a chord AB.

a Label:

i the minor segment

ii the major segment.

A chord CD is such that the circle is divided into two segments that are the same size.

b What can you say about the chord CD?

2

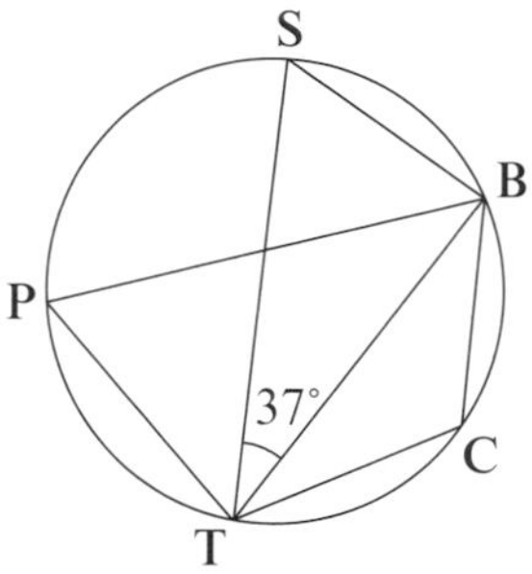

a Name an angle in the minor segment.

b Calculate the size of TBS when $T\hat{P}B = 75°$. Explain your answer.

3 In the diagram PT is a chord and O the centre of the circle.

Calculate the size of $T\hat{P}O$. Give reasons for your answer.

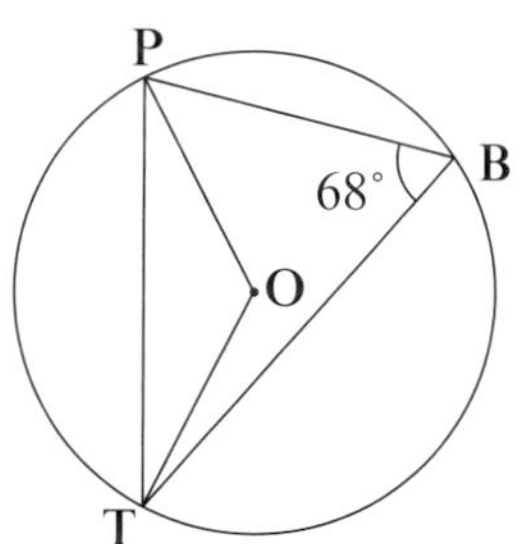

4 The circle has a centre O.

a Calculate the size of $P\hat{T}B$. Explain your answer.

b What is the size of $A\hat{C}B$? Explain your answer.

c Explain why $A\hat{B}P$ is 76°.

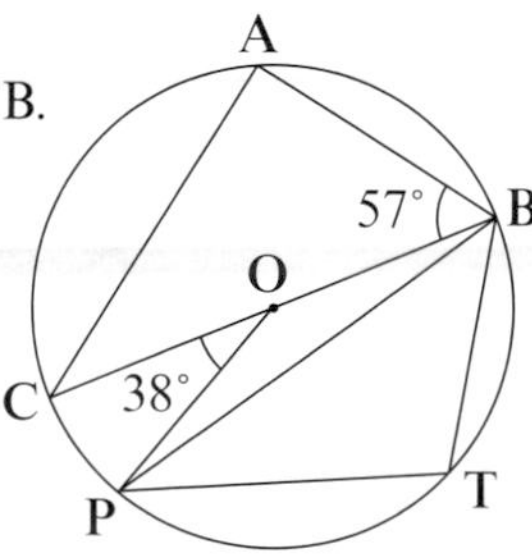

5 The circle has a centre O.

a Explain why $P\hat{A}T$ is 42°.

b Calculate $P\hat{C}T$. Explain your answer.

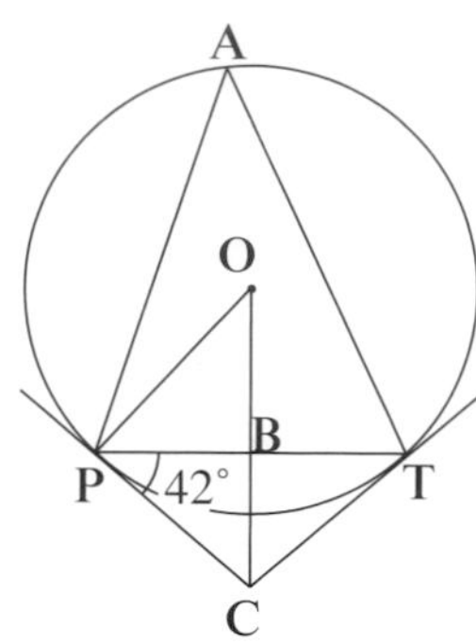

6 a Explain why $T\hat{B}P$ is 35°.

b Calculate $B\hat{P}T$. Give reasons for your answer.

c Explain why $A\hat{C}B$ is 73°.

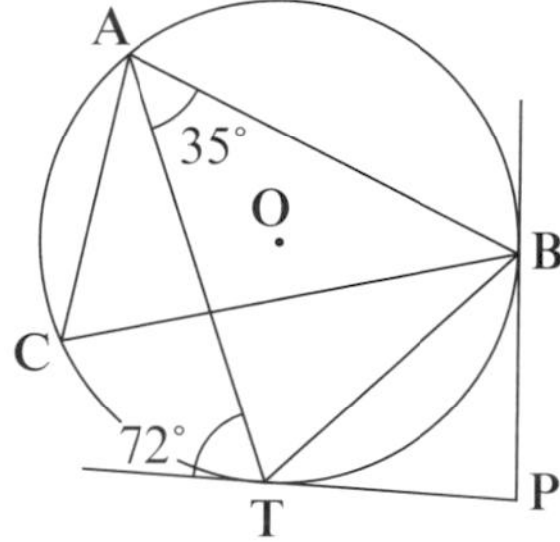

7 a Explain why $E\hat{C}D = 31°$.

b Calculate the size of $A\hat{B}C$. Explain your answer.

c $E\hat{A}B = 52°$ Give two different explanations to show this.

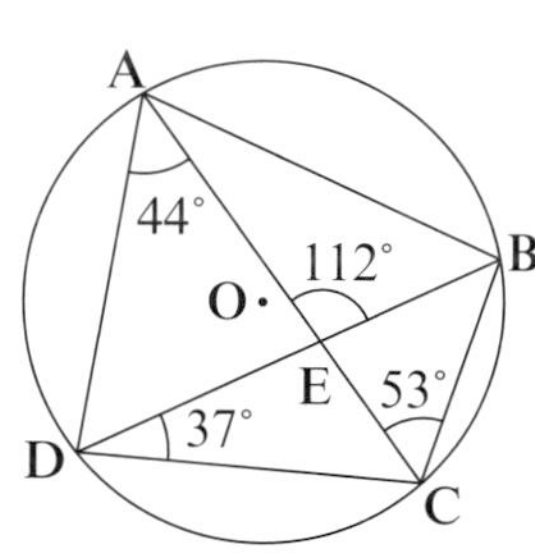

8 Show why an expression for $D\hat{A}B$ is given as:

$180 - a$

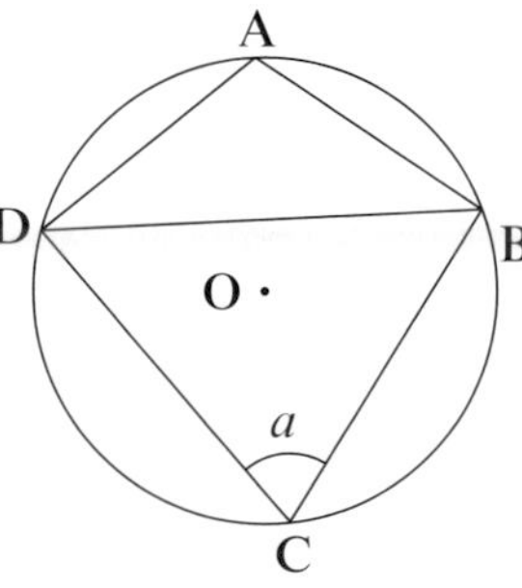

For your GCSE Maths you will need to produce two pieces of coursework:

1. Investigative
2. Statistical

Each piece is worth 10% of your exam mark.
This unit gives you guidance on approaching your coursework and will tell you how to get good marks for each piece.

Investigative coursework

In your investigative coursework you will need to:

- Say how you are going to carry out the task and provide a plan of action
- Collect results for the task and consider an appropriate way to represent your result
- Write down observations you make from your diagrams or calculations
- Look at your results and write down any observations, rules or patterns – try to make a general statement
- Check out your general statements by testing further data – say if your test works or not
- Develop the task by posing your own questions and provide a conclusion to the questions posed
- Extend the task by using techniques and calculations from the content of the higher tier
- Make your conclusions clear and link them to the original task giving comments on your methods.

Assessing the task

Before you start, it is helpful for you to know how your work will be marked. This way you can make sure you are familiar with the sort of things which examiners are looking for.

Investigative work is marked under three headings:

1. **Making and monitoring decisions to solve problems**
2. **Communicating mathematically**
3. **Developing skills of mathematical reasoning**

Each strand assesses a different aspect of your coursework. The criteria are:

1. Making and monitoring decisions to solve problems

This strand is about deciding what needs to be done then doing it. The strand requires you to select an appropriate approach, obtain information and introduce your own questions which develop the task further.

For the higher marks you need to analyse alternative mathematical approaches and, possibly, develop the task using work from the higher tier GCSE syllabus content.

For this strand you need to:

- solve the task by collecting information
- break down the task by solving it in a systematic way
- extend the task by introducing your own *relevant* questions
- extend the task by following through alternative approaches

- develop the task by including a number of mathematical features
- explore the task extensively using higher level mathematics

2. Communicating mathematically

This strand is about communicating what you are doing using words, tables, diagrams and symbols. You should make sure your chosen presentation is accurate and relevant.

For the higher marks you will need to use mathematical symbols accurately, concisely and efficiently in presenting a reasoned argument.

For this strand you need to:

- illustrate your information using diagrams and calculations
- interpret and explain your diagrams and calculations
- begin to use mathematical symbols to explain your work
- use mathematical symbols consistently to explain your work
- use mathematical symbols accurately to argue your case
- use mathematical symbols concisely and efficiently to argue your case

3. Developing skills of mathematical reasoning

This strand is about testing, explaining and justifying what you have done. It requires you to search for patterns and provide generalisations. You should test, justify and explain any generalisations.

For the higher marks you will need to use provide a sophisticated and rigorous justification, argument or proof which demonstrates a mathematical insight into the problem.

For this strand you need to:

- make a general statement from your results
- confirm your general statement by further testing
- make a justification for your general statement
- develop your justification further
- provide a sophisticated justification
- provide a rigorous justification, argument or proof

This unit uses a series of investigative tasks to demonstrate how each of these strands can be achieved.

Task 1

TRIANGLES

Patterns of triangles are made as shown in the following diagrams:

Pattern 1

Pattern 2

Pattern 3

What do you notice about the pattern number and the number of triangles?

Investigate further.

Note: You are reminded that any coursework submitted for your GCSE examination must be your own. If you copy from someone else then you may be disqualified from the examination.

Planning your work

A straightforward approach to this task is to continue the pattern further. You can record your results in a table and then see if you can make any generalisations.

Collecting information

A good starting point for any investigation is to collect information about the task.

You can see that:

- Pattern 1 shows 1 triangle
- Pattern 2 shows 4 triangles
- Pattern 3 shows 9 triangles

You can then extend this by considering further patterns.

Moderator comment

It is a good idea not to collect too much data as this is time consuming ... aim to collect 4 or 5 items of data to start with.

Pattern 4

16 triangles

Pattern 5

25 triangles

Drawing up tables

The information is not easy to follow so it is **always** a good idea to illustrate it in a table.

A table to show the relationship between the pattern number and the number of triangles

It is important to give your table a title and to make it quite clear what the table is showing.

Pattern number	1	2	3	4	5
No. of triangles	1	4	9	16	25

Now you try ...

Using the table:

- Can you see anything special about the number of triangles?
- Ask yourself: are they odd numbers, even numbers, square numbers, triangle numbers, are they multiples, do they get bigger, do they get smaller ...
- What other relationships might you look for?

Being systematic

Another important aspect to your work is to be systematic.

This task shows you what that means.

Task 2

ARRANGE A LETTER

How many different ways can you arrange the letters ABCD?

Investigate further.

Note: You are reminded that any coursework submitted for your GCSE examination must be your own. If you copy from someone else then you may be disqualified from the examination.

You could haphazardly try out some different possibilities:

ABCD
BCDA
CDAB etc

You may or may not find all the arrangements this way. To be sure of finding them all you should list the arrangements systematically.

To be systematic you group letters starting with A:

Notice the system here:
A followed by B,
then A followed by C,
then A followed by D
and so on.

ABCD
ABDC
ACBD
ACDB
ADBC
ADCB

Now you try ...

Now see if you can systematically produce all of the arrangements starting with B (you should find six of them) then complete the other arrangements.

You should also be systematic about the way you collect your data. In task 2, you were asked to find the arrangements of 4 letters: A, B, C and D.

To approach the whole task systematically, you should start by finding the arrangements for:

- 1 letter: A
- 2 letters: A and B
- 3 letters: A, B and C

This will give you more information about the task.

For different numbers of letters you should find:

Arrangements of different numbers of letters

Number of letters	1	2	3	4
Number of arrangements	1	2	6	24

You should now try and explain what your table tells you:

From my table I can . . .

Finding a relationship – making a generalisation

To make a generalisation, you need to find a relationship from your table of results. A useful method is to look at the differences between terms in the table.

Here is the table of results from task 1:

A table to show the relationship between the pattern number and the number of triangles

Pattern number	1		2		3		4		5	
No of triangles	1		4		9		16		25	
		+3		+5		+7		+9		+11
			+2		+2		+2		+2	

In this table the first "differences" are going up in two's.

The "second differences" are all the same ... this tells you that the relationship is quadratic, so you should compare the numbers with the square numbers

Here are some generalisations you could make:

From my table I notice that the number of triangles are all square numbers.

From my table I notice that the number of triangles are the pattern numbers squared.

Moderator comment

Remember that the idea is to identify some relationship from your table or, graph so that there is some point in you using it in your work ... a table or graph without any comment on the findings is of little use.

A graph may help see the relationship:

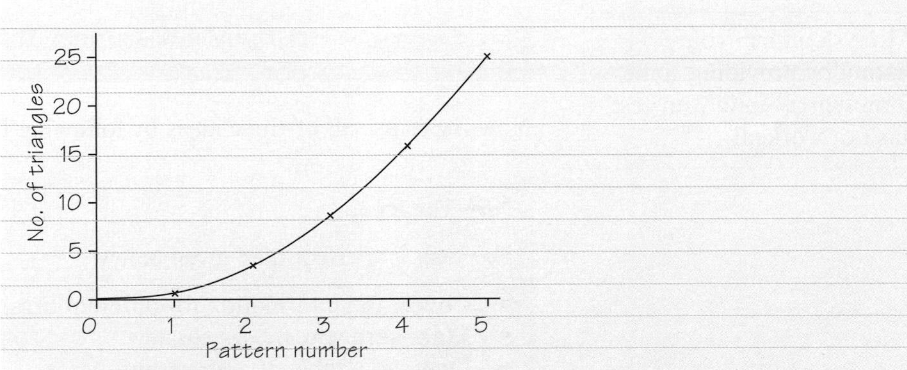

From my graph I notice that there is a quadratic relationship between the number of triangles and the pattern number.

Using algebra

You will gain marks if you can express your generalisation or rule using algebra.

The rule:

I notice that the number of triangles are the pattern numbers squared.

can be written in algebra:

Remember that you must explain what t and p stand for.

$t = p^2$ where t is the number of triangles and p is the pattern number

Testing generalisations

You need to confirm your generalisation by further testing.

From my table I notice that the number of triangles are the pattern numbers squared so that the sixth pattern will have $6^2 = 36$ triangles.

You can confirm this with a diagram:

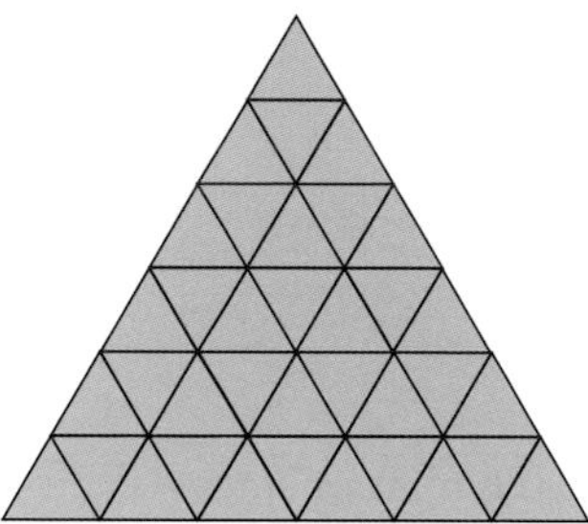

Always confirm your testing by providing some comment, even if your test has not worked!

My diagram confirms the generalisation for the sixth pattern.

Now try to use all of these ideas by following task 3.

Now you try ...

For task 3: 'Perfect tiles' given below:

- Collect the information for different arrangements
- Make sure you are systematic
- Draw up a table of your results.

Task 3

PERFECT TILES

Floor spacers are used to give a perfect finish when laying tiles on the kitchen floor.

Three different spacers are used including

L spacer
T spacer
+ spacer

Here is a 3 × 3 arrangement of tiles

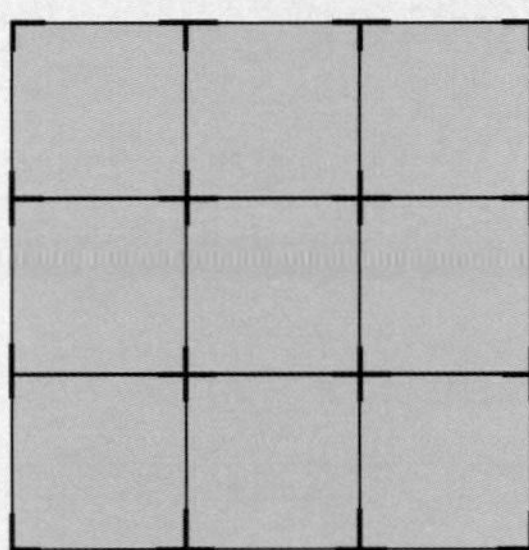

Draw picture of 3 × 3 arrangement

This uses 4 **L** spacers
8 **T** spacers and
4 **+** spacers

Investigate different arrangements of tiles.

Note: You are reminded that any coursework submitted for your GCSE examination must be your own. If you copy from someone else then you may be disqualified from the examination.

You should be able to produce this table:

Number of spacers for different arrangements of tiles

Size of square	1 × 1	2 × 2	3 × 3	4 × 4
Number of L spacers	4	4	4	4
Number of T spacers	0	4	8	12
Number of + spacers	0	1	4	9

Now you try ...

What patterns do you notice from the table?
What general statements can you make?
Now test your general statements.

Generalisations for the 'Perfect tiles' task might include:

$L = 4$ where L is the number of L spacers

$T = 4(n - 1)$ where T is the number of T spacers and n is the size of the arrangement ($n \times n$)

The formula $T = 4(n - 1)$ can also be written as $T = 4n - 4$.

Test:

For a 5 × 5 arrangement (ie $n = 5$)

$T = 4(n - 1)$
$T = 4(5 - 1)$
$T = 4 \times 4$
$T = 16$

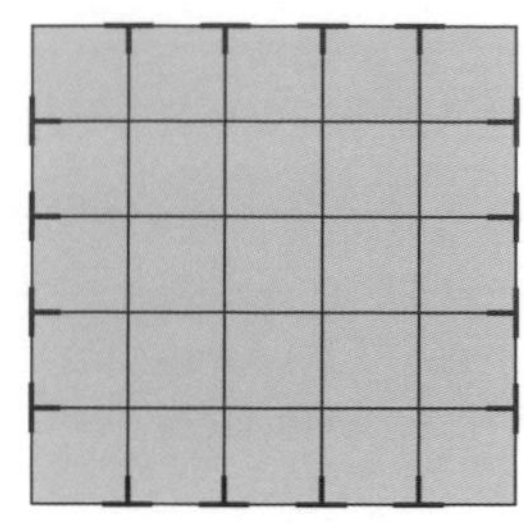

The number of T spacers is 16 so the formula works for a 5 × 5 arrangement.

$+ = (n - 1)^2$ where + is the number of + spacers and n is the size of the arrangement ($n \times n$)

Test:
For a 6 × 6 arrangement (ie $n = 6$)

$+ = (n - 1)^2$
$+ = (6 - 1)^2$
$+ = (5)^2$
$+ = 25$

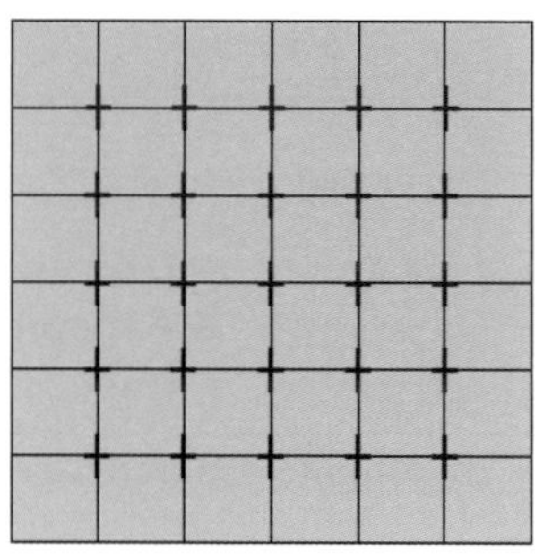

The number of + spacers is 25 so the formula works for a 6 × 6 arrangement.

Making justifications

Once you have found the general statement and tested it then you need to justify it. You justify the statement by explaining WHY it works.

For example, here are possible justifications for the rules in task 3:

WHY is the number of L spacers always 4?

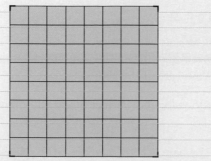

The number of L spacers is always 4 because there are 4 corners to each arrangement.

WHY are all of the T spacers multiples of 4?

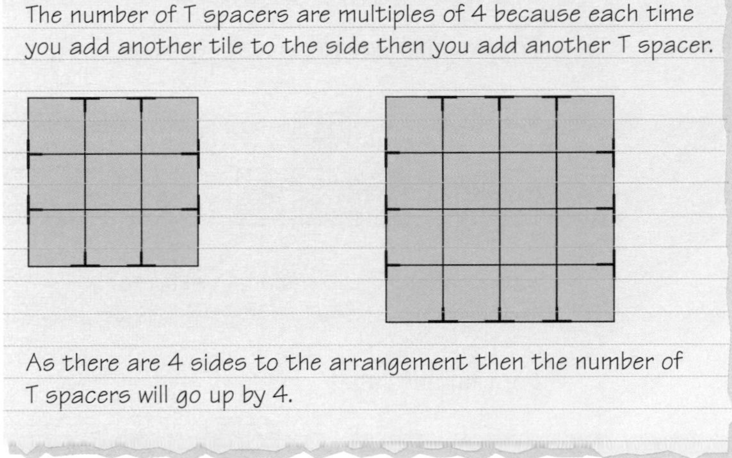

The number of T spacers are multiples of 4 because each time you add another tile to the side then you add another T spacer.
As there are 4 sides to the arrangement then the number of T spacers will go up by 4.

WHY are all of the + spacers square numbers?

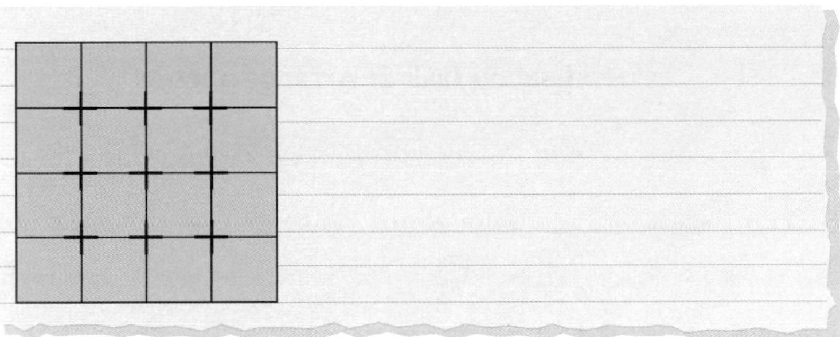

Now you try ...

Explain why the number of + spacers are **always** square numbers.

Extending the problem – investigating further

Once you have understood and explained the basic task, you should extend your work to get better marks.

To extend a task you need to pose your own questions.
This means that you must think of different ways to extend the original task.

Extending task 1: The triangle problem

You could ask and investigate:

What about different patterns?

For example, triangles

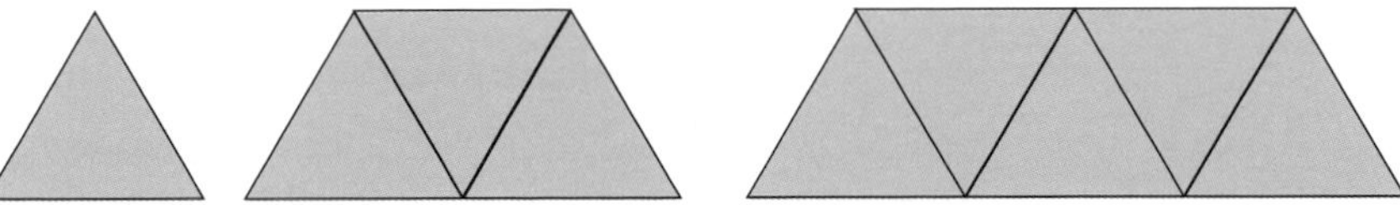

What about patterns of different shapes?

For example, squares

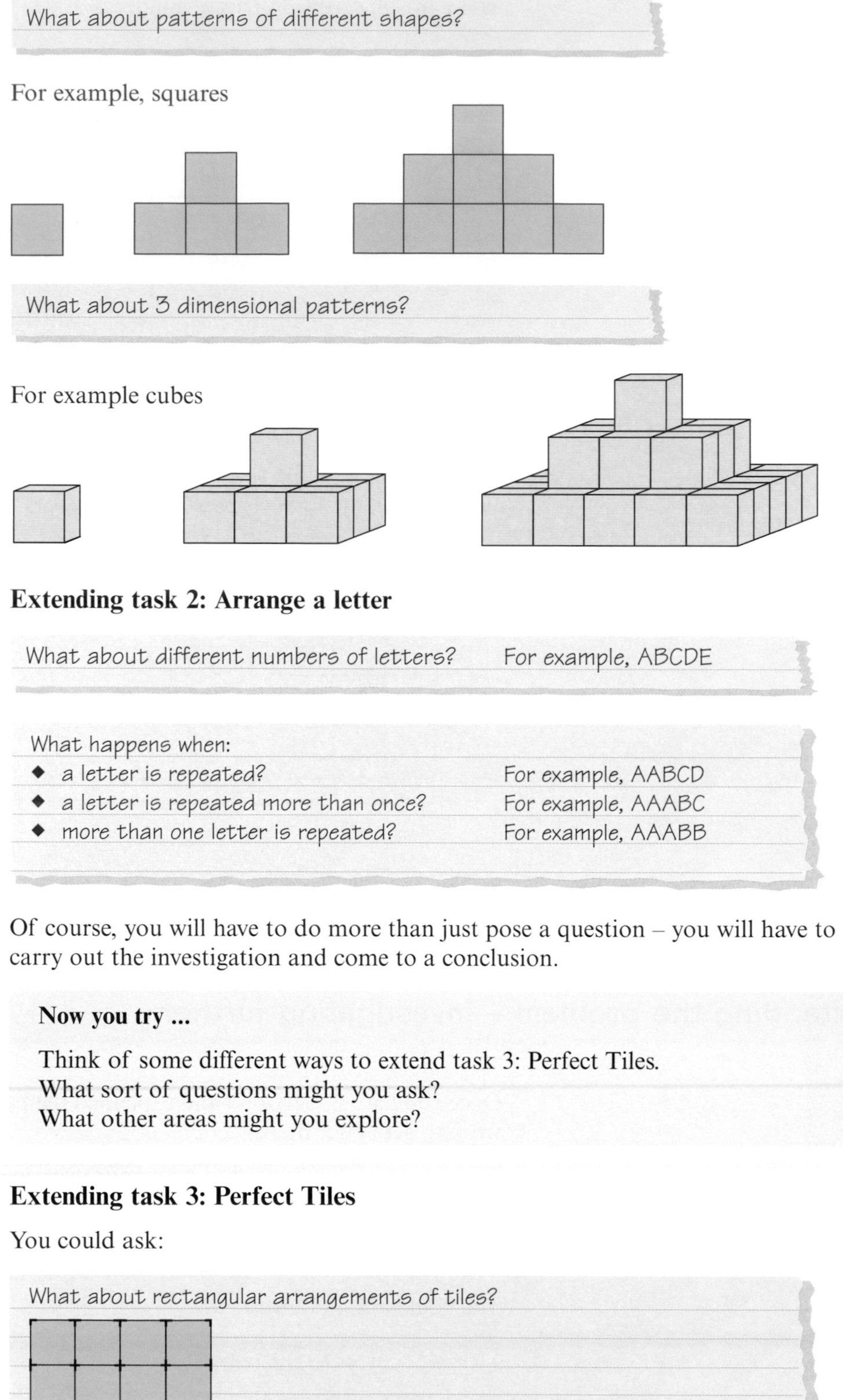

What about 3 dimensional patterns?

For example cubes

Extending task 2: Arrange a letter

What about different numbers of letters? For example, ABCDE

What happens when:

- a letter is repeated? For example, AABCD
- a letter is repeated more than once? For example, AAABC
- more than one letter is repeated? For example, AAABB

Of course, you will have to do more than just pose a question – you will have to carry out the investigation and come to a conclusion.

Now you try ...

Think of some different ways to extend task 3: Perfect Tiles.
What sort of questions might you ask?
What other areas might you explore?

Extending task 3: Perfect Tiles

You could ask:

What about rectangular arrangements of tiles?

What about different shaped tiles?
For example, triangles

Remember to work systematically.

Consider rectangular arrangements of tiles:

Number of spacers for different arrangements of tiles

Size of arrangement	2 × 1	2 × 2	2 × 3	2 × 4	2 × 5
Number of L spacers	4	4	4	4	4
Number of T spacers	2	4	6	8	10
Number of + spacers	0	1	2	3	4
Number of spacers	6	9	12	15	18

Number of spacers for different arrangements of tiles

Size of arrangement	3 × 1	3 × 2	3 × 3	3 × 4	3 × 5
Number of L spacers	4	4	4	4	4
Number of T spacers	4	6	8	10	12
Number of + spacers	0	2	4	6	8
Number of spacers	8	12	16	20	24

Rules for the 'Perfect tiles' task extension might include:

Note that the formula
$T = 2(x - 1) + 2(y - 1)$
can also be written
$T - 2(x \quad 1) + 2(y \quad 1)$
$T = 2x - 2 + 2y - 2$
$T = 2x + 2y - 4$

It is sensible to replace + by P so that P stands for the number of + spacers.

$L = 4$	where L is the number of L spacers
$T = 2(x - 1) + 2(y - 1)$	where T is the number of T spacers and x is the length of the arrangement and y is the width of the arrangement $(x \times y)$
$P = (x - 1)(y - 1)$	where P is the number of + spacers and x is the length of the arrangement and y is the width of the arrangement $(x \times y)$
$N = (x + 1)(y + 1)$	where N is the total number of spacers and x is the length of the arrangement and y is the width of the arrangement $(x \times y)$

Moderator comment

The use of higher order algebra can result in high marks awarded on the middle strand as well as an opportunity to provide a sophisticated justification

The total number of spacers should equal the number of L spacers plus the number of T spacers plus the number of + spacers

Proof: $N = L + T + P$

$$= 4 + [2(x - 1) + 2(y - 1)] + [(x - 1)(y - 1)]$$
$$= 4 + 2x - 2 + 2y - 2 + xy - y - x + 1$$
$$= xy + x + y + 1$$
$$= (x + 1)(y + 1)$$

Since $N = (x + 1)(y + 1)$ is true, then my theory is proved

Task 4

SQUARE SEA SHELLS

Square sea shells are formed from squares whose pattern of growth is shown as follows:

Day 1

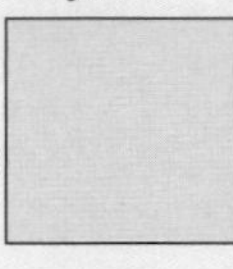

Day 2

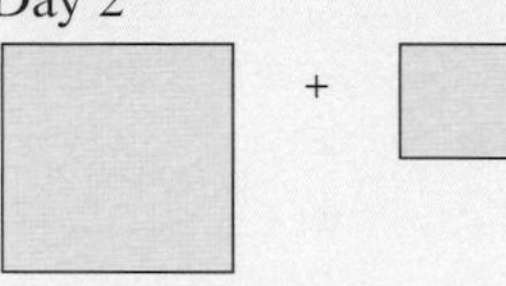

The length of the new square is half that of the previous day

Day 3

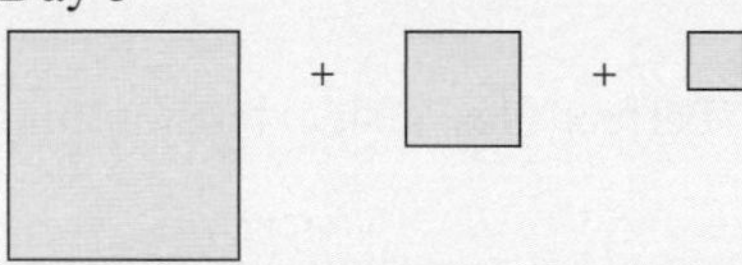

The length of the new square is half that of the previous day

Day 4

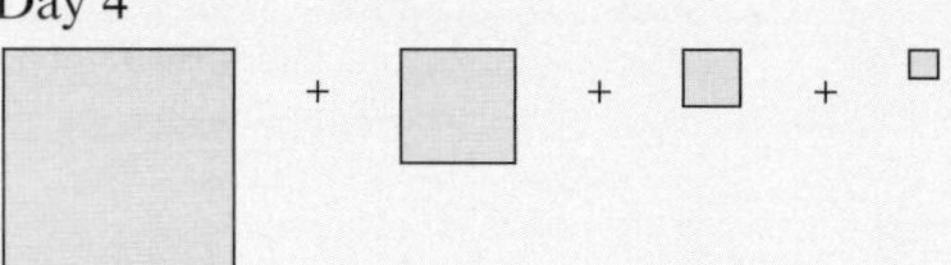

The length of the new square is half that of the previous day

This pattern of growth continues

Investigate the area covered by square sea shells on different days.

Investigate further.

For a square of side a

Day 1 Area $= a^2$

Day 2 Area $= a^2 + \frac{a^2}{4}$

Day 3 Area $= a^2 + \frac{a^2}{4} + \frac{a^2}{16}$

Day 4 Area $= a^2 + \frac{a^2}{4} + \frac{a^2}{16} + \frac{a^2}{64}$

The proof of the formula is beyond GCSE and the rest of the work is left as a challenge for the reader.

In this unit we have tried to give you some hints on approaching investigative coursework to gain your best possible mark.

This mathematics is often useful in investigative tasks:

- Creating tables
- Drawing and interpreting graphs
- Recognising square numbers
- Recognising triangular numbers
- Finding the nth term of a linear sequence

Summary

These are the grade criteria your coursework will be marked by:

Identifying information and making statements (grade E/F)

To achieve this level you must:

- solve the task by collecting information
- illustrate your information using diagrams and calculations
- make a general statement from your results

Testing general statements (grade D)

To achieve this level you must:

- break down the task by solving it in an orderly manner
- interpret and explain your diagrams and calculations
- confirm your general statement by further testing

Posing questions and justifying (grade C)

To achieve this level you must:

- extend the task by introducing your own relevant questions
- begin to use mathematical symbols to explain your work
- make a justification for your general statement
- provide a sophisticated justification

Making further progress (grade B)

To achieve this level you must:

- extend the task by following through alternative approaches
- use mathematical symbols consistently to explain your work
- develop you justification further

Justifying a number of mathematical features (grade A)

To achieve this level you must:

- develop the task by including a number of mathematical features
- use mathematical symbols accurately to argue your case
- provide a sophisticated justification

Statistical coursework

In your statistical coursework you will need to:

- Provide a well considered hypothesis and provide a plan of action to carry out the task
- Decide what data is needed and collect results for the task using an appropriate sample size and sampling method
- Consider the most appropriate way to represent your results and write down any observations you make
- Consider the most appropriate statistical calculations to use and interpret your findings in terms of the original hypothesis
- Develop the task by posing your own questions – you may need to collect further data to move the task on
- Extend the task by using techniques and calculations from the content of the higher tier
- Make your conclusions clear – always link them to the original hypothesis recognising limitations and suggesting improvements.

Assessing the task

It may be helpful for you to know how the work is marked. This way you can make sure you are familiar with the sort of thing that examiners are looking for.

Statistical work is marked under three headings:

1. Specifying the problem and planning
2. Collecting, processing and representing the data
3. Interpreting and discussing the results

Each strand assesses a different aspect of your coursework as follows:

1. Specifying the problem and planning

This strand is about choosing a problem and deciding what needs to be done then doing it. The strand requires you to provide clear aims, consider the collection of data, identify practical problems and explain how you might overcome them. For the higher marks you need to decide upon a suitable sampling method, explain what steps were taken to avoid possible bias and provide a well structured report.

2. Collecting, processing and representing the data

This strand is about collecting data and using appropriate statistical techniques and calculations to process and represent the data. Diagrams should be appropriate and calculations mostly correct. For the higher marks you will need to accurately use higher level statistical techniques and calculations from the higher tier GCSE syllabus content.

3. Interpreting and discussing the results

This strand is about commenting, summarising and interpreting your data. Your discussion should link back to the original problem and provide an evaluation of the work undertaken. For the higher marks you will need to provide sophisticated and rigorous interpretations of your data and provide an analysis of how significant your findings are.

This unit uses a series of statistical tasks to demonstrate how each of these strands can be achieved.

Planning your work

Statistical coursework requires careful planning if you are to gain good marks. Before undertaking any statistical investigation, it is important that you plan your work and decide exactly what you are going to investigate – do not be too ambitious!

Before you start you should:

- decide what your investigation is about and why you have chosen it
- decide how you are going to collect the information
- explain how you intend to ensure that your data is representative
- detail any presumptions which you are making

Getting started – setting up your hypothesis

Moderator comment

Your statistical task must always start with a 'hypothesis' where you state exactly what you are investigating and what you expect to find.

A good starting point for any statistical task is to consider the best way to collect the data and then to write a clear hypothesis you can test.

Task 1

WHAT THE PAPERS SAY

Choose a passage from two different newspapers and investigate the similarities and differences between them.

Write down a hypothesis to test.

Design and carry out a statistical experiment to test the hypothesis.

Investigate further

Note: You are reminded that any coursework submitted for your GCSE examination must be your own. If you copy from someone else then you may be disqualified from the examination.

First consider different ways to compare the newspapers:

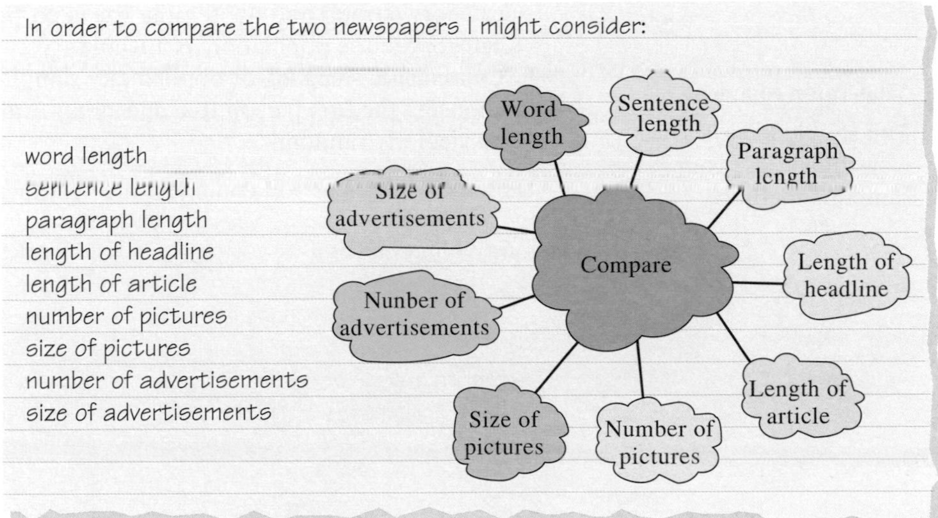

Now formulate your hypothesis.

Here are some possible hypotheses:

For my statistics coursework, I am going to investigate the hypothesis that 'tabloid' papers use shorter words than 'broadsheet' newspapers.

My hypothesis is that word lengths in the tabloid newspaper will be shorter than word lengths in the broadsheet newspaper.

My hypothesis is that sentence lengths in the tabloid newspaper will be shorter than sentence lengths in the broadsheet newspaper.

My hypothesis is that the number of advertisements in the tabloid newspaper will be greater than the number of advertisements in the broadsheet newspaper.

Remember:
it does not matter whether your hypotheses are true or false and you will still gain marks if your hypothesis turns out to be false.

Now you try ...

See if you can add some hypotheses of your own.

Explain how you would proceed with the task.

Choosing the right sample

Moderator comment

It is important to choose an appropriate sample, give reasons for your choice and explain what steps you will take to avoid bias.

Moderator comment

You should always say why you choose your sampling method.

Once you have decided your aims and set up a hypothesis then it is important to consider how you will test your hypothesis.

Sampling techniques include:

- **Random sampling** is where each member of the population has an equally likely chance of being selected. An easy way to do this would be to give each person a number and then choose the numbers randomly.
- **Systematic sampling** is the same as random sampling except that there is some system involved such as numbering each person and then choosing every 20th number to form the sample.
- **Stratified sampling** is where each person is divided into some particular group or category (strata) and the sample size is proportional to the size of the group or category in the population as a whole.
- **Convenience sampling** or opportunity sampling is one which involves simply choosing the first person to come along ... although this method is not particularly random!

Here are some ways you could test the hypotheses for task 1:

Sampling method	Reason
To investigate my hypothesis I am going to choose a similar article from each type of newspaper and count the lengths of the first 100 words.	I decided to choose similar articles because the words will be describing similar information and so will be better to make comparisons.
To investigate my hypothesis I will choose every tenth page from each type of newspaper and calculate the percentage area covered by pictures.	I decided to choose every tenth page from each type of newspaper because the types of articles vary throughout the newspaper (for example headlines at the front and sports at the back of the paper).

Collecting primary data

Moderator comment

It is important to carefully consider the collection of reliable data. Appropriate methods of collecting primary data might include observation, interviewing, questionnaires or experiments.

Moderator comment

It is always a good idea to undertake a small scale 'dry run' to check for problems. This 'dry run' is called a pilot survey and can be used to improve your questionnaire or survey before it is undertaken.

If you are collecting primary data then remember:

- **Observation** involves collecting information by observation and might involve participant observation (where the observer takes part in the activity), or systematic observation (where the observation happens without anyone knowing).
- **Interviewing** involves a conversation between two or more people. Interviewing can be formal (where the questions follow a strict format) or informal (where they follow a general format but can be changed around to suit the questioning).
- **Questionnaires** are the most popular way of undertaking surveys. Good questionnaires are
  - simple, short, clear and precise,
  - attractively laid out and quick to complete

  and the questions are
  - not biased or offensive
  - written in a language which is easy to understand
  - relevant to the hypothesis being investigated
  - accompanied by clear instructions on how to answer the questions.

Avoiding bias

You must be very careful to avoid any possibility of bias in your work. For example, in making comparisons it is important to ensure that you are comparing like with like.

Jean undertook task 1: 'What the papers say'.
She collected this data from two newspapers by measuring (observation).

	Tabloid	Broadsheet
Number of advertisements	35	30
Number of pages	50	30
Area of each page	1000cm^2	2000cm^2

Her hypothesis is:

The tabloid newspaper has more advertisements than the broadsheet newspaper.

A quick glance at the table may make her claim look true: the broadsheet has 30 adverts but the tabloid has 35, which is more.

However, the sizes of the newspapers are different so Jean is not really comparing like with like.
To ensure Jean compares like with like she should take account of:

- the area of the pages
- the number of pages

and so on.

Percentage coverage per page:
Tabloid $= 35 \div 50 \times 100\% = 70\%$
Broadsheet $= 30 \div 30 \times 100\% = 100\%$

This shows that:

The broadsheet newspaper has more advertisements per page than the tabloid newspaper.

The total area of the pages is:
Tabloid = $50 \times 1000\text{cm}^2 = 50\,000\text{cm}^2$
Broadsheet = $30 \times 2000\text{cm}^2 = 60\,000\text{cm}^2$

Note: this still doesn't take account of the size of the adverts!

So the percentage coverage is:
Tabloid = $35 \div 50\,000 \times 100\% = 0.07\%$
Broadsheet = $30 \div 60\,000 \times 100\% = 0.05\%$

This shows:

The tabloid newspaper has more advertisements per area of coverage than the broadsheet newspaper.

Methods and calculations

You can represent the data using statistical calculations such as the mean, median, mode, range and standard deviation.

Once you have collected your data, you need to use appropriate statistical methods and calculations to process and represent your data.

Task 2

GUESSING GAME

Dinesh asked a sample of people to estimate the length of a line and weight of a packet.

Write down a hypothesis about estimating lengths and weights and carry out your own experiment to test your hypothesis.

Investigate further

Note: You are reminded that any coursework submitted for your GCSE examination must be your own. If you copy from someone else then you may be disqualified from the examination.

Maurice and Angela decide to explore the hypothesis that:

Students are better at estimating the length of a line than the weight of a package.

To test the hypothesis they collect data from 50 children, detailing their estimations of the length of a line and the weight of a parcel.

Here are their findings:

A table to show the estimations for the length of a line and the weight of a parcel

	Length (cm)	Weight (g)
Mean	15.9	105.2
Median	15.5	100
Mode	14	100
Range	8.6	28
SD	1.2	2.1

Moderator comment

It is important to consider whether information on all of these statistical calculations is essential.

Note: the actual length of the line is 15cm and the weight of the parcel is 100g.

Now you try ...

Using the table:

- What do you notice about the average of the length and weight?
- What do you notice about the spread of the length and weight?
- Does the information support the hypothesis?

Graphical representation

Other statistical representations might include pie charts, bar charts, scatter graphs and histograms.

Moderator comment

You should only use appropriate diagrams and graphs.

Remember to consider the possibility of bias in your data:

The percentage error for the lengths is $0.9/15 \times 100\% = 6\%$

The percentage error for the weights is $5.2/100 \times 100\% = 5.2\%$

The data for task 2 was sorted into different categories and represented as a table.

It may be useful to show your results using graphs and diagrams as sometimes it is easier to see trends.

Graphical representations might include stem and leaf diagrams and cumulative frequency graphs and box-and-whisker diagrams.

Once you have drawn a graph you should say what you notice from the graph:

From my representation, I can see that the estimations for the line are generally more accurate than the estimations for the weight because:

- the mean is closer to the actual value for the lengths and
- the standard deviation is smaller for the lengths

However, on closer inspection:

- the percentage error on the mean is smaller for the weights
- so the median and mode value are better averages to use for the weights

Using secondary data

Moderator comment

If you use secondary data there must be enough 'to allow sampling to take place' – about 50 pieces of data.

You may use secondary data in your coursework.

Secondary data is data that is already collected for you.

Task 3

GENDER DIFFERENCES IN EXAMINATIONS

Jade is conducting a survey in the GCSE examination results for Year 11 students at her school.

She has collected data on last year's results, and wants to compare the performance of boys and girls at the school.

Jade explores the hypothesis that:

Year 11 girls do better in their GCSE examinations than boys

To test the hypothesis she decided to concentrate on the core subjects and her findings are shown in the table:

A table to show the performance of Year 11 girls and boys in their GCSE examinations

		%A*-C	%A*-G	APS
English	Girls	62	93	4.9
	Boys	46	89	4.3
	All	54	91	4.6
Mathematics	Girls	47	91	4.2
	Boys	45	89	4.2
	All	46	90	4.2
Science	Girls	48	91	4.4
	Boys	45	88	4.3
	All	47	90	4.4

Now you try ...

Using the table:

- What do you notice about the percentages of A*-C grades?
- What do you notice about the percentages of A*-G grades?
- What do you notice about the average point scores?
- Does the information support the hypothesis?

The data has been sorted into different categories and represented as a table.

Jade could use comparative bar charts to represent the data as it will allow her to make comparisons more easily.

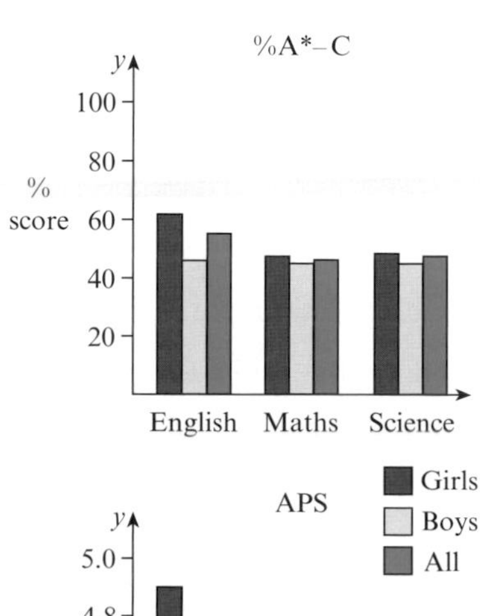

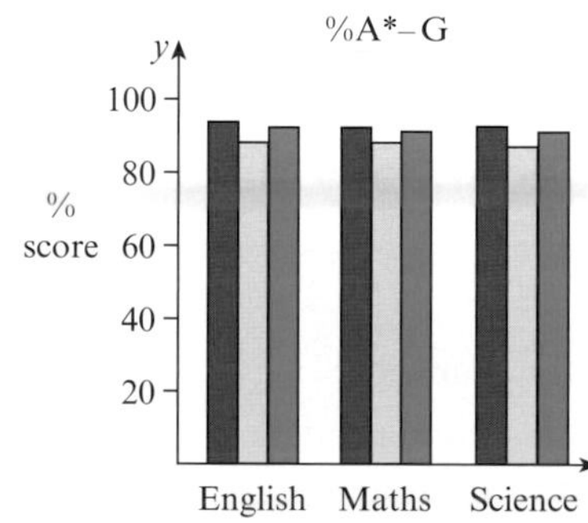

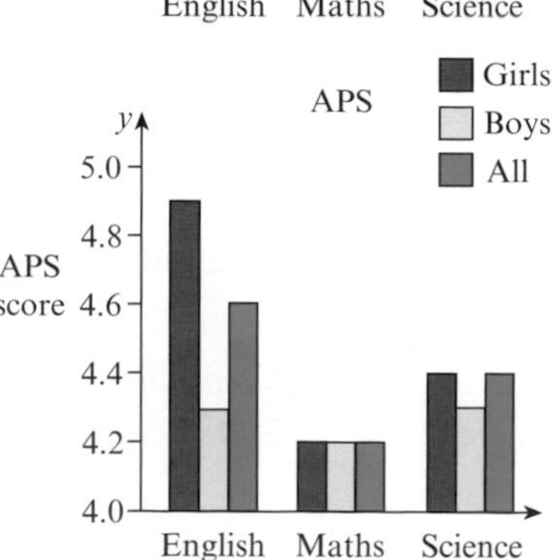

Summarising and interpreting data

Moderator comment

You should refer to your original hypothesis when you summarise your results

Hint:
You need to appreciate that the data is more secure if the sample size is 500 rather than 50.

Jade summarises her findings like this:

> From my graph, I can see that Year 11 girls do better in their GCSE examinations than boys in terms of A*-C grades, A*-G grades and average point scores.
>
> The performance of Year 11 girls is significantly better than boys in English although less so in mathematics.

Moderator comment

In your conclusion you should also suggest limitations to your investigation and explain how these might be overcome.

You may wish to discuss:

- sample size
- sampling methods
- biased data
- other difficulties

Extending the task

To gain better marks in your coursework you should extend the task in light of your findings

In your extension you should:

- Give a clear hypothesis
- Collect further data if necessary
- Present your findings using charts and diagrams as appropriate
- Summarise your findings referring to your hypothesis

Extending task 3: Gender differences in examinations

Jade extends the task by looking at the performance of individual students in combinations of different subjects.

> I am now going to extend my task by looking at the performance of individual students in English and mathematics. My hypothesis is that there will be no correlation between the two subjects.

She presents the data on a scattergraph.

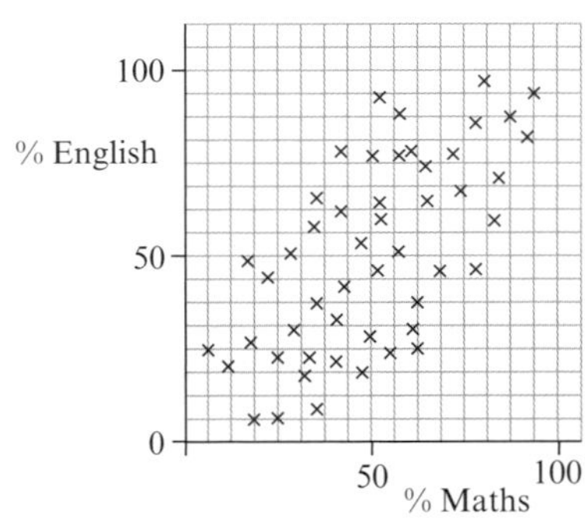

Note:
You should only draw a line of best fit on the diagram to show the correlation if you make some proper use of it (for example to calculate a students' likely English mark given their mathematics mark.)

She summarises her findings:

From my scattergraph, I can see that there is some correlation between the performance of individual students in English and mathematics.

She extends the task further:

I shall now look at the performance of individual students in mathematics and science. My hypothesis is that there will be a correlation between the two subjects.

She presents the data on a scattergraph.

Note:
The strength of the correlation could be measured using higher level statistical techniques such as Spearman's Rank Correlation.

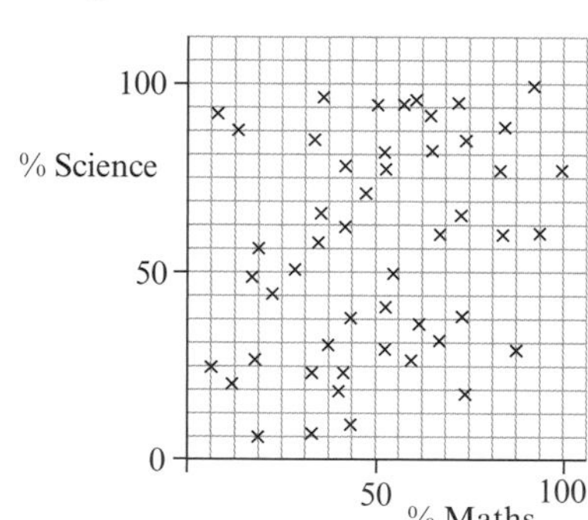

From my scattergraph, I can see that there is no correlation between the performance of individual students in mathematics and science.

Extending task 2: Guessing game

Maurice extends the 'Guessing game' like this:

I am now going to extend my task by looking to see whether people who are good at estimating lengths are also good at estimating weights. My hypothesis is that there will be a strong correlation between peoples' ability at estimating lengths and estimating weights.

He draws a scattergraph:

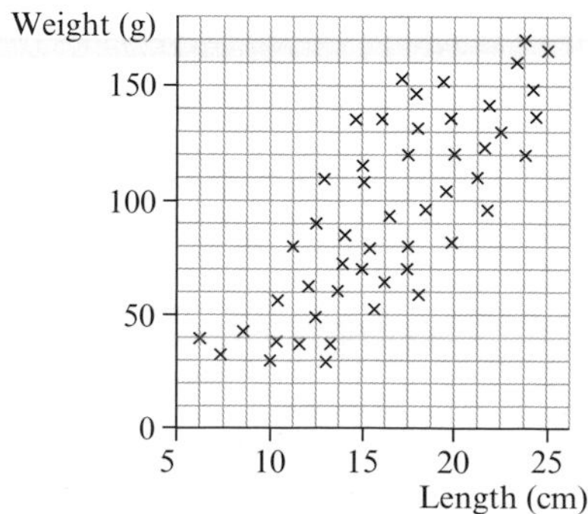

He summarises his findings:

From my scattergraph I can see that there is a strong correlation between peoples' ability at estimating lengths and estimating weights.

Moderator comment

The development of the task to include higher level statistical analyses must be 'appropriate and accurate'.

50 people were in the survey so $n = 50$

The strength of the correlation can be measured using higher level statistical techniques such as Spearman's Rank Correlation.

I am now going to extend my investigation by using Spearman's Rank Correlation to calculate the rank correlation coefficient.

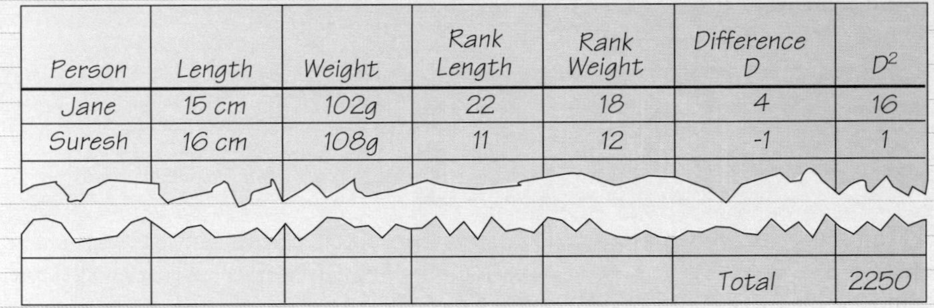

Person	Length	Weight	Rank Length	Rank Weight	Difference D	D^2
Jane	15 cm	102g	22	18	4	16
Suresh	16 cm	108g	11	12	-1	1
					Total	2250

$$r = 1 - \frac{6(\sum D^2)}{n(n^2 - 1)}$$

$$r = 1 - \frac{6(2250)}{50(50^2 - 1)}$$

$$r = 1 - \frac{13500}{50(2499)}$$

$$r = 1 - \frac{13500}{124950}$$

$$r = 1 - 0.108043.....$$

$$r = 0.891956.....$$

$$r = 0.89 \text{ (2dp)}$$

The value for the rank correlation coefficient is quite close to 1 so that there is a strong positive correlation between my results.

This tells me that there is a strong correlation between peoples' ability at estimating lengths and estimating weights and confirms my original hypothesis.

Using a computer

It is quite acceptable that calculations and representations are generated by a computer, as long as any such work is accompanied by some analysis and interpretation.

Remember:
make sure that your computer generated scattergraph has labelled axes and a title to make it quite clear what it is showing

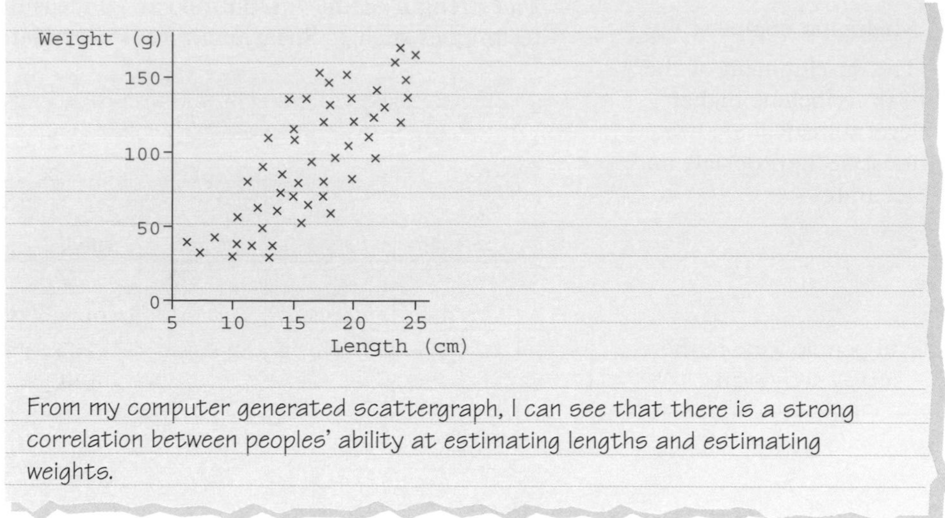
Weight (g)
150
100
50
0
5
10
15
20
25
Length (cm)
From my computer generated scattergraph, I can see that there is a strong correlation between peoples' ability at estimating lengths and estimating weights.

In this unit we have tried to give you some hints on approaching statistical coursework to gain your best possible mark.

This statistics is often useful in investigative tasks:

- Calculating averages (mean, median and mode)
- Finding the range
- Pie charts, bar charts, stem and leaf diagrams
- Constructing a cumulative frequency graph
- Finding the interquartile range
- Histograms
- Calculating the standard deviation
- Drawing a scatter graph and line of best fit
- Sampling techniques
- Discussing bias

Summary

These are the grade criteria your coursework will be marked by:

Foundation statistical task (grade E/F)

To achieve this level you must:

- set out reasonably clear aims and include a plan
- ensure that the sample size is of an appropriate size (about 25)
- collect data and make use of statistical techniques and calculations

  For example: pie charts, bar charts, stem and leaf diagrams, mean, median, mode and scattergraphs
- summarise and interpret some of your diagrams and calculations

Intermediate statistical task (grade C)

To achieve this level you must:

- set out clear aims and include a plan designed to meet those aims
- ensure that the sample size is of an appropriate size (about 50)
- give reasons for your choice of sample
- collect data and make use of statistical techniques and calculations

  For example: pie charts, bar charts, stem and leaf diagrams, mean, median, mode (of grouped data), scatter graphs and cumulative frequency
- summarise and correctly interpret your diagrams and calculations
- consider your strategies and how successful they were

Higher statistical task (grade A)

To achieve this level you must:

- set out clear aims for a more demanding problem
- include a plan which is specifically designed to meet those aims
- ensure that sample size is considered and limitations discussed
- collect relevant data and use statistical techniques and calculations

  For example: pie charts, bar charts, stem and leaf diagrams, mean, median, mode (of grouped data), scatter graphs, cumulative frequency, histograms and sampling techniques

- summarise and correctly interpret your diagrams and calculations
- use your results to respond to your original question
- consider your strategies, limitations and possible improvements

Number

Algebra

Shape, space and measures

Handling data

Formula sheet

In the GCSE examination you will be given a formula sheet like this one.
You will be required to memorise all other formulae, such as area of a circle, Pythagoras' theorem and trigonometry.
The formula sheet is the same for all Examining Groups.

Area of trapezium = $\frac{1}{2}(a + b)h$

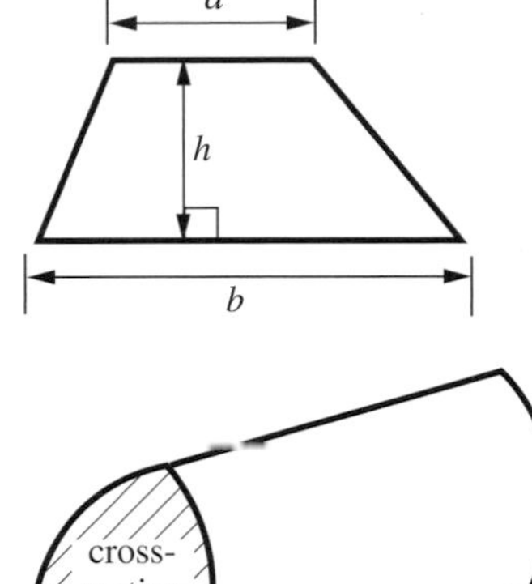

Volume of prism = area of cross-section × length

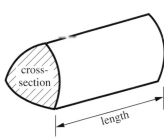

Number

N1
Units, conversions, and compound measures

N1.1 Do not use a calculator to answer this question.
John's dog eats two meals a day, every day for a year.
Each meal costs 75p.
How much money will John need to feed his dog for a year?
(3 marks) (NEAB, 1999)

N1.2 The weights and prices of two different sizes of tomato soup are shown.

Which size of soup gives more grams per penny?
You must show all your working.
(3 marks) (SEG, 1998)

N1.3 Sweets are sold in two sizes of jars.

The large jar holds 450 grams of sweets and costs £1.49.
The small jar holds 310 grams of sweets and costs 99p.
Which jar of sweets is better value for money?
You must show all your working.
(4 marks) (NEAB, 1998)

N1.4 This is a list of ingredients to make 12 rock cakes.

Rock cakes (makes 12)
240 g flour
75 g margarine
125 g sugar
150 g fruit
$\frac{1}{4}$ teaspoon spice

You have plenty of margarine, sugar, fruit and spice but only half a kilogram of flour.
What is the largest number of rock cakes you can make?
(3 marks) (NEAB, 2000)

N1.5 A shop in Dover sells gifts.
The gifts can be bought in either English or French currency.
English and French prices are shown for the following gifts.

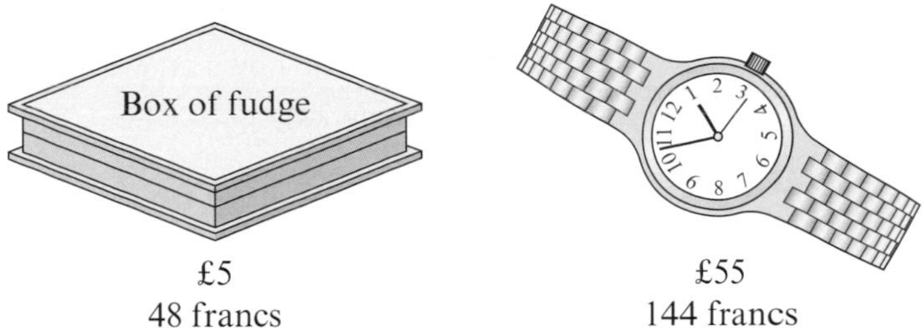

£5
48 francs

£55
144 francs

a Use this information to draw a conversion graph for pounds (£) and francs on a copy of this grid.

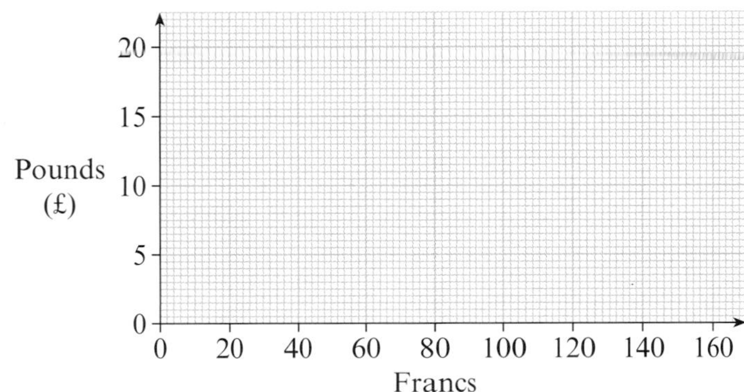

b A clock costs 120 francs.

How much is the clock in pounds (£)?

c A camera costs £60.

How much is the camera in francs?

(5 marks) (SEG, 1999)

N1.6 Tomato sauce is sold in two sizes.
A large bottle contains 681 g and costs 66p.
A small bottle contains 408 g and costs 38p.
Which size is better value for money?
You must show all your working.

(3 marks) (SEG, 1999)

N1.7 A recipe for 18 buns needs
150 g flour
100 g butter
50 g sugar

a Calculate the amount of flour needed for 45 buns.

Give your answer in kilograms.

b Lucy has plenty of flour and butter but only one kilogram of sugar.

How many buns can she make with this?

(7 marks) (NEAB, 1999)

N2
Types of number

N2.1 **a** Write down the multiple of 6 that is larger than 105, but smaller than 111.

b Write down all the factors of 111.

c Write down the largest prime number that is smaller than 107.

d Write down an even prime number.

(5 marks) (NEAB, 1999)

N2.2 **a** Write down the value of 3×2^4.

b Write 36 as a product of prime factors.

c p and q are whole numbers.
The lowest common multiple of p and q is 36.
Write down the values of p and q when $p + q = 13$.

(4 marks) (SEG, 1999)

N2.3 Jenny and David are both investigating number patterns.

Jenny

(1)	2	3	4	5	6
7	8	9	10	11	12
13	14	15	16	17	18
19	20	21	22	23	(24)

David

1	(2)	(3)	4	(5)	6
(7)	8	9	10	11	12
13	14	15	16	17	18
(19)	20	21	22	(23)	24

a **i** Jenny is looking for factors of 24.
She circles 1 and 24.
On a copy of Jenny's grid circle all the other factors of 24.

ii David is looking for prime numbers.
He has circled some.
On a copy of David's grid circle all the other prime numbers less than 24.

iii Write 24 as a product of its prime factors.

b **i** Ahmed has circled the first five numbers of a number pattern on his grid.

Ahmed

(1)	2	(3)	4	5	6
(7)	8	9	10	11	12
(13)	14	15	16	17	18
19	20	(21)	22	23	24

Write down the next number in this pattern.

ii Explain how you found your answer.

(6 marks) (SEG, 1999)

N2.4

3, 9, 20, 25, 29, 75, 92, 100

Which of the numbers in the box are:

a square numbers

b factors of 100

c prime numbers?

(6 marks) (NEAB, 1998)

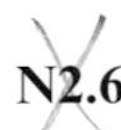

N2.5 **a** **i** Write down the highest common factor of 10 and 15.

ii Write down the lowest common multiple of 10 and 15.

p, q and r are prime numbers.
$x = pq$ and $y = qr$.
The prime numbers p, q and r have the following values:

$1 < p \leqslant 4$

$7 \leqslant q \leqslant 9$

$10 \leqslant r < 15$

b Write down all the possible values of x and y.

(5 marks) (SEG, 1999)

N2.6 Which is bigger: 2^6 or 3^4?
Show all your working.

(2 marks) (NEAB, 2000)

N2.7 Work out

a the cube of 5

b 2^6.

(2 marks) (NEAB, 1998)

N2.8 **a** Work out $\frac{2}{5}$ of 12.

b Write down the value of $3^2 + \sqrt{16}$.

(4 marks) (SEG, 1999)

N2.9 **a** Evaluate $32^{0.4}$

b Work out $32^{0.4}(4 - \sin 45°)$

Give your answer correct to 3 significant figures.

(3 marks) (NEAB, 2000)

N3 Fractions, decimals and percentages

N3.1 Joe earns £650 in May.
In June he earns 20% more.
How much does he earn in June?

(3 marks) (NEAB, 1998)

N3.2 On a computer keyboard there are 104 keys.

a 26 of the keys have letters on them.
What fraction of the keys on the keyboard have letters on them?
(Give your answer in its simplest form.)

b 13 of the keys have arrows on them.
What percentage of the keys on the keyboard have arrows on them?

(4 marks) (NEAB, 1998)

N3.3

All types of Cabins for hire		
Summer	Spring	Winter
Summer Price	Pay $\frac{3}{4}$ of Summer Price	Pay $\frac{1}{3}$ of Summer Price

a The summer price of a Standard Cabin is £240.
How much do you pay for a Standard Cabin in Spring?

b The winter price of a de luxe Cabin is £120.
How much do you pay for a de luxe Cabin in Spring?

(6 marks) (NEAB, 1999)

N3.4 The land area of a farm is 385 acres.
96 acres is pasture.

What percentage of the total land is pasture?
Give your answer to the nearest 1%.

(3 marks) (NEAB, 2000)

N3.5 **a** Write down a decimal that lies between $\frac{1}{3}$ and $\frac{1}{2}$.
b Which of these two fractions is the bigger?

$$\frac{3}{4} \text{ or } \frac{2}{3}$$

Show your working.

(3 marks) (NEAB, 2000)

N3.6 Clyde wins £120.
He gives his daughter one quarter.
He gives his son one third.
He keeps the remainder.

What fraction does he keep?

(4 marks) (NEAB, 2000)

N3.7 When a ball is dropped onto the floor, it bounces and then rises. This is shown in the diagram.

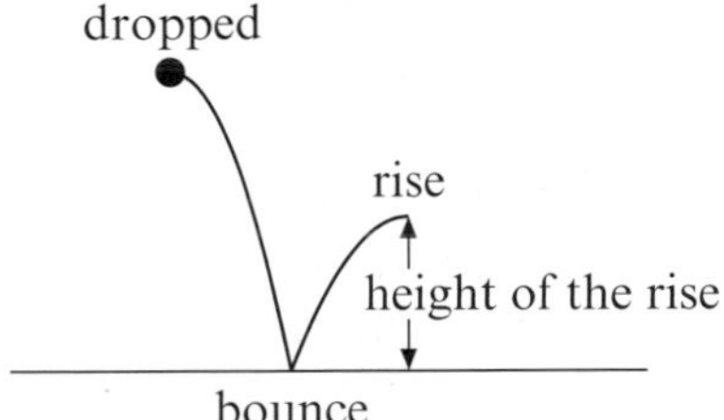

The ball rises to 80% of the height from which it was dropped.
It was dropped from a height of 3 metres.

a Calculate the height of the rise after the first bounce.

The ball bounces a second time.
It rises to 80% of the height of the first rise.

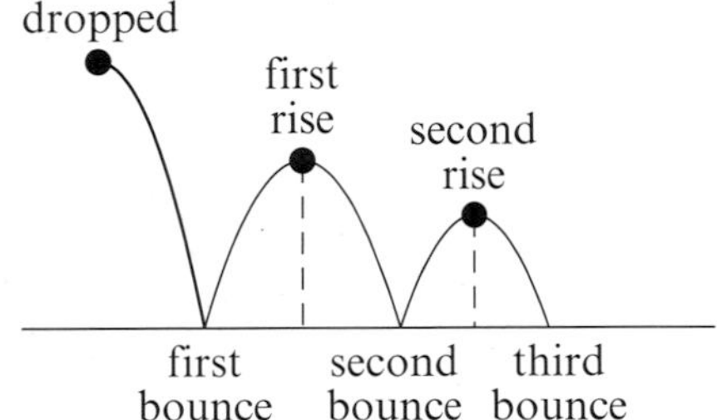

b Calculate the height of the second rise.

c The ball carries on bouncing in this way.

Each time it rises to 80% of the last rise.
For how many bounces does it rise to a height greater than 1 metre?

(4 marks) (NEAB, 1998)

N3.8

In a sale a dress costs £32.40.
The original price has been reduced by 10%.

What was the original price?

(3 marks) (NEAB, 1998)

N3.9 In a sale, all the prices are reduced by 20%.
I bought a coat for £68 in the sale.
What was the price of the coat before the sale?

(3 marks) (NEAB, 1999)

N4
Number patterns and sequences

N4.1 A sequence begins 1, –2, ...
The next number in the sequence is found by using the rule:

ADD THE PREVIOUS TWO NUMBERS AND MULTIPLY BY TWO

Use the rule to find the next **two** numbers in the sequence.
(2 marks) (SEG, 1999)

N4.2 **a** What is the next number in this sequence?

3, 7, 11, 15, ...

One number in the sequence is x.

b **i** Write, in terms of x, the next number in the sequence.

ii Write, in terms of x, the number in the sequence before x.

(3 marks) (SEG, 1999)

N4.3 Look at this pattern:

$15^2 - 14^2 = 29$ *row 1*
$14^2 - 13^2 = 27$ *row 2*
$13^2 - 12^2 = 25$ *row 3*
$12^2 - 11^2 = 23$ *row 4*

a Write down *row 6* of the pattern.

b Copy and complete this line to give the general rule for this pattern.

r^2 – .... = ...

(4 marks) (NEAB, 2000)

N4.4 **a** Write down the next two terms in the sequence

3, 5, 9, 15, 23, ...

b **i** Write down the next term in the sequence

3, 5, 9, 17, 33, ...

ii Explain how you got your answer.

c Write down the nth term for the sequence

3, 5, 7, 9, 11, ...

(6 marks) (NEAB, 1998)

N4.5 **a** Here are the first three lines of a number pattern.

Line 1 $1^2 + 2 = 3^2 - 6$
Line 2 $2^2 + 4 = 4^2 - 8$
Line 3 $3^2 + 6 = 5^2 - 10$

Write down the fourth line of this pattern.

b Here is another pattern.

Line 1 $1^2 + 2 = 1 \times 3$
Line 2 $2^2 + 4 = 2 \times 4$
Line 3 $3^2 + 6 = 3 \times 5$

i Write down the fourth line of this pattern.

ii Write down the nth line of this pattern.

c Expand and simplify

$$(n + 2)^2 - 2(n + 2)$$

(9 marks) (NEAB, 2000)

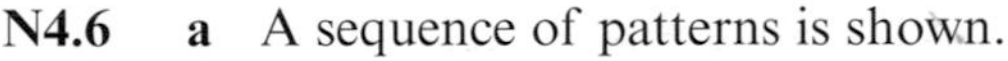

N4.6 **a** A sequence of patterns is shown.

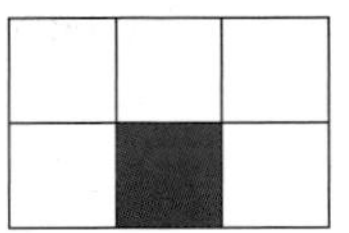

Pattern 1

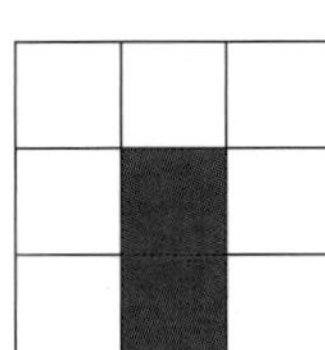

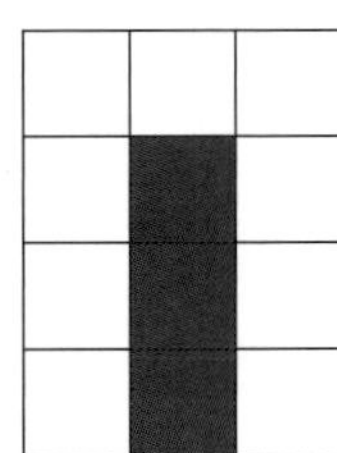

Pattern 2 Pattern 3

Write an expression, in terms of n, for the number of white squares in the nth pattern of the sequence.

b A number sequence begins

3, 6, 11, 18, 27, ...

Write an expression, in terms of n, for the nth term of this sequence.

(4 marks) (SEG, 1999)

N4.7 **a** Sticks are arranged to form a sequence of patterns as shown.

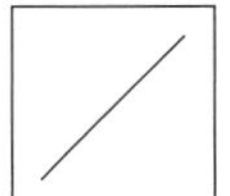

Pattern 1
(5 sticks)

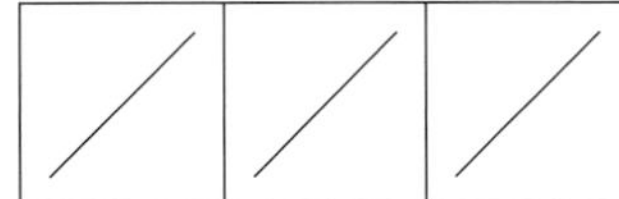

Pattern 2 Pattern 3

Write an expression, in terms of n, for the number of sticks needed to form the nth pattern.

b Squares are arranged to form a sequence of rectangular patterns as shown.

Pattern 1

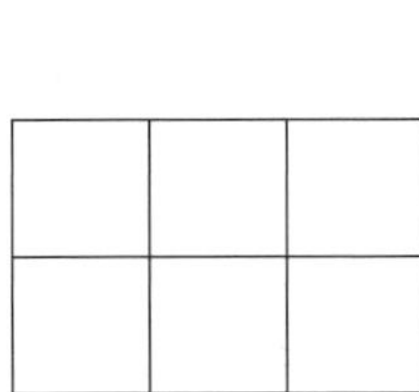

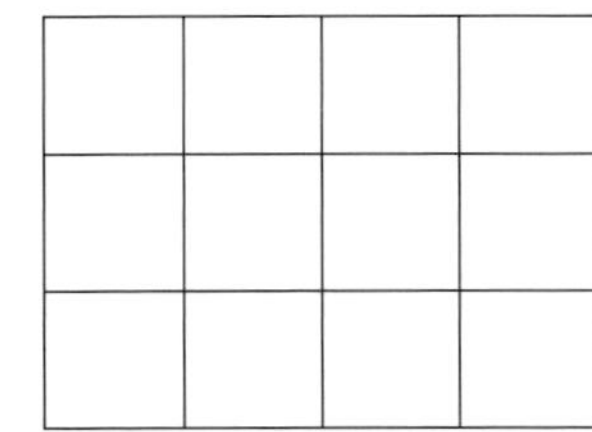

Pattern 2 Pattern 3

Write an expression, in terms of n, for the number of squares needed to form the nth pattern.

(4 marks) (SEG, 1999)

N5
Indices and standard form

N5.1 In the box are six numbers written in standard form.

8.3×10^{4}	3.9×10^{5}	6.7×10^{-3}
9.245×10^{-1}	8.36×10^{3}	4.15×10^{-2}

a **i** Write down the largest number.

ii Write your answer as an ordinary number.

b **i** Write down the smallest number.

ii Write your answer as an ordinary number.

(6 marks) (NEAB, 2000)

N5.2 **a** Write 34 500 000 000 in standard form.

b Write 0.000 000 543 in standard form.

c Work out $\dfrac{7.2 \times 10^{5}}{6.4 \times 10^{3}}$

Give your answer in standard form.

(5 marks) (NEAB, 1999)

N5.3 **a** $p = 3.7 \times 10^{-4}$
Write p as an ordinary number.

b Work out each of the following.
Give your answers in standard form.

i $5.2 \times 10^{4} + 9.6 \times 10^{5}$

ii $\dfrac{8.2 \times 10^{4}}{2.5 \times 10^{5}}$

(5 marks) (NEAB, 1998)

N5.4 The area of the United Kingdom is 244 018 square kilometres.

a Write 244 018 in standard form.

b The area of the Earth is 5.09×10^{8} square kilometres.
Calculate the area of the United Kingdom as a percentage of the area of the Earth.

(5 marks) (SEG, 1999)

N5.5 Very large distances in the Universe are measured in **parsecs** and **light-years**.

One parsec is 3.0857×10^{13} kilometres.
One parsec is 3.26 light-years.

How many kilometres are in 1 light-year?
Give your answer in standard form to an appropriate degree of accuracy.

(4 marks) (NEAB, 2000)

N6
Ratio

N6.1 A bar of Fruit & Nut chocolate normally weighs 200 g.
The ratio by weight of a special offer bar to a normal bar is 5 : 4.
What is the weight of a special offer bar?

(2 marks) (SEG, 2000)

N6.2 Fiona is given £24 for her birthday.
She spends 5 times as much as she saves.
How much does she save?

(2 marks) (NEAB, 1999)

N6.3 80 million pounds of lottery funds was given to the sports of swimming and athletics.
This money was shared between these two sports in the ratio

swimming	:	athletics
3	:	1

Calculate the amount that was given to swimming.

(2 marks) (NEAB, 1998)

N6.4 Clare has 200 sweets to share between her two grandchildren.
She shares them in the ratio 3 : 5
Calculate the larger share.

(3 marks) (NEAB, 2000)

N6.5 **a** Phil has 50 birds.
Some are blue, the rest are yellow.
The ratio of blue birds to yellow birds is 3 : 7.
How many yellow birds are there?

b Phil sells some of his birds and buys some others.
The new ratio of blue birds to yellow birds is 5 : 2.
There are 16 yellow birds.
How many blue birds are there?

(5 marks) (NEAB, 1998)

N6.6 Max shares £420 with a friend in the ratio 5 : 3.
How much does each receive?

(3 marks) (NEAB, 2000)

N7
Miscellaneous number

N7.1 A farmer has 175 sheep.

a $\frac{4}{5}$ of the sheep had lambs.
How many sheep had lambs?

b Of the sheep which had lambs, 35% had two lambs.
How many sheep had two lambs?

(3 marks) (SEG, 1999)

N7.2 **a** Paul can write shorthand at the rate of 110 words per minute.
How long will it take him to write 484 words?
Give your answer in minutes and seconds.

b Susan types 280 words in 5 minutes.
Paul types 5% faster than Susan.
How many words does Paul type in 5 minutes?

(5 marks) (SEG, 1999)

N7.3 A Munch Crunch bar weighs 21 g.
Each bar contains the following nutrients.

Protein	1.9 g
Fat	4.7 g
Carbohydrate	13.3 g
Fibre	1.1 g

a What percentage of the bar is fat?
Give your answer to an appropriate degree of accuracy.

b What is the ratio of protein to carbohydrate?
Give your answer in the form 1 : n.

(5 marks) (SEG, 1999)

N7.4 A survey counted the number of visitors to a website on the Internet.
Altogether it was visited 30 million times.
Each day it was visited 600 000 times.
Based on this information, for how many days did the survey last?
(2 marks) (NEAB, 2000)

N7.5 **a** This table gives information about the sailing times of Hovercrafts.

Hovercraft Sailing Times — **First sailing 7 am**

A Hovercraft leaves Dover
every 70 minutes.
The first sailing is at 7 a.m.

How many sailings from Dover are there before midday?

b

London–Paris by Plane
Normal Price: £91
Special Offer: 15% discount

An airline company has flights from London to Paris for £91.
They offer a 15% discount.
Calculate the cost of the flight after the discount.

(5 marks) (NEAB, 1999)

N7.6 **a** Estimate the values of:

i 25.1×19.8

ii $119.6 \div 14.9$

b Jonathan uses his calculator to work out the value of 42.2×0.027

The answer he gets is 11.394

Use approximation to show that his answer is wrong.

(6 marks) (NEAB, 2000)

N7.7 **a** John is 8.4 kg heavier than Alan.
The sum of their weights is 133.2 kg.

How heavy is Alan?

b Before starting a diet Derek weighed 80 kg.
He now weighs 8 kg less.

Calculate his weight loss as a percentage of his previous weight.

c Sarah weighs 54 g.

i Sarah's weight has been given to the nearest kilogram.
What is the minimum weight she could be?

ii What is Sarah's weight in pounds?

(8 marks) (SEG, 1999)

N7.8 **a** A load of 4000 bricks weighs 9200 kilograms.

i What is the weight of one brick in pounds?

The bricks are moved by lorry.

ii The ratio of the weight of the bricks to the weight of the lorry is 4 : 3.

Calculate the weight of the lorry, in kilograms.

b Bags of cement weigh 25 kilograms.
The weight of each bag is accurate to the nearest kilogram.

What is the minimum weight of a bag of cement?

(7 marks) (SEG, 1999)

N7.9 Jack invests £2000 at 7% per annum compound interest.

Calculate the value of his investment at the end of 2 years.

(3 marks) (SEG, 2000)

N7.10 The cost of hiring a coach is £40 plus £2 for every mile travelled.
For example, a journey of 25 miles would cost

$$£40 + 25 \times £2 = £90$$

a How much will it cost for a journey of 72 miles?

b Write down an expression for the cost in £ of a journey of M miles.

c A journey costs £124.

i Use your answer to part **b** to form an equation using this information.

ii How many miles did the coach travel on this journey?

(6 marks) (NEAB, 1999)

Algebra

A1 Formulae

A1.1 **a**

Southern Rental
Van Hire charges
£24 per day plus 12 pence per mile

i Anita hires a van for one day.
She drives 68 miles.
How much is the hire charge?

ii John hires a van for two days.
The total hire charge is £66.
How many miles did he drive?

Cars can also be hired.
When a car is hired for d days and driven m miles, the hire charge, C pounds, is calculated by using the formula

$$C = 0.06(300d + m)$$

b Use the formula to calculate the cost of hiring a car for 7 days and driving 458 miles.

(8 marks) (SEG, 1999)

A1.2 **a** These polygons are similar.

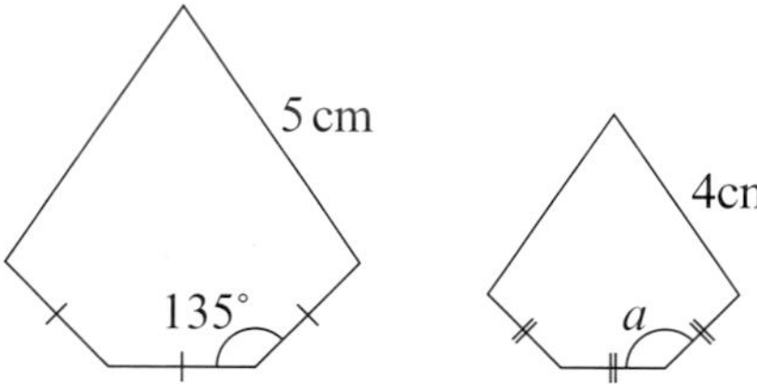

What is the size of angle a?

b Part of a regular polygon is shown.

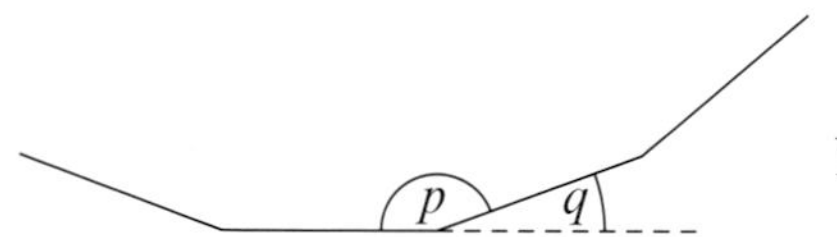

The complete polygon has n sides, where $n = \dfrac{360}{q}$ and $p + q = 180°$.

i Calculate the value of n, when $p = 168°$.

ii Write down a formula for n terms of p.

(5 marks) (SEG, 1999)

A1.3 A formula is given as $V = 4h + p^2$
Find the value of V when $h = 0.5$ and $p = 8$.

(2 marks) (NEAB, 1998)

A1.4 s is given by the formula

$$s = ut + \tfrac{1}{2}at^2$$

Find the value of s when $u = 2.8$, $t = 2$ and $a = -1.7$.

(2 marks) (NEAB, 2000)

A1.5 You are given the formula $P = \dfrac{V^2}{R}$.

a Work out the value of P when $V = 3.85$ and $R = \frac{8}{5}$.

b Rearrange the formula to give V in terms of P and R.

(4 marks) (SEG, 1999)

A1.6 You are given the formula $y = 3x^2$.

a Calculate y when $x = -3$.

b Find a value of x when $y = 147$.

c Rearrange the formula to give x in terms of y.

(6 marks) (SEG, 2000)

A1.7 Find the exact value of

$$\frac{1}{a} + \frac{1}{b} \quad \text{when} \quad a = \tfrac{1}{2} \quad \text{and} \quad b = \tfrac{1}{3}$$

(3 marks) (NEAB, 1999)

A1.8 A magic square is shown below.
Every row, column and diagonal adds up to the same total of 15.

8	3	4
1	5	9
6	7	2

For the magic square below, every row, column and diagonal should add up to 3.

2	−3	4
3	1	
−2		0

a Fill in the missing numbers.

b Paul says, 'If I multiply each number in the square by −6, the new total for each row, column and diagonal will be −18.'

Show **clearly** that this is true for the first row of numbers.

(NEAB, 2000)

A2 Solving linear and simultaneous equations

A2.1

Think of a number.
Multiply it by three.
Now add seven.
My answer is 94.
Teacher
Mabel

What is the number Mabel thinks of?
(3 marks) (NEAB, 1998)

A2.2 Solve the equation $7x - 6 = 15$.
(2 marks) (NEAB, 2000)

A2.3 Solve these equations.

a $5x - 2 = 13$

b $3(2x - 1) = 9$

(5 marks) (SEG, 2000)

A2.4 Solve the equations.

a $4x + 7 = 13$

b $3x + 7 = 3 - x$

(5 marks) (NEAB, 1998)

A2.5 Solve the equations.

a $3x - 14 = 4$

b $5x + 7 = x + 9$

c $\dfrac{x - 7}{8} = 2$

(8 marks) (NEAB, 1999)

A2.6 **a** Solve the equation $2x + 3 = 15 - x$

b Solve the simultaneous equations

$$2x + y = 9$$
$$x - 2y = 7$$

You **must** show all your working.
(5 marks) (SEG, 1998)

A2.7 Solve the simultaneous equations

$$x - 3y = 7$$
$$2x + y = 0$$

(3 marks) (SEG, 2000)

A2.8 **a** On a copy of this grid, draw the line $y = 2x$.

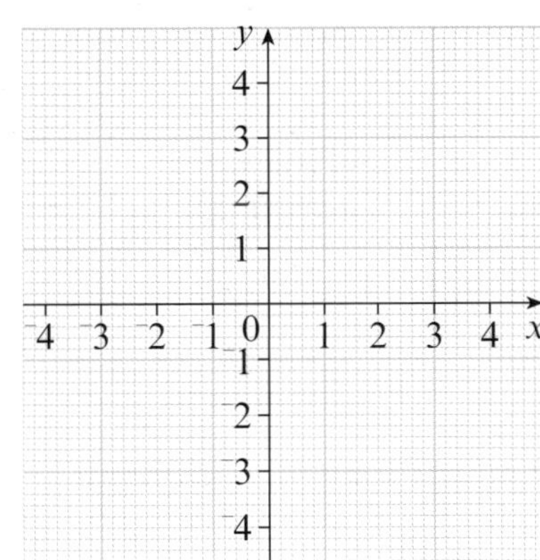

b The line $y = 2x$ crosses the line $x = -1$ at P.

Give the coordinates of P.

(3 marks) (SEG, 2000)

A2.9 Copy the grid. Use a **graphical method** to solve the simultaneous equations

$$y = x + 7 \quad \text{and} \quad y + 3x = 5$$

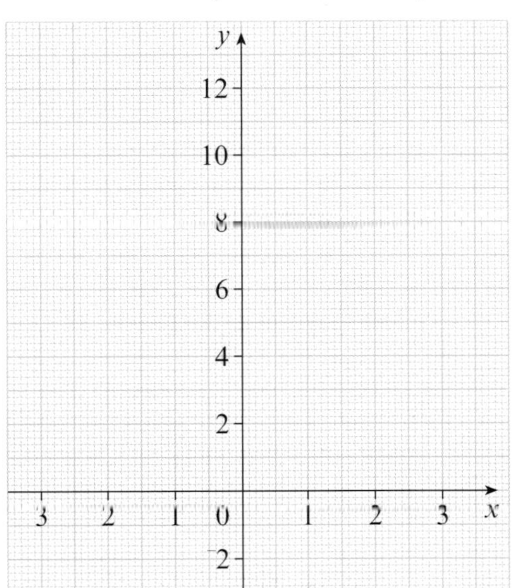

(4 marks) (NEAB, 2000)

A3
Quadratic equations and higher order equations (trial and improvement)

A3.1 **a** Factorise the expression $x^2 + 8x + 15$

b Hence solve the equation $x^2 + 8x + 15 = 0$

(3 marks) (NEAB, 1998)

A3.2 Solve the quadratic equation $x^2 + 3x - 10 = 0$.

(3 marks) (SEG, 2000)

A3.3 The diagram shows a square $ABCD$ and a rectangle $RBPQ$.

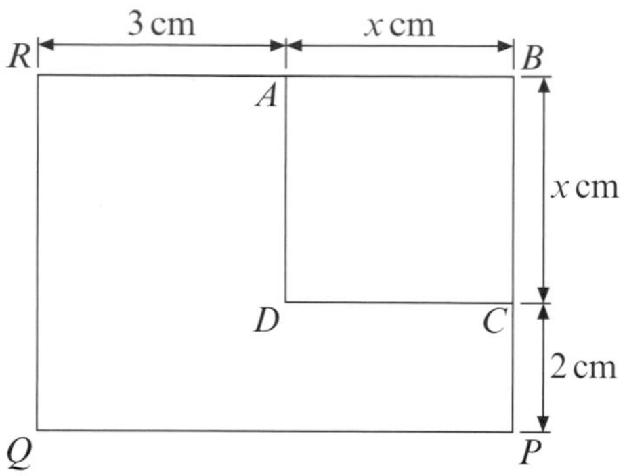

Not to scale

$CP = 2$ cm and $RA = 3$ cm.

The length of the side of square $ABCD$ is x cm.
The rectangle $RBPQ$ has an area of 42 cm^2.

a Form an equation, in terms of x, for the area of the rectangle.
Show that it can be written in the form $x^2 + 5x - 36 = 0$.

b Solve $x^2 + 5x - 36 = 0$, and hence calculate the area of the square $ABCD$.

(6 marks) (SEG, 1999)

A3.4 The diagram shows a shape, in which all the corners are right angles.

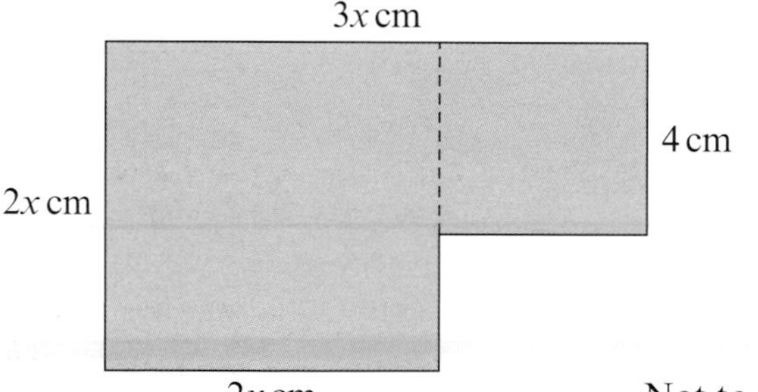

Not to scale

The area of the shape is 48 cm^2.

a Form an equation, in terms of x, for the area of the shape.
Show that it can be simplified to $x^2 + x - 12 = 0$.

b By solving the equation $x^2 + x - 12 = 0$, calculate the value of x.

(6 marks) (SEG, 1999)

A3.5 **a** Copy and complete the table of values for $y = x^2 - x - 1$.

x	-2	-1	0	1	2	3
y		1	-1		1	

b On a copy of the axes below, draw the graph of $y = x^2 - x - 1$ for values of x from -2 to 3.

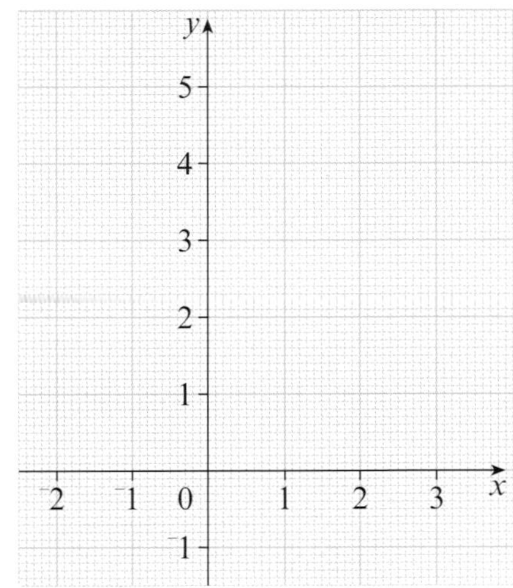

c Use your graph to solve the equation $x^2 - x - 1 = 0$.
(6 marks) (SEG, 1999)

A3.6 Nico is using trial and improvement to find a solution to $x^2 + x = 765$.

This table shows his first try.

x	$x^2 + x$	Comment
20	420	too small

Copy and continue the table to find a solution to the equation.
(3 marks) (NEAB, 1999)

A3.7 Use trial and improvement to solve the equation $x^3 + x^2 = 500$.

One trial has been completed for you.

Show all your trials.

Give your answer correct to 1 decimal place.

x	$x^3 + x^2$	
7	$7^3 + 7^2 = 392$	too small

Copy and complete this sentence:

The answer is $x =$ .... (correct to 1 decimal place)
(4 marks) (NEAB, 1998)

A4
Linear and quadratic graphs

A4.1 **a** On a copy of the diagram draw and label the following lines.

$y = 2x$ and $x + y = 5$

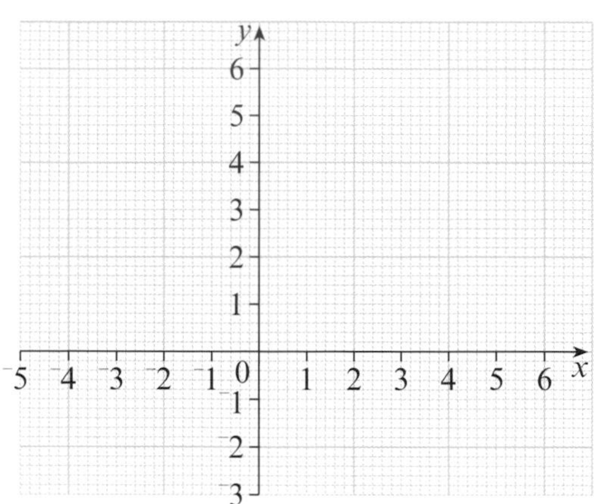

b Explain how to use your graph to solve the equation $2x = 5 - x$.

c Show clearly the single region that is satisfied by **all** of these inequalities.

$x + y \leqslant 5$ $y \geqslant 2x$ $x \geqslant 0$

Label this region R.

(6 marks) (SEG, 1998)

A4.2 The graph shows the cost of printing wedding invitation cards.

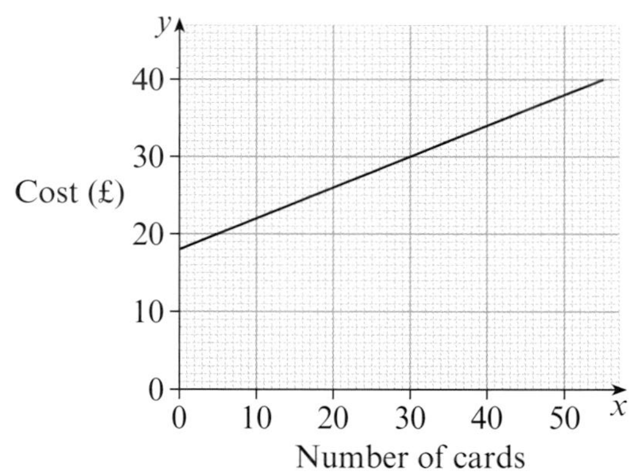

a Find the equation of the line in the form $y = mx + c$.

b For her wedding Charlotte needs 100 cards to be printed. How much will they cost?

(5 marks) (SEG, 2000)

A4.3 A graph of the equation $y = ax + b$ is shown.

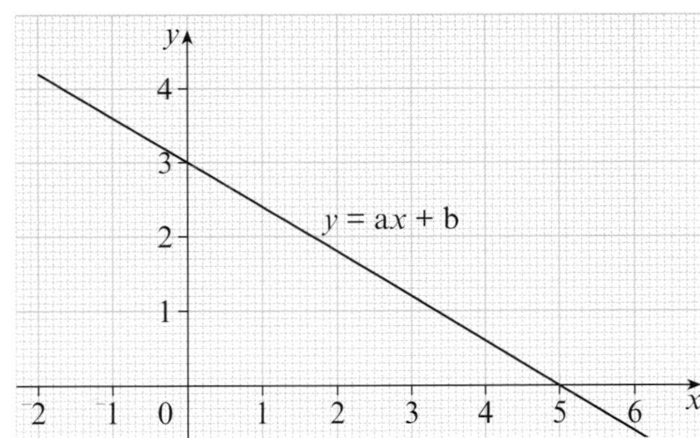

Find the values of a and b.
(3 marks) (SEG, 1998)

A4.4 **a** Below are three graphs.
Match each graph with one of the following equations.
Equation A: $y = 3x - p$
Equation B: $y = x^2 + p$
Equation C: $3x + 4y = p$
Equation D: $y = px^3$
In each case *p* is a positive number.

i

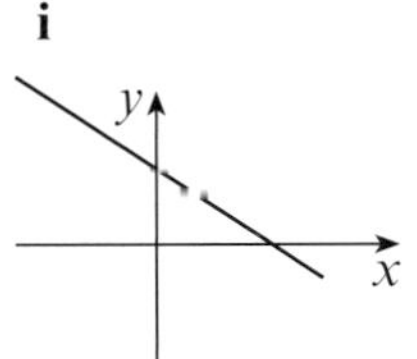

ii

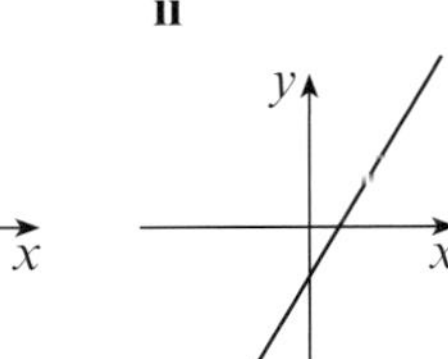

iii

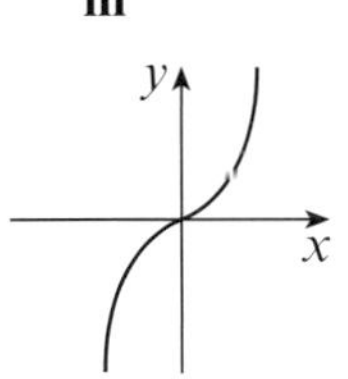

b Sketch a graph of the equation you have not yet chosen.
(5 marks) (NEAB, 1998)

A4.5 The graph shows the line AB.

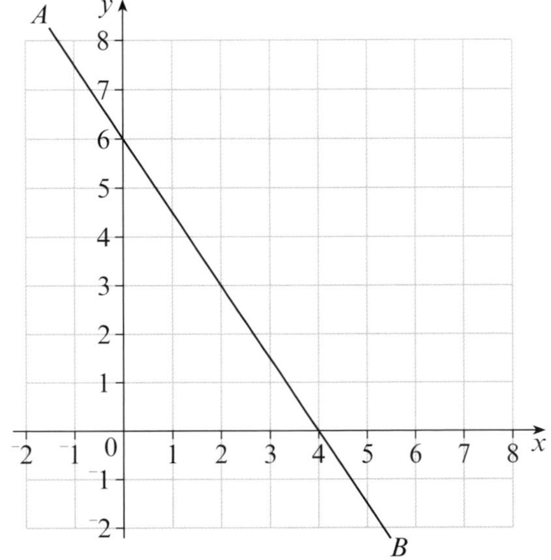

Work out the equation of the line AB.
(4 marks) (NEAB, 2000)

A4.6 **a** **i** Copy and complete the table for the graph of $y = x^2 - 7$

x	−3	−2	−1	0	1	2	3
y	2	−3	−6				2

ii Draw the graph on a the copy of the grid.

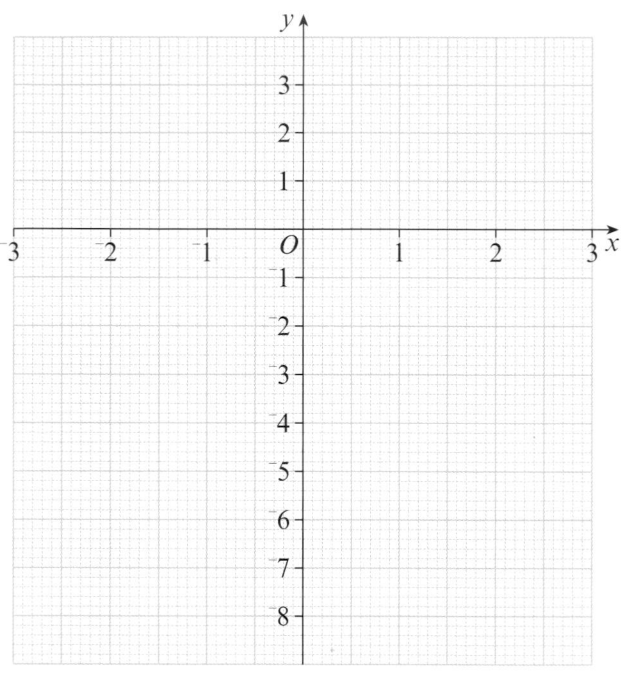

b **Use your graph** to solve the equation $x^2 - 7 = 0$.

c Use your graph to state the minimum value of y.

(6 marks) (NEAB, 2000)

A4.7 **a** Copy and complete the table of values for $y = 5 - x^2$.

x	−3	−2	−1	0	1	2	3
y		1	4	5			−4

b Draw the graph of $y = 5 - x^2$ for values of x from −3 to 3.

c Use your graph to solve the equation $5 - x^2 = 0$.

(6 marks) (SEG, 2000)

A5 Inequalities

A5.1 List the values of n, where n is an integer, such that $3 \leqslant n + 4 < 6$
(2 marks) (SEG, 1998)

A5.2 Solve the inequality $3 < 2x + 1 < 5$.
(2 marks) (SEG, 2000)

A5.3 List the values of n, where n is an integer, such that $1 \leqslant 2n - 3 < 5$.
(3 marks) (SEG, 1999)

A5.4 **a** Solve the inequality $3x + 4 \leqslant 7$.
b **i** Solve the inequality $3x + 11 \geqslant 4 - 2x$.
ii If x is an integer what is the smallest possible value of x?
(5 marks) (NEAB, 1999)

A5.5 Solve the inequality $x + 20 < 12 - 3x$.
(2 marks) (NEAB, 2000)

A5.6 **a** On a copy of the diagram draw and label the lines

$x = 1$ and $x + y = 4$

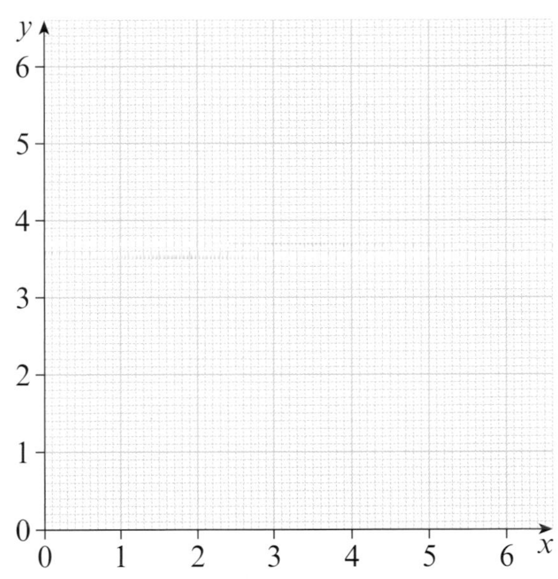

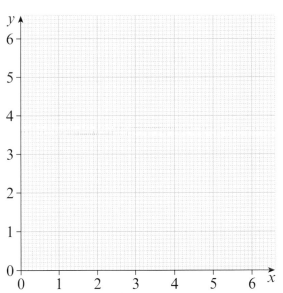

b Show clearly on the diagram the single region that is satisfied by all of these inequalities.

$y \geqslant 0$ $x \geqslant 1$ $x + y \leqslant 4$

Label this region R.
(4 marks) (SEG, 1999)

A6 Manipulation

A6.1 **a** Simplify $2(x-3)+5$.

b Factorise completely $4a-2ab$.

c Solve $5x-3=3x+7$.

(6 marks) (SEG, 1999)

A6.2 Simplify

$$\frac{a^6c^4}{a^2c^5}$$

(2 marks) (NEAB, 1998)

A6.3 Simplify:

a $t^3 \times t^5$

b $p^6 \div p^2$

c $\frac{a^3 \times a^2}{a}$

(3 marks) (NEAB, 1998)

A6.4 **a** Expand $x(3x^2-5)$.

b Expand and simplify $(2x+1)(3x-2)$.

(5 marks) (NEAB, 1998)

A6.5 Simplify:

a $p \times p \times p \times p$

b $2a \times 3b \times 4c$

c $x^3 \div x^3$

(3 marks) (NEAB, 2000)

A6.6 **a** Simplify $x^2-2x-3+x-5$.

b Factorise completely $3a^2-6a$.

c Multiply out and simplify $(2x-1)(x-3)$.

(6 marks) (SEG, 1998)

A6.7 **a** Simplify the following expressions.

i $a^5 \times a^3$

ii $a^5 \div a^3$

iii $(a^5)^3$

b **i** Which expression in part **a** is negative when $a=-1$?

ii Which expression in part **a** has the greatest value when $a=0.1$?

(5 marks) (NEAB, 2000)

A6.8 **a** Simplify:

i $2a^3 \times 3a^2$

ii $4a^6 \div 2a^3$

b Factorise $2x^2+4x$.

(4 marks) (SEG, 2000)

A6.9 **a** Simplify the expression $2(x-3)-x$.

b Solve the equation $4x-1=x+5$.

(4 marks) (SEG, 1999)

A7 Miscellaneous algebra

A7.1 **a** Factorise completely $3x^2 - 6x$.

b Expand and simplify $(3x + 2)(x - 4)$.

c Make t the subject of the formula

$$W = \frac{5t + 3}{4}$$

(8 marks) (NEAB, 1998)

A7.2 **a** Simplify $2x - x + 1$.

b Find the value of $3x + y^2$ when $x = -2$ and $y = 3$.

(3 marks) (SEG, 2000)

A7.3 **a** Expand and simplify $(2x + 5)(3x - 4)$.

b Factorise completely $5x^2y + 15xy^3$.

(5 marks) (NEAB, 1999)

A7.4 **a** Solve the equation $\frac{1}{3}x = 2$.

b Solve the simultaneous equations

$$x + 3y = 5$$

and $$x - y = 3$$

You **must** show all your working.

(4 marks) (SEG, 1999)

A7.5 Adrian has three regular polygons: A, B and C.

A has x sides.
B has $(2x - 1)$ sides.
C has $(2x + 2)$ sides.

a Write an expression, in terms of x, for the total number of sides of these three polygons.

Write your answer in its simplest form.

The three polygons have a total of 16 sides.

b **i** Form an equation and hence find the value of x.

ii Use your value of x to find the number of sides of polygon B.

(5 marks) (SEG, 1999)

A7.6 The diagram shows a sketch of the line $2y + x = 10$.

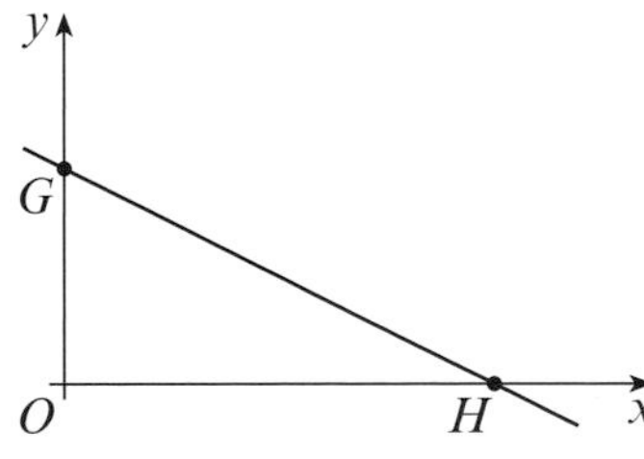

a Find the coordinates of points G and H.

b **i** On a copy of the diagram, sketch the graph of $y = 2x$.

ii Solve the simultaneous equations

$$2y + x = 10$$

$$y = 2x$$

c The equation of a straight line is $y = ax + b$.

Rearrange the equation to give x in terms of y, a and b.

(8 marks) (SEG, 1999)

Shape, space and measures

S1 Angles and polygons

S1.1 The diagram shows an isosceles triangle with two sides extended.

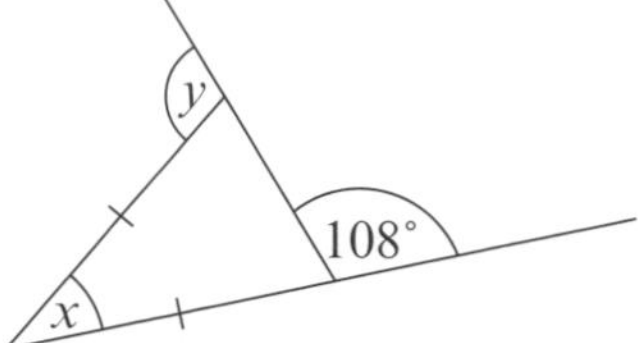

Not to scale

a Work out the size of angle x.

b Work out the size of angle y.

(4 marks) (SEG, 2000)

S1.2 The diagram shows a regular octagon with centre O.

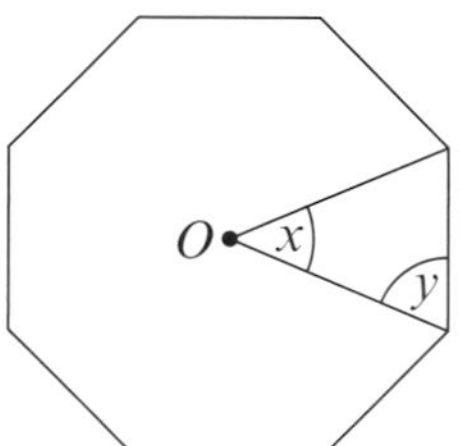

Not to scale

a Work out the size of angle x.

b Work out the size of angle y.

(4 marks) (SEG, 1998)

S1.3

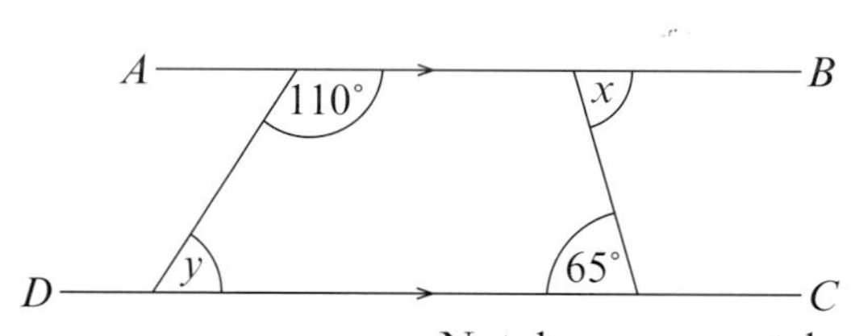

Not drawn accurately

AB is parallel to DC.

a Work out the size of angle x.
Give a reason for your answer.

b Work out the size of angle y.
Give a reason for your answer.

(4 marks) (NEAB, 1999)

S1.4

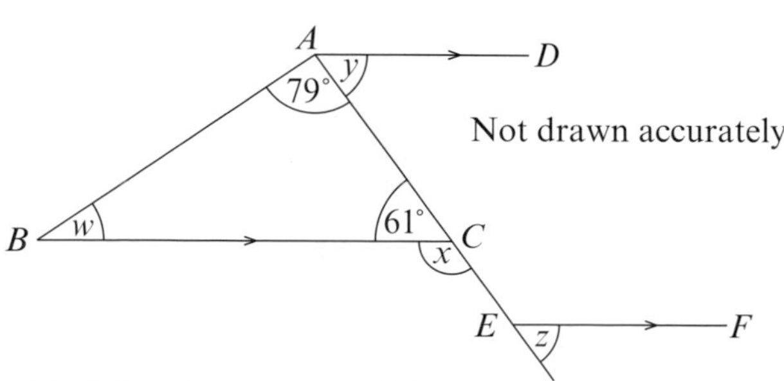

Not drawn accurately

AD, BC and EF are parallel.
Angle $BAC = 79°$
Angle $ACB = 61°$

Work out the size of the angles marked w, x, y and z.
(5 marks) (NEAB, 2000)

S1.5 **a** What is the sum of the angles in a triangle?

b Write down an expression, in terms of x, for the sum of the three angles in this triangle.

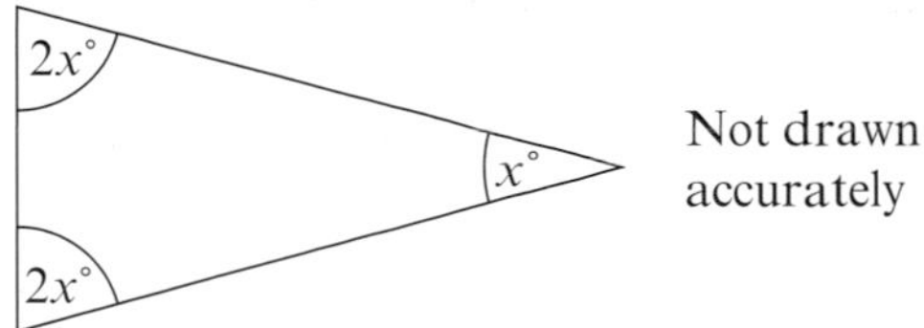

c Use your answers to parts **a** and **b** to write down an equation in x. Hence find the size of **each** angle in the triangle.

d What special name is given to this kind of triangle?

(6 marks) (NEAB, 1999)

S1.6 The four angles of a quadrilateral are $3x$, x, $4x$ and $2x$.

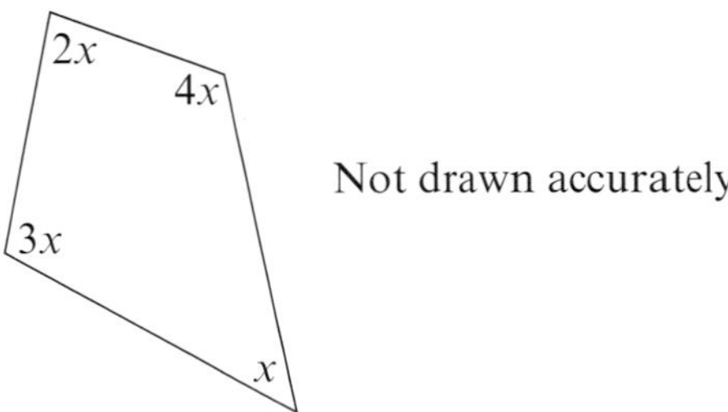

Calculate the value of x.

(3 marks) (NEAB, 1998)

S1.7

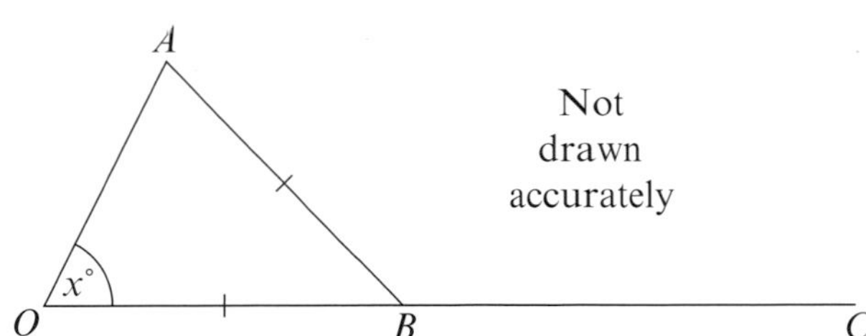

OBC is a straight line.
AOB is an isosceles triangle with $OB = AB$.
Angle $AOB = x°$.

a Write down, in terms of x,

i angle OAB

ii angle ABC

b Angle OBA is $(x - 12)$ degrees.
Find the value of x.

(4 marks) (NEAB, 2000)

S2
Length, area and volume

S2.1 Calculate the area of this trapezium.
(Remember to state the units in your answer).

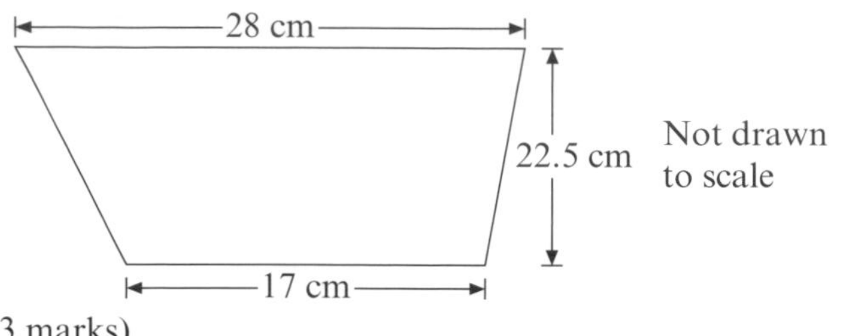

(3 marks) (NEAB, 1998)

S2.2

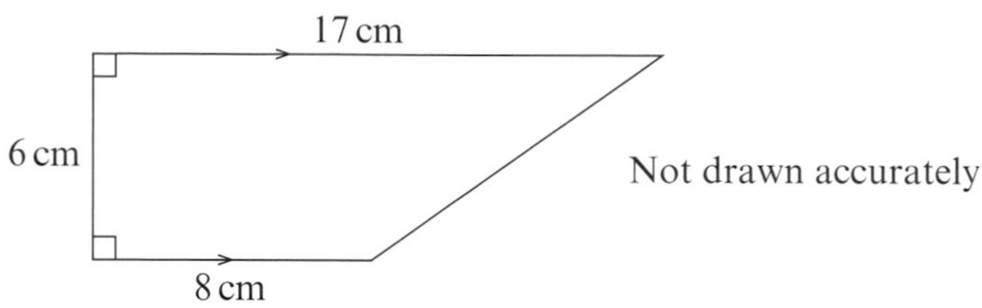

a Calculate the area of this trapezium.
(Remember to state the units in your answer.)

b Calculate the perimeter of this trapezium.
(Remember to state the units in your answer.)

(7 marks) (NEAB, 1999)

S2.3 A rectangular carpet measures 1.4 m by 0.75 m.
60% of the carpet is red.

Calculate the area of red carpet.
State your units.
(5 marks) (SEG, 1999)

S2.4 The diagram shows a length of plastic guttering.

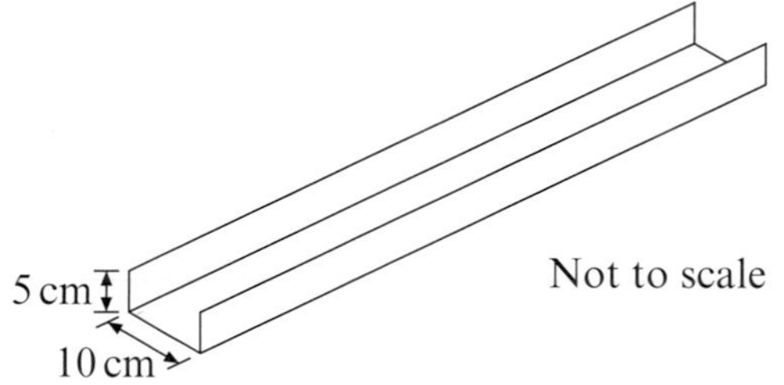

The cross-section of the guttering is a rectangle measuring 10 cm by 5 cm.

a Calculate the area of plastic needed to make a 200 cm length of guttering.

b Calculate the volume of water a 200 cm length of guttering could contain if the ends were sealed and it was full of water.
State your units.

(5 marks) (SEG, 1999)

S2.5 A girder is 5 metres long.
Its cross-section is L-shaped as shown below.

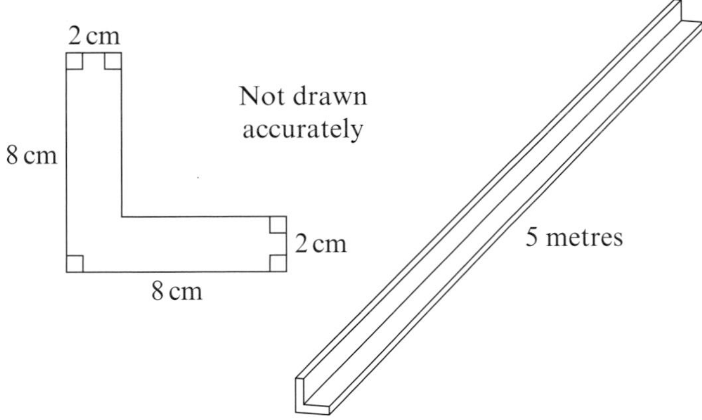

Find the volume of the girder.
Remember to state the units in your answer.
(5 marks) (NEAB, 2000)

S2.6 **a** Water is being poured at a constant rate into a container which has the shape of a prism.
The diagram shows the cross-section of the container.
Copy the axes and sketch a graph to show the height of water in the container as it is being filled.

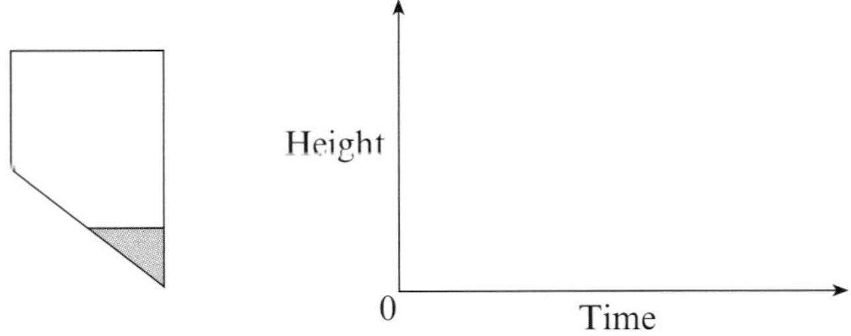

b Another view of the container is shown.

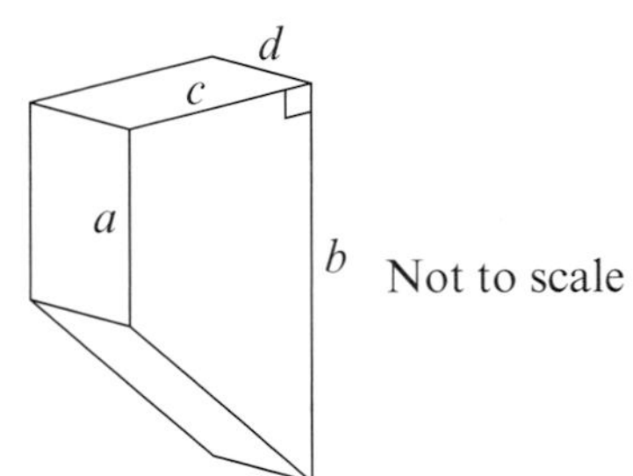

The following formulae represent certain quantities connected with the container.

$$2(a + b) \qquad \tfrac{1}{2}(a + b)c \qquad \tfrac{1}{2}(a + b)cd$$

Which of these formulae represents area?
(3 marks) (SEG, 1998)

S2.7 The following formulae represent certain quantities connected with containers, where a, b and c are dimensions.

$$\pi a^2 b \qquad 2\pi a(a + b) \qquad 2a + 2b + 2c \qquad \tfrac{1}{2}(a + b)c \qquad \sqrt{a^2 + b^2}$$

a Which of these formulae represent area?

b Which of these formulae represent volume?

(2 marks) (SEG, 1999)

S2.8 The diagram shows a prism.

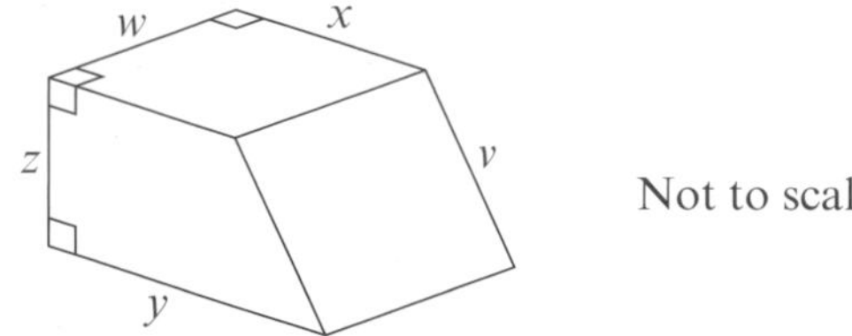

Not to scale

The following formulae represent certain quantities connected with the prism.

$$wx + wy \qquad \tfrac{1}{2}z(x + y)w \qquad \frac{z(x + y)}{2} \qquad 2(v + 2w + x + y + z)$$

a Which of these formulae represents length?

b Which of these formulae represents volume?

(2 marks) (SEG, 2000)

S2.9 The diagram shows a bale of straw.
The bale is a cylinder with radius 70 cm and height 50 cm.

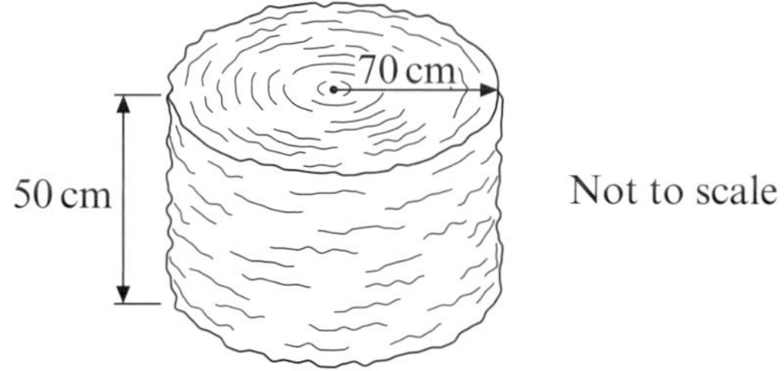

Not to scale

a Calculate the circumference of the bale.
Give your answer to an appropriate degree of accuracy.

b Calculate the volume of the bale.
State your units.

(7 marks) (SEG, 2000)

S2.10 The diagram shows a circular paddling pool with a vertical side.

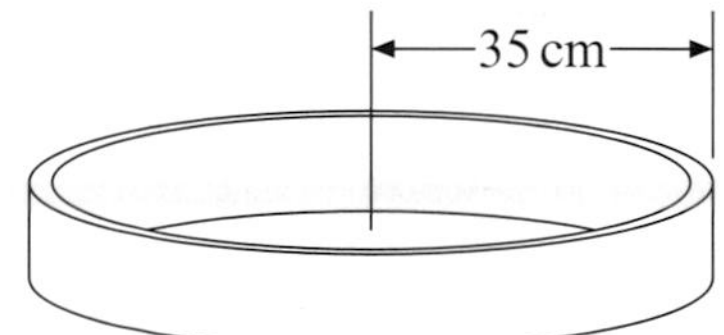

Not to scale

The radius of the pool is 35 cm.

a Calculate the circumference of the pool.

b The water in the pool is 8 cm deep.
Calculate the volume of water in the pool.

(5 marks) (SEG, 1999)

S3
Distance, speed and time

S3.1 **a** A train takes 3 hours to travel 150 miles.
What is its average speed?

b Another train travels 50 miles at an average speed of 37.5 mph.
How long does this journey take?
Give your answer in hours and minutes.

(5 marks) (NEAB, 1998)

S3.2 **a** Brian travels 225 miles by train.
His journey takes $2\frac{1}{2}$ hours.
What is the average speed of the train?

b Val drives 225 miles at an average speed of 50 mph.
How long does her journey take?

(4 marks) (NEAB, 2000)

S3.3 The diagram shows a map of a group of islands.
The map has been drawn to a scale of 1 cm to 5 km.

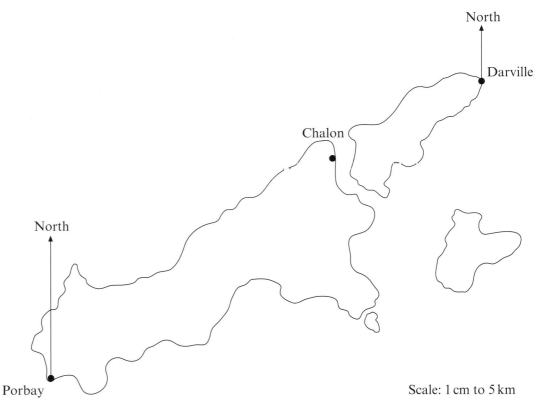

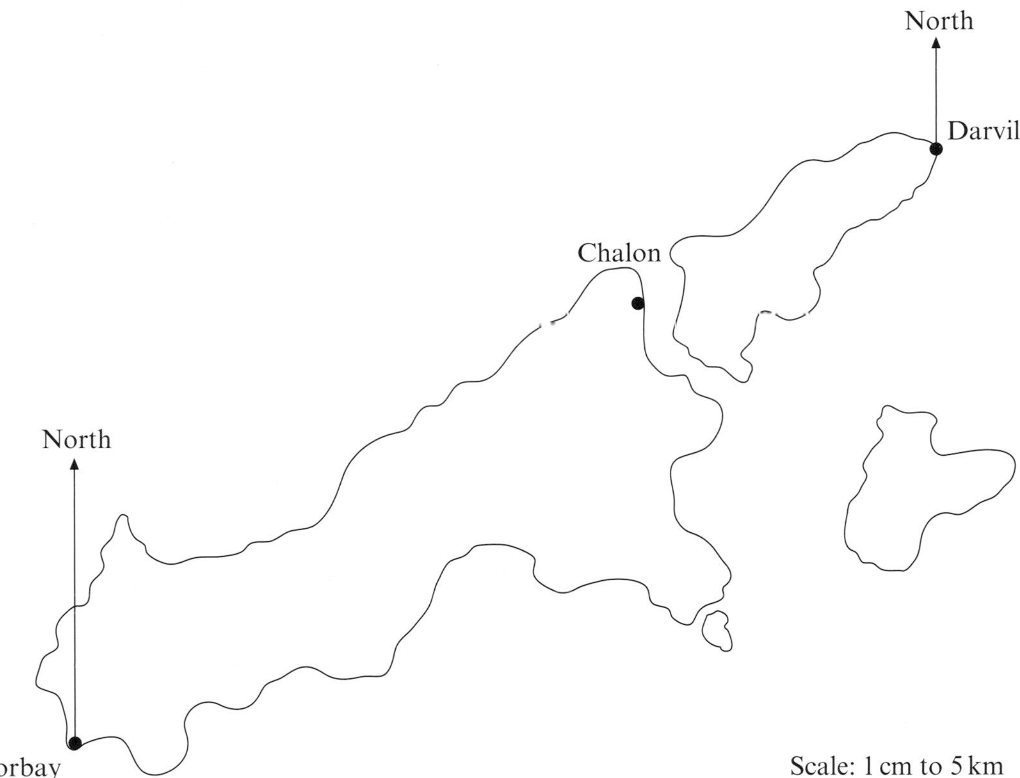

A straight road joins Porbay to Chalon.

a Use the map to find the length of this road in kilometres.

b Brian cycles from Porbay to Chalon along this road.
He sets off at 0930 and cycles at an average speed of 18 kilometres per hour.
At what time does he arrive in Chalon?

(6 marks) (SEG, 1998)

S3.4 **a** The train from London to Manchester takes 2 hours 30 minutes. This train travels at an average speed of 80 miles per hour.

What is the distance from London to Manchester?

b The railway company is going to buy some faster trains. These new trains will have an average speed of 100 miles per hour.

How much time will be saved on the journey from London to Manchester?

(5 marks) (NEAB, 1998)

S3.5 The Grand National horse race is $4\frac{1}{2}$ miles long.

In 1839 the winning horse ran this in 15 minutes.

What was the average speed of this horse in miles per hour?

(2 marks) (NEAB, 1999)

S3.6 The distance between Southampton and London is 120 km.

A coach leaves Southampton at 0800 to travel to London.

The coach travels 60 km at an average speed of 80 kilometres per hour and then stops for 30 minutes. It then continues its journey arriving in London at 1100.

At 1115 the coach leaves London and returns to Southampton, without stopping. It arrives in Southampton at 1230.

a Copy the grid and draw a travel graph for the journey of the coach.

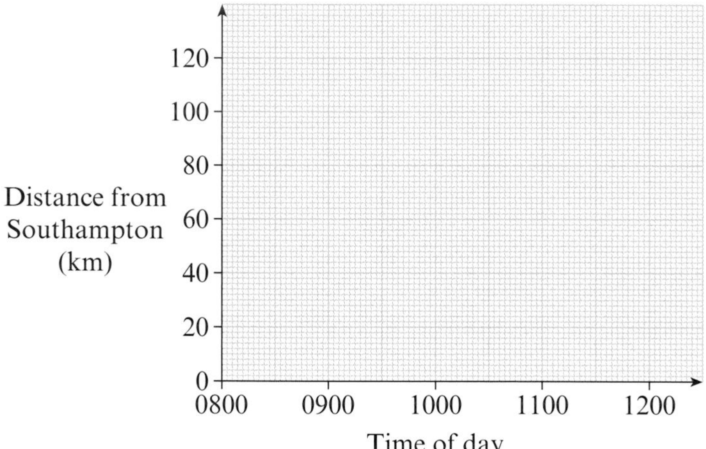

b What was the average speed of the coach on the return journey?

(5 marks) (SEG, 1998)

S4
Maps and bearings

S4.1 The diagram shows an accurate plan of a race.

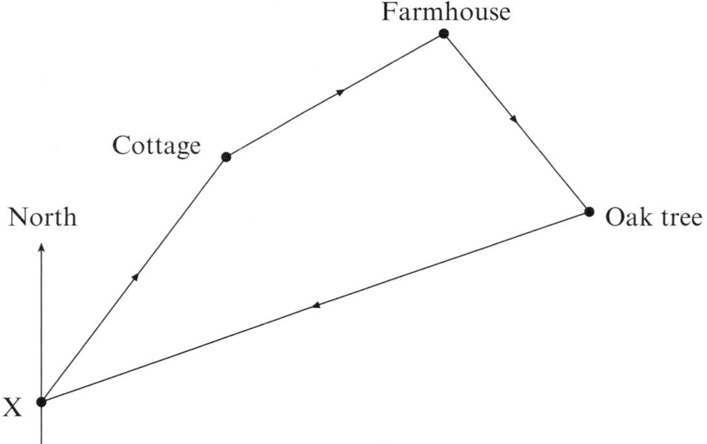

a The start and finish of the race is at **X**.

i What is the bearing of the cottage from **X**?

ii What is the bearing of **X** from the oak tree?

b The plan has been drawn using a scale of 1 mm to represent 20 m.
Use the map to estimate the length of the race in kilometres.
Give your answer to the nearest tenth of a kilometre.

(5 marks) (SEG, 1999)

S4.2 The map of an island is shown.

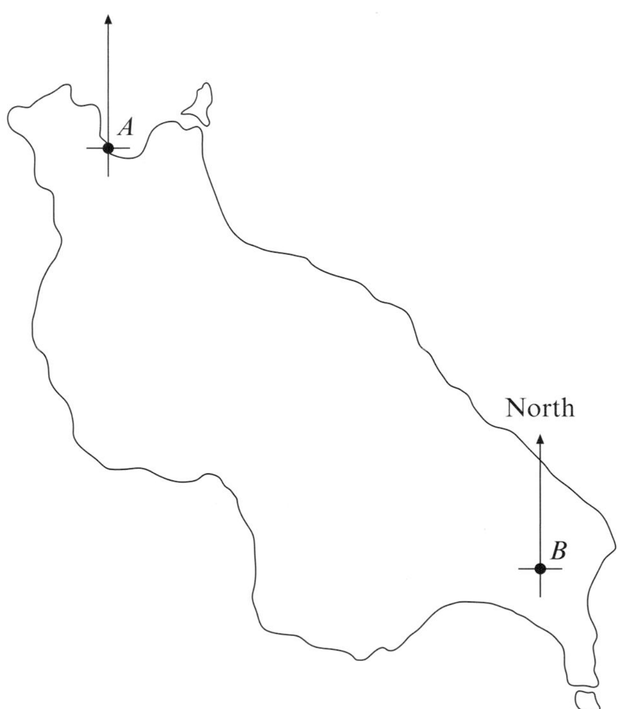

The map has been drawn to a scale of 5 cm to 4 miles.

a The port is at A and the airport is at B.
Use the map to find distance AB in miles.

b The Hotel Central is equidistant from A and B and on a bearing of 150° from A.
On a copy of the diagram draw loci to represent this information.
Mark, with a cross, the position of the hotel on the map.

(6 marks) (SEG, 1999)

S4.3 The diagram shows the course sailed by a yacht from O to A and then from A to B.

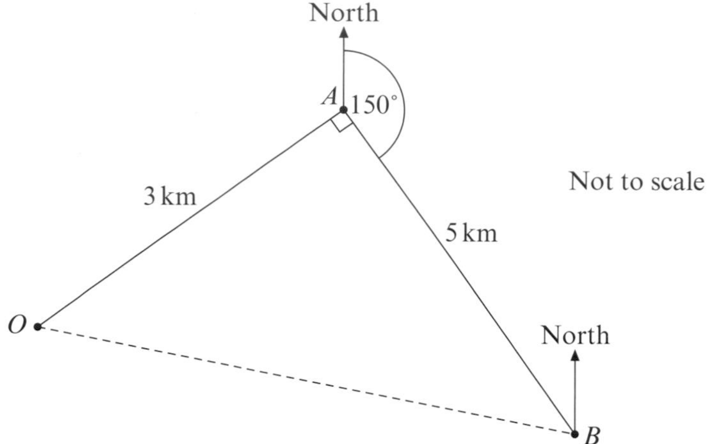

A is 3 km from O.
B is 5 km from A on a bearing of 150°.
Angle OAB is 90°.

Calculate the bearing on which the yacht must sail to return directly from B to O.

No marks will be given for a scale drawing.
(5 marks) (SEG, 1998)

S4.4 The map shows part of Shropshire.

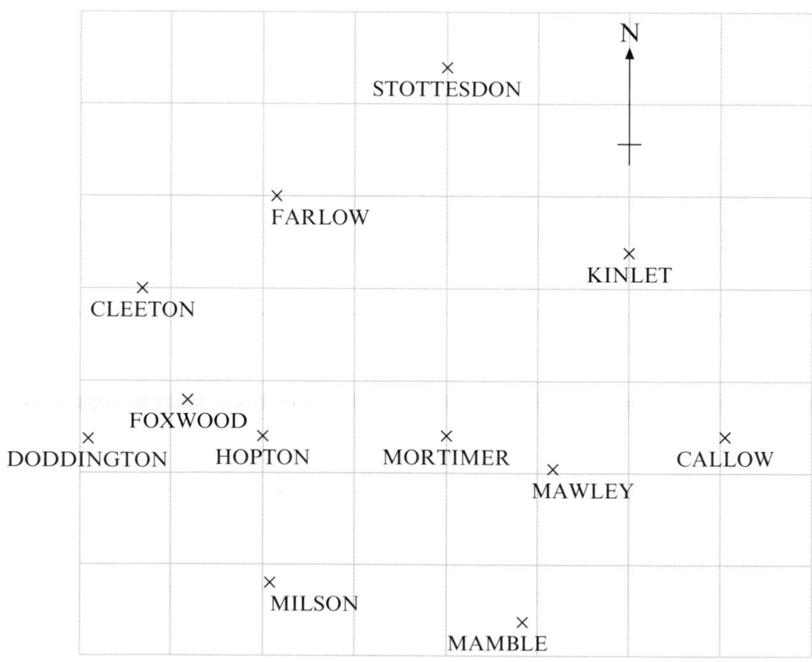

What is the bearing of Cleeton from **Mortimer**?
(2 marks) (NEAB, 1998)

S5
Pythagoras and trigonometry

S5.1 **a** Calculate angle x in triangle ABC.

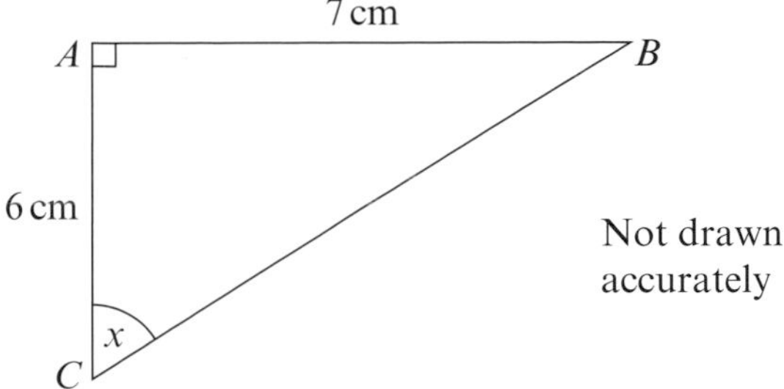

b Calculate length QR in triangle PQR.

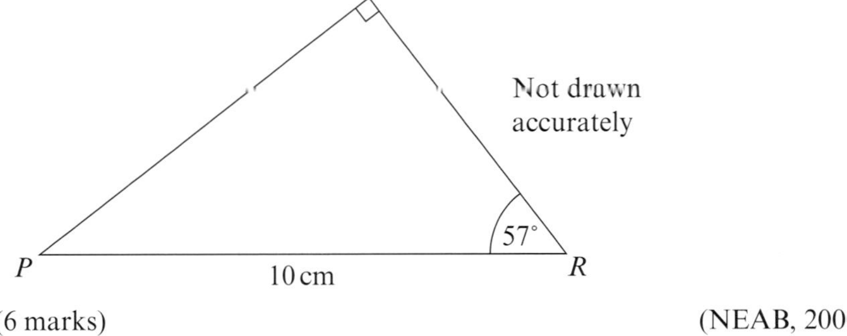

(6 marks) (NEAB, 2000)

S5.2 A lift at the seaside takes people from sea level to the top of a cliff, as shown.

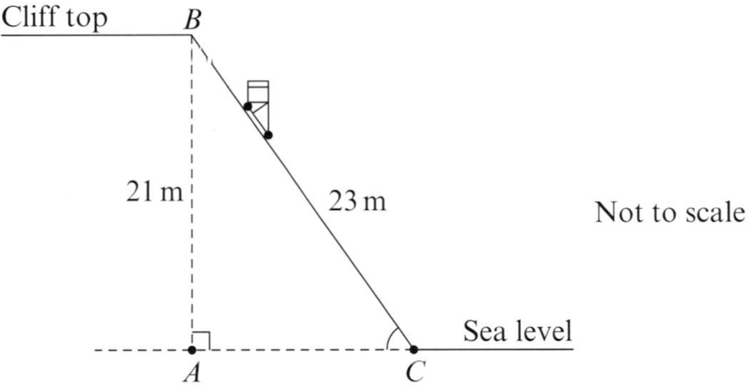

From sea level to the top of the cliff, the lift travels 23 m and rises a height of 21 m.

a Calculate the distance AC.

b Calculate angle BCA.

(6 marks) (SEG, 2000)

S5.3 The diagram shows a kite, $ABCD$.

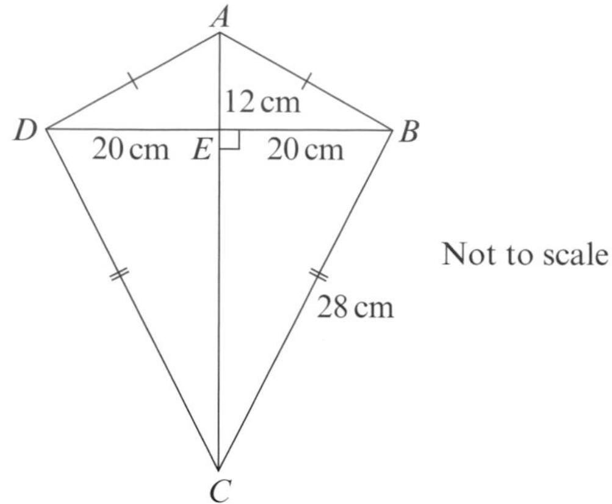

$AE = 12$ cm, $DE = EB = 20$ cm and $BC = 28$ cm.

a Calculate the size of angle EBC.

b Calculate the length of EC and hence find the area of the kite.

(8 marks) (SEG, 1998)

S5.4 The diagram shows a right-angled triangle, ABC.

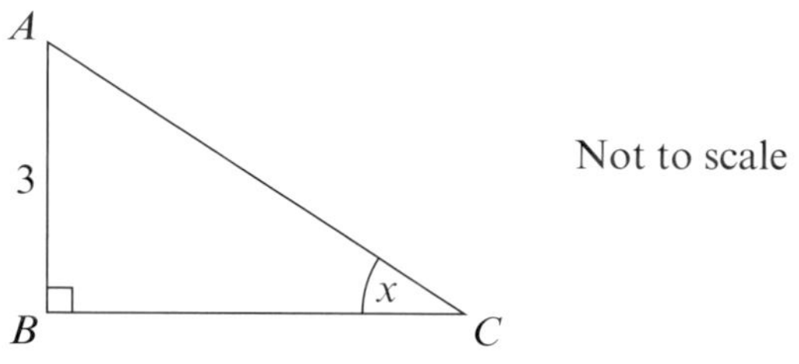

Angle $ABC = 90°$.
Tan $x = \frac{3}{4}$.

a Calculate sin x.

ABC and PQR are similar triangles.

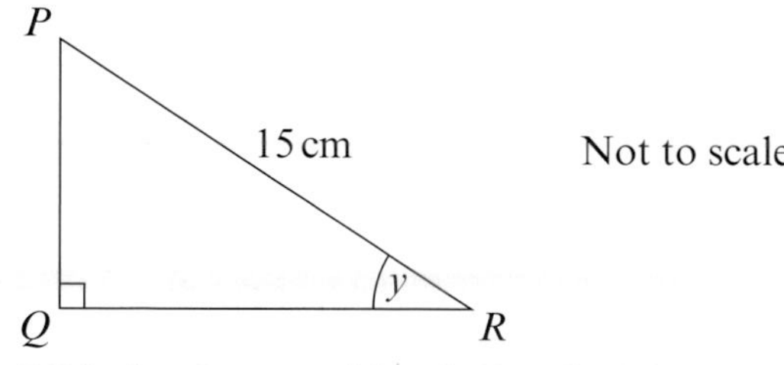

PQ is the shortest side of triangle PQR.

Angle $PQR = 90°$ and $PR = 15$ cm.

b **i** What is the value of cos y?

ii What is the length of PQ?

(7 marks) (SEG, 2000)

S5.5

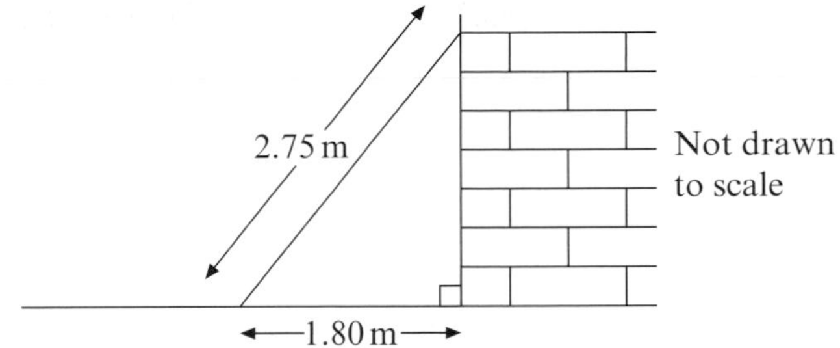

A ladder, 2.75 m long, leans against a wall.
The bottom of the ladder is 1.80 m from the wall, on level ground.

Calculate how far the ladder reaches up the wall.
Give your answer to an appropriate degree of accuracy.
(4 marks) (NEAB, 1998)

S5.6 The sketch shows triangle ABC.
$AB = 40$ cm, $AC = 41$ cm and $CB = 9$ cm.

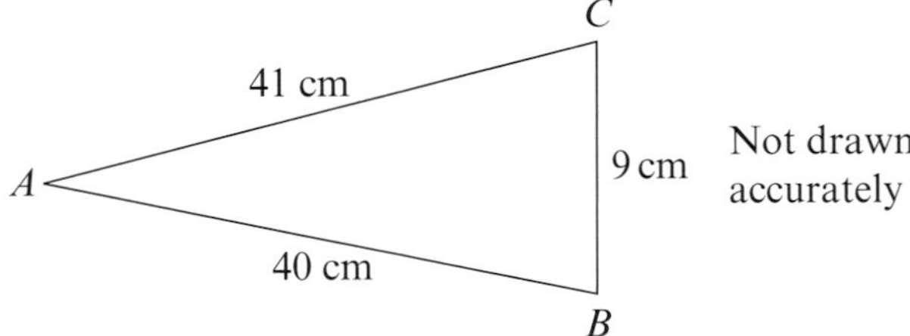

By calculation, show that triangle ABC is a right-angled triangle.
(2 marks) (NEAB, 2000)

S5.7 The diagram shows part of a framework for a roof.

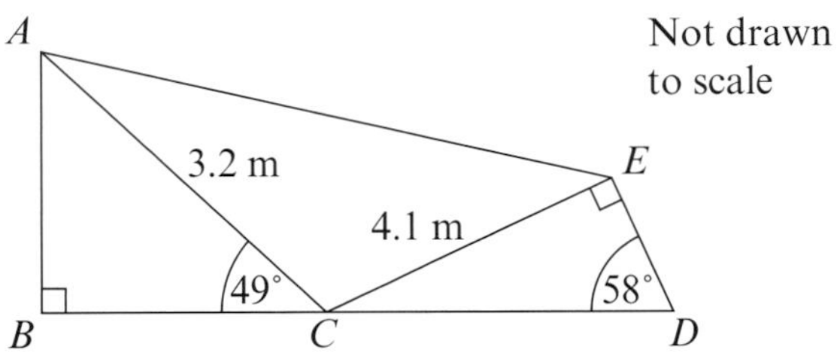

Triangles ABC and CED are right angled.
$AC = 3.2$ m. $CE = 4.1$ m.
Angle ACB is 49°. Angle EDC is 58°.

a Calculate the length BC.

b Calculate the length CD.
(6 marks) (NEAB, 1998)

S6 Constructions, loci and scale drawings

S6.1 Construct an accurate drawing of this triangle.

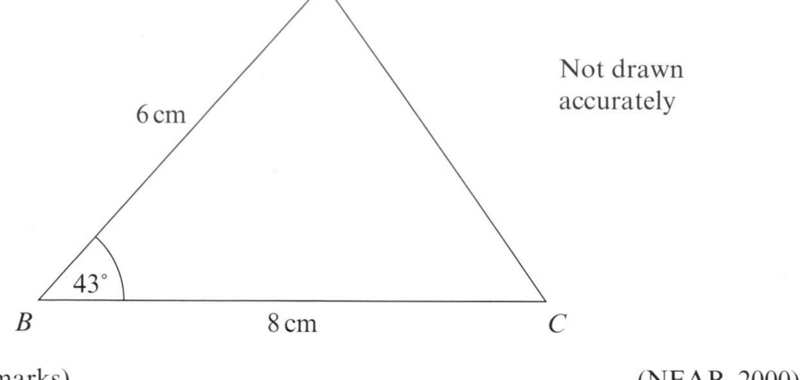

(3 marks) (NEAB, 2000)

S6.2 The scale diagram below shows a plan of Paul's garden.
Paul has an electric lawn mower.
The lawn mower is plugged in at point *P*. It can reach a maximum distance of 12 metres from *P*.
Copy the diagram and, using the same scale, show the area of the garden which the lawn mower can reach.

Scale: 1 cm represents 1 m

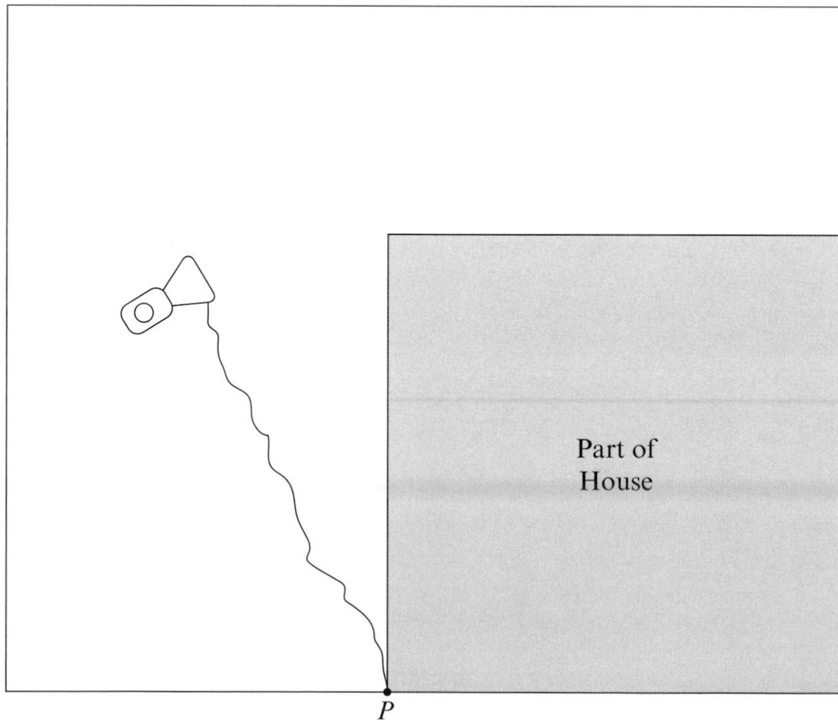

(3 marks) (NEAB, 1998)

S6.3 The scale diagram below shows a plan of a room.
The dimensions of the room are 9 m and 7 m.

Two plug sockets are fitted along the walls.
One is at the point marked *A*. The other is at the point marked *B*.
A third plug socket is to be fitted along a wall.
It must be equidistant from *A* and *B*.

Copy the diagram and, **using ruler and compasses**, find the position of the new socket.
Label it C.

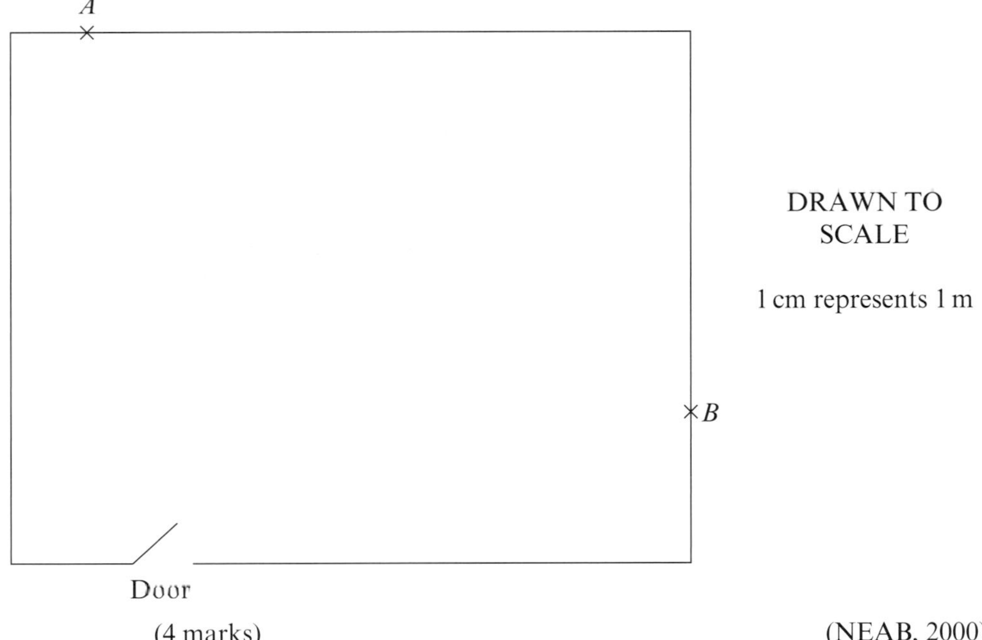

(4 marks) (NEAB, 2000)

S6.4 The plan shows the landing area, *ABCD*, for a javelin event.
AD is the throwing line.
The arc *BC* is drawn from the centre *X*.
The plan has been drawn to a scale of 1 cm to 5 m.

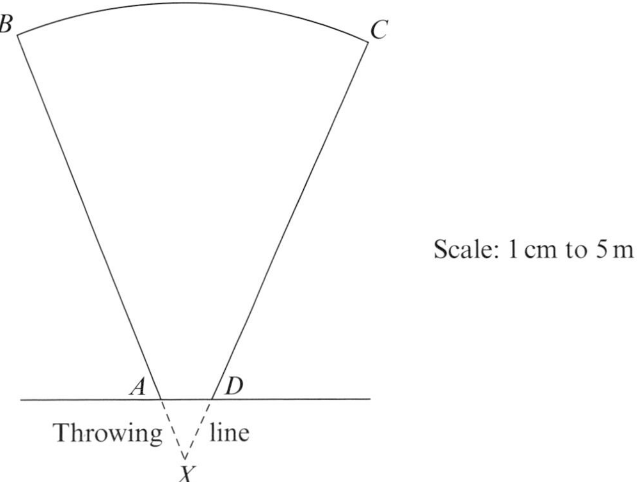

The landing area is fenced off in front of the throwing line.
The position of the fence is always 10 m from the boundaries *AB*, *BC* and *CD* of the landing area.
Copy the diagram and draw accurately the position of the fence on the plan.
(4 marks) (SEG, 2000)

S7 Nets

S7.1 Nets are to be made using different sizes of the following shapes.

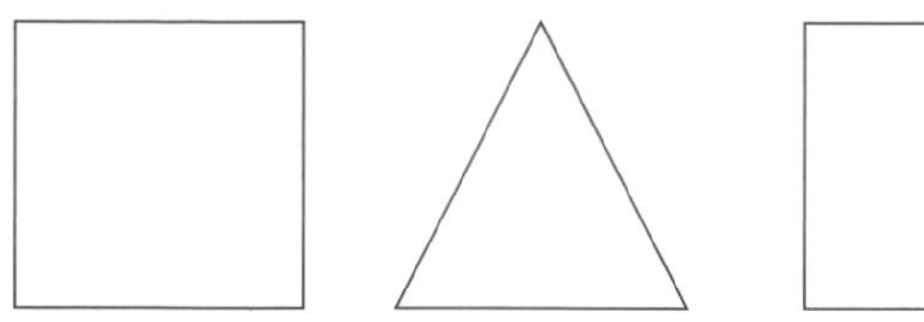

Example: a cuboid has 2 squares, 0 triangles and 4 rectangles.

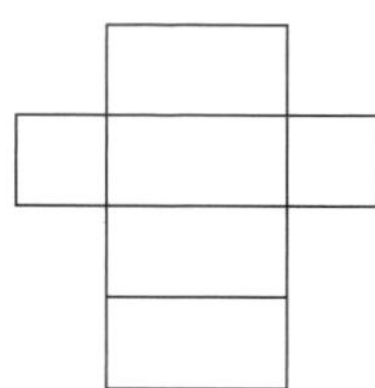

What number of each shape is required to complete a net of

a a cube

b a square-based pyramid

c a triangular prism?

(3 marks) (NEAB, 2000)

S7.2 **a** Which of the following is a net of a cube?

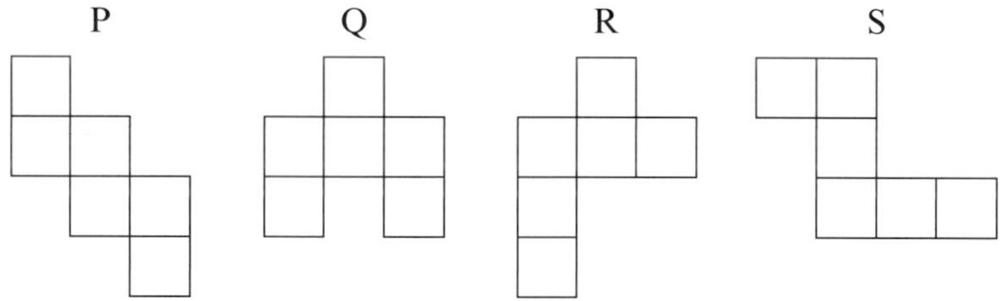

b A cube has edges of length 3 cm.
Calculate the surface area of the cube.

c Cubes of edge 3 cm are stored in a box, as shown.

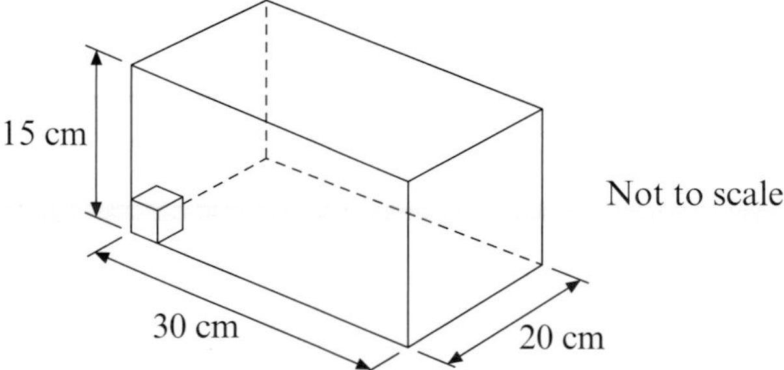

The box is a cuboid with dimensions 15 cm by 20 cm by 30 cm.
What is the maximum number of cubes that can be stored inside the box?

(6 marks) (SEG, 2000)

S7.3 The diagram shows a triangular prism.

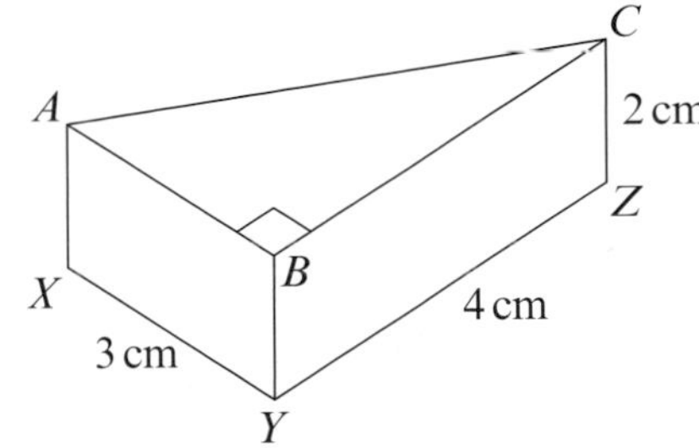

Angle $ABC = 90°$.
$XY = 3$ cm, $YZ = 4$ cm and $CZ = 2$ cm.

a Calculate the volume of the prism, stating your units.

b One face of the prism, $BCZY$, has been drawn below.
Copy the diagram and complete an accurate net of the prism.

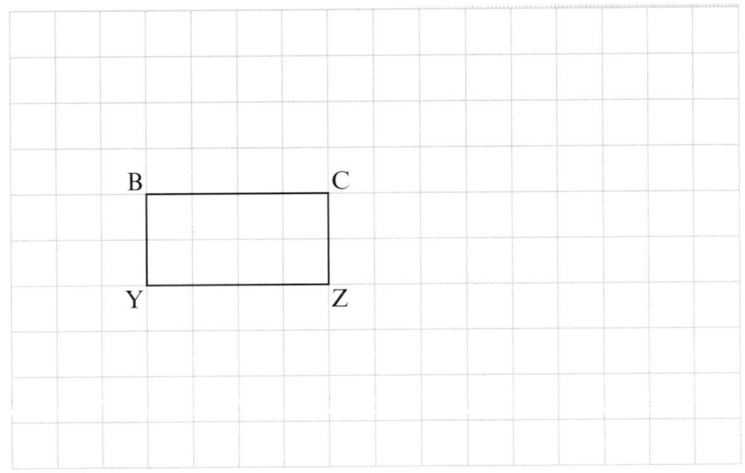

(7 marks) (SEG, 1998)

S7.4 This chocolate box is a triangular prism.

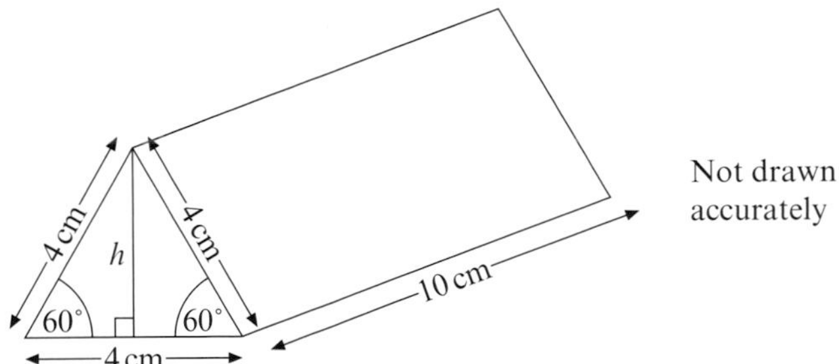

a Draw an accurate net for this triangular prism.
Copy the base which has been drawn for you below.

b Use your net to find the height, h, of the prism.
(4 marks) (NEAB, 2000)

S8 Transformations

S8.1 Copy the triangle and enlarge it with scale factor $\frac{1}{3}$ centre P.

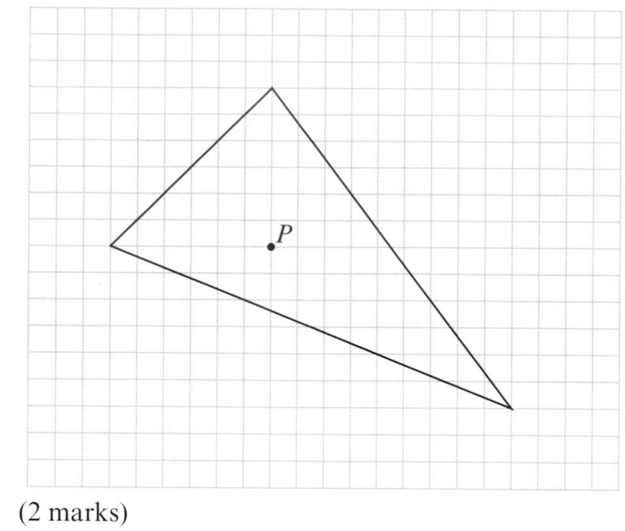

(2 marks) (SEG, 2000)

S8.2 The diagram shows three triangles P, Q and R.

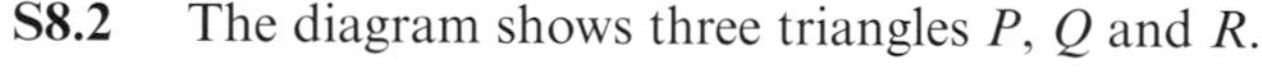

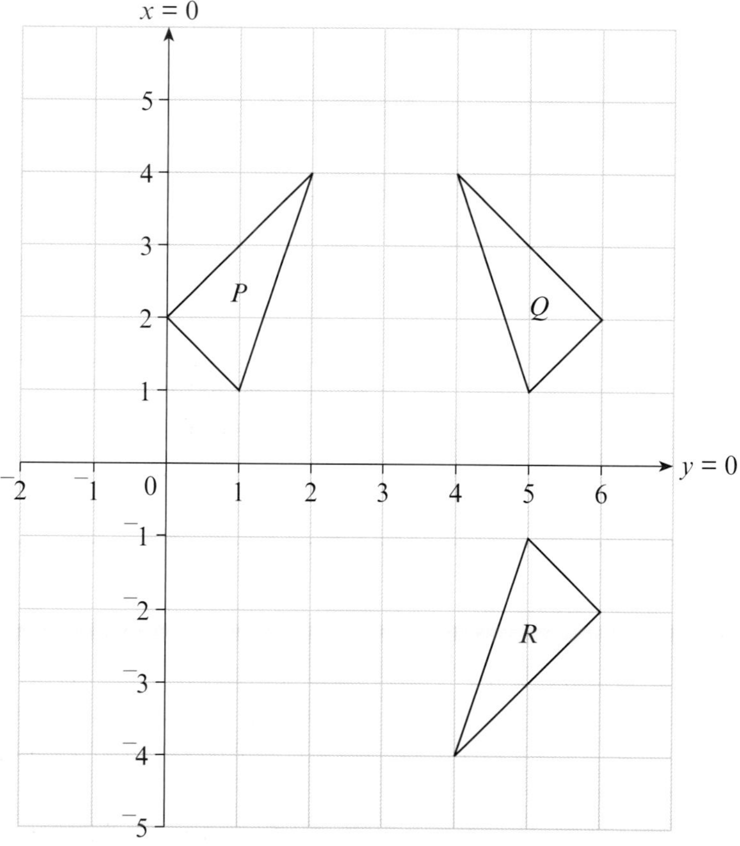

a Describe fully the single transformation which takes P onto Q.

b Describe fully the single transformation which takes P onto R.

(5 marks) (SEG, 1999)

S8.3

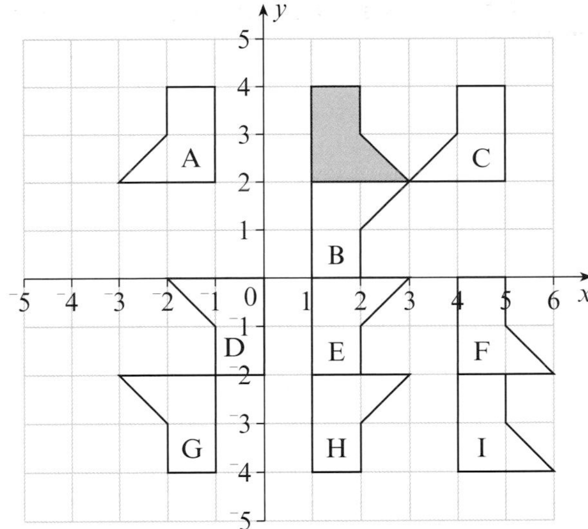

a Give the letter of the finishing position

i after the shaded shape is reflected in the x-axis

ii after the shaded shape is rotated $\frac{1}{2}$ turn about (0, 0)

iii after the shaded shape is translated 3 units right and 4 units down.

b Describe fully the single transformation which will map shape G onto shape H.

(5 marks) (NEAB, 1998)

S8.4 The star $ABCDEFGHIJKL$ is made up of 12 equilateral triangles.

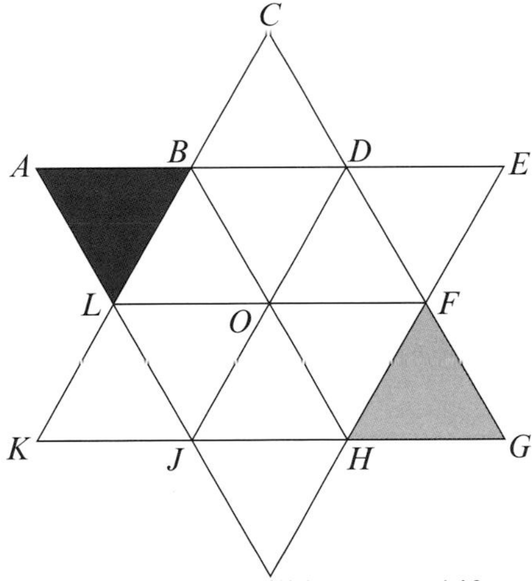

a Which triangle will be covered if

i triangle BCD is rotated by 60° clockwise about the point O

ii triangle BCD is enlarged by a scale factor of 2 from the point C?

b Describe two different **single** transformations that take the black triangle to the grey triangle.

(6 marks) (NEAB, 2000)

S8.5 The diagram shows a rectangle $ABCD$.
The coordinates of A, B and C are given.

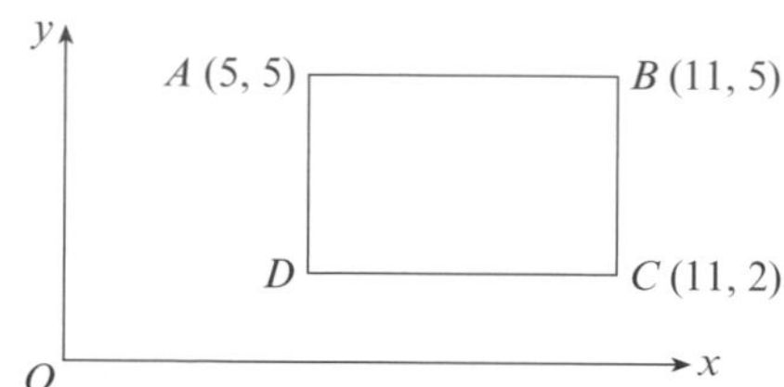

a Write down the equations of the lines of symmetry of the rectangle $ABCD$.

b The rectangle $ABCD$ has been translated 9 units to the left and 4 units down.
Its new position is shown by the rectangle $EFGH$.

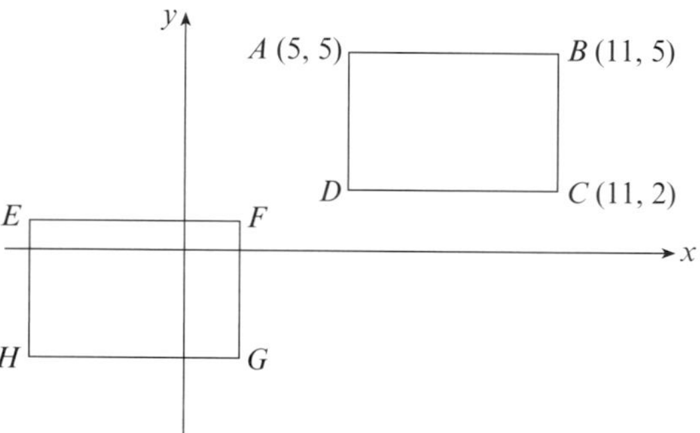

i Write down the coordinates of E.

ii Describe the translation that would move the rectangle back to its original position.

(5 marks) (NEAB, 1999)

S8.6 The diagram shows shapes Q and R which are transformations of shape P.

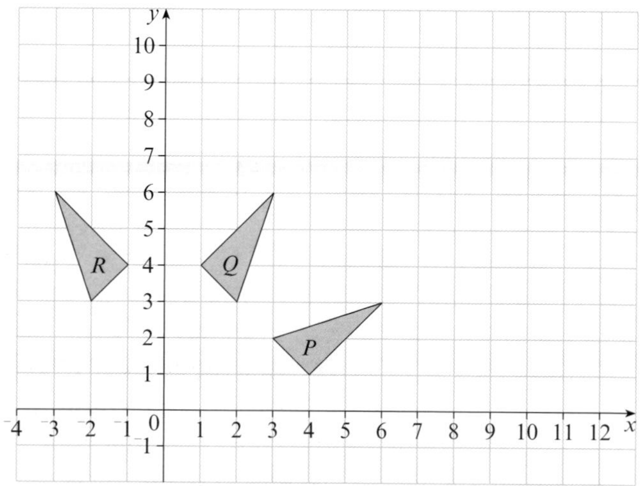

a Describe fully the **single** transformation which takes P onto R.

b Describe fully the **single** transformation which takes P onto Q.

c Copy the diagram and draw an enlargement of shape P with scale factor 2, centre (3, 2).

(7 marks) (SEG, 1999)

S9
Similarity

S9.1 These irregular polygons are similar.

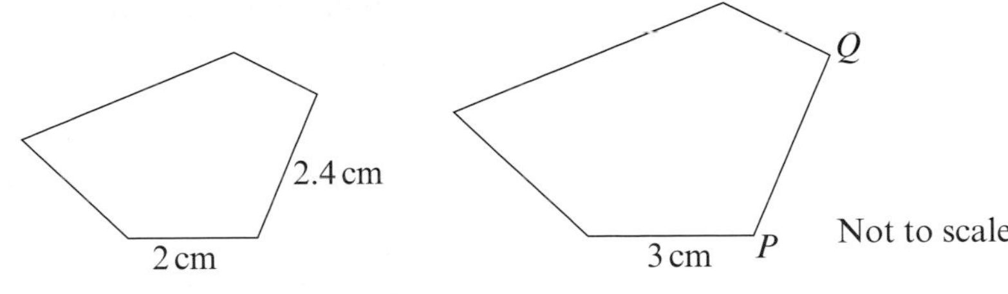

Calculate the length of PQ.
(2 marks) (SEG, 1999)

S9.2 In the diagram AB is parallel to CD.

$AB = OC = 12$ cm. $OB = 10$ cm.

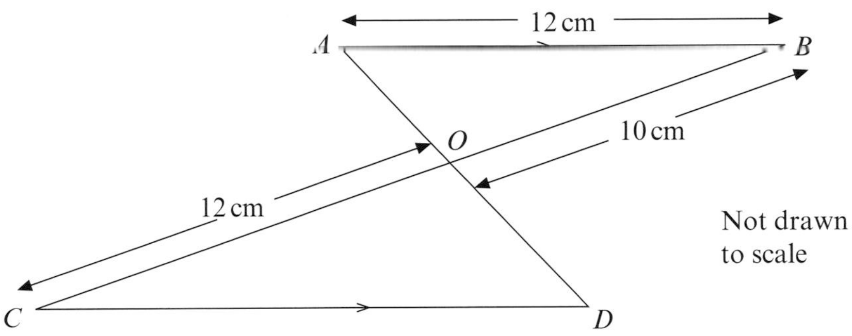

Use similar triangles to calculate the length of CD.
(3 marks) (NEAB, 1998)

S9.3 In the diagram, WA is a straight line and ZX is parallel to BD.

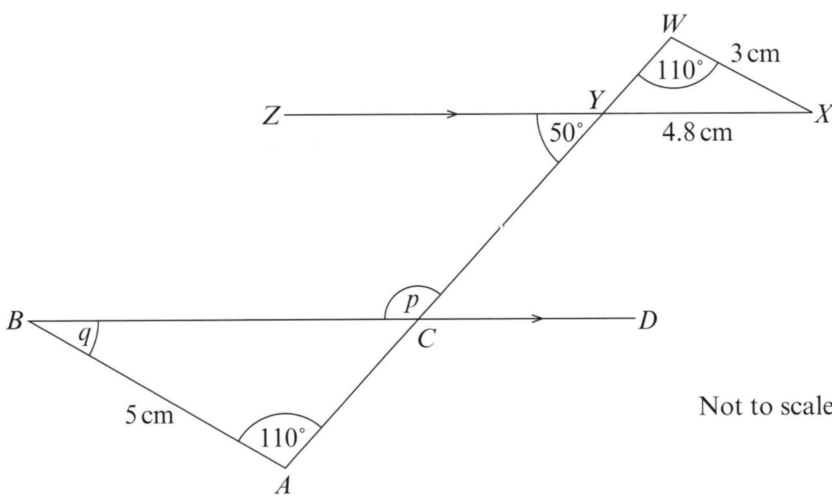

a Work out the size of angle p and the size of angle q.

Triangle ABC is similar to triangle WXY.
$AB = 5$ cm, $WX = 3$ cm and $YX = 4.8$ cm.

b Calculate the length of BC.
(6 marks) (SEG, 1999)

S10
Miscellaneous shape

S10.1 The diagram shows a rectangle which has been partly shaded.

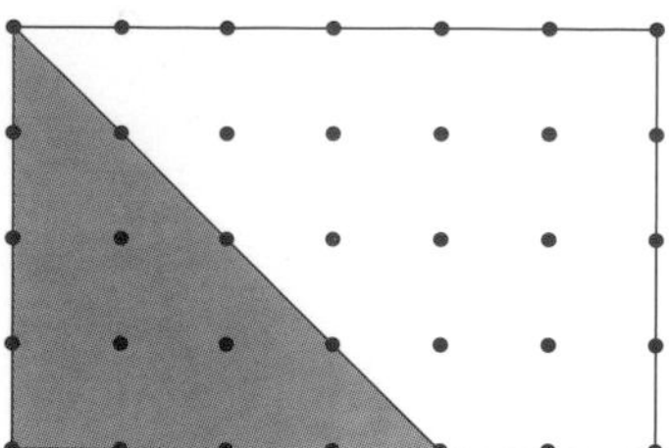

a What fraction of the rectangle is shaded?

b Copy the rectangle and shade more of it so that the final diagram has rotational symmetry of order 2 but no lines of symmetry.

(3 marks) (SEG, 1999)

S10.2 *PQRS* is a parallelogram with vertices at *P*(1, 1), *R*(7, 8) and *S*(5, 3).

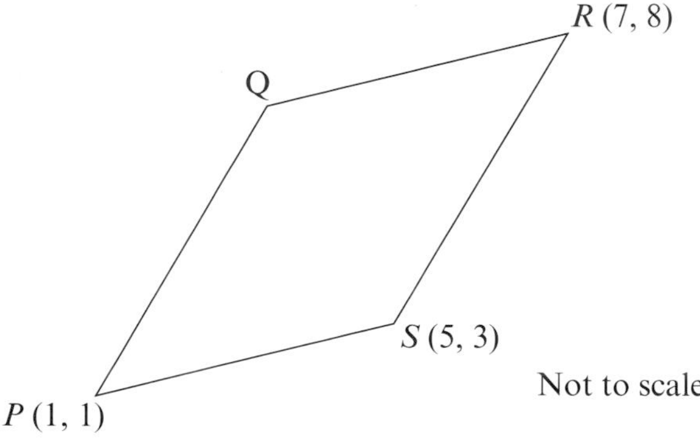

a Write down the coordinates of *Q*.

b The parallelogram is enlarged with scale factor $\frac{1}{2}$, centre *P*(1, 1).
Write down the new coordinates of *S*.

(4 marks) (SEG, 1999)

S10.3 **a** The diagram shows a rectangle with its diagonals drawn.

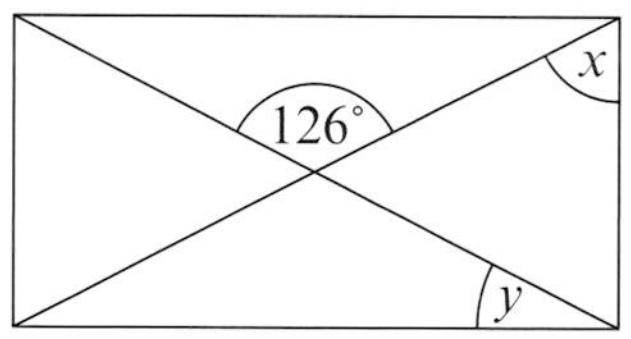

Work out the size of angle x and the size of angle y.

b The diagram shows a triangle *ABC*.

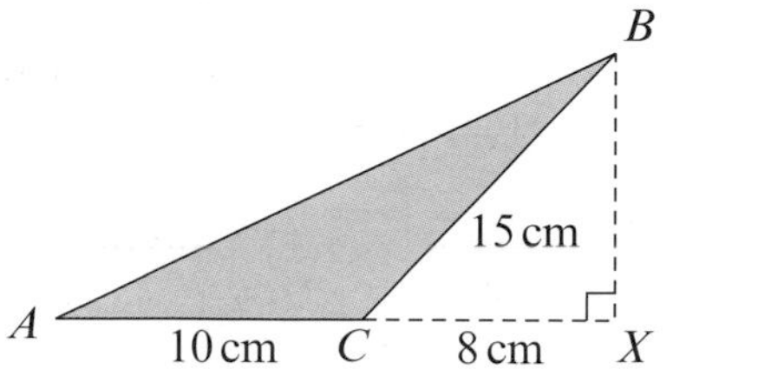

Angle $BXA = 90°$, $BC = 15$ cm and $CX = 8$ cm.

i Calculate *BX*.

ii $AC = 10$ cm.
Calculate the area of triangle *ABC*.

(8 marks) (SEG, 1999)

S10.4 The diagram shows a right-angled triangle ABC and a trapezium $ACDE$.

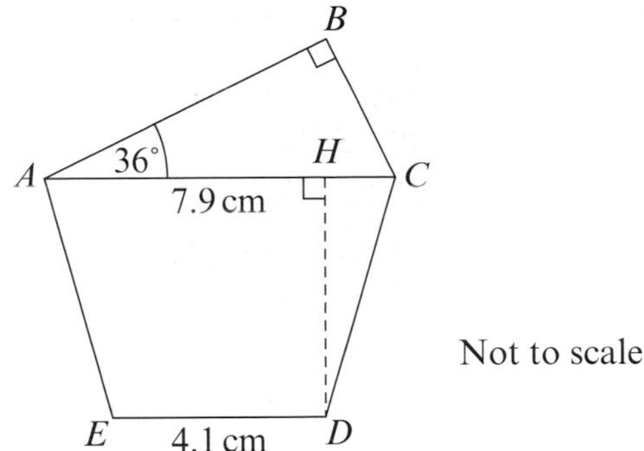

Not to scale

Angle $BAC = 36°$ and $AC = 7.9$ cm.

a Calculate the length of AB.

The area of the trapezium is 52 cm^2.

$ED = 4.1$ cm.

b Calculate DH, the height of the trapezium.

(7 marks) (SEG, 1999)

S10.5 A piece of land is bounded by three straight roads PQ, QR and RP, as shown.

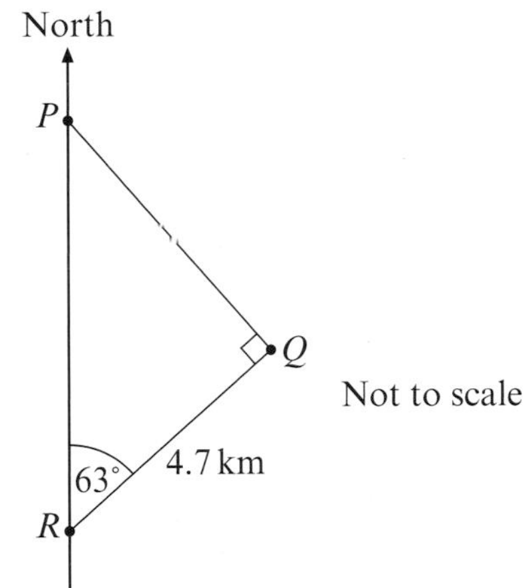

Not to scale

P is due north of R.
Q is on a bearing of 063° from R.
PQR is a right angle.

a What is the bearing of P from Q?

The length of QR is 4.7 km.

b Calculate the area of the piece of land.

(6 marks) (SEG, 1999)

S10.6 Sketch a tetrahedron.
Show any hidden edges as dotted lines.
(2 marks) (NEAB, 2000)

Handling data

D1 Probability

D1.1 Two fair spinners are used for a game.

The scores from each spinner are added together.

For example: The total score from these two spinners is $4 + 5 = 9$.

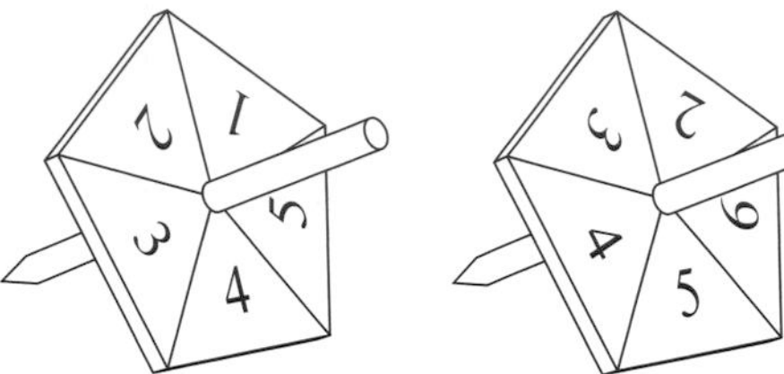

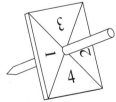

a Copy and complete this table to show all the possible totals for the two spinners.

	1	2	3	4	5
2	3	4			
3	4	5			
4					
5					
6					

b What is the probability of scoring

i a total of 3

ii a total of more than 8?

(5 marks) (NEAB, 1998)

D1.2 Two fair spinners are used for a game.

The score is the **difference**.

For example, the score for these two spinners is $6 - 4 = 2$

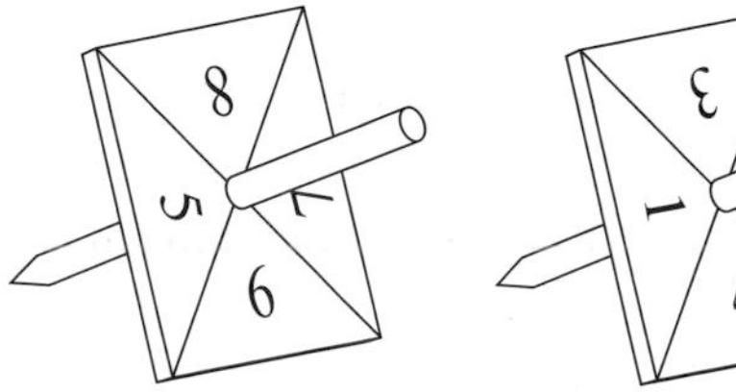

a Copy and complete this table to show all the possible scores for the two spinners.

	5	6	7	8
1			6	
2	3			
3				5
4		2		

b What is the probability of the score being an odd number?
(4 marks) (NEAB, 1999)

D1.3 A fair six-sided dice and a fair coin are thrown at the same time.

a List all the possible outcomes.

b What is the probability of getting a head and an even number?

c What is the probability of getting a tail **or** an odd number **or** both?
(5 marks) (NEAB, 2000)

D1.4 One person is to be chosen at random from four men and two women.

Four events are defined as

Event J: Someone with a name beginning with J is chosen.
Event M: A man is chosen.
Event N: Someone reading a newspaper is chosen.
Event W: A woman is chosen.

What is the probability that, if **one person** is chosen at random:

a both J and M are true

b both J and N are true

c either N or W is true?
(3 marks) (NEAB, 2000)

D1.5 Boxes P and Q each contain five numbered balls.
The balls in each box are numbered as shown.

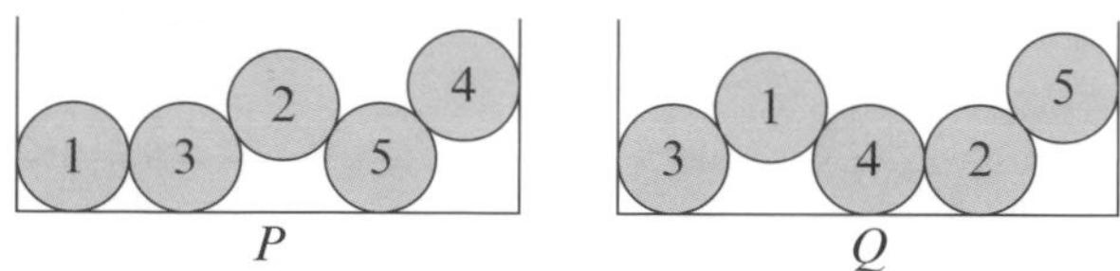

A ball is taken from each box at random.

a What is the probability that both balls are numbered 2?

b What is the probability that both balls have the same number?

c What is the probability that the number on the ball from box P is greater than the number on the ball from box Q?

(6 marks) (SEG, 1998)

D1.6 Penny has a green jacket, a black jacket and a white jacket.
She also has a red skirt and a yellow skirt.

a List all the possible colour combinations if Penny wears one jacket with one skirt.

b Penny chooses a jacket and skirt at random.

i What is the probability that Penny wears the yellow skirt with the black jacket?

ii What is the probability that Penny wears the green jacket?

(5 marks) (NEAB, 1998)

D1.7 An office has two photocopiers, A and B.
On any one day,

the probability that A is working is 0.8,
the probability that B is working is 0.9.

a Calculate the probability that, on any one day, both photocopiers will be working.

b Calculate the probability that, on any one day, only one of the photocopiers will be working.

(5 marks) (SEG, 2000)

D1.8 The table shows information about a group of children.

		Boys	Girls
Wears glasses	Yes	5	3
	No	14	10

a A boy in the group is chosen at random.
What is the probability that he wears glasses?

b A child in the group is chosen at random.
The probability that the child wears glasses is 0.25.

What is the probability that the child does **not** wear glasses?

(3 marks) (SEG, 1998)

D1.9 Andy travels to work by bus on two days.
The probability that the bus is late on any day is 0.6.

a Copy and complete the tree diagram to show the possible outcomes for the two days.

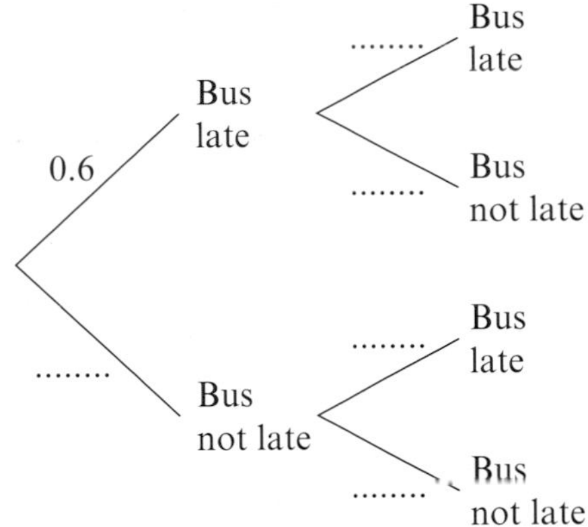

b **i** What is the probability that the bus is 'not late' on both days?

ii What is the probability that the bus is late on at least one of the two days?

(6 marks) (NEAB, 1998)

D1.10 A bag contains 20 coins.
There are 6 gold coins and the rest are silver.

A coin is taken at random from the bag.
The type of coin is recorded and the coin is then **returned** to the bag.
A second coin is then taken at random from the bag.

a The tree diagram shows all the ways in which two coins can be taken from the bag.

Copy the diagram and write in the probabilities.

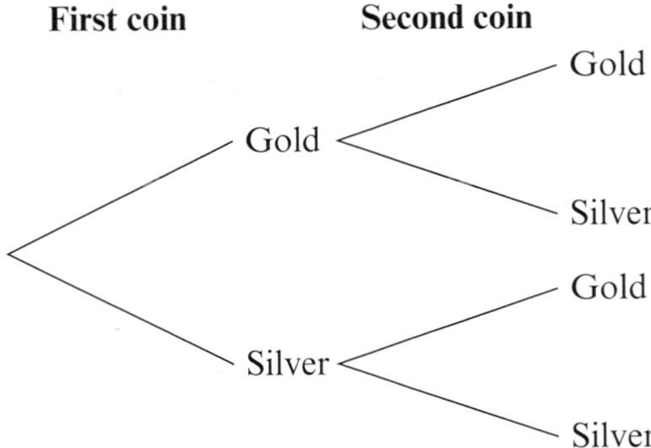

b Use your diagram to calculate the probability that one coin is gold and one coin is silver.

(5 marks) (SEG, 2000)

D1.11 Geoff throws a coin 70 times.
He plots the relative frequency of the number of tails after every 10 throws.

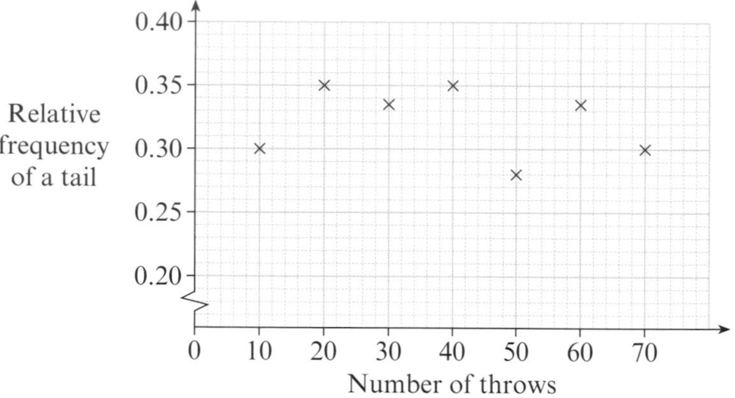

a How many tails were obtained in 50 throws?

b **Use the diagram** to estimate the probability of obtaining a tail.

c Do you think the coin was biased?
Give a reason for your answer.

(6 marks) (NEAB, 1999)

D1.12 Ann and Paul each make an ordinary six-sided dice.

a Paul throws his dice 12 times.
The number '2' occurs 7 times.
Paul says, 'My dice is no good!'
Explain why Paul may be wrong.

b Ann throws her dice 100 times.
The number '2' occurs 19 times.
What is the relative frequency of getting a '2' from Ann's results?

c Ann's dice is now thrown 1000 times.
How many times do you expect the number '2' to occur?

(4 marks) (NEAB, 1999)

D1.13 A fair spinner is labelled as shown.

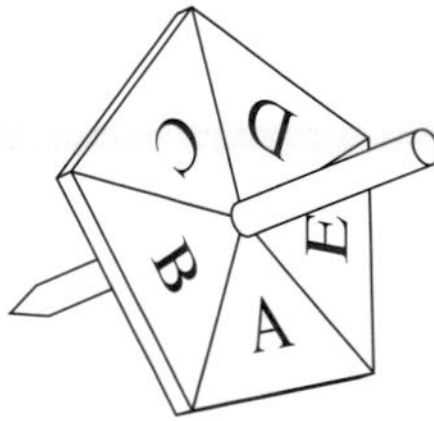

The results of the first 20 spins are shown below.

A B C D B C A A D E
E E B C D C D E A A

a What is the relative frequency of the letter A for these results?
The results of the next 10 spins are shown below.

B C D C A B C D B E

b What is the relative frequency of the letter A after these 30 spins?
The spinner continues to be spun.

c Estimate the number of times the letter A will occur in 1000 spins.

(4 marks) (SEG, 2000)

D2
Drawing and interpreting graphs and charts

D2.1 Karan did a survey to find the most popular cereal.

a The results for adults are shown in the table.

Cereal	Muesli	Weeta Bites	Cornflakes	Other cereals
Number of adults	20	15	30	25

Draw a clearly labelled pie chart to illustrate this information.

b The results for children are shown in this pie chart.

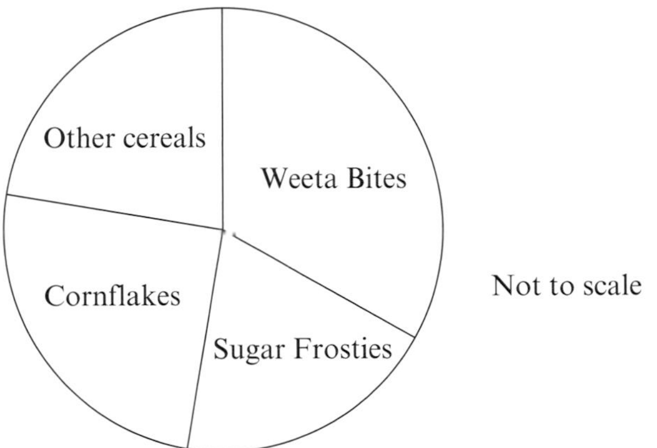

Not to scale

A third of the children eat Weeta Bites.
A quarter of the children eat cornflakes.
Weeta Bites are eaten by 40 children.

How many children eat cornflakes?
(7 marks) (SEG, 2000)

D2.2 One Saturday a newsagent sells the following:

National daily newspapers	510
Echo	360
Magazines and comics	210

a Draw a clearly labelled pie chart to represent these sales.

The frequency polygon shows the sales of national daily newspapers.

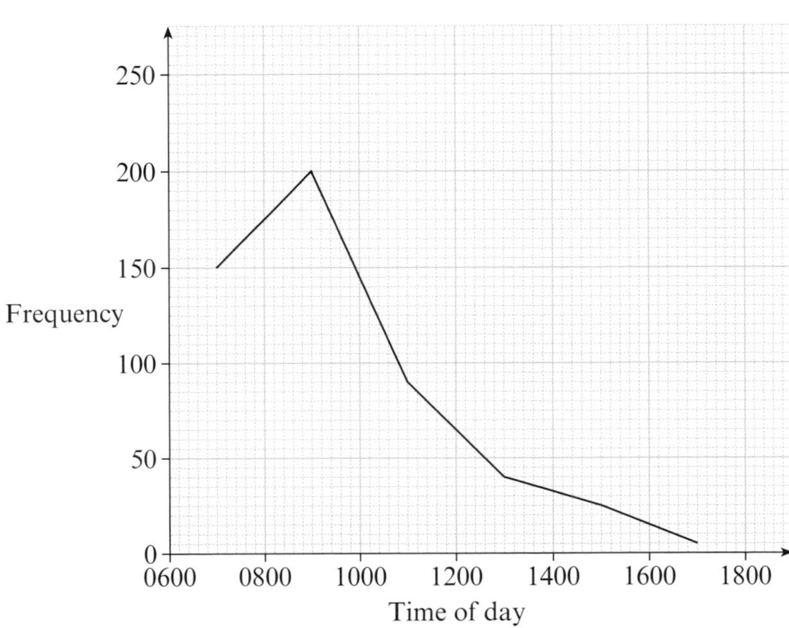

b **i** How many of these papers were sold between 1000 and 1400?
The table shows the sales of the Echo.

Time of day	0600–	0800–	1000–	1200–	1400–	1600–1800
Frequency	0	0	0	20	125	215

ii On a copy of the graph on page 409 draw a frequency polygon to show the sales of the Echo.

iii Compare and comment on the sales of these two types of paper.

(9 marks) (SEG, 1998)

D2.3 Sophie conducts a survey in her class to find out about computer use.
Here are her results.

Have own	9
Share	6
use only in school	10
Never use	5
	30

a Show this information in a pie chart.
Label each sector.

b Sophie decides to find out the main use of the computer.
The table shows her results.

Main use	**Number of people**	**Percentage of people asked**
e-mail	11	
Internet	5	
Word-processing	3	
Games	6	
Total	25	

i Copy and complete the table to show the percentages.

ii Draw a percentage bar chart to show this data.

(8 marks) (NEAB, 2000)

D2.4 A school entered pupils for GCSE maths as shown in the table below.

Tier	Number of pupils
Higher	21
Intermediate	57
Foundation	30

Draw a pie chart for the school GCSE maths entry.

(4 marks) (NEAB, 1999)

D2.5 The pie chart shows information from a survey about the holiday destinations of a number of people.

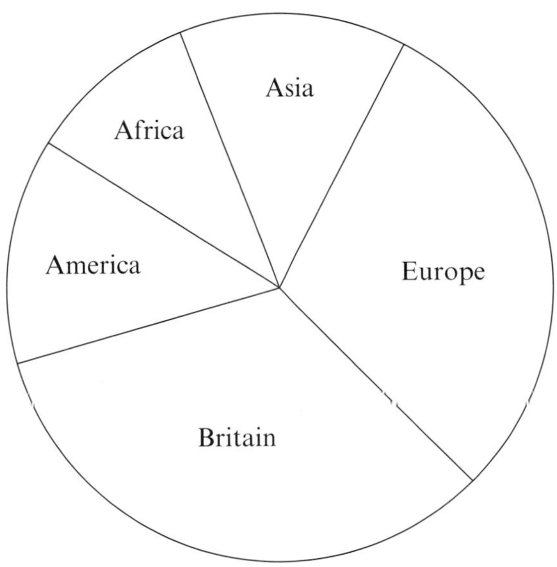

a **i** Which holiday destination is the mode?

ii America is the holiday destination of 24 people.
How many people go to Africa?

In another survey it was found the America is the holiday destination of 21 people out of 180 people asked.

b What percentage of all the people asked in these two surveys gave America as their holiday destination?

(7 marks) (SEG, 1998)

D3 Scatter diagrams and correlation

D3.1 The scatter diagram shows the heights of sixteen Year 9 boys and their fathers.

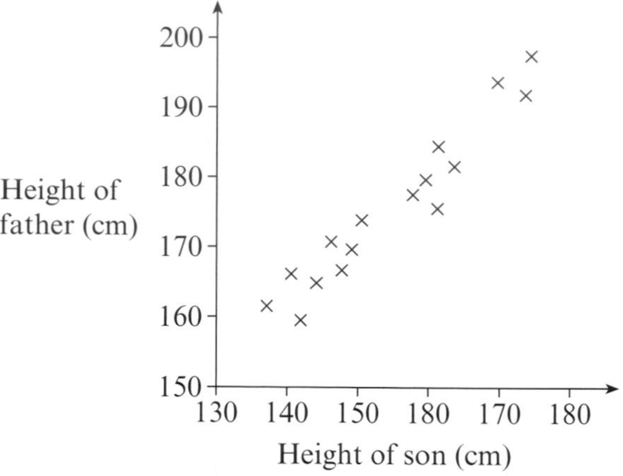

a What does the scatter diagram tell you about the relationship between the heights of these boys and their fathers?

b Copy the diagram and draw a line of best fit.

c Bill, another Year 9 boy, is 155 cm tall.
Use the diagram to estimate the height of Bill's father.
Explain clearly how you obtained your answer.

(4 marks) (NEAB, 1998)

D3.2 A hospital carries out a test to compare the reaction times of patients of different ages.

The results are shown.

Age in years	17	21	24	25	31	15	18	29	20	26
Time (hundredths of a second)	29	40	45	65	66	21	33	62	32	53

a Plot the results as a scatter graph on a copy of the grid.

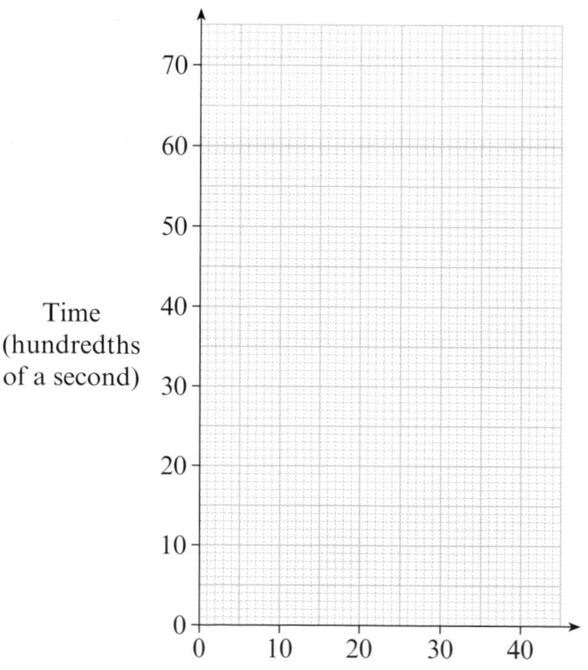

b What does the scatter graph tell you about the reaction times of these patients?

c Draw a line of best fit on the scatter graph.

d The hospital is worried about the reaction time of one patient.

  i How old is the patient?

  ii Using the line of best fit, what should the reaction time be for this patient?

e Hospital records for this reaction test give the following information.

	15-year-olds	30-year-olds
	Time (hundredths of a second)	Time (hundredths of a second)
Lower quartile	20	61
Median	22	65
Upper quartile	25	76

Look at the information for two age groups.
Write down two comments comparing the two sets of information.

(9 marks) (NEAB, 2000)

D3.3 The table gives information about the age and mileage of a number of cars. The mileages are given to the nearest thousand miles.

Age (years)	1	3	5	3	5	4	7	$4\frac{1}{2}$
Mileage (nearest 1000)	9 000	26 000	46 000	27 000	41 000	39 000	62 000	40 000

a Use this information to draw a scatter graph.

b What type of correlation is there between the age and mileage of cars?

c By drawing a line of best fit estimate the age of a car with a mileage of 54 000.

(5 marks) (SEG, 1998)

D3.4 Gordon wants to compare the cost of 10 paperback books with the number of pages in each book.

Cost (£)	4	5.50	3	9	8	2.50	6	10	7.50	5
Number of pages	120	240	75	100	500	80	350	550	400	220

The grid shows the first five pairs of values.

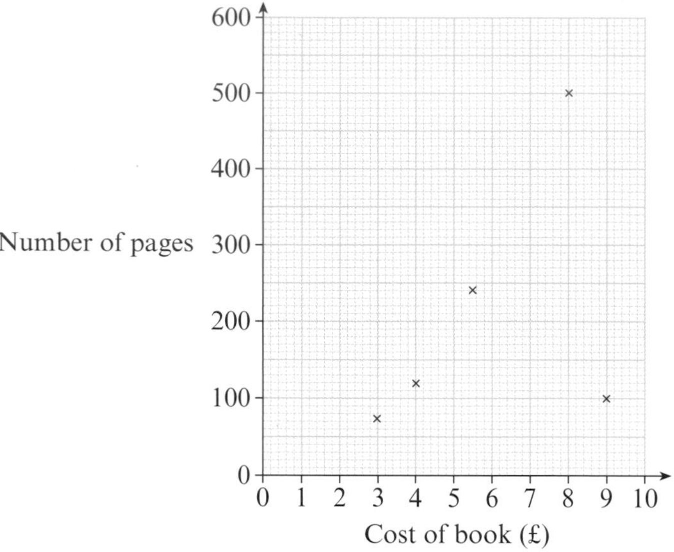

a Copy the grid and plot the other five pairs of values.

b Which book does **not** follow the general pattern?

c Describe the relationship between the cost of a book and the number of pages.

(4 marks) (NEAB, 2000)

D3.5 A teacher asked the pupils in his maths class how long they had spent revising for a maths test.

He drew a scatter diagram to compare their test results and the time they had spent revising.

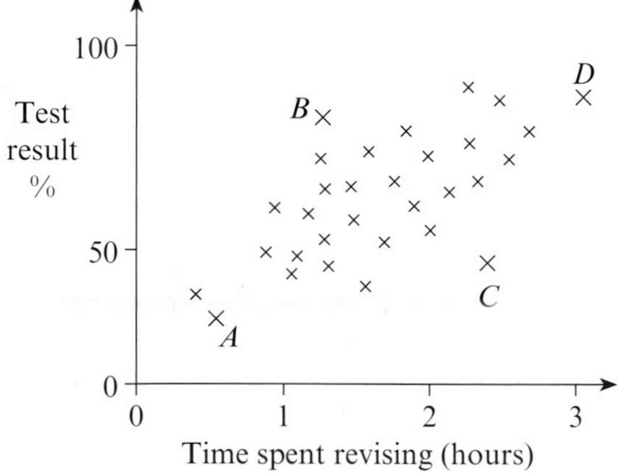

a State which point *A*, *B*, *C* or *D* represents the statement:

i Keith 'Even though I spent a long time revising, I still got a poor test result.'

ii Val 'I got a good test result despite not doing much revision.'

iii Jane 'I revised for ages and got a good test result.'

b Make up a statement which matches the point you have **not** used in your answer to part **a**.

c What does the scatter diagram tell you about the relationship between the time the pupils spent revising and their test results?

(5 marks) (NEAB, 1999)

D4 Averages

D4.1 The temperatures at midnight on the last five nights in December were,

−2° −1° 2° 0° −4°

What is the mean (average) of these temperatures?

(2 marks) (NEAB, 1999)

D4.2 **a** The lateness of 12 trains is recorded.
The results, in minutes, are shown.

1 1 2 2 3 3 3 5 7 7 9 15

i What is the range in lateness for these trains?

ii What is the mean lateness for these trains?

The lateness of 12 buses is also recorded.
The range in lateness for these buses is 7 minutes and mean lateness is 12 minutes.

b Give **one** difference in the lateness for these trains and buses.

(5 marks) (SEG, 1999)

D4.3 **a** Write down four whole numbers whose mode is 10 and whose range is 8.

b Write down four whole numbers whose mode is 10 and whose mean is 9.

c Write down four whole numbers whose mode is 10, whose mean is 9 and whose range is 8.

(6 marks) (NEAB, 2000)

D4.4 The pupils in a class are asked how many goals each one scored in a netball lesson.
The results are shown in the table.

Number of goals	Number of pupils
0	1
1	8
2	11
3	5
4	4
5	1

Calculate the mean number of goals per pupil.

(3 marks) (NEAB, 2000)

D4.5 The graph shows the weekly pocket money of 100 children aged 10 years.

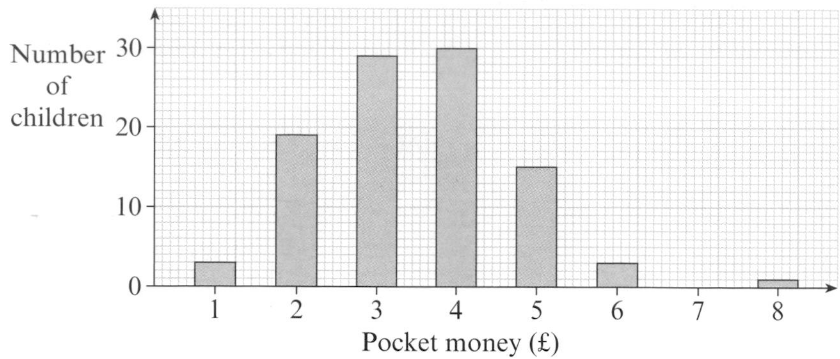

a **i** What is the median amount of pocket money?

ii Calculate the mean amount of pocket money.

The table shows the weekly pocket money of 100 students aged 16 years.

Pocket money (£)	0–	5–	10–	15–	20–25
Number of students	19	18	21	19	23

b Do the 10-year-olds or the 16-year-olds have the larger range in pocket money?
Give a reason for your answer.

(7 marks) (SEG, 1999)

D4.6 The table shows the age distribution of females taking part in a marathon.

Age (a years)	**Female**
$10 \leqslant a < 20$	0
$20 \leqslant a < 30$	15
$30 \leqslant a < 40$	27
$40 \leqslant a < 50$	14
$50 \leqslant a < 60$	4
$60 \leqslant a < 70$	0

Calculate an estimate of the mean age of these females.

(4 marks) (SEG, 1999)

D4.7 The table shows the time taken by 40 pupils to do their homework.

Time, t (minutes)	**Number of pupils**
$0 < t \leqslant 10$	2
$10 < t \leqslant 20$	3
$20 < t \leqslant 30$	12
$30 < t \leqslant 40$	11
$40 < t \leqslant 50$	8
$50 < t \leqslant 60$	4

Calculate an estimate for the mean time taken to do the homework.

(4 marks) (NEAB, 1999)

D5
Cumulative frequency

D5.1 The table gives the age distribution of the population of the United Kingdom.

Age, A, years	Number of people (millions)	Cumulative frequency (millions)
$0 \leqslant A < 10$	9	
$10 \leqslant A < 20$	7	
$20 \leqslant A < 30$	8	
$30 \leqslant A < 40$	9	
$40 \leqslant A < 50$	8	
$50 \leqslant A < 60$	7	
$60 \leqslant A < 70$	6	
$70 \leqslant A < 80$	4	
$80 \leqslant A < 90$	2	
$90 \leqslant A < 100$	1	
	60	

a **i** Copy the table and complete the cumulative frequency column.

ii Draw a cumulative frequency diagram on a copy of the grid.

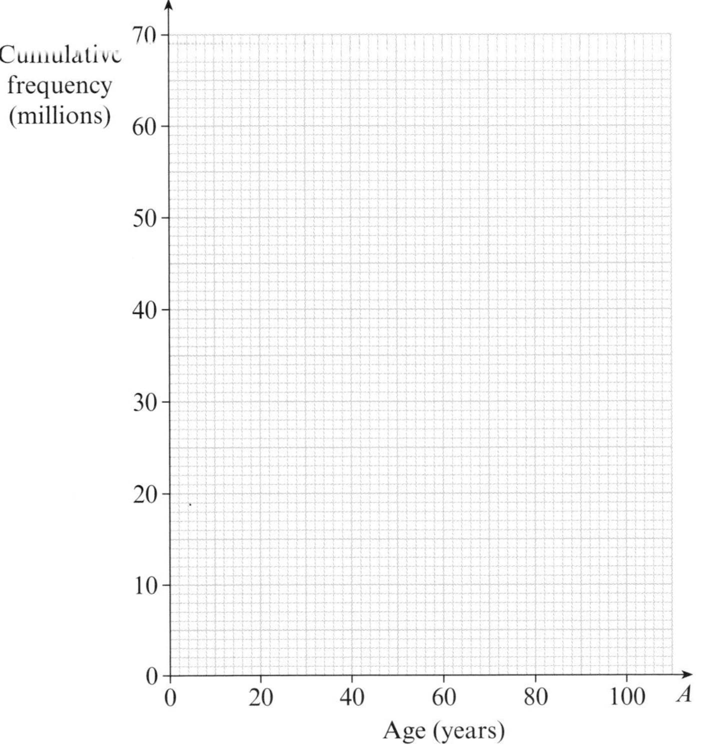

b Use your diagram to estimate

i the median age

ii the interquartile range.

c Use your diagram to estimate the number of people who are 65 years or over.

(10 marks) (NEAB, 1999)

D5.2 The length of life of 100 batteries of a certain make was recorded. The table shows the results.

Length of life (hours)	<10	<15	<20	<25	<30	<35	<40
Cumulative frequency	0	2	9	50	86	96	100

a Draw a cumulative frequency graph to illustrate these data.

b How many batteries had a life of more than 32 hours?

c Use your graph to estimate:

i the median

ii the interquartile range.

d Another make of battery has a median length of life of 25 hours and an interquartile range of 7 hours.
Is this make of battery likely to be more reliable than the first?

Give a reason for your answer.

(8 marks) (SEG, 1998)

D5.3 The ages of 500 people attending a concert are given in the table below.

Age, A, years	**Number of people**	**Cumulative frequency**
$0 \leqslant A < 10$	20	
$10 \leqslant A < 20$	130	
$20 \leqslant A < 30$	152	
$30 \leqslant A < 40$	92	
$40 \leqslant A < 60$	86	
$60 \leqslant A < 80$	18	
$80 \leqslant A < 100$	2	

a **i** Copy the table and complete the cumulative frequency column.

ii Draw a cumulative frequency graph.

b Use your diagram to estimate:

i the median age

ii the interquartile range.

c Use your diagram to estimate the percentage of people at the concert who are under the age of 16 years.

(10 marks) (NEAB, 1999)

D6 Surveys and questionnaires

D6.1 This graph shows the time of day when accidents happened.

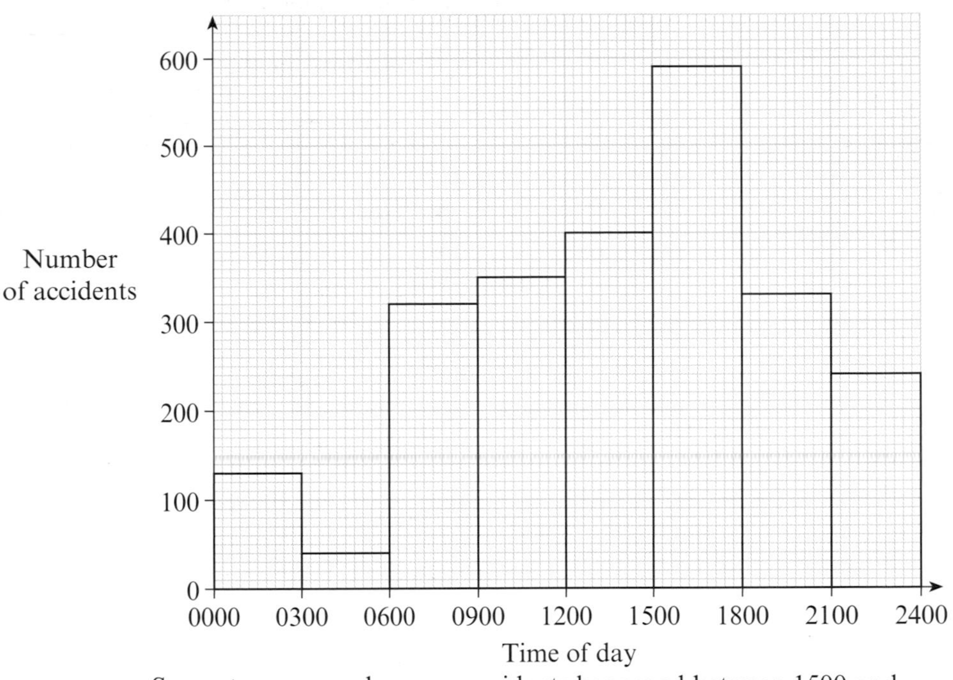

a Suggest a reason why more accidents happened between 1500 and 1800 than at other times of the day.

b **i** How many accidents happened between 2100 and 2400?

$12\frac{1}{2}\%$ of the accidents which happened between 2100 and 2400 involved drivers who had been drinking.

ii How many accidents between 2100 and 2400 involved drivers who had been drinking?

(4 marks) (SEG, 1999)

D6.2 Brian does a survey on what pupils eat for breakfast.
Here are three questions from his questionnaire.

1. What time do you get up?
   - Before 7.00 ☐
   - Between 7.00 and 8.00 ☐
   - After 8.00 ☐
2. Which of these do you have for breakfast?
   Bacon ☐ Eggs ☐ Cereal ☐
3. How long does it take to eat your breakfast?

   ....

Write down one **criticism** of each question.
(3 marks) (NEAB, 2000)

D6.3 Paul helps in his school library.
He notices that some pupils who go into the library do not borrow books.
Paul knows those pupils are using books, magazines, careers information, the CD ROM, or the photocopier.

Paul decides to use a questionnaire to find out how pupils at the school use the library.
Here is part of Paul's questionnaire.

> ***School Library Questionnaire***
>
> 1. *Please tick the box to show which year you are in.*
>
> ☐ *Year 7* ☐ *Year 8* ☐ *Year 9* ☐ *Year 10* ☐ *Year 11*

a Write down another question Paul could ask in his questionnaire.
Make sure you include a section for the pupil's response.

Paul gives the pupils a questionnaire when he stamps their books.

b Is this a suitable way to give out the questionnaires?
Give a reason for your answer.

(3 marks) (NEAB, 1998)

D6.4 This statement is made on a television programme about health.

'Three in every eight pupils do not take any exercise outside school.'

a A school has 584 pupils.
According to the television programme, how many of these pupils do not take any exercise outside school?

b Clare says, 'I go to the gym twice a week after school.'
She decides to do a survey to investigate what exercise other pupils do outside school.

Write down two questions that she could ask.

c Matthew decides to do a survey in his school about the benefits of exercise.
He decides to ask the girls' netball team for their opinion.
Give **two** reasons why this is **not** a suitable sample to take.

d This is part of Matthew's question.

> **Question** *Don't you agree that adults who were sportsmen when they were younger suffer more from injuries as they get older?*
>
> **Response** *Tick one box*
>
> ☐ *Yes* ☐ *Usually* ☐ *Sometimes* ☐ *Occasionally*

i Write down one criticism of Matthew's question.

ii Write down one criticism of Matthew's response section.

(8 marks) (NEAB, 2000)

D7 Miscellaneous handling data

D7.1 A school wants to open a snack bar.
They want to sell sandwiches, pizzas, soup, jacket potatoes and salad.

a Design a data collection sheet that could be used to find out from pupils in the school what they would buy.

b Invent the first 30 entries on your data sheet.

(3 marks) (NEAB, 1999)

D7.2 An automatic car wash cleans approximately 10 000 cars a year.
About 100 cars a year get scratched paintwork.

a A car uses the war wash.
Estimate the relative frequency that it gets scratched paintwork.

The probability that a car is green is 0.1.

b A car uses the car wash.
Calculate the probability that it is green and gets scratched paintwork.

(3 marks) (SEG, 1999)

D7.3 A school holds a 'mini marathon'.
Every pupil who finishes the marathon gains points.
The table shows the times taken, the number of pupils and the points gained.

Time interval	Number of pupils	Points gained
1 hour or less	0	60
More than 1 hour but less than or equal to 2 hours	10	50
More than 2 hours but less than or equal to 3 hours	15	40
More than 3 hours but less than or equal to 4 hours	40	30
More than 4 hours but less than or equal to 5 hours	57	20
More than 5 hours but less than or equal to 6 hours	23	10

a What was the most common time interval?

b Ann and her brother Ben both take part.
Ann finishes only 5 minutes before Ben but gains twice as many points.
Explain how this could have happened.

c Draw a frequency diagram to show the number of pupils and the time taken.

(4 marks) (NEAB, 2000)

Do NOT use a calculator for this exam paper.

1 A window cleaner charges £1.50 plus 25p per window.

a Roy has seven windows cleaned.
How much is the total charge?

(3 marks)

b Hayley has x windows cleaned.
Write down an expression for the amount she pays.

(2 marks)

2 a Here is a number machine.
Copy and complete the answer box.

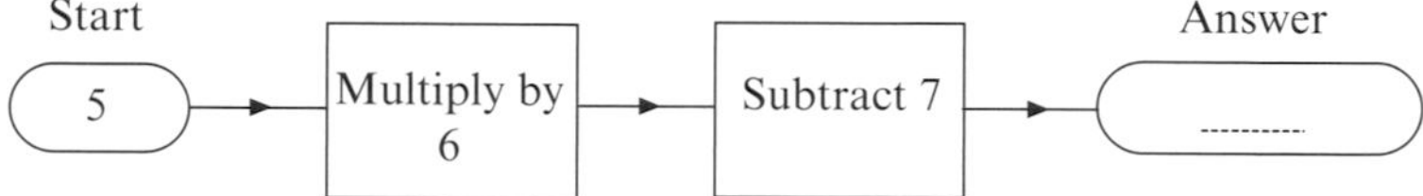

(1 mark)

b Here is an algebra machine.
Copy and complete the answer box.

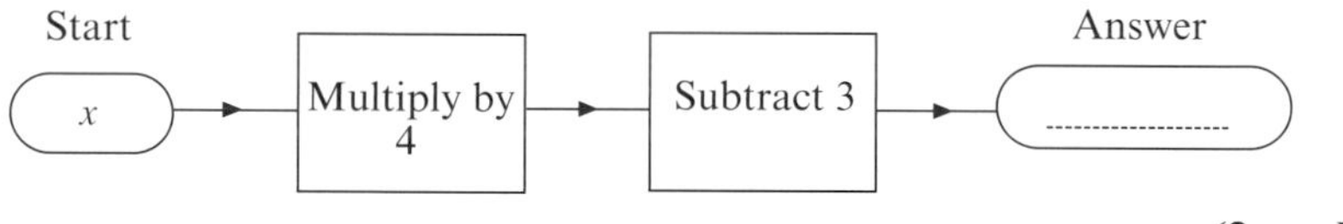

(2 marks)

3 a On a copy of the grid, reflect the shape in the dotted line.

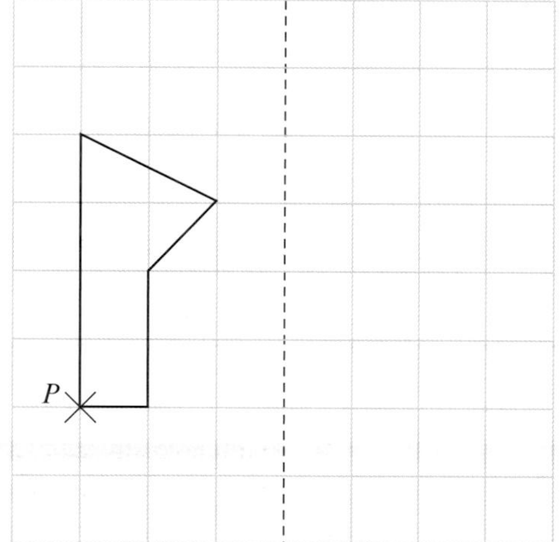

(2 marks)

b Rotate your original shape through 180° about the point P.

(2 marks)

4 120 customers in a bread shop were asked which type of bread they prefer.
Here are the results.

Type of bread	White	Brown	Rolls	other
Number of customers	25	43	36	16

a Draw a pie chart to show the results. **(4 marks)**

b Calculate the percentage of customers who prefer rolls. **(2 marks)**

5 The triangle ABC is isosceles.

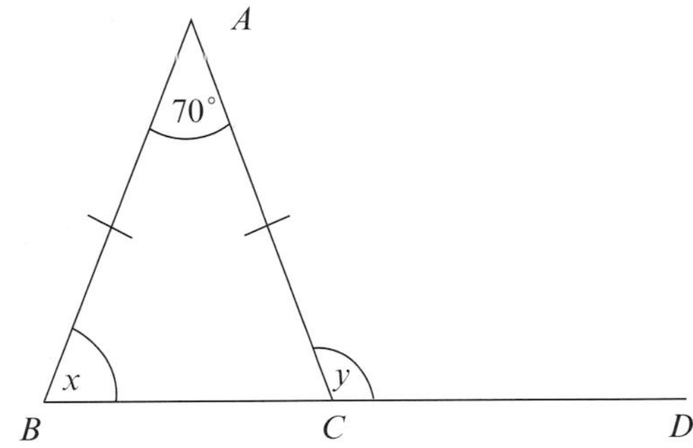

a Work out the value of angle x. **(2 marks)**

b Work out the value of angle y. **(2 marks)**

6 **a** Find the value of $2x + 3y$ when $x = \frac{1}{2}$ and $y = 4$ **(2 marks)**

b Solve the equation

$$4x + 5 = 29$$

(2 marks)

c Solve the equation

$$3(x - 2) = x + 7$$

(3 marks)

7 A motorist travelled 100 miles in 3 hours.

a Calculate his average speed.
Give your answer to a suitable degree of accuracy. **(4 marks)**

b He then travels a further 60 miles at 30 miles per hour.
Calculate how long the whole journey takes. **(3 marks)**

8 A group of 15 pupils decide to record how long it takes to complete a Mathematics homework. Here are the results to the nearest minute.

11	15	20	23	35	38	31	25
17	15	12	23	15	21	22	

a Draw a stem and leaf diagram to show this information.

(3 marks)

b Find the median of the recorded times.

(3 marks)

c The homework was expected to take 20 minutes to complete.
Do you think that this was reasonable?
Explain your answer.

(1 mark)

9 The weights of two parcels differ by 500 grams.
The combined weights of the two parcels is 4.6 kg.
Calculate the weight of the heavier parcel.

(4 marks)

10 The diagram shows a triangle ABC.

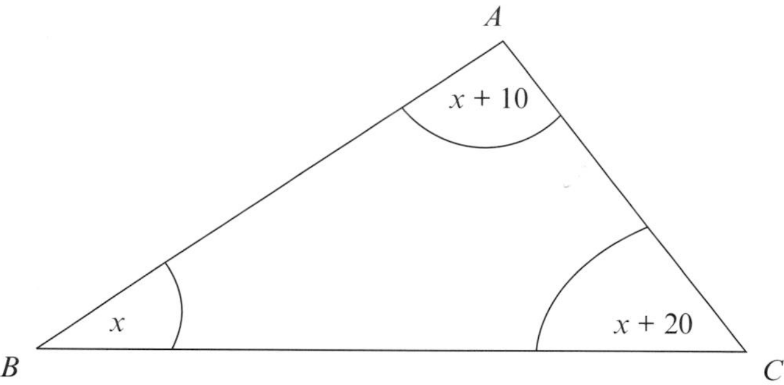

Work out the size of x.

(4 marks)

11 **a** Express 72 as a product of its prime factors.

(3 marks)

b What is the highest common factor of 16 and 72?

(1 mark)

c What is the least common multiple of 6 and 8?

(2 marks)

12 Measure this angle and copy it onto paper. Using ruler and compasses only construct the bisector of the angle.

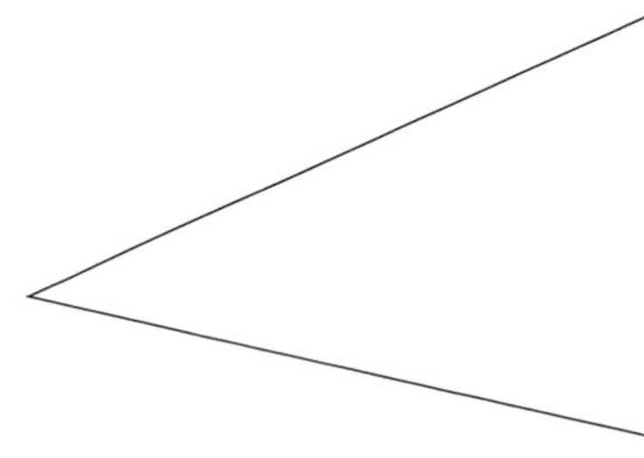

(3 marks)

13 Simplify

a $x^2 \times x^3$

(1 mark)

b $\dfrac{x^7}{x^5}$

(1 mark)

c $\dfrac{x^2 \times x^4}{x^3}$

(2 marks)

14 The grid shows a line segment AB.

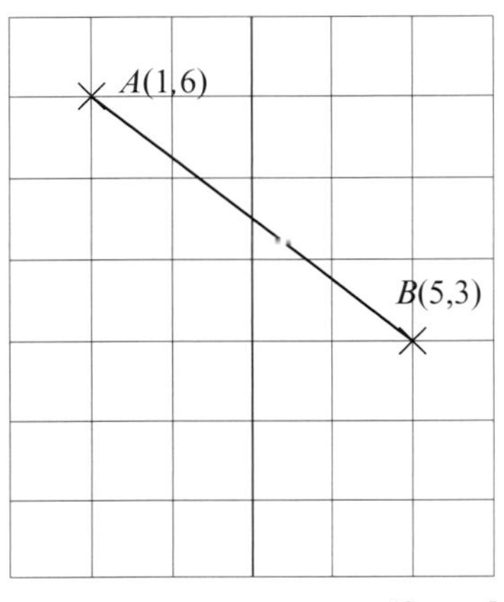

Find the length of the line segment AB.

(3 marks)

15

Which savings scheme makes the most money after two years?
Explain your answer.

(4 marks)

16 A logo is made up of a rectangle and a semicircle as shown.
Find the perimeter of the logo.
Give your answer in terms of π.

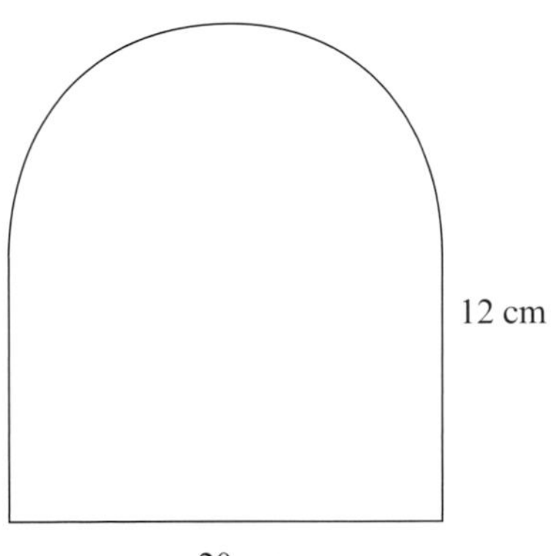

(4 marks)

17 Clare tiles a wall with 180 tiles.
There are plain tiles and patterned tiles.
The ratio of plain tiles to patterned tiles is 7 : 2
How many plain tiles does she use?

(3 marks)

18 Solve the simultaneous equations

$$2x + y = 7$$
$$x - 3y = 0$$

(4 marks)

19 The graph shows a straight line passing through the points $A\ (0, 2)$ and $B\ (6, 5)$.

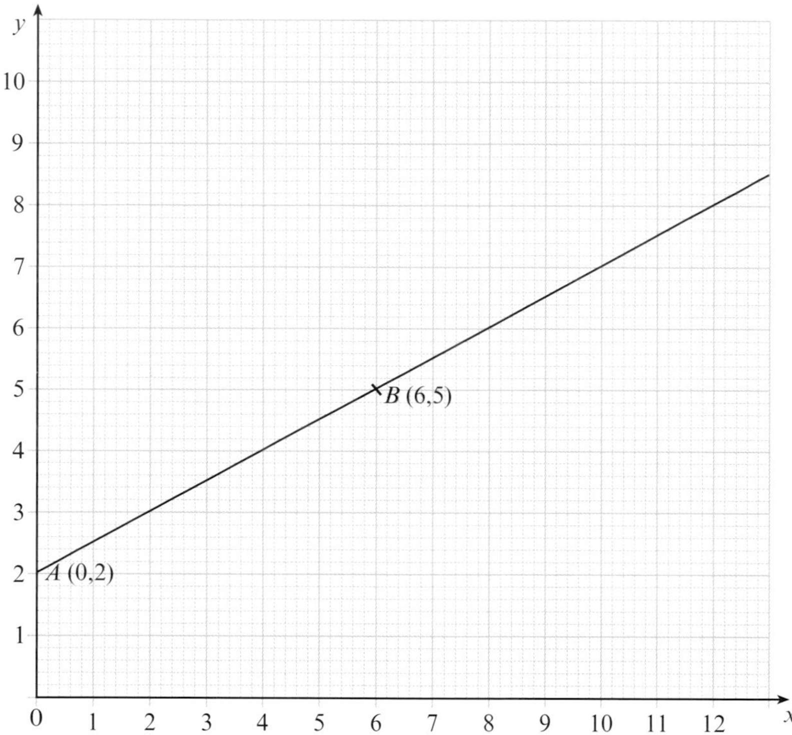

Find the equation of the line.

(4 marks)

20 When man first walked on the moon 600 million people worldwide watched it on television.

a Write 600 million as a number in standard form.

(2 marks)

b 600 million was $\frac{1}{5}$ of the world's population.
Calculate the world population.
Give your answer in standard form.

(3 marks)

21 The box plot shows the results of a Science examination.

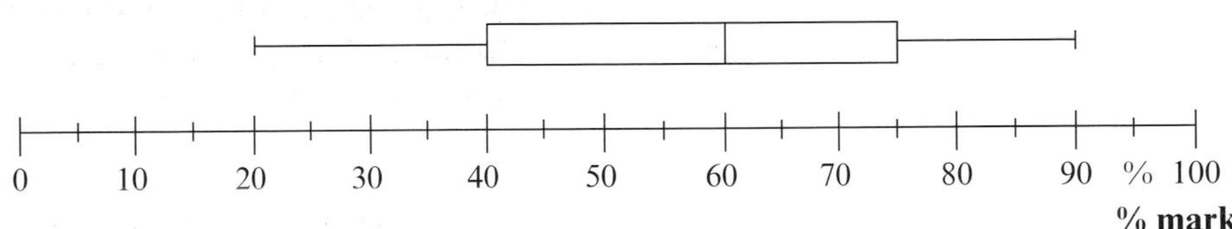

The following data is collected for an English examination.

Lowest score	30
Lower quartile	48
Median	60
Upper quartile	72
Highest score	80

a Draw the box plot for the English data.

(3 marks)

b Make two comparisons between the Science data and the English data.

(2 marks)

22 The diagram shows a circle.

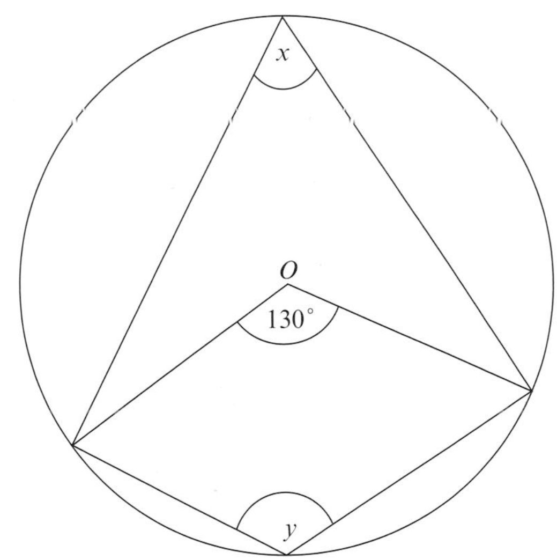

O is the centre of the circle.

a Work out the value of x.

(2 marks)

b Work out the value of y.

(2 marks)

100 marks total

1 A magazine costs £2.40
The price is increased by 15%.
Calculate the new cost.

(3 marks)

2 **a** Simplify

$$3a + 2b - 2a + 4b$$

(2 marks)

b $x = 5$ and $y = -2$
Find the value of $3x + 4y$

(2 marks)

c Solve

$$9x + 1 = 4x + 25$$

(3 marks)

3

Country	£1 is equal to
France	9.90 Francs
Germany	2.96 Marks
Greece	511 Drachma
Spain	250.7 Pesetas
U.S.A.	1.43 Dollars

a Belinda goes to Germany.
How many marks should she get if she changes £150?

(2 marks)

b Jonathan comes home from the U.S.A. with 200 dollars.
How much money should he get when he changes his money back into pounds?

(2 marks)

4 A garage has x cars.
Each car has 4 wheels.

a Write down an expression, in terms of x, for the total number of wheels.

(1 mark)

b Each wheel uses 4 nuts.
Write down an expression, in terms of x, for the total number of nuts.

(1 mark)

c The total number of nuts used is 144.
How many cars are there?

(2 marks)

5

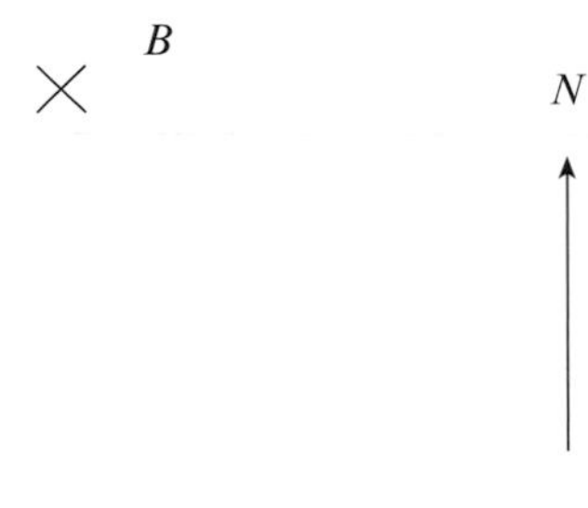

A

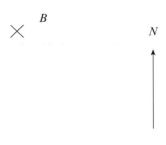

a Using tracing paper copy the diagram.
Measure the bearing of *B* from *A*.

(2 marks)

b *C* is due east of *A*.
C is also on a bearing of 160° from *B*.
Mark accurately the position of *C* on the diagram.

(3 marks)

6 A teacher says that anyone who scores 60% or more in a test will pass.
Matthew scores 38 out of 60.
Did he score enough to pass?
Give a reason for your answer.

(4 marks)

7 A bus leaves the station at 9.00 a.m.
The graph shows the journey.

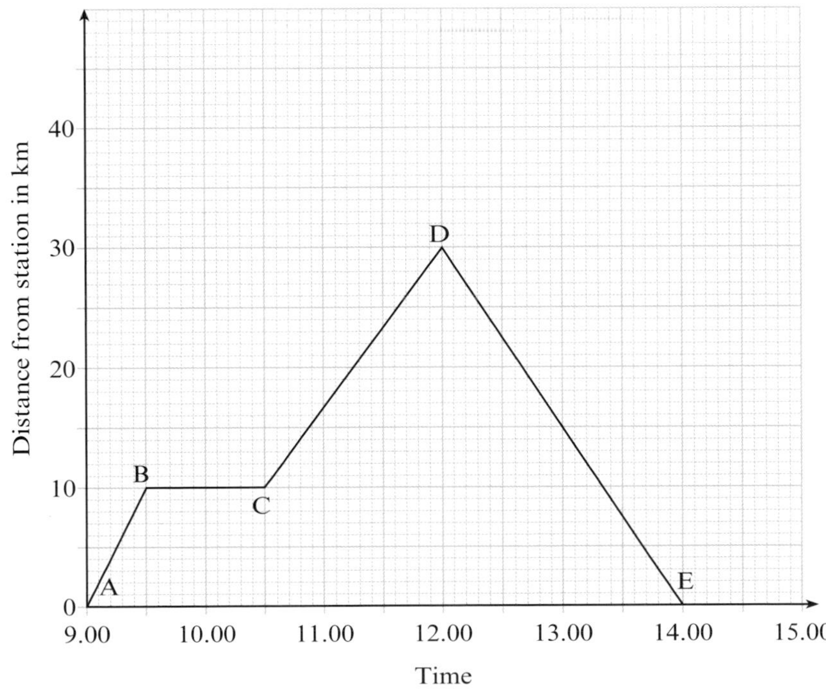

a Describe each stage of the journey.

(3 marks)

b What was the average speed of the bus, in km/h, when travelling from *C* to *D*?

(2 marks)

8 The table shows the scores awarded by 10 judges for a diving competition between two competitors.

Competitor 1	6	8	7	9	9	7	8	8.5	6.5	7
Competitor 1	6.5	4	7	9.5	8.5	6.5	7	9.5	6.5	8

a Plot the scores on a scatter diagram.

(2 marks)

b What do you notice about the correlation between the scores awarded to the two competitors?

(1 mark)

c **i** Draw a line of best fit.

(1 mark)

ii Use your graph to estimate the score awarded to Competitor 2 if Competitor 1 is awarded 7.5.

(1 mark)

9 A circle has a radius of 6.4 cm.
Calculate the area of the circle.
Give your answer to a suitable degree of accuracy.

(3 marks)

10 Here is a number pattern

Line 1: $1 \times 2 = 2^2 - 2$

Line 2: $2 \times 3 = 3^2 - 3$

Line 3: $3 \times 4 = 4^2 - 4$

Line 4: $4 \times 5 = 5^2 - 5$

a Write down the fifth line of the pattern.

(2 marks)

b Write down the *nth* line of the pattern.

(2 marks)

11 Use your calculator to work out

a $\dfrac{(0.3 + 0.51)}{(0.27 + 0.34)}$

(2 marks)

b $\sqrt{\dfrac{8}{0.4}}$

(2 marks)

12 The times taken for a group of pupils to get to school one day is shown.

Time taken T minutes	Number of people
$0 < T \leqslant 5$	10
$5 < T \leqslant 10$	18
$10 < T \leqslant 15$	15
$15 < T \leqslant 20$	7
$20 < T \leqslant 25$	6
$25 < T \leqslant 30$	4
	Total = 60

a Write down the modal class.

(1 mark)

b Estimate the mean time taken to get to school.

(4 marks)

13 Estimate the value of

$$\frac{593 \times 4.98}{0.52}$$

(3 marks)

14 Solve the equation

$$\frac{21 \quad x}{3} = 5$$

(3 marks)

15 Use trial and improvement to solve the equation

$$x^3 - 11x = 12$$

Give your answer to 1 decimal place.

(3 marks)

16 The diagram shows an airport car park.

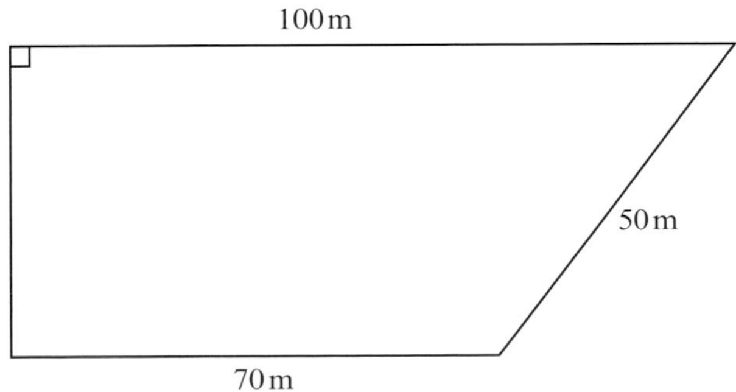

Calculate the area of the car park.

(7 marks)

17 Work out

a $\sqrt{25}$ **b** 4^0 **c** $\sqrt[3]{27}$ **d** 2^{-3}

(4 marks)

18 The diagram shows a circle.

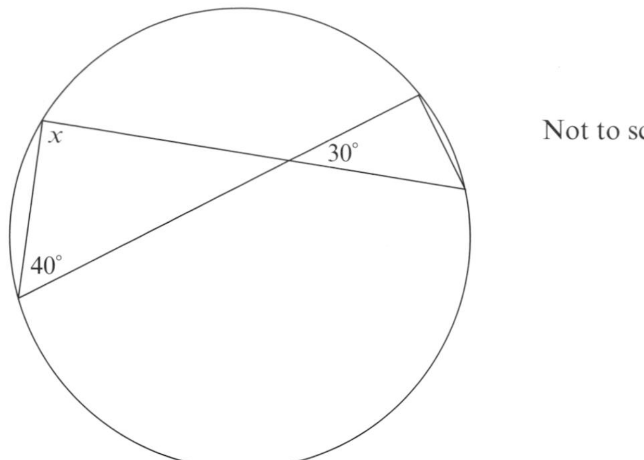

Calculate the value of x.
Show your working.

(3 marks)

19 In the diagram $AB = 7.2$ cm, $CD = 2.3$ cm and angle $B = 48°$

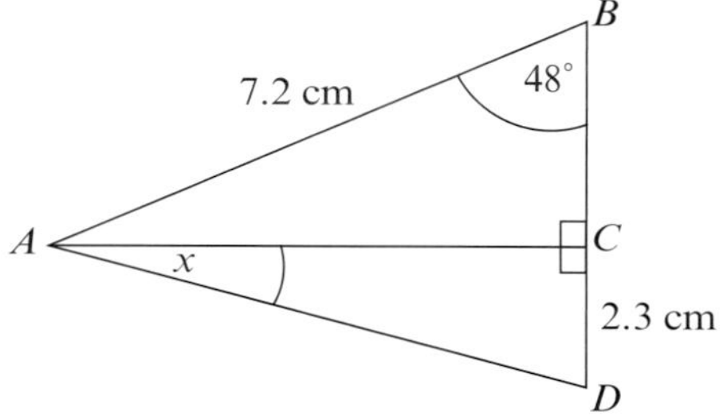

Calculate the size of the angle marked x.
Give your answer to 1 decimal place.

(5 marks)

20 A rectangle is drawn so that the length is 4 cm more than the width as shown.

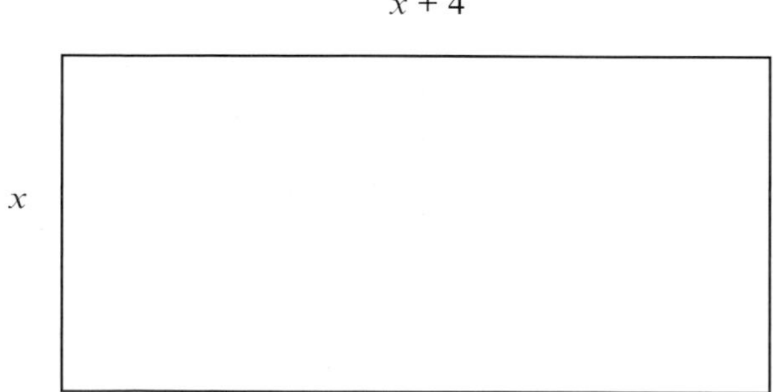

The area of the rectangle is 45 cm^2.

a Form an expression, in terms of x, for the area of the rectangle.

(1 mark)

b Show that the equation can be simplified to

$$x^2 + 4x - 45 = 0$$

(2 marks)

c Solve the equation $x^2 + 4x - 45 = 0$ and hence write down the perimeter of the rectangle.

(3 marks)

20 **d** A different rectangle has length 11 cm and width 9 cm.
All measurements are to the nearest centimetre.
What is the minimum value of the area of the rectangle?

(3 marks)

21 The probability that the school bus is late on a Monday morning is 0.4.

a Write down the probability that the school bus is not late on a Monday morning.

(1 mark)

b The probability that the bus is late on Tuesday morning is 0.3.

i Copy and complete the tree diagram.

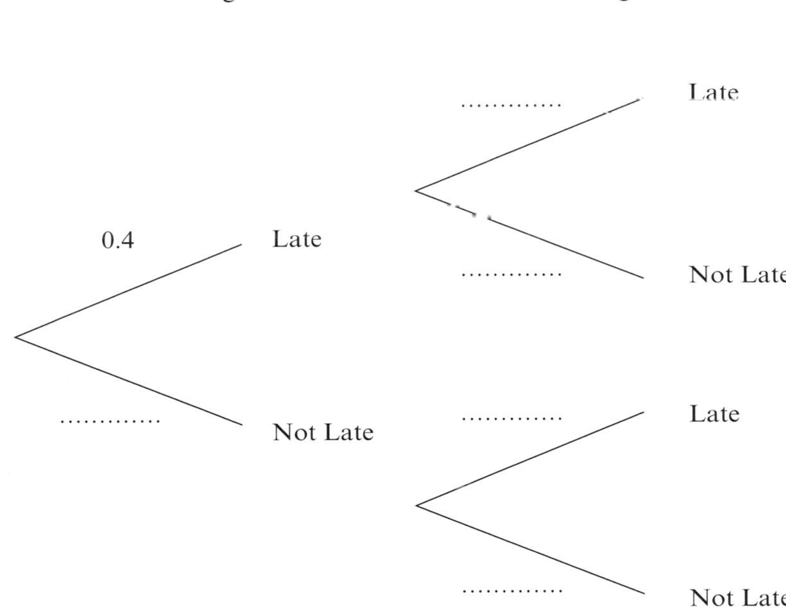

(2 marks)

ii Calculate the probability that the bus is late on both Monday and Tuesday mornings.

(2 marks)

22 **a** Given that $2^n = 64$
Work out the value of n.

(1 mark)

b $3^p = 3^2 \times 3^4$
Write down the value of p.

(1 mark)

23 Rearrange the formula to make x the subject.

$$ax = bx + c$$

(3 marks)

100 marks total

Answers

Section 1 Number bites

Starting points

A1 a $^{-}2$ b $^{-}7$ c 4 d 10 e $^{-}4$ f 5

A2 a $^{-}2$ b $^{-}2$ c 9 d $^{-}9$ e $^{-}3$ f 2 g $^{-}4$ h $^{-}5$ i $^{-}4$ j $^{-}6$

B1 a 5, 10, 15, 20, 25, 30, 35 for example
b 8, 16, 24, 32, 40, 48, 56 for example
c 12, 24, 36, 48, 60, 72, 84 for example

B2 a 40, 80, 120 for example b 24, 48, 72 for example

B3 a 1, 2, 3, 4, 6, 8, 12, 24 b 1, 2, 3, 5, 6, 10, 15, 30 c 1, 5, 25 d 1, 17

B4 a 1, 2, 3, 6 b 1, 5

B5 It has only one factor

B6 a 2, 3, 5, 7, 11, 13, 17, 19

C1 a 5^2 b 2^5

C2 a 64 b 81 c 2.25 d 1.61051

D1 a 2, 3, 7 b 2, 7, 11

D2 a $2^3 \times 3 \times 5$ b $2 \times 5^2 \times 7$

E2 a $1\frac{1}{3}$ b $2\frac{1}{6}$ c $5\frac{3}{4}$

E3 a $\frac{7}{5}$ b $\frac{10}{3}$ c $\frac{5}{1}$ d $\frac{37}{13}$ e $\frac{11}{1}$

F1 a $\frac{4}{10}, \frac{6}{15}, \frac{8}{20}$ b $\frac{1}{3}, \frac{2}{6}, \frac{8}{24}$ c $\frac{2}{3}, \frac{4}{6}, \frac{8}{12}$

G1 a 0.4375 b $0.\dot{5}$ c 0.63 d $0.4\dot{2}5\dot{9}$

H1 a £90 b 92 litres c 99 m

Exercise 1.1

1 A, B, D, F, H, I

2 a $3 \times {}^{-}5 = {}^{-}15$ b $^{-}2 \times 6 = {}^{-}12$ c $5 \times {}^{-}4 = {}^{-}20$ d $^{-}3 \times {}^{-}6 = 18$
e $^{-}12 \div {}^{-}3 = 4$ f $^{-}8 \div 4 = {}^{-}2$ g $24 \div {}^{-}6 = {}^{-}4$ h $^{-}40 \div 5 = {}^{-}8$

3 a $4 \times {}^{-}7 = {}^{-}28$ b $^{-}15 \div {}^{-}5 = 3$ c $^{-}4 \times 8 = {}^{-}32$ d $^{-}3 \times {}^{-}6 = 18$
e $^{-}35 \div {}^{-}7 = 5$ f $36 \div {}^{-}6 = {}^{-}6$ g $^{-}3 \times {}^{-}12 = 36$ h $14 \div 7 = 2$

4 a 9.43 b 9.43 c $^{-}9.43$ d $^{-}7.2$ e 7.2 f $^{-}7.2$

5 If you multiply an odd number of negative numbers you get a negative number. If you multiply an even number of negative numbers you get a positive number.

Exercise 1.2

2 a 5.5, $^{-}5.5$ b 8.6, $^{-}8.6$ c 3.5, $^{-}3.5$
d 6.7, $^{-}6.7$ e 7.7, $^{-}7.7$ f 9.7, $^{-}9.7$

3 a 6.93, $^{-}6.93$ b 8.49, $^{-}8.49$ c 6.25, $^{-}6.25$
d 5.18, $^{-}5.18$ e 30.00, $^{-}30.00$ f 0.1, $^{-}0.1$

4 a 13, $^{-}13$ b 14, $^{-}14$ c 10, $^{-}10$
d 9, $^{-}9$ e 15, $^{-}15$

Exercise 1.3

2 a 729 b 27 000 c 42.875 d $^{-}729$ e $^{-}1.331$ f 0.064

4 a 2 b 4 c 10 d 3 e 9 f 7 g 2.5 h 20 i 5 j 0.7

5 a $^{-}125, {}^{-}64, {}^{-}27, {}^{-}8, {}^{-}1, 0, 1, 8, 27, 64, 125$

Exercise 1.4

1 a 0.25 b 10 c 0.4 d $0.\dot{3}$ e 1.5 f 1.2 g Not defined

2 a $\frac{5}{3}$ b $\frac{4}{7}$ c $\frac{2}{3}$ d $\frac{3}{10}$ e $\frac{5}{6}$

3 Graph passing through (1, 1), $(2, \frac{1}{2})$, $(4, \frac{1}{4})$ and $(10, \frac{1}{10})$.

Exercise 1.5

1 a 3.8×10^5 b 4.51×10^7 c 9.2×10^{-4} d 2.62×10^{-5}

2 a 94 200 b 0.0025 c 741 400 000 d 0.000 000 627

3 a **P:** 13 is greater than 10. **Q:** 1000 isn't written as a power of 10.
R: Should be a multiplication not a division.
S: 0.58 is less than 1.
b **P:** 1.3×10^5, **Q:** 5.7×10^3, **R:** 6.42×10^{-3}, **S:** 58×10^{-5}

5 a 8.76×10^6 b 1.7×10^5 c 1.98×10^4
d 7.88×10^{10} e 7.68×10^{-15} f 3.972×10^4

Exercise 1.6

1 a 3^7 b 4^2 c 2^3 d 7^{-6} e 6^{-2} f 7^{-6} g 5^0 h 2^5 i 6^{-1} j 7^{-3}

2 a $2^3 \times 2^4 = 2^7$ b $4^5 \times 4^{-2} = 4^3$ c $3^5 \div 3^3 = 3^2$ d $7^7 \times 7^{-3} = 7^4$
e $8^4 \div 8^6 = 8^{-2}$ f $5^4 \div 5^2 = 5^2$ g $2^{-6} \times 2^5 = 2^{-1}$ h $3^4 \div 3^1 = 3^3$

3 a 3^8 b 5^9 c 4^{-10} d 4^{-10} e 2^{-16} f 7^0

4 a $2^3 \times 2^2 = 32$ b $3^6 \times 3^{-3} = 27$ c $2^8 \div 2^2 = 64$
d $4^2 \div 4^{-1} = 64$ e $(3^2)^2 = 81$ f $(5^3)^1 = 125$
g $(2^{-3})^1 = 0.125$ h $(5^{-1})^2 = 0.04$

Exercise 1.7

1 a 2.4×10^4 b 2×10^{-6} c 2×10^3 d 6×10^{-11}

Exercise 1.8

1 a 36 b 420 c 420 d 3150 e 630 f 1890 g 14 850

2 45, 90, 315 or 630

Exercise 1.9

1 a $\frac{5}{6}$ b $\frac{13}{20}$ c $\frac{1}{6}$ d $\frac{11}{20}$ e $\frac{31}{40}$ f $\frac{3}{4}$ g $\frac{1}{2}$ h $\frac{17}{20}$

2 a $1\frac{1}{6}$ b $1\frac{1}{20}$ c $1\frac{1}{12}$ d $1\frac{5}{8}$

3 a $2\frac{3}{10}$ b $1\frac{3}{10}$ c $3\frac{5}{6}$ d $\frac{9}{10}$ e $3\frac{23}{24}$ f $\frac{13}{20}$

Thinking ahead to ...

A a 8 b 8 c 12.8 d 12.8

B Same as multiplying by its reciprocal

Exercise 1.10

1 a $\frac{5}{12}$ b $\frac{1}{2}$ c $\frac{12}{35}$ d $\frac{3}{7}$ e $\frac{2}{5}$ f $\frac{7}{9}$ g $\frac{8}{15}$ h $\frac{5}{12}$

2 a $2\frac{2}{5}$ b 6 c 12 d $10\frac{1}{2}$ e 21 f $31\frac{1}{2}$ g $4\frac{2}{7}$ h $4\frac{3}{13}$

3 a $4\frac{1}{2}$ b $2\frac{11}{12}$ c $2\frac{4}{7}$ d $7\frac{1}{3}$ e 1 f $1\frac{1}{9}$ g 6 h $\frac{20}{33}$

Exercise 1.11

1 a 36 b 15 c 77 d 3

End points

A1 a $7 \times {}^{-}2 = {}^{-}14$ b $^{-}3 \times {}^{-}5 = 15$ c $^{-}12 \div 4 = {}^{-}3$ d $20 \div {}^{-}5 = {}^{-}4$
e $6 \times {}^{-}3 = {}^{-}18$ f $^{-}16 \div {}^{-}2 = 8$ g $^{-}4 \times {}^{-}9 = 36$ h $^{-}21 \div 7 = {}^{-}3$

B1 a 1.2 b 35.937 c 6 d $^{-}1.5$

B2 a $\frac{1}{8}$ b 5 c 1.8 d $2\frac{1}{2}$ e $\frac{4}{5}$

C1 a 3^7 b 4^3 c 7^6

C2 a $2^3 \times 2^4 = 128$ b $3^5 \times 3^{-2} = 27$ c $(4^2)^{-1} = 0.0625$

C3 a 2^2 b 3^{-9} c 5^{10} d 4^{-6}

C4 a $3^4 \times 3^6 = 3^{10}$ b $4^3 \div 4^{-4} = 4^7$ c $7^{-2} \times 7^7 = 7^5$ d $(6^{-2})^0 = 6^0$

D1 a 7.062×10^7 b 3.75×10^{-6}

D2 a 0.000 000 069 b 10 300 000 000

D3 a 9.984×10^{14} b 9.4×10^{-8} c 4.8×10^{27}

D4 a 4.2×10^{-3} b 2×10^4

E1 a $\frac{13}{15}$ b $\frac{13}{30}$ c $\frac{5}{8}$ d $3\frac{37}{42}$

E2 a $\frac{3}{10}$ b $\frac{1}{4}$ c 6 d $1\frac{3}{7}$

F1 a 84 b 360

F2 a 7 b 35

Section 2 Sequences

Starting points

A1 a 21 b 3 c 0 d 18

A2 a 19 b 36 c 4 d 12

B1

B2 1, 9, 25, 49 for example

B3 1, 3, 6, 10, 15, 21

B4 5, 25, 125, 625

B5 34

B6 29, 47, 76, 123

C1 33, 39

C2 31

C3 **a** 31, 42, 55 **b** 32, 38, 44 **c** 47, 65, 86 **d** 57, 83, 114

D1 **a**

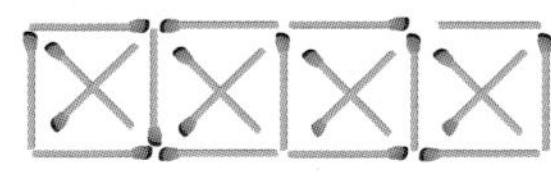

b **i** 16 **ii** 26

c

Pattern number (n)	Number of matches (m)
1	6
2	11
3	16
4	21
5	26

d $m = 5n + 1$

Exercise 2.1

1

Pattern number		Number of matches
1	→	3
2	→	4
3	→	5
4	→	6
20	→	22
n	→	$n + 2$

2 **a**

Pattern number		Number of matches
1	→	4
2	→	8
3	→	12
4	→	16
5	→	20
6	→	24

b $n \rightarrow 4n$ **c** 400

Thinking ahead to ...

A 301 matches; $3n + 1$

Exercise 2.2

1 **a** $m = 2n + 1$ **b** 17

2 **a** $m = 4n + 1$ **b** 161 **c** 32nd

3 **a** $m = 6n - 2$ **b** 598

4 **b** $m = 4n + 2$

5 **a** $50 \rightarrow 249$, $n \rightarrow 5n - 1$ **b** $36 \rightarrow 293$, $p \rightarrow 8n + 5$
c $2 \rightarrow 5$, $3 \rightarrow 9$, $4 \rightarrow 13$, $25 \rightarrow 97$

Thinking ahead to ...

A **a** 19, 21 **b** 29 **c** 105

Exercise 2.3

1 **a** A: $3n + 3$, B: $5n - 4$, C: $10n + 3$, D: $8n - 6$
b A: 153, B: 246, C: 503, D: 394

2 **a** He should have had $3n$ instead of $n + 3$. **b** $3n + 2$

3 $n + 7$

4 $22 - 2n$

Exercise 2.4

1 **a** 5th number, $40 = 5 \times 8$ **b** 130

2 **a** 5th triangle number, $15 = \frac{(5 \times 6)}{2}$ **b** 78 **c** $\frac{\{n \times (n + 1)\}}{2}$

3 **a** 5th power, $32 = 2 \times 2 \times 2 \times 2 \times 2 = 2^5$ **b** 256 **c** 2^n

Exercise 2.5

1 **c** 204 **d** 80th

2 **a** $2n + 6$, $2(n + 3)$

End points

A1 **a** $m = 2n + 3$ **b** 203

B1 **a**

1	→	2
2	→	5
3	→	8
4	→	11
10	→	29
p	→	$3p - 1$

b

1	→	7
2	→	12
3	→	17
4	→	22
20	→	102
n	→	$5n + 2$

C1 **a** A: $3n + 4$, B: $5n - 3$, C: $4n + 1$, D: $n + 5$, E: $3n + 6$, F: $7n + 4$
b A: 64, B: 97, C: 81, D: 25, E: 66, F: 144

SECTION 3 Properties of shapes

Starting points

A1 **a** **i** reflex **ii** acute **iii** right angle
b **i** acute **ii** right angle **c** Triangles AED, BAD

B1 **a** 60° **b** 300° **c** 45

B2 **a** 35°, 72.5°
b The three angles in a triangle add up to 180°, so if one angle is more than 90°, the other two must be less than 90°.

C1 116°, 116°, 64°, 64°

D1 trapezium; rectangle or square; square or rhombus; rectangle, square, parallelogram or rhombus

E1 1080°

E2 142°

E3 129°

E4 **a** 1800° **b** 150°

Exercise 3.1

1 **b** 74°, 53°, 53°, 74°, 53°, 53°

2 62°, 120°, 117°

Exercise 3.2

1 **a** The corresponding angles are: any two out of a, c and e; any two out of b, d and f; any two out of g, i and k; any two out of h, j and l
b The alternate angles are: b and i or k, d and k, h and c or e, j and e

2 72°, 135°, 45°

3 **a** **i** The corresponding angles are:
any pair out of angles ADC, DHG, HLK and LPO;
any pair out of angles EDH, IHL, MLP and QPS;
any pair out of angles ADE, DHI, HLM and LPQ;
any pair out of angles CDH, GHL, KLP and OPS
ii The corresponding angles are:
any pair out of angles BDC, DGF, GKJ and KON;
any pair out of angles BDE, DGH, GKL and KOP;
any pair out of angles CDG, FGK, JKO and NOR;
any pair out of angles EDG, HGK, LKO and POR
b BR and AS are not parallel

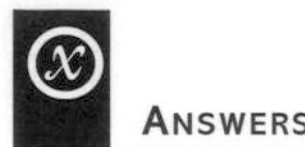

c The alternate angles are:
angles CDG and DGH, GKL or KOP;
angles FGK and GKL or KOP;
angles JKO and KOP;
angles EDH and DHG, HLK or LPO;
angles IHL and HLK or LPO;
angles MLP and LPO;
angles EDG and DGF, GKJ or KON;
angles HGK and GKJ or KON;
angles LKO and KON;
angles CDH and DHI, HLM or LPQ;
angles GHL and HLM or LPQ;
angles KLP and LPQ

d **i** 65° **ii** 54° **iii** 61° **iv** 54° **v** 126° **vi** 61°

4 42°, 77°, 61°, 72°, 72°, 72°, 68.5°, 68.5°, 68.5°

Exercise 3.3

1 **a** 56°, 42°, 69°, 104°, 89°

2 **a** The 12 exterior angles are 12 equal turns which are equivalent to one full turn, 360°.
b 30° **c** 150°

3 40°, 140°, 70°, 40°

Exercise 3.4

1 **a** AMD is isosceles, so angle AMD = 180° − 2 × 22.5° = 135°
b **i** 67.5°, 157.5°

2 **a** **i** 67.5° at A, 67.5° at C, 45° at F
ii 112.5° at A, 112.5° at C, 135° at F
b 5 **c** 45°, 67.5°, 90°, 112.5°, 135°, 157.5°

End points

A1 42°, 67°, 109°

B1 C is regular, B and E have only one obtuse angle

C1 **a** B 120°, 30°, 30°; C 60°, 60°, 60°; D 90°, 90°, 90°, 90°; E 60°, 90°, 120°, 90°
b B 60°, 150°, 150°; C 120°, 120°, 120°; D 90°, 90°, 90°, 90°; E 120°, 90°, 60°, 90°

C2 45°

Section 4 Linear graphs

Starting points

A1 **d** Straight line passing through ($^-3$, 2), and (3, 2), and another straight line passing through ($^-1$, $^-3$) and ($^-1$, 3).
e ($^-1$, 2)

A2 **a** (0, 2)
b No, the lines are parallel.
c The x-coordinates will always be 0.

A3 **a** horizontal **b** vertical **c** vertical
d horizontal **e** horizontal **f** vertical

B1 **a**

x	$^-2$	$^-1$	0	1	2	3
y	$^-7$	$^-4$	$^-1$	2	5	8

b

x	$^-2$	$^-1$	0	1	2	3
y	0	1	2	3	4	5

c

x	$^-2$	$^-1$	0	1	2	3
y	$^-2$	$^-1$	0	1	2	3

d

x	$^-2$	$^-1$	0	1	2	3
y	$^-5$	$^-4$	$^-3$	$^-2$	$^-1$	0

e

x	$^-2$	$^-1$	0	1	2	3
y	$^-2$	$^-0.5$	1	2.5	4	4.5

B2 **a** Straight line passing through ($^-1$, $^-4$) and (1, 2).
b Straight line passing through ($^-2$, 0) and (0, 2).
c Straight line passing through ($^-1$, $^-1$) and (1, 1).
d Straight line passing through ($^-1$, $^-4$) and (3, 0).
e Straight line passing through ($^-2$, $^-2$) and (1, 2).

B3 ($^-2$, $^-7$), (2, 1), (3, 3) and ($^-0.25$, $^-3.5$) all lie on the line $y = 2x - 3$

Exercise 4.1

1 **a** 30p **b** 10p

2 **a** more **b** 26p

3 34p

4 20 seconds at low call charge is cheaper

5 Low call charge

6 26p

7 1 minute 40 seconds

8 **c** Peak charge **d** Weekend, Low call, Infotel, Peak
e The higher the charge rate the greater the steepness.

9 16p

10 **a** $c = t \div 2$ **b** 45p

Exercise 4.2

1 **a** 3 **b** 2 **c** 1 **d** 2 **e** 2 **f** 1 **g** 3 **h** 1

3 The higher the gradient the greater the steepness.

Exercise 4.3

1 **a** **i** $\frac{3}{2}, \frac{4}{3}, \frac{4}{5}, \frac{5}{4}, \frac{3}{2}, \frac{5}{2}, \frac{1}{2}$ **ii** 1.5, 1.3, 0.8, 1.25, 1.5, 2.5, 0.5
b f = 2.5, a = 1.5, e = 1.5, b = 1.3, d = 1.25, c = 0.8, g = 0.5

2 Lines that are parallel have the same gradient.

Exercise 4.4

1 **a** a, c, e, f, h
b **i** $\frac{^-1}{2}, \frac{5}{2}, \frac{^-3}{2}, \frac{3}{2}, \frac{^-3}{4}, \frac{^-1}{2}, \frac{1}{2}, {^-1}$ **ii** $^-0.5$, 2.5, $^-1.5$, 1.5, $^-0.75$, $^-0.5$, 0.5, $^-1$

3 Positive gradient

4 A

5 G

6 F

7 $^-1$

8 C

9 gradient of E = $\frac{^-3}{2}$

10 gradient of D = $\frac{2}{3}$

11 **a–d** Straight line passing through ($^-5$, 5) and (3, 3), and another passing through ($^-3$, $^-5$) and (3, 3).
e (3, 3) **f** $^-5$ **g** ($^-1$, 4)

Exercise 4.5

1 2

2 $\frac{^-1}{2}$

3 h

4 **a** 3 **b** $\frac{^-1}{3}$ **c** $^-1$ **d** yes

5 Their gradients multiply to give $^-1$

6 l and g

7 yes

8 d

9 e

10 gradient of g × gradient of f = 3 × $\frac{^-1}{3}$ = $^-1$

Exercise 4.6

1 **a** Multiply the x-coordinate by 3 then add 2 to get the y-coordinate.

b

x	$^-2$	$^-1$	0	1	2	3
y	$^-4$	$^-1$	2	5	8	11

c ($^-2$, $^-4$), ($^-1$, $^-1$), (0, 2), (1, 5), (2, 8), (3, 11)
d & e Straight line passing through ($^-1$, $^-1$) and (1, 5).

2 **a** Multiply the x-coordinate by 2 then subtract 3 to get the y-coordinate.

b

x	$^-1$	0	1	2	3
y	$^-5$	$^-3$	$^-1$	1	3

c Straight line passing through ($^-1$, $^-5$) and 3, 3)

3 **a**

x	$^-2$	$^-1$	0	1	2	3
y	$^-3$	$^-1$	1	3	5	7

b Straight line passing through ($^-2$, $^-3$) and (1, 3).

4 Straight line passing through ($^-2$, $^-9$) and (1, 3).

5 **a** Straight line passing through ($^-2$, $^-5$) and (1, 1) and another passing through ($^-1$, $^-6$) and (2, 3).
b (2, 3)

Exercise 4.7

1 **a** 2 **b** $^-5$

2 **a** 3 **b** $^-1$

3 $y = 4x + 2$ and $y = 2 + 4x$

4 **a** 3, $^-2$ **b** 1, 2 **c** 4, 0 **d** 5, $^-4$ **e** 2, 3 **f** 3, $^-\frac{1}{2}$ **g** 5, 1 **h** $\frac{1}{2}$, 0
i $^-5$, 5 **j** 1, 0 **k** 1, $^-2$ **l** $^-4$, 0

5 K: $y = 3x + 5$, M: $y = \frac{1}{2}x - 1$, P: $y = 2x + 1.5$, R: $y = -3x + 2$, T: $y = x$

Exercise 4.8

1 **a** $y = 4x - 3$ passes through (0, $^-3$) and (1, 1).
$y = x - 1$ passes through (0, $^-1$) and (1, 0).
$y = 2x + 1$ passes through ($^-1$, $^-1$) and (0, 1).

2 **a** $y = 3x$ passes through (0, 0) and (1, 3).
$y = x + 2$ passes through (0, 2) and (1, 3).
$y = 3x - 3$ passes through (0, $^-3$) and (1, 0).
b $y = 3x$, $y = x + 2$
c $y = x + 2$

3 **a** $^-3$, 2 **b** Straight line passing through (0, 2) and (1, $^-1$).

Exercise 4.9

1 **a** 2 **b** $^-4$ **c** $y = 2x - 4$

2 **a** $^-3$ **b** $y = 3x - 3$

3 C: $y = 3x + 1$, D: $y = 2x + 3$, E: $y = {}^-3x + 1$

Exercise 4.10

1 **a** 4 **b** $y = 2x - 5$

2 **a** 2 **b** $y = \frac{3}{2}x + 2$

3 **a** $y = 2x + 4$ **b** $y = \frac{4}{3}x - 3$ **c** $y = x - 2$ **d** $y = \frac{3}{5}x$
e $y = \frac{2}{5}x + 1$ **f** $y = 3x - 2$ **g** $y = \frac{3}{2} - 2x$ **h** $y = \frac{1}{4}x$
i $y = 1 - x$ **j** $y = 2x + 2$ **k** $y = \frac{1}{5}x + \frac{1}{5}$ **l** $y = \frac{2}{3}x$

4 $3y = 6x - 9$

Exercise 4.11

1 **a** $y = \frac{3}{2}x + 1$ **b** $\frac{3}{2}$ **c** 1
d Straight line passing through (0, 1) and (2, 4).

2 **a** $\frac{2}{3}$ **b** -1
c Straight line passing through (0, −1) and (3, 1).

3 Straight line passing through (0, −3) and (1, 0).

4 **c** (4, 2).

5 Straight line passing through (−1, −2) and (2, 2).

6 Yes, $2 \times 3 = (5 \times 2) - 4$

Exercise 4.12

1 **a** (2.5, 3) **b** $5y = 2x + 10$, $y = 2x - 2$

2 **b** (2, 3)

3 (1.2, 2.2)

End points

B1 1

B2 **a** $\frac{7}{4}$ **b** 1.75

B3 $\frac{2}{3}$, $^-1$

C1

x	$^-1$	0	1	2	3
y	$^-7$	$^-4$	$^-1$	2	5

C2 Straight line passing through (−1, −7) and (1, 0).

D1 $y = \frac{3}{4}x + \frac{3}{2}$

E1 2, $^-5$

E2 Straight line passing through (0, 1) and (2, 2).

E3 Straight line passing through (0, $^-2$) and (3, 3).

F1 $3y = 2x + 6$, $y = 2x - 2$

SECTION 5 Comparing Data

Starting points

B1

Number of children in car	Tally	Number of cars
0	~~IIII~~ ~~IIII~~ I	11
1	~~IIII~~ II	7
2	IIII	4
3	I	1
4	I	1

B2

Colour of car	Frequency
Red	5
Blue	6
Green	2
White	3
Silver	2
Black	4
Brown	2

C1 bars with heights 5, 6, 2, 3, 2, 4, 2.

C2 bars with heights 11, 7, 4, 1, 1.

D1

Colour of car	Percentage
Red	20.8%
Blue	25%
Green	8.3%
White	12.5%
Silver	8.3%
Black	16.7%
Brown	8.3%

E1 **a** 15 and 42 **b** 39.5 **c** 42.5 **d** 69

E2 **a** 25 **b** 38 **c** 48.8 **d** 93

E3 2, 3, 3, 5, 7, 7, 10, 11 for example

F1 blue

F2 **a** 0 **b** 4

G1 **a** £340 **b** £340 **c** £314.89
d The mean is not sensible, because there is an extreme value, £112.

H1 **a** 54.4% **b** 34.8% **c** 10.8%

H2 **a** 38.8% **b** 33.0% **c** 28.2%

Exercise 5.1

1 **a** 72, 48, 100 **b** 5°, 7.5°, 3.6°

c

	Caring for Children	Child Action	Children in Crisis
Child Care	306°	252°	252°
Fund raising	32°	66°	54°
Administration	8°	18°	36°
Other costs	14°	24°	18°

Thinking ahead to ...

A **a** 7 **b** 4

Exercise 5.2

1 **a** 2 **b** 5, 5, 5, 5, 6, 6, 7, 8, 8, 9 **c** 6

2 **a** 15 **b** 5, 5, 5, 6, 6, 7, 7, 7, 7, 9, 9, 9, 10, 10, 10 **c** 7

3 **a** Peta fired 18 arrows, so the median is halfway between the 9th and 10th values.
b 5

4 For example

Ring score	2	3	4	5	6	7	8	9	10
Frequency	6	0	0	0	1	1	0	0	6

Thinking ahead to ...

A **a** 24 points

Exercise 5.3

1 **a** Jodie 84, Geeta 156 **b** Jodie 16, Geeta 20
c Jodie 5.25, Geeta 7.8

2 **a** They have not multiplied the frequency by the ring score to find the total score.
b 7.4

Exercise 5.4

1 **a** Javed 6, Lisa 5.5 **b** Javed 6, Lisa 3
c Javed's scores are higher on average. Lisa is more consistent.

2 **a** Amy's mean = 5.8, range = 4; Paul's mean = 5.7, range = 5
b Amy's scores are slightly higher on average. She is also more consistent.

3 **a** Imran's mode = 5, range = 8; Viv's mode = 7, range = 8
b Most of her scores are between 4 and 7, but the range is distorted by a couple of extreme values.

Exercise 5.5

1 **a** William: 2, 4, 4, 4, 5, 5, 6, 6, 6, 6, 6, 7, 7, 7, 8, 8
Bryony: 4, 4, 5, 5, 5, 5, 6, 6, 6, 7, 7, 7, 9, 9
Daniel: 2, 3, 4, 4, 5, 5, 5, 6, 6, 6, 6, 6, 7, 7, 7, 9, 9, 9, 10, 10
b William's lower quartile = 4.5, upper quartile = 7
Bryony's lower quartile = 5, upper quartile = 7
Daniel's lower quartile = 5, upper quartile = 8
c William's interquartile range = 2.5, Bryony's = 2, Daniel's = 3

2 Sheera's interquartile range = 2
Dave's interquartile range = 3.5
Sheera is more consistent

Thinking ahead to ...

A **a** 5 **b** 8 **b** 8

Exercise 5.6

1

Ring score	Cumulative Frequency
⩽5	2
⩽6	7
⩽7	19
⩽8	26
⩽9	31

2

Ring score	2	3	4	5	6	7	8
Frequency	3	7	12	19	13	9	4
Cumulative frequency	3	10	22	41	54	63	67

3 **a** 15 **b** 36 **c** 64

4

Age	Cumulative frequency
10 and under	8
11 and under	15
12 and under	25
13 and under	36
14 and under	44
15 and under	53
16 and under	64

5

Distance to target in metres	10	15	20	25	30	35	40
Number of arrows hitting target	20	19	17	16	15	13	11
Cumulative frequency	20	39	56	72	87	100	111

6

	Ring score									
	⩽1	⩽2	⩽3	⩽4	⩽5	⩽6	⩽7	⩽8	⩽9	⩽10
April	4	11	18	26	31	37	39	42	47	48
May	6	11	15	24	29	32	40	42	44	48
June	3	9	15	19	24	28	36	40	42	48

b She is improving: she is getting more high scores and fewer low ones.

Thinking ahead to ...

A **b** 8

Exercise 5.7

1 **a** Lee

Ring score	3	4	5	6	7	8
Cumulative Frequency	7	17	22	23	29	40

Faith

Ring score	4	5	6	7	8
Cumulative Frequency	11	18	26	35	45

Aqib

Ring score	1	2	3	4	5	6	7	8	9
Cumulative Frequency	2	6	8	11	12	16	23	32	39

Tegan

Ring score	1	2	3	4	5	6	7	8	9	10
Cumulative Frequency	6	15	20	24	26	31	37	41	44	48

c Lee's median score is 5
Faith's median score is 6
Aqib's median score is 7
Tegan's median score is 4.5

2 **a** Median is between 40th and 41st score i.e. halfway between 7 and 8
b Upper quartile – lower quartile = 9 – 6 = 3

Exercise 5.8

1 **a**

```
4 | 0 6
3 | 2 3 8
2 | 0 3 4 6
1 | 2 2 4 9
0 | 7
    1 2 3 4 5 6
```

b

```
4 | 0
3 | 0 1 2
2 | 0 2 4 4 4 6 6
1 | 0 2 5
0 |
    1 2 3 4 5 6 7
```

c

```
4 | 1 2 4
3 | 0 6 0
2 | 4 5 5
1 | 6 6 8
0 | 6 6 7 9
    1 2 3 4 5 6
```

d

```
5 | 6
4 | 0 0 6
3 | 0 2 7 8
2 | 2 4 8 8 8
1 | 4 8
0 |
    1 2 3 4 5 6 7
```

2 **a**

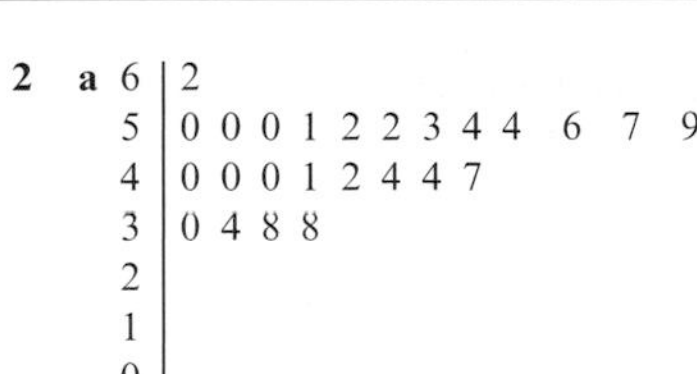

b 50

Thinking ahead to ...

A **b** **i** X38 Y 36 Z 38
ii X3.5 Y 4 Z 2.5

End points

A1 Pie chart with angles: child care 230°, fundraising 85°, administration 25°, other 20°.

B1 **a** 5.5

b **i**

Ring score	1	2	3	4	5	6	7	8
Cumulative Frequency	2	3	8	16	23	28	31	33

ii 5

C1 **a** 5.3 **b** 4.6

D1 Manoj's mean is higher with a smaller range.

D2 Kim's mode is higher with a smaller range.

E1 **a** LQ3 UQ 7 **b** LQ5.5 UQ 8

SECTION 6 Working in 2D

Starting points

A1 23.2 cm

A2 57.5 cm

A3 44.4 cm

A4 **a** 16.3 cm **b** 2.2 cm **c** 15.1 cm

B1 121.6 cm²

B2 56.3 cm²

B3 15.1 cm²

B4 **a** 58.1 cm² **b** 176.7 cm² **c** 483.1 cm²

C1 **a** 128.6 cm² **b** 53.0 cm²

D1 4.1 cm, 9.7 cm, 8.7 cm

Exercise 6.1

1 **a** 102.5 cm² **b** 5.6 cm²

2 **a** 92 cm² **b** 120 cm² **c** 238 cm² **d** 299 cm² **e** 344 cm² **f** 22.5 cm² **g** 592 cm² **h** 222 cm **i** 2450 mm² **j** 558 cm²

3 **a** 101 cm² **b** 303 cm²

4 **b** trapezium **c** 12 797 cm²

5 **b** 675.8 cm²

6 **b** 2827.5 cm²

Exercise 6.2

1 **a** 162.9 cm² **b** 120.8 cm² **c** radius = 1.5, area = $1.5^2\pi = 2.25\pi$ **d** 8π **e** $\sqrt{53}$

2 **a** 18 mm **b** 1017.9 mm² **c** 180 mm **d** 3010.6 mm²

3 120 424.8 mm²

4 **a** triangle **b** 61.4 cm² **c** 52.4 cm²

5 **a** 183.7 cm² **b** less **c** 85.4 cm

6 **a** 154 mm **b** 136 mm **c** 504 mm² **d** 928 mm² **e** 16 288 mm² **f** 4656 mm² **g** between $\frac{1}{5}$ and $\frac{1}{4}$

7 **a** 18 mm **b** $\frac{1}{4}$ **c** 254.5 mm² **d** 763.4 mm² **e** 278.1 mm²

Exercise 6.3

1 **a** 2.9 cm **b** 4.7 cm **c** 13.0 cm²

2 30.0 cm²

3 **a** 6.5 cm **b** 21.5 cm²

4 **a** 9.5 cm² **b** 21.6 cm²

5 10.4 cm

6 62.4 cm²

7 8.6 cm²

8 49.8 cm²

9 262.6 cm²

10 413.4 cm²

11 73.4 cm

Exercise 6.4

1 **a** 59.7 cm **b** 77.0 cm **c** $2 \times \text{radius} \times \pi = 2 \times 4.8 \times \pi = 9.67\pi$ **d** 7.5 cm **e** 55.3 cm

2 **a** 99.0 m **b** 33.0 m

3 32.5

4 23 cm

5 6240 tins

Exercise 6.5

3 **a** triangle **b** rectangle **c** rectangle

5 9600 mm²

7 **b** 18

End points

A1 **a** 27.1 cm² **b** 29.4 cm² **c** 46.3 cm²

B1 Composite shapes are made up from more than one shape.

B2 **a** 109 cm² **b** 147 cm²

B3 41.4 cm²

C1 18.3 m

SKILLS BREAK 1A

1 In 2001 it is 44 years.

2 9.5 hours

3 £2.70

4 Malnourished

5 12.3 kg

6 1000 kg

7 0.55 kg

8 50%

9 7319

10 In 2001, 26 years.

11 35 minutes

12 $\frac{1}{3}$

13 60%

14 8

15 $\frac{1}{3}$

16 5%

17 2

18 2

19 0.85 kg

20 770 g

21 14.25 kg

22 Tony

23 24.5%

24 **a** 19 kg **b** 18.5 kg **c** 19 kg

25 23 kg, 17 kg, 12 kg, 24 kg, 19 kg

26 65 000 g, 56 000 g, 66 000 g, 98 000 g, 85 000 g

27 150 kg

28 Outer Hebrides

29 2100, 9500, 7000, 900, 900

30 2000, 10 000, 7000, 1000, 1000

32 80 000

33 **a** 40 km **b** $1\frac{3}{4}$ hours **c** 3 minutes **d** $\frac{1}{3}$ km

34 24 000

35 45 minutes

36 **a** $\frac{1}{6}$ **b** $\frac{1}{6}$

37 **a** 46p **b** £1.44 **c** 81p

SKILLS BREAK 1B

1 17 miles

2 4 square miles

3 32 square miles

4 20 square miles

5 right

6 Silchurch

7 **a** Robridge **b** 286°

8 **a** 382 594 **b** 345 585 **c** 362 589, 378 593 **d** 354 585 **e** 422 580

10 **a** 48 km **b** 88 km **c** 128 km **d** 40 km **e** 208 km **f** 456 km

11 **a** 34 miles **b** 60 miles **c** 25 miles **d** 75 miles
e 147 miles **f** 234 miles

12 $\frac{5}{8}$

13 27 km

14 43 minutes

15 **a** 25 minutes **b** 50 minutes

16 2 hours 5 minutes

17 13 minutes

18 **a** 61 days **b** 183

19 **a** 15 625 **b** 9375

20 No, some of the adults may be senior citizens and pay a cheaper fare so ticket sales will be less.

21 410 passengers

22 5469

23 No, the running costs will be nearly £120 000.

24 Adult £5.80, Child £3.20, Senior Citizen £3.70

25

	Frequency
Adult	67
Child	19
Senior Citizen	93

26 £1844.50

27 £10.30

29 2.2 tons

30 500 gallons of water

31 58.7 tons

32 $p + n = 8$

33 **a** $27p + 37n$ **b** $55 + 27p + 37n$

SECTION 7 Solving Equations

Starting points

A1 **a** 17 **b** 6 **c** 24 **d** $^-3$

A2 **a** 13 **b** 7 **c** 49 **d** $^-11$

B1 C

B2 **a** $9t$ **b** $10s + 9k$ **c** $8b + 3c$ **d** $11x + 2$ **e** $7v + 16$

C1 C

C2 **a** $z = 1.5$ **b** $y = 3$ **c** $x = ^-2$ **d** $w = 4$ **e** $v = 8$ **f** $t = ^-1$ **g** $s = 2$

D1 A and C

D2 **a**

x	$^-2$	$^-1$	0	1	2
y	1	3	5	7	9

b

x	$^-2$	$^-1$	0	1	2
y	8	7	6	5	4

Thinking ahead to ...

A P and U, Q and S, R and T

Exercise 7.1

1 **a** $8t$ **b** $5x$ **c** $3h + 3$ **d** a **e** $1 - 5y$ **f** $6 - 3f$ **g** $x + 4y - 5$
h $2x - 5y$ **i** $2x + 2y - 4$ **j** $6ax + 3x$ **k** $2xy + 2x + y$ **l** $4x - 2y - 10$
m $4x + 5y - 3$ **n** $3y - 4x$ **o** $4 - x - 7y$ **p** $12 - 5x - 15y$
q $2y - x - 3$ **r** $4ax + ay + y$

2 **b** It is a magic square

13	1	10
5	8	11
6	15	3

c $6x - 3y$

Thinking ahead to ...

A A and F, B and E, C and G, D and H

Exercise 7.2

1 16

2 0

3 $2(a + 4)$ and $2a + 8$, $8a - 12$ and $4(2a - 3)$, $2(a + 2)$ and $2a + 4$, $2a - 12$ and $2(a - 6)$

4 **b** $12 + 3x$

5 $5(2n - 1) = (5 \times 2n) - (5 \times 1) = 10n - 5$

6 **a** $4n + 4$ **b** $5m - 15$ **c** $6c - 54$ **d** $6p + 4$ **e** $16s - 40$
f $12 + 4t$ **g** $15 - 3k$ **h** $14n - 10$ **i** $60f - 390$ **j** $900 + 540h$
k $30 - 70q$ **l** $10y + 15z$

7 $2(n + 2) + 2n = 2n + 4 + 2n = 4n + 4$

8 **a** $8p + 10$ **b** $5q - 4$ **c** $26 - x$ **d** $54r + 30$ **e** $6s + 2t$ **f** $7x - 1$
g $^-2x - 2$ **h** $5x - 9$ **i** $^-2x - 23$

Exercise 7.3

1 **a** $x = 4$ **b** $y = 6$ **c** $x = 7$ **d** $w = 18$ **e** $v = 4$ **f** $u = 1.2$ **g** $t = 7.25$
h $s = 5.375$ **i** $r = 9$ **j** $q = 1.8$ **k** $p = 2$ **l** $n = 1.5$ **m** $m = 4$
n $l = 3.6$ **o** $k = 2.5$ **p** $j = 10$ **q** $h = 5$ **r** $g = 3.5$

2 **a** $z = ^-5$ **b** $y = ^-1$ **c** $x = -2$ **d** $w = ^-3$ **e** $v = ^-4$ **f** $u = ^-6$
g $t = ^-0.5$ **h** $s = ^-1.2$ **i** $r = ^-3.5$

3 **a** $z = \frac{1}{3}$ **b** $y = \frac{5}{6}$ **c** $x = 3.5$ **d** $w = 5$ **e** $v = 2$ **f** $u = \frac{2}{5}$

Exercise 7.4

1 **a** 5 cm, 10 cm, 11 cm **b** 20 cm **c** $(6x - 4)$ cm
d 18 cm, 36 cm, 50 cm

2 a square 80 m, rectangle 50 m b $8p$ m c $(4p + 10)$ m
d 20 m, 25 m e $p = 2.5$ m

3 $t = 1$ cm

4 a $x = 11$
b

22	9	32
31	21	11
10	33	20

c $x = 20$
d i

26	11	38
37	25	13
12	39	24

e Sum of corners $= 2x + 3x - 1 + x - 1 + 2x - 2 = 8x - 4 = 4(2x - 1)$ $= 4 \times$ centre number

Exercise 7.5

1 A

2 A $3n + 5 = 13 - n$, $n = 2$
B $4(n - 2) = 5(n - 3)$, $n = 7$
C $3(n - 1) = 4(8 - n)$, $n = 5$
D $10 - 6n = 7 - 3n$, $n = 1$
E $2n - 11 = 15 - 2n$, $n = 6.5$

3 b $n = 7$

Exercise 7.6

1 a $t + 2b = 70$, $3t + b = 130$
b $t + 2b = 70$

t	0	10	20	30	40
b	35	30	25	20	15

$3t + b = 130$

t	0	10	20	30	40
b	130	100	70	40	10

d tea = 38p, biscuit = 16p

2 a 2, $^-3$ d $x = 1.5$, $y = 4.5$

3 a 2,1 d $x = {}^-1$, $y = 3$

4 c A: $x = 3$, $y = {}^-2$, B: $x = {}^-2.5$, $y = 1$, C: $x = 1.8$, $y = 3.6$

Exercise 7.7

1 A: £9.20, C: £2.30

2 a £3.04 b £1.24 c £4 d 96p e 28p f 68p

3 The woman is 160 cm tall, each twin is 120 cm tall.

Thinking ahead to ...

A a i 33 ii $4q$ iii 34 iv 12 v 6
b $p = 1$, $q = 5$

Exercise 7.8

1 a $x = 6$, $y = 9$ b $x = 8$, $y = 1$ c $x = 1.5$, $y = 3.5$

2 A false B true C false

3 $m = {}^-1$, $n = 10$

4 a $v = 2$, $w = 12$ b $v = {}^-1$, $w = 3$ c $v = 2.5$, $w = {}^-0.5$

Exercise 7.9

1 a $a = 10$, $b = 2$ b $a = 2$, $b = 5$ c $a = 5.5$, $b = 1.5$
d $a = {}^-2$, $b = 8$ e $a = 3$, $b = 1$ f $a = 6$, $b = {}^-3$

2 a $m = 5$ b $n = 10$

3 $p = 3$, $q = 1$

4 $m = 6$, $n = 1$

5 a $x = 4$, $y = 2$ b $x = 3$, $y = 5$ c $x = 2$, $y = 1$
d $x = 2.5$, $y = 1.5$ e $x = 5$, $y = {}^-1$ f $x = 7$, $y = {}^-4$

Exercise 7.10

1 38 and 35

2 26 large, 63 small

End points

A1 a $2k + 9m$ b $2p + 8$

B1 a $z = 6.5$ b $y = 3$ c $x = 1.5$
d $w = 1$ e $v = {}^-3$ f $t = 1.5$

B2 a $3x + 2$ b $3x + 2 = 56$, 18 cm, 12 cm, 26 cm

B3 $2n - 1 = 14 - 4n$, $n = 2.5$

C1 $x = 0.5$, $y = {}^-2$

D1 a $m = 1.5$, $n = 0.25$ b $m = {}^-2$, $n = 3$ c $m = 12$, $n = 2$

D2 Fifty-seven 50p coins, eight-three 20p coins

SECTION 8 Estimation and Approximation

Starting points

A1 a 2000 b 2200 c 2180 d 2176

A2 a 45.64 b 45.6

A3 a 34.60 b 3 c 38.5 d 3500

A4 c, d, e, f

B1 a 177.753 b 43.463 c 3.984 d 3442.8895

B2 a 27.98 b 11.17 c 149.86 d 232.236

B3 23.4 has been added instead of 2.34.

B4 The hundredths and thousandths digits of 2.278 have been added instead of subtracted.

Exercise 8.1

1 a Round up, in case it costs more than expected.
b Do not round at all, or you will sit in someone else's seat.
c Round up, in case you make any mistakes when cutting the paper.
d Do not round at all, or you will visit someone else.
e Round down, in case you do not earn as much as you expect.

3 a Round down to 95 cm to make sure it fits.
b Round down to 67 500 miles, because lower mileage cars are more saleable.
c Round up to 80, just in case a few more people come.
d Round up to 15.5 metres, in case there is a slight error in your calculations.

Exercise 8.2

1 a 10:43 should be 11:00, 71 934 should be 70 000, 63 479.6 cm should be 635 m, 19.6 cm should be 20 cm, 1452.56 metres^2 should be 1500 metres^2, 56 minutes should be 1 hour, 492 or so should be about 500
b 10.84 sec, because the record will only have been broken by a fraction of a second; 100 m and 2003.
c 19.6, because it is a bigger percentage of the number.

3 1000 km or 600 miles

4 a 10 243 b 24.6

5 At least to the nearest metre.

Exercise 8.3

1 a 1450 b 21 700 c 143 d 2 130 000 e 150 000

2 a 70 000 b 72 000 c 71 900 d 71 930

3 Because the unit 9 is rounded up.

4 3200, 3210 or 3150 for example

5

Number	45 287	2395	302 604.32	14.823
to 2 sf	45 000	2400	300 000	15
to 3 sf	45 300	2400	303 000	14.8
to 4 sf	45 290	2395	302 600	14.82

6 The length measurements are only given to 3 sf, so the area measurement should not be given more accurately than that. Round to 1500 m^2.

7 419 500 for example

8 3 to 6 sf

Exercise 8.4

1 0.008 cm

2 The last two zeros indicate that the number has been rounded to 4 sf.

3

to 1 sf	to 2 sf	to 3 sf
0.03	0.025	0.0254
0.009	0.0091	0.00914
0.04	0.039	0.0394
0.6	0.62	0.621
0.002	0.0018	0.00176
0.06	0.061	0.0610
0.03	0.028	0.0283
0.005	0.0045	0.00454

4 Japan

5 Countries with similar areas, for example Bangladesh and Cambodia, could end up with very different looking areas. Other countries with different areas, for example Cambodia and United Kingdom, would end up with exactly the same area.

6 **a** 244 020 **b** 460 **c** 370 000 **d** 9 561 000 **e** 2

7 Congo and Bangladesh

Exercise 8.5

1 **a** $30 \times 40 = 1200$ **b** $700 \times 4 = 2800$ **c** $900 \times 80 = 72\ 000$
d $500 \times 50 = 25\ 000$ **e** $60\ 000 \div 80 = 750$ **f** $60\ 000 \div 500 = 120$
g $500 \div 2 = 250$ **h** $30\ 000 \div 1 = 30\ 000$

2 **a** £18 000 **b** Both numbers get rounded up.

3 120 people per km^2

4 In the first calculation both numbers are rounded down a lot, in the second they are both rounded up a lot.

5 **a** too large **b** too small **c** about right **d** too small **e** too small **f** too large **g** about right **h** too small

6 The size of the numbers is too different. You would be rounding the first number to the nearest hundred-thousand, then adding on a multiple of one hundred.

Thinking ahead to ...

A 9.234, smaller

B 12 666.67, larger

C **a** it gets smaller **b** it gets bigger

Exercise 8.6

1 **a** 17 784 **b** 17.784

2 estimate 0.1, actually 0.1035

3 **a** 7500, 8657.5 **b** 18, 15.738 **c** 0.18, 0.17524 **d** 1000, 1360
e 4,4.2558 **f** 1 000 000, 1 189 915.556

4 **a** 0.6 m^2, 70 m^2 **b** 0.822 m^2, 97.35042 m^2

5 40 m × 0.5 m, 50 000 m × 0.04 m

6 36 000 cm^3, 20 000 cm^3, 84 cm^3

7 0.5 cm

Exercise 8.7

1 **a** The £5 has reduced in size each time except in 1971 when the length increased.

b

Year	Method 1	Method 2
1925	286	294
1957	142	144
1963	118	112
2971	114	120
1994	93	98

d Method 1

End points

A1 Alez tuned into her favourite station. Channel 162 on the infrawave. She knew the slot lasted for about 90 minutes so she would need a compulsory meal before the end. She dined on 27 of her favourite food pills with 785 ml of ice-cold isophoric delight. Jeq materialised in 11 minutes raving about some antique maths book with about 400 pages that he'd found and dated as 1997.

A2 **a** Round up, to make sure you get enough paint.
b Round down, just in case the maximum is inaccurate.

B1 **a** 346 **b** 345.68 **c** 300

B2 **a** 34.7 **b** 1.97 **c** 195 000 **d** 0.00319 **e** 144 **f** 6.99

C1 **a** 10 m^3 **b** 8.90 m^3

SECTION 9 Probability A

Starting points

A1 **a** $\frac{1}{4}$ **b** $\frac{1}{12}$ **c** $\frac{1}{12}$ **d** $\frac{1}{6}$ **e** $\frac{1}{12}$

A2 **a** $\frac{5}{12}$ **b** $\frac{1}{3}$ **c** $\frac{2}{3}$ **d** $\frac{1}{6}$ **e** 1 **f** 0

A4 **a** $\frac{1}{6}$ **b** $\frac{1}{2}$ **c** $\frac{1}{3}$

B1 $\frac{1}{5}$

C1 **a** $\frac{3}{16}$ **b** $\frac{5}{16}$ **c** $\frac{2}{7}$ **d** $\frac{3}{32}$

D1

Outcome of Spinner A					
	X	X1	X2	X3	X4
	Y	Y1	Y2	Y3	Y4
	Z	Z1	Z2	Z3	Z4
		1	2	3	4

Outcome of Spinner B

D2 **a** $\frac{1}{12}$ **b** $\frac{1}{6}$

E2 $\frac{1}{8}$

Exercise 9.1

1 **a** $\frac{1}{18}$ **b** $\frac{1}{36}$ **c** $\frac{1}{6}$ **d** $\frac{1}{6}$ **e** $\frac{5}{6}$ **f** $\frac{1}{12}$ **g** $\frac{5}{18}$ **h** $\frac{1}{4}$ **i** $\frac{5}{12}$

2

Number on red dice											
	9	9, 0	9,1	9, 2	9, 3	9, 4	9, 5	9, 6	9, 7	9, 8	9, 9
	8	8, 0	8,1	8, 2	8, 3	8, 4	8, 5	8, 6	8, 7	8, 8	8, 9
	7	7, 0	7,1	7, 2	7, 3	7, 4	7, 5	7, 6	7, 7	7, 8	7, 9
	6	6, 0	6, 1	6, 2	6, 3	6, 4	6, 5	6, 6	6, 7	6, 8	6, 9
	5	5, 0	5, 1	5, 2	5, 3	5, 4	5, 5	5, 6	5, 7	5, 8	5, 9
	4	4, 0	4, 1	4, 2	4, 3	4, 4	4, 5	4, 6	4, 7	4, 8	4, 9
	3	3, 0	3,1	3, 2	3, 3	3, 4	3, 5	3, 6	3, 7	3, 8	3, 9
	2	2, 0	2,1	2, 2	2, 3	2, 4	2, 5	2, 6	2, 7	2, 8	2, 9
	1	1, 0	1,1	1, 2	1, 3	1, 4	1, 5	1, 6	1, 7	1, 8	1, 9
	0	0, 0	0, 1	0, 2	0, 3	0, 4	0, 5	0, 6	0, 7	0, 8	0, 9
		0	1	2	3	4	5	6	7	8	9

Number on blue dice

3 **a** $\frac{1}{10}$ **b** $\frac{3}{20}$ **c** $\frac{1}{4}$

Exercise 9.2

1 **a** $\frac{2}{9}$ **b** $\frac{7}{27}$ **c** $\frac{2}{9}$ **d** $\frac{2}{3}$ **e** $\frac{19}{27}$ **f** $\frac{1}{27}$ **g** $\frac{1}{9}$ **h** $\frac{2}{3}$

3 **a** $\frac{1}{8}$ **b** $\frac{1}{4}$ **c** $\frac{7}{8}$ **d** $\frac{3}{4}$ **e** 0 **f** $\frac{3}{8}$

Exercise 9.3

1 **a**

EHNW	HEWN	NEHW	WEHN
EHWN	HENW	NEWH	WENH
ENHW	HWEN	NHEW	WHEN
ENWH	HWNE	NHWE	WHNE
EWNH	HNEW	NWEH	WNEH
EWHN	HNWE	NWHE	WNHE

c $\frac{1}{24}$

2 NOT, NTO, ONT, OTN, TON, TNO

3 2

4 **a** and **c**

No. of letters	1	2	3	4	5	6
No. of arrangements	1	2	6	24	120	720

b 2, 3, 4; the number you multiply by increases by 1 each time.

5 **a** NNN, NNO, NON, ONN, NNT, NTN, TNN, OOO, OON, ONO, NOO, OOT, OTO, TOO, TTT, TTO, TOT, OTT, TTN, TNT, NTT, NOT, NTO, ONT, OTN, TON, TNO
b 27
c By showing that there is a choice of 3 letters for each of the 3 spaces in the word, giving $3 \times 3 \times 3 = 27$ possibilities.

Exercise 9.4

1

Rolls of dice	Number of sixes	Relative frequency
20	7	0.34
50	17	0.34
70	24	0.34

2 **a** 0.23 **b** 2 **c** 0.77

3 **a** 0.25 **b** 0.75 **c** 150

4 0.26

5 0.49

End points

A1

Outcome of Spinner A					
	R	RR	RY	RB	RP
	Y	YR	YY	YB	YP
	B	BR	BY	BB	BP
		R	Y	B	P

Outcome of Spinner B

A2 **a** $\frac{1}{4}$ **b** $\frac{1}{12}$ **c** $\frac{1}{12}$ **d** 0

A4 **a** $\frac{1}{8}$ **b** $\frac{3}{4}$ **c** $\frac{1}{4}$ **d** $\frac{3}{8}$

B1 **a** 24 **b** $\frac{1}{2}$

C1 0.36

SECTION 10 Using algebra A

Starting points

A1 **a** $3a + 12$ **b** $2b - 12$ **c** $6c + 15$ **d** $8d - 8$ **e** $32 - 8e$ **f** $15 + 30f$

B1 **a** $5a + 4b - 6$ **b** $4p + 6q - 10$

B2 **a** $8a + 10$ **b** $14a + 10$

C1 **a** $p = 3$ **b** $q = 10$ **c** $t = 0.75$ **d** $x = 18.75$ **e** $x = {}^{-}4.5$

D1 A and E, B and F, C and D

E1 **a** 31.28 **b** 12.6 **c** 4.12 **d** 27.92 **e** 29.76 **f** 45.08

Exercise 10.1

1 **a** **i** 16.3 m **ii** 3.4 m **iii** 11.9 m
b 54.1 m **c** 8.2 m × 10.5 m **d** 26.9 m **e** 108 m² **f** 306 m²

2 **a** 27.9 m **b** 19.4 m **c** 21.4 m² **d** 86.12m²

Exercise 10.2

1 **b** **i** $13.4 + w$ **ii** $16.2 - 2w$ **d** **i** $32.4 - 2w$ **ii** 28.4 m
e **ii** $131.22 - 16.2w$ **f** 2.7 m

2 **a** **i** $46 + p$ **ii** $24 - p$ **iii** $138.24 - 9.6p$ **iv** $24p$
b 48 m **c** 2.4

Exercise 10.3

1 **a** $2a + 8$ **b** $3b - 15$ **c** $8c + 12$ **d** $d^2 + 7d$ **e** $6e - e^2$ **f** $4f + f^2$
g $2p^2 + 9p$ **h** $4r^2 - 2r$

2 **a** $a(a + 5)$, $b(b - 2)$, $c(c - 6)$, $d(12 - d)$, $4(e + 8)$, $7(8 - f)$
b $a^2 + 5a$, $b^2 - 2b$, $c^2 - 6c$, $12d - d^2$, $4e + 32$, $56 - 7f$

3 **a** $5(2p + 2)$, $4(3q - 2)$, $r(2r + 2)$ **b** $10p + 10$, $12q - 8$, $2r^2 + 2r$

4 $2a$, b, c

5 $d + 3$, $e + 2$, $5 + f$, $2g + 3$

6 **a** $4(x + 3)$ **b** $2(3p - 2)$ **c** $2(a + 4)$ **d** $5(2 - q)$ **e** $b(b - 6)$
f $y(y + 4)$ **g** $r(2r + 3)$ **h** $d(3d - 5)$ **i** $t(3t + 4)$ **j** $s(6s + 7)$

Exercise 10.4

1 **a** $(a + 8)$ cm **b** $(4a + 16)$ cm **c** 209 cm²

2 **a** $5d$ cm **b** 23.2 cm

3 225 m²

4 84.5 cm²

Exercise 10.5

1 **a** $6ab$ **b** $3pq$ **c** $20xy$ **d** $30pq$ **e** $2x^2$ **f** a^2b
g $2xy^2$ **h** $6a^2b$ **i** $6p^2q$ **j** a^5 **k** $6b^3$ **l** $10bc^3$

2 A and E, B and F, C and H, G and I

3 **a** $ab + 4a$ **b** $xy + xz$ **c** $2mn + 3mp$ **d** $6xy + 4xz$ **e** $ac + c^2$
f $p^2 - pq$ **g** $3bc + 3b^2$ **h** $4a^2 - 4ab$ **i** $3p^2 - 4p$ **j** $2ab - 4ac$
k $6a^2 + 8ab$ **l** $8pq - 12p^2$ **m** $6p^2q + 4pq^2$ **n** $4x^2y - 8xy^2$
o $3abx^2 - 3aby^2$

4 **a** $9a - 22b$ **b** $2x^2 + x$ **c** $4a + 5ab$ **d** $x^2 + x^3 - 2x$ **e** $11a + 2b$
f $6x^2 + 4xy$ **g** $4p^2 + 6q$ **h** $6a$

5 **a** $9a + 26b$ **b** $10x + 5y$ **c** $2x^2 + x$ **d** $10x^2 + 3xy$ **e** $2a^2b$
f $3a^2b + 3ab + 2ab^2 - 2b^2$

6 **a** $6a^2 + 6ab + 12b^2$ **b** $16x^2y + 20xy^2 - 6xy$ **c** $8p^2q - 7pq^2$
d $14m^2n - 8mn^2 - 12m^3$

Exercise 10.6

1 **a** $5(ax + 2y)$ **b** $3(2ay - 5bc)$ **c** $3(3xy - 4a)$ **d** $2x(2y + 5)$
e $7(3xy + 5c)$ **f** $9(4ab - 5y)$ **g** $3(2ax + 4y + 5x)$
h $4(2ac + 5cy - 6ax)$ **i** $3(3x + 5ay - 7xy)$ **j** $6(5kx + 3y + 7ac)$
k $14(2xy + 4y - 3c)$ **l** $12(12ac - 5y + 9x)$

2 **a** $a(3x - 5y)$ **b** $x(6y + 7a)$ **c** $y(9x - 5a)$ **d** $x(6a - 17y)$
e $a(3bc + 5x)$ **f** $b(9a + 4)$ **g** $y(7x - 9)$ **h** $ax(6 + 7y)$ **i** $c(7ab - 8)$
j $x(3a + 5y - 7c)$ **k** $y(5x - 7a + 6)$ **l** $y(12x - 19a + 15)$
m $c(6ab + 5b + 7)$ **n** $x(8 + 7ay + 9by)$ **o** $ab(7c + 1)$
p $y(16x + 25 - 13ax)$ **q** $3bx(7a - 5c)$

3 **a** $2a(x - 4y)$ **b** $5x(5y + 3a)$ **c** $4y(4a + 7b)$ **d** $8x(4y - 3a)$
e $7b(6ac - 5x)$ **f** $14b(4a + 3y)$ **g** $3a(2b - 3y + 4)$
h $2y(2x + 3a - 5c)$ **i** $3x(7y - 5 + 4a)$ **j** $5xy(3k + 5a)$
k $3x(6c + 4ay + 5)$ **l** $7ac(3b + 8x)$ **m** $3x(5ay + 6a - 10y)$
n $bc(a + 5 - 15x)$ **o** $8x(3y - 2x + a)$ **p** $2x(x + 2 - 3y)$
q $5x(1 - 2y - 5x)$ **r** $3ab(2 - 3a + c)$

End points

A1 53.8 m

A2 16.7 m

A3 **a** 15.9 m **b** 796 m²

B1 **a** $20f - 16$ **b** $m^2 + mn$ **c** $2pr - 2p^2$ **d** $a^2b + ab^2$
e $2m^2n + 3mn^2$ **f** $6xy^2 - 10x^2y$

B2 **a** $22b + 16$ **b** $26b + 14$ **c** $10ab^2 + 4a^2b$ **d** $3x^2y + 7xy^2 + 19xy$

C1 **a** **i** $(p - 8.4)$ cm **ii** $(p - 6.1)$ cm **b** **i** $4p$ cm **ii** 14 cm

C2 **a** $(a + 5)$ cm **b** $(4a + 10)$ cm **c** $a(a + 5) = a^2 + 5a$
d 11.5 cm **e** 51.51 cm²

D1 **a** $4(2p + q)$ **b** $6(a - 2b)$ **c** $a(a + b)$ **d** $m(7 + 3n)$ **e** $2y(4x + 5)$
f $2a(2b + 3a)$ **g** $xy(3y - 5x)$ **h** $2pq(2 - 3p)$ **i** $h(2g^2 - 5h)$
j $2x(15y + 12x - 8y^2)$ **k** $5ab(5a - 3b)$ **l** $7x(2ay - 3x + 5xy)$
m $50y(2x + ay - 5x^2)$

SECTION 11 Constructions and loci

Starting points

A1 **a** AI **b** EL **c** EF, EL or EC **d** triangle ECL
e triangle AGI **f** triangle HIJ **g** AK or CL **h** GD **i** FC

B1 F

Exercise 11.1

2 Rydon and Barton

3 Rydon

7 about 20 miles

Exercise 11.2

2 isosceles

3 a right-angled b scalene c equilateral d scalene

4 This triangle does not exist. The sum of the lengths of the two shorter sides must always be greater than the length of the longest side.

6 104 miles

Exercise 11.3

2 angle SQR = 77°

3 Triangles LMN and PQR look identical.

Thinking ahead to ...

E b The locus is a line perpendicular to the line AB.

Exercise 11.5

1 c It is always the same distance from A and B.

3 c The three bisectors intersect at one point.
d yes

Thinking ahead to ...

A A, C and H

Exercise 11.6

2 Each angle is 60°. When you bisect one of the angles you make an angle of 30°.

Exercise 11.7

1 a 2 cm b 20 m

End points

B1 The gradients are the same.

C2 The bisectors intersect at one point.

SECTION 12 Ratio

Starting points

A1 a 3 : 1 b 1 : 3

B1 a 12 : 3 b 1 : $2\frac{1}{2}$: 2 c 3 : 4, 6 : 8

B2 2 : 5, 1 : $2\frac{1}{2}$ and 10 : 25, 1 : 3 : 2 and 3 : 9 : 6, 10 : 20 and $2\frac{1}{2}$: 5

B3 a 3 : 2 b 2 : 3 c 2 : 4 : 1 d 10 : 1 e 4 : 3 : 6

C1 a 0.625 b 1.8 c $0.58\dot{3}$ d $1.\dot{3}\dot{6}$

C2 a $61.\dot{1}$% b 50% c $70.8\dot{3}$% d 87.5%

D1 a 1, 3, 4, 7, 11, 18, 29, 47 b 2, 2, 4, 6, 10, 16, 26, 42
c 1, 5, 6, 11, 17, 28, 45, 73 d 3, 6, 9, 15, 24, 39, 63, 102

E1 a GI b GHI c HGI and GIH d GH

F3 JKL and PQR

Exercise 12.1

1 a 480 ml b 1200 ml

2 a 315 ml b 525 ml

3 a 50 ml b 175 ml

4 a 250 kg b i 2 : 1 : 9 ii 2400 kg

5 a 8 m b 9 cm

6 a 15 cm b 2.25 m

7 a 6.3 cm b 1 : 5000

Exercise 12.2

1 Sarah gets 20, Denzil gets 15

2 £3, £6, £9

3 £75, £45, £30

4 a $1\frac{1}{4}$ hours b $1\frac{1}{2}$ hours

Exercise 12.3

1 345 g caster sugar, 450 ml water, 75 g cocoa, 1800 ml chilled milk

2 a 60 g b 75 ml
c 300 g white sugar, 100 ml milk, 100 ml water, 50 g butter, 20 g cocoa

3 a 7 drops b 1.75, which is not practical

Exercise 12.4

1 a 2 : 1 b $\frac{2}{1}$

2 2 : 3, $\frac{2}{3}$

3 a $\frac{5}{3}$ b $\frac{3}{5}$

4 a $\frac{1}{5}$ b $\frac{4}{5}$

5 a $\frac{2}{3}$ b That would be height as a fraction of height and diameter.

Exercise 12.5

1 a 1.5 b 7.5 cm

2 a 2.5 b 12.75 cm

3 a 0.67 b 7.2 cm

End points

A1 a 450 ml b 630 ml

B1 £160, £128, £112

C1 a i 375 g ii 10 tablespoons b 3 eggs

D1 a 3 : 4 b $\frac{3}{4}$

D2 $\frac{2}{5}$

SKILLS BREAK 2A

1 £1.27, £1.69, £2.54, £4.24

2 8 litre

3 12 litre

4 yes

5 2 gallon

6 5105 g

7 a 0.315 kg b Less, it will weigh closer to 7 kg.

8 a 33 g b $2\frac{1}{4}$ scoops

9 a 66 gallons b 10 g c i $\frac{1}{100}$ ii 0.01 iii 1%

10 Yes, 300 litres is about the same as 66 gallons.

11 £3502

12 a 24.75 m^2 b 15p

13 4913 cm^2

14 30 metre

15 a three 50 m rolls and one 30 m roll, or six 30 m rolls, or for 28p per metre
b six 30 m rolls, which costs £44.94

16 2.4 m^2

17 a rectangle b 5.5 m by 1.9 m

18 6.3 m^2

19 34.4 m^2

20 No, nearly 9 litres will be required.

21 76.0°

22 a rectangle b 2.9 m c 16.5 m^2 d 72.4 kg

23 a £180 b £14.40 c £10.80

24 £24, £26.67, £33.33

SKILLS BREAK 2B

1 12

2 12

3 40.8 m

4 132.7 m^2

5 a octagon b 45° c 135°

6 696 cm^2

7 4

8 1110 tons

9 50 m

10 16 cm

11 370

12 a 1.01×10^4 b 5.0×10^5

13 a 1650 b 1700 c 2000

14 c

15 a 225° b 45°

16 £7.43

17 a 308m FF b £41m

18 a 2 200 000 b 2 190 000 c 2 189 000 d 2 188 900

19

Vowel	French	English
a	41	42
e	73	51
i	26	35
o	21	36
u	27	12

21 a EP, EV, EM, PV, PM, VM b $\frac{1}{2}$ c $\frac{1}{6}$ d i $\frac{1}{3}$ ii $\frac{1}{2}$ e loss

22 a £851 b £99 c £89 d £277

SECTION 13 Distance, speed and time

Starting points

A1 a 5000 m b 3.4 m

A2 a 4800 m b 1.5 m c 1.8 m

A3 a 4 cm b 400 000 cm c 390 cm d 18 cm

A4 a 5.13 km b 11.36 km

A5 12.5 miles

B1 a 15.4 pounds b 8.4 pounds

B2 a 7.3 kg b 2.3 kg c 3.6 kg d 2.1 kg

B3 2.2 pounds

B4 a

Gallons	1	2	3	4	5	6	7	8
Litres	4.55	9.1	13.65	18.2	22.75	27.3	31.85	36.4

c i 11.4 litres ii 3.3 gallons

C1 a 9:30 am b 2:21 pm

C2 a 02:23 b 17:25

C3 75 minutes

C4 6 hours 40 minutes

D1 13:00

D2 0625: 4 hours 40 minutes, 0825: 4 hours 45 minutes, 0955: 4 hours 45 minutes, 1145: 4 hours 50 minutes, 1345: 4 hours 55 minutes, 1700: 4 hours 45 minutes

D3 The 1400 bus from Torquay take the longest to get to Exeter.

D4 5:15 pm

D5 3 hours 12 minutes

D6 9:45 pm

Exercise 13.1

1 a i 5.1 m ii 25.5 m iii 11.48 m b i 153 m ii 9180 m
c i 413 seconds ii 6 minutes 53 seconds

2 b Alperton: 29.8 seconds, The Angel: 80 seconds, Chancery Lane: 15.2 seconds: Kentish Town: 66.8 seconds c 2:02 pm

Exercise 13.2

1 520 mph

2 a 1 hour and 50 minutes is not the same as 1.5 hours b 608.2 mph

3 a 10.33 miles per minute b 103.3 miles

4 a 6.72 miles per minute b 17.9 minutes

5 a 35.5 minutes b 3487.5 miles

Exercise 13.3

1 b 175 m c 2.7 seconds d 120 m
f i 20 m/s, 10 m/s ii red iii red has the steepest graph
g i 3.5 seconds ii It is the point where their graphs cross.

2 a Purple 90 m, black 100 m, blue 118 m, red 140 m
b Black c i 4 seconds ii 3.3 seconds
d Purple 20 m/s, black 40 m/s, blue 30 m/s, red 0 m/s
e It broke down!

3 b 200 m

Exercise 13.4

1 a 30 miles b 50 minutes c 18.5 miles
d i 9 minutes ii 21 minutes e 11 miles f A3, B2, C4, D5, E1
g It is a graph of Geeta's distance from home, not a diagram of the roads.

Exercise 13.5

1 a 0.2 miles per minute b 12 miles per hour

2 a 0.875 miles per minute b 52.5 miles per hour

3 a 58.9 mph

4 1:40 pm

5 16 miles

6 a i 20 minutes ii 33 miles per hour
b 25.44 mph, 29.4 mph c 45 miles d 1:46 pm

Exercise 13.6

1 a 1500 m b A c 10 minutes
d A: 1500 m, B: 1200 m, C: 0 m, D: 700 m e 18 km/h
f i A is Sharon, B is Jane, C is Jason, D is Lorna
ii 500 m iii Walked, she was the slowest.
g i A: 150 m/min, B: 100 m/min, C: 300 m/min, D: 60 m/min
ii A: 9000 m/h, B: 6000 m/h, C: 18 000 m/h, D: 3600 m/h
iii A: 9 km/h, B: 6 km/h, C: 18 km/h, D: 3.6 km/h

2 a i 13:40 ii 20 minutes b i 14:00 to 14:16 ii 45 mph
c The motorbike passed the car in the opposite direction
e i The motorbike travelled 60 miles, the car travelled 30 miles
ii motorbike 33 mph, car 18 mph

Exercise 13.7

1 a ii 11.1 km/h b ii 3:16 pm iii 2:44 pm

2 a ii 9 mph iii 9.2 mph b i 10 miles iii 12.8 mph
iv Alice was buying cakes as Emily passed c ii 20 mph

Exercise 13.8

1 a A: 5.6 gallons, B: 3.6 gallons b 35 miles c twice
d 6.4 gallons e 3 gallons, 2.8 gallons
f i 9.2 gallons, 12.6 gallons ii 43.5 mpg, 31.7 mpg

2 a 0.89 cars per household b 55 cars per kilometre c No

3 a 0.45 cars per household

End points

A1 225.25 m

A2 4 minutes 49 seconds

A3 45.6 mph

A4 16:32

B1 a 2.5 miles b 26 minutes c twice d 10.3 mph
e i 20:40 and 21:08 ii steepest part f 50 minutes g 60 mph
h i 21:26 ii 15.5 miles

C1 b 7.7 mph

D1 3000 people

D2 0.51 litres per second

D3 £1.95 per person

SECTION 14 Trigonometry

Starting points

A1 **a** GI **b** angle GHI **c** angle GIH **d** GH

B1 A and C, B and E

B2 **a** XZ **b** FG

Thinking ahead to ...

A **b** 0.55 **c** they are the same

B **a** 1.83 and 1.83 **c** they are the same

Exercise 14.1

1 **a** **i** BC/AC, ED/EF **ii** 1.18
b **i** AC/BC, EF/ED **ii** 0.85
c AB/AC = DF/EF = HI/GI = 0.81, AC/AB = EF/DF = GI/HI = 1.22
AB/BC = DF/ED = HI/GH = 0.69,
BC/AB = ED/DF = GH/HI = 1.44

2 **a** All corresponding angles are equal.
b 0.75, 1.33, 0.6, 1.67, 0.8, 1.25

3 6

Exercise 14.2

1

adj	opp	hyp	adj/hyp	adj/opp	opp/hyp	opp/adj	hyp/adj	hyp/opp
40	23	46	0.87	1.74	0.5	0.56	1.15	2

Exercise 14.4

1 **b** A: $\sin x = \frac{21.6}{22.5} = 0.96$, $\cos x = \frac{6.3}{22.5} = 0.28$, $\tan x = \frac{21.6}{6.3} = 3.43$
B: $\sin x = \frac{3}{5} = 0.6$, $\cos x = \frac{4}{5} = 0.8$, $\tan x = \frac{3}{4} = 0.75$
C: $\sin x = \frac{4.5}{11.7} = 0.38$, $\cos x = \frac{10.8}{11.7} = 0.92$, $\tan x = \frac{4.5}{10.8} = 0.42$
D: $\sin x = \frac{1.05}{1.75} = 0.6$, $\cos x = \frac{1.4}{1.75} = 0.8$, $\tan x = \frac{1.05}{1.4} = 0.75$
E: $\sin x = \frac{5}{13} = 0.38$, $\cos x = \frac{12}{13} = 0.92$, $\tan x = \frac{5}{12} = 0.42$
F: $\sin x = \frac{8}{10} = 0.8$, $\cos x = \frac{6}{10} = 0.6$, $\tan x = \frac{8}{6} = 1.33$
G: $\sin x = \frac{12}{12.5} = 0.96$, $\cos x = \frac{3.5}{12.5} = 0.28$, $\tan x = \frac{12}{3.5} = 3.43$
H: $\sin x = \frac{9.6}{10.4} = 0.92$, $\cos x = \frac{4}{10.4} = 0.38$, $\tan x = \frac{9.6}{4} = 2.4$

Exercise 14.5

2 EF = 3.9 cm, GH = 5.7 cm, JK = 3.5 cm, MN = 20.7 cm

3 74.9 cm

4 AB = 6.9 cm, DE = 6.5 cm, KL = 3.8 cm, GI = 7.1 cm

5 JK = 3.1 cm, NO = 8.1 cm, AC = 3.7 cm, PR = 11.5 cm

Exercise 14.6

2 **a** 21° **b** 46° **c** 69°

3 **a** 21° **b** 65° **c** 59°

4 28°

5 $a = 42°$, $b = 56°$, $c = 46°$

6 $a = 37°$, $b = 52°$, $c = 30°$, $d = 41°$, $e = 61°$,
$f = 60°$, $g = 38°$, $h = 44°$, $i = 60°$, $j = 51°$

7 **a** Wayne must be incorrect
b The cosine of an angle cannot be greater than 1.

Thinking ahead to ...

A **a** less than
b RS is opposite the smallest angle

C 5.41 cm

Exercise 14.7

1 **a** A: greater than hyp, B: isosceles triangle, C: opposite 50°, but almost as large as hyp
D: opposite smallest side but greater than 45°, E: smaller than opp, F: isosceles triangle
b 3.79 cm, 4.32 cm, 9.46 cm, 42.1°, 10.2 cm, 45°

2 42.0°, 1.33 cm, 13.2 cm, 31.3°, 72.4°, 22.6°

Exercise 14.8

1 **a** 75.52° **b** 4.84 m

2 **a** 1.39 m **b** 1.09 m **c** 4.28 m

3

	72°	76°
4 m	3.80 m	3.88 m
6 m	5.71 m	5.82 m
8 m	7.61 m	7.76 m

4 **a** 312 mm **b** 140 mm **d** 24.2°

5 **a** 0.42 m **b** 8.03 m **c** 8.83 m **d** 8.01 m **e** 10.41 m

6 **a** 2.86°
b A: 3.33°, unsafe B: 2.01°, safe C: 1.51°, safe D: 1.06°, safe

7 2.80°

8 **a** 32.74° **b** 21.66°

End points

A1 ABC and MNO

B1 48.59°, 35.54°, 59.74°

C1 10.7 cm, 3.75 m, 6.63 cm

SECTION 15 Using algebra B

Starting points

A1 48 cm^2

A2 **a** 19 **b** 2.5 **c** 49 **d** 20

B1 **a** 12 **b** 39

C1 **a** $n = 20$ **b** $n = 2$

C2 **a** $7n + 9 = 65$ **b** 8

D1 **a** $5x + 8$ **b** $9a + 2b$ **c** $5x^2 + 2x$ **d** $5a^2 + 6a - 7$
e $3t^2$ **f** $10cd$ **g** $12k^2$ **h** $7w^2$

Thinking ahead to ...

A **a** 3 hours 20 minutes **b** 3 lb

Exercise 15.1

1 **a** $p = t - 1$ **b** $p = \frac{s}{7}$ **c** $p = \frac{(y + 2)}{5}$ **d** $p = 10 - v$ **e** $p = \frac{(12 - h)}{9}$
f $p = r - q$ **g** $p = \frac{t}{q}$ **h** $p = \frac{(x - f)}{2}$ **i** $p = \frac{(k + m)}{l}$ **j** $p = 9g$
k $p = mn$ **l** $p = 2(s - 1)$

2 **a** £10.25 **b** $n = \frac{(c - 50)}{15}$ **c** 129 **d** 63

3 **a** $t = 5d$ **b** 7.5 seconds

4 **a** 720° **b** $n = \frac{S}{180} + 2$ **c** 20
d Putting 600° into the formula does not give a whole number of sides.

5 **a** 58 **b** 3 **c** $d = \frac{(100 - n)}{3}$ **d** 30 days

6 **a** 550 **b** $x = \frac{(800 - n)}{5}$ **c** 60p
d The formula would give a negative value for the number of ice-creams she would sell.

7 **a** 94 miles **b** 600 mph **c** $t \longrightarrow \times s \longrightarrow d$
d $d \longrightarrow \div s \longrightarrow t$, $t = \frac{d}{s}$ **e** $2\frac{1}{2}$ hours

Exercise 15.2

1 **a** $w = \frac{5k}{3}$ **b** $w = \frac{4x}{3}$ **c** $w = \frac{5(k - 1)}{3}$ **d** $w = 2(3x - 4)$ **e** $w = 3(3y - 3)$
f $w = 3(2k - 4)$ **g** $w = \frac{5(4 - x)}{3}$ **h** $w = 3k - 1$ **i** $w = 6k + 2$
j $w = \frac{3}{4}k + 1\frac{1}{2}$ **k** $w = -\frac{4k}{3}$ **l** $w = k + 4$

2 **a** 17.6 lb **b** $k = \frac{5}{11}p$ **c** 1.4 kg

3 **a** $h = \frac{2A}{b}$ **b** 80 cm

4 **a** 20 °C
b $F \longrightarrow [-32] \longrightarrow F - 32 \longrightarrow [\times 5] \longrightarrow 5(F - 32) \longrightarrow [\div 9] \longrightarrow \frac{5(F - 32)}{9} = C$
c $C \longrightarrow [\times 9] \longrightarrow 9C \longrightarrow [\div 5] \longrightarrow \frac{9C}{5} \longrightarrow [+32] \longrightarrow \frac{9C}{5} + 32 = F$
d 75.2 °F

5 **a** $y = \frac{4z}{3}$ **b** $x = \frac{3m}{x}$ **c** $y = 2(d - 4)$ **d** $y = \frac{7(k + h)}{4}$

6 **a** 36.14 km/h **b** $d = \frac{5st}{18}$ **c** $t = \frac{18d}{5s}$

Exercise 15.3

1 **a** 403.44 cm^2 **b** $x\sqrt{(\frac{S}{6})}$ **c** 4.47 cm

2 **a** 7.07 cm^2 **b** B **c** 7.98 cm

3 **a** 207.03 m **b** $t=\sqrt{(\frac{10d}{49})}$ **c** 9.51 seconds

4 **a** $k=\sqrt{(m-1)}$ **b** $k=\sqrt{(\frac{m-1}{3})}$ **c** $k=\sqrt{(m-n)}$

5 **a** $y=\sqrt{(\frac{k-2}{4})}$ **b** $y=\sqrt{(\frac{1-3x}{2})}$ **c** $y=\sqrt{(\frac{ax}{5})}$ **d** $y=\sqrt{(\frac{ax+4}{3})}$
e $y=\sqrt{(2(x+4))}$ **f** $y=x\sqrt{(\frac{a}{b})}$ **g** $y=\sqrt{\left(\frac{(a^2-3x)}{5}\right)}$
h $y=\sqrt{(3-a^2-x^2)}$ **i** $\sqrt{\left(\frac{(x^2-4a^2+3x)}{2}\right)}$ **j** $y=\sqrt{(\frac{4}{a-3})}$ **k** $y=\sqrt{(\frac{2w}{3-x})}$
l $y=\sqrt{(\frac{4}{5+a})}$ **m** $y=x\sqrt{\left(\frac{(b)}{(c-a)}\right)}$ **n** $y=\sqrt{\left(\frac{(3-x^2)}{(a-5)}\right)}$ **o** $y=\sqrt{\left(\frac{(2w+3x^2)}{x}\right)}$
p $y=\sqrt{(\frac{4x}{1+4a})}$ **q** $y=\sqrt{(3-2x^2)}$ **r** $y=\sqrt{\left(\frac{(4a^2)}{x^2}\right)}$

Exercise 15.4

1 **a** 42, 42, 42, 42 **b** 16, 20, 20, 12 **c** 6, 12, 12, 0
d 10.75, 15.75, 15.75, 5.75

2 82.5 m

3 **a** 121.5 cm^3 **b** 64 cm^3 **c** $x=2$

4 87.29 m/s

5 **a** 31.11 m/s **b** 21.14 m/s

Exercise 15.5

1 **a** 1500 cm^2, 180 cm^2, 150 cm^2, 18 cm^2 **b** 1848 cm^2

2 **b** 378 cm^2

3 **a**

×	40	+5
60	2400	300
+1	40	5

So 45 × 61 = 2745

b

×	20	+5
20	400	100
+7	140	35

So 25 × 27 = 675

c

×	30	+2
810	2400	160
+4	120	8

So 32 × 84 = 2688

Thinking ahead to ...

A **a** n^2, $3n$, $2n$, 6 **b** n^2+5n+6

Exercise 15.6

1 **a** a^2+4a+3 **b** b^2+6b+8 **c** $c^2+13c+40$ **d** $d^2+14d+45$
e e^2+8e+7 **f** $f^2+13f+22$ **g** $x^2+7x+12$ **h** $x^2+11x+24$
i $x^2+18x+81$ **j** $x^2+9x+14$ **k** $x^2+24x+135$ **l** $x^2+26x+160$
m x^2+4x+3 **n** $x^2+15x+56$ **o** $x^2+27x+126$

2 **a** **i** x^2+2x+1 **ii** x^2+4x+4 **iii** x^2+6x+9

3

×	$2x$	+4
$3x$	$6x^2$	$12x$
+12	$2x$	4

So $(2x+4)(3x+1)=6x^2+14x+4$

4 **a** $2a^2+9a+4$ **b** $3b^2+14b+15$ **c** $6c^2+25c+14$
d $12d^2+25d+12$ **e** $4e^2+12e+9$ **f** $25f+10f+1$
g $10x^2+13x+4$ **h** $12x^2+43x+35$ **i** $9x^2+12x+4$
j $56x^2+43x+5$ **k** $20x^2+9x+1$ **l** $16x^2+20x+6$

5 B

6 **a** $5\times8=40$, $6\times7-2=40$
b $(n+1)(n+2)-2=n^2+3n+2-2=n^2+3n=n(n+3)$

7 $(2x+1)(2x+3)-3=4x^2+8x+3-3=4x^2+8x=4x(x+2)$

Exercise 15.7

1 **a** x^2-x-12 **b** x^2+x-20 **c** $x^2-5x-14$ **d** $x^2-2x-24$
e x^2-x-56 **f** $x^2+4x-32$ **g** x^2-x-42 **h** $x^2+7x-18$
i $x^2+4x-21$ **j** $x^2+4x-96$ **k** $x^2-3x-54$ **l** x^2+7x-8

2 **a** x^2+x-6 **b** x^2+x-30 **c** $x^2-4x-32$ **d** $x^2-2x-35$
e x^2+x-72 **f** $x^2-5x-14$ **g** $x^2-5x-36$ **h** $x^2-3x-28$
i $x^2-4x-45$ **j** $x^2-2x-80$ **k** $x^2-7x-60$ **l** x^2-x-56

3 **a** $x^2-11x+24$ **b** $x^2-10x+24$ **c** $x^2-13x+40$ **d** $x^2-10x+9$
e $x^2-10x+21$ **f** $x^2-13x+36$ **g** $x^2-19x+84$ **h** $x^2-19x+90$
i $x^2-11x+30$ **j** $x^2-27x+180$ **k** $x^2-19x+84$ **l** $x^2-29x+100$

4 **a** $6x^2+11x+4$ **b** $12x^2+23x+10$ **c** $30x^2+39x+12$ **d** $8x^2+2x-3$
e $15x^2+8x-12$ **f** $54x^2-3x-12$ **g** $18x^2-2$ **h** $28x^2-13x-6$
i $32x^2-8x-12$ **j** $42x^2-46x+12$ **k** $18x^2-60x+18$
l $20x^2-41x+20$

Thinking ahead to ...

A $(x+2)(x+6)$

B **a**

×	p	−2
p	p^2	$-2p$
+7	$7p$	−14

b $(p-2)(p+7)$

Exercise 15.8

1 **a** $(x+1)(x+7)$ **b** $(x+1)(x+2)$ **c** $(x+1)(x+3)$
d $(x+2)(x+6)$ **e** $(x+3)(x+3)$ **f** $(x+4)(x+9)$

2 **a** $(x+1)(x+5)$ **b** $(x+8)(x-6)$ **c** $(x+7)(x-2)$
d $(x+7)(x-3)$ **e** $(x+9)(x-3)$ **f** $(x+9)(x-2)$
g $(x+10)(x-1)$ **h** $(x+12)(x-5)$ **i** $(x+10)(x-2)$

3 **a** $(x+1)(x-7)$ **b** $(x+5)(x-9)$ **c** $(x+5)(x-8)$
d $(x+6)(x-10)$ **e** $(x+1)(x-5)$ **f** $(x+5)(x-12)$
g $(x+7)(x-9)$ **h** $(x+2)(x-7)$ **i** $(x+3)(x-12)$

4 **a** $(x-2)(x-7)$ **b** $(x-5)(x-4)$ **c** $(x-2)(x-4)$
d $(x-2)(x-5)$ **e** $(x-4)(x-7)$ **f** $(x-5)(x-9)$
g $(x-6)(x-5)$ **h** $(x-7)(x-8)$ **i** $(x-7)(x-13)$

End points

A1 **a** 100 m **b** 175 m **c** 100 m

A2 52.4 cm^3

B1 **a** $n=\frac{(C-20)}{30}$ **b** 15 days

B2 **a** $m=5(C-10)$ **b** 100 miles

B3 **a** $k=\sqrt{(\frac{h}{5})}$ **b** $k=5m-n$ **c** $k=\sqrt{(A-9)}$

C1 C

C2 **a** a^2+9a+8 **b** $b^2+8b+16$ **c** $5c^2+17c+6$
d d^2+d-2 **e** $6e^2-5e-50$ **f** $3f^2-23f+14$

D1 $(k-2)(k+7)$

D2 **a** $(x+1)(x+11)$ **b** $(x+2)(x+7)$ **c** $(x+5)(x-1)$
d $(x+3)(x-2)$ **e** $(x+4)(x-5)$ **f** $(x-3)(x-2)$

SECTION 16 Grouped data

Starting points

A1 2.305 m

A2 2.25 m

B1 1.97 m

B2 1.91 m

C1 **a** 14 cm **b** 12 cm

C2 **a** 7 cm **b** 6 cm

E1 **a** 6 years **b** 4 years

E2 **a** bars with heights 8, 13, 10, 3
b bars with heights 3, 17, 14, 8, 5, 1

Exercise 16.1

1 a 70–72 b 70–72

2 a 1st round
b The frequency polygon for the 1st round scores is higher in the low score section of the graph than the frequency polygon for the 4th round.

3 a 17, 18, 19, 20, 21, 22, 23, 24, 25, 26, 27 b 22 c 22, 33, 44, 55

5 No, you need roughly the same number to compare.

Thinking ahead to ...

A a 64, 65, 66 b 79, 80, 81

B 17, 16, 15, 14, 13

C 15, the middle value and happens most often

Exercise 16.2

1 a 5 b 69, 74, 79, 84 c 15

2 a 4 b 68.5, 72.5, 76.5, 80.5 c 12

3 a 128, 129, 130, 131, 132 b 130
c The mid-class value is the best estimate of the score in a class, so the mid-class value multiplied by the frequency is the best estimate of the total.

Exercise 16.3

1 43 900

2 252 750

3 8153.2

4 a 3089 m b 2837 m

Exercise 16.4

1 63.0 m

2 About the same as the mean throw is the same

3 18.7 m

4 20.1 m

5 The men threw further than the women

Exercise 16.5

Reaction time	0.120–	0.140–	0.160–	0.180–	0.200–	0.220–	0.240–	0.260–	Total
Frequency	5	8	23	5	4	2	0	0	47

1

2 a i 0.14 seconds ii 0.1 seconds
b i 0.1738 seconds ii 0.1704 seconds

3 true, the mean is lower and the range is narrower

Reaction time	0.120–	0.140–	0.160–	0.180–	0.200–	0.220–	0.240–	0.260–	Total
Frequency	10	16	46	10	8	4	0	0	94

4 a

d No, the finalists may have reacted exactly the same in the semifinals, the athletes that didn't make it to the finals perhaps had the slowest reaction times in the semifinals.

Thinking ahead to ...

A 12.73, 12.68, 12.749

Exercise 16.6

1 a 9.835 to 9.845 seconds b 7.115 to 7.125 metres
c 0.815 to 0.825 metres d 0.1735 to 0.1745 seconds

2 a 12.375 to 12.385 seconds b 9.15 to 9.25 metres
c 7.25 to 7.35 seconds d 7.295 to 7.305 seconds

Exercise 16.7

1 10.93 to 10.94 seconds

2 7.00 to 7.01 metres

3 So that the time/distance recorded isn't better than they performed.

Exercise 16.8

1 a

Points	< 700	< 800	< 900	< 1000	< 1100
Cumulative frequency	0	8	31	47	50

2 a

Distance (metres)	< 76	< 78	< 80	< 82	< 84	< 86	< 88	< 90
Cumulative Frequency	0	2	6	20	33	41	47	48

Exercise 16.9

1 a

Distance (metres)	< 72	< 74	< 76	< 78	< 80	< 82
Cumulative Frequency	0	1	12	35	48	52

2 a 77.2 b 2.3

3 The javelin was, on average, thrown further but the hammer was throw more consistently.

4 b 840 c 180

5 a 875 b 115

6 More points were scored on day 1 with a smaller spread.

7 a

Distance (metres)	< 56	< 58	< 60	< 62	< 64	< 66	< 68
Cumulative Frequency	0	7	18	28	38	45	46

8 a at 11.5, 23, 34.5 b 61, 4.5

Exercise 16.10

1 5

2 23 3 b 15

Exercise 16.11

1 a 12, 10.7, 9, 12, 8, 11, 9, 10, 14
b 12, 13, 10, 9, 6, 9, 5, 12, 9, 15
c 9, 12, 10, 11, 6, 15, 14, 17.3, 7, 10

2 a 23, 20.8, 27.2, 27.8, 25.8, 25, 25.2, 16
b 12.8, 12, 12.6, 11, 14.6, 21.4, 21.2
c 19.8, 17.4, 14.2, 11.8, 12.6, 15.2
d 12.4, 10.8, 10.6, 18, 17, 16.4, 18.8

3 a 16.25, 15, 10.25, 11.25, 13.25, 20.25, 23.25, 20.5, 18
b 20.5, 19.75, 18.25, 19, 14.75, 19.5, 16.5, 24.75
c 10, 20.75, 27.75, 32, 38.25, 29.75, 32.75, 44
d 31.5, 19, 17.25, 16.5, 8, 17, 29.5, 40.25, 48

4 a 8, 7.5, 6, 5, 6, 6.75, 9, 8, 8
c Sales of boots are quite steady over each 4 week period.

5 a 20.25, 16.75, 15.5, 14, 13.5, 16.5, 15.5, 14.25, 15.25, 14.75, 10.25, 11.75, 10.5, 7.75
c Watch sales declined over the period

End points

A1 a bars with heights 7, 22, 23, 10, 3, 2, 2, 1
b bars with heights 5, 10, 30, 17, 5, 3, 1

A2 a 0.16– b 0.16–

B1 a

Reaction Time (s)	< 0.12	< 0.14	< 0.16	< 0.18	< 0.20	< 0.22	< 0.24	< 0.26	< 0.28
Cumulative Frequency	0	7	29	52	62	65	67	69	70

c i 0.165 seconds ii 0.03 seconds

C1 a 0.12 seconds b 0.14 seconds

C2 **a** 0.175 seconds **b** 0.170 seconds

D1 The men were, on average, quicker but the women were more consistent

E1 **b i** 8 **ii** 15

F1 **a** 7, 9, 10, 11, 7, 9, 10, 12, 7, 10, 12, 13
c The number of lates rises then falls in cycles.

SECTION 17 Working with percentages

Starting points

A1 **a** 0.75, 75% **b** 0.625, 62.5% **c** 1.6, 160% **d** 1.25, 125%
e 1.2, 120% **f** 0.65, 65% **g** 0.25, 25% **h** 1.5, 150%
i 0.4375, 43.75% **j** 2.2857, 228.57% **k** 1.125, 112.5% **l** 1.8, 180%

B1 **a** 325 miles **b** £300 **c** 2 marks **d** 2484 **e** 10 marks **f** 171 kg
g £240.70 **h** 360 g **i** 18p **j** 16.2 m **k** £1.71 **l** 1.8 cm **m** £6.25
n 2.4 miles **o** 10 km **p** £480 **q** 110.7 miles **r** 9 cm **s** 8.4 m
t 438 m

C1 **a** £13.60 **b** 5.6 mm **c** 1540 g **d** 20p **e i** 50.7 mm **ii** 5.07 cm
f i 9p **ii** £0.09 **g i** 1.4 tonnes **ii** 1400 kg

Exercise 17.1

1 50%

2 400%

3 a little more

4 **i** 8 **ii** $80 \times \frac{10}{100}$

5 50%

6 **a** 16 **b** 21

7 **a** 9 **b** His number is a little more than 8 but not as much as 10.

8 25%, 25%, 20%, 25%, 75%, 10%

9 **a** December and March **b** January **c** 62.5%

Exercise 17.2

1 **a** 44.44% **b** 65.00% **c** 29.17% **d** 46.20%
e 80.00% **f** 12.50% **g** 3.75% **h** 133.33%
i 166.67% **j** 187.50%

2 **a** 69.09% **b** Pretty accurate, rounded to the nearest 5%.

3 14%

4 **a** 170%

Thinking ahead to ...

A 45 ml, less **B** $\frac{3}{20}$ **C** 345 ml

Exercise 17.3

1 **a** 110% **b** 484 g

2 **a** 420 kg **b** 514.05 km **c** 5.49 m **d** £37 800

3 206 000 000

4 26g

Exercise 17.4

1 **a** 266 kg **b** 312 km **c** 20.9 m **d** £287 300
e £18.36 **f** 16 588 tonnes **g** 4 717 000 **h** 1.2 cm

2 1466 kg

3 **a** 1125
b Round down to the whole number below, because fractions of cars are irrelevant.

4 **a** 26 650 **b** 14 350

5 7644 kg

6 **a** £19.60 **b** £4.90

Thinking ahead to ...

A Divide by 10, halve the answer, halve the answer again, and add the three amounts.

B £235

C **i** £14.69 **ii** Round to the nearest penny.

Exercise 17.5

1 **a** £18.80 **b** £51.99 **c** £217.38 **d** £22.80
e 85p **f** £432.89 **g** £307.85 **h** £28.85 **i** £10.99

2 **a** £287.88 **b** £42.88

Exercise 17.6

1 £387.60

2 **a** £155 **b** £181.35 **c** £301.35

3 **a** £2612.50 **b** £112.50

4 £89 979

Exercise 17.7

1 **a** £324 **b** £67.20 **c** £1350 **d** £10 200 **e** £90

2 **a** £5950 **b** £9450

3 £231

4 £66 440 000

Exercise 17.8

1 **a** £103.26 **b** £175.71 **c** £1217.61
d £1529.58 **e** £1030.32 **f** £8967.60

2 **a** £2346.64 **b** £7346.64 **c i** £7346.64 **ii** Round each year.

3 **a** £1900.16 **b** £7542.60 **c** £9881.86 **d** £10 438.16

4 £5898.58

5 £30 653.86

Exercise 17.9

1 **a** £115.32 **b** £29.78 **c** £63.82 **d** £56.00
e £158.94 **f** £12.33 **g** £42.54 **h** £680.83 **i** £84.25

2 **a** £56.25 **b** £11.25

3 £26

4 444 g

End points

A1 25%

A2 300%

A3 28.13%

A4 7.7%

B1 556.8 km

B2 426 ml

C1 £52.45

C2 39 203 breakdowns

D1 17.5%

D2 **a** £40.83 **b** £186.12

E1 £221.27

F1 £115.20

F2 **a** £180 **b** £680

G1 **a** £809.89 **b** £2898.19 **c** £4926.85 **d** £9067.93
e £4810.70 **f** £330.91

SECTION 18 Transformations

Starting points

A1 $x = 3, y = {}^-1$

B1 **b i** Triangle with coordinates (3, 2), (3, 0), (2, 0).
ii Triangle with coordinates (⁻2, 2), (⁻2, 3), (⁻4, 3).

C1 **a** 90° **b** (0, 1)

D1 U: (2, 2), (4, 3), (2, 3).
V: (⁻1, 0), (1, 0), (⁻1, ⁻1)

E1 **a** triangle: (1, 0), (1, 2), (5, 2)
b triangle: (0, 3.5), (0, 4), (1, 4)

Exercise 18.1

1 **a i** 0 **ii** 3 **b i** 1 **ii** 3

2 **a** rotation of 60° clockwise **b** translation, reflection
c A and F **d** Z, N, V **e i** C, R, X **ii** E, J, T

Exercise 18.2

1 **a i** triangle 8 **ii** triangle 1 **iii** triangle 3 **b i** B **ii** 90°
c i rotation of 90° about D **ii** rotation of 180° about P
d i BD **ii** HD
e triangle 7 **f** triangle 4 **g** rotation of -90° about F
h reflection in AE or rotation of 180° about P
i rotation of 360° about any point, or reflection in GP

Exercise 18.3

1 **a** $\begin{pmatrix}0\\-4\end{pmatrix}$ **b** (-2, -1) **c** $y = -1$

2 **a i** C and E, D and H, F and G

ii

Object	Mirror line	Image
C	$y = x$	E
D	$y = 0$	H
F	$x = -1$	G

b i C and F, D and G, E and H

ii

Object	Vector	Image
C	$\begin{pmatrix}-4\\-4\end{pmatrix}$	F
D	$\begin{pmatrix}2\\-2\end{pmatrix}$	G
E	$\begin{pmatrix}-4\\-2\end{pmatrix}$	H

3 **a** $\begin{pmatrix}1\\-2\end{pmatrix}$ **b** $\begin{pmatrix}4\\-1\end{pmatrix}$ **c** $\begin{pmatrix}4\\2\end{pmatrix}$ **d** $\begin{pmatrix}-3\\-1\end{pmatrix}$ **e** $\begin{pmatrix}4\\-3\end{pmatrix}$ **f** $\begin{pmatrix}5\\3\end{pmatrix}$ **g** $\begin{pmatrix}-6\\3\end{pmatrix}$ **h** $\begin{pmatrix}-5\\5\end{pmatrix}$ **i** $\begin{pmatrix}-5\\-3\end{pmatrix}$

4 H

5 H

6 **a** B **b** $\begin{pmatrix}-4\\3\end{pmatrix}$

7 **a** $\begin{pmatrix}1\\-3\end{pmatrix}, \begin{pmatrix}-1\\-2\end{pmatrix}, \begin{pmatrix}-4\\-1\end{pmatrix}, \begin{pmatrix}0\\4\end{pmatrix}, \begin{pmatrix}3\\1\end{pmatrix}$ **b** $\begin{pmatrix}-1\\-1\end{pmatrix}$

8 $\begin{pmatrix}6\\0\end{pmatrix}$

Exercise 18.4

1 B: (1, 2), (2, 2), (1, 4)
C: (0, 1), (0, 2), (-2, 1).
D: (-2, -1), (-4, -1), (-2, -2)
E: (2, -1), (3, -1), (2, -3).
c Rotation 90° anticlockwise about (1, 1)
d i Rotation 90° anticlockwise about (0, 1)
ii Reflection in $y = x - 1$

2 B: (2, 2), (2, 3), (3, 4), (3, 2)
C: (0, 3), (0, 4), (-1, 4), (-2, 3)
D: (0, -1), (0, 0), (-1, 1), (-1, -1).

3 B: (-1, 3), (-1, 4), (-3, 4), (-2, 3), (-2, 2)
C: (2, 0), (2, -1), (3, -2), (1, -2), (1, -1)
D: (2, 2), (3, 3), (4, 3), (4, 1), (3, 2)
E: (0, 0), (-1, -1), (-1, -2), (-2, -1), (-2, 0)

4 B: (0, 5), (-1, 5), (-2, 4), (-1, 3), (0, 3), (-1, 4)
C: (1, 2), (2, 1), (3, 2), (3, 1), (2, 0), (1, 1)
D: (0, -1), (1, -2), (1, -3), (0, -2), (-1, -3), (-1, -2).
b They are the same
d They are the same
f They are the same

5 28 cm^2, 23 cm

6 6 cm^2, 12 cm

7 **a** area = 4 $units^2$, perimeter = 10 units
d area = 4 $units^2$, perimeter = 10 units
g They all have the same area and perimeter

Exercise 18.5

1 F: (2, 2), (3, 3), (4, 2), (4, 1)
H: (4, 1), (5, 1), (6, 2), (5, 3)

2 B: (-1, 0), (-2, -1), (-3, -1), (-3, -2), (-2, -2), (-1, -3)
C: (2, -3), (3, -3), (3, -4), (4, -5), (1, -5), (2, -4)
c The area and perimeter are all the same

3 **b i** translation [vector(-2, -2)] **ii** 180° rotation about (2, 0)
iii 180° rotation about (3, 1)

Note: there are a number of answers to the following questions. These are examples.

4 **a** rotation 90° clockwise about (2, 0) **b** reflection in $x = 5$.

5 A rotation of 90° clockwise about (0, 0) followed by a reflection in $x = 3$.

6 **a** [vector (6 0)] **b** reflection in $y = x - 8$

7 **a** reflection in $y = 2$ **b** rotation 90° anticlockwise about (7, 3)

8 A reflection in the x axis and a reflection in $x = 2$, in either order.

Exercise 18.6

1 **b** B

c

Object	Image	SF	Centre
S	Y	2	(4, 8)
Y	S	$\frac{1}{2}$	(4, 8)
X	S	$\frac{1}{2}$	(8, 8)
X	U	$\frac{1}{2}$	(12, 8)

d They are not in the same orientation.
e i Rotate -90° about (8, 8) **ii** Enlarge SF $\frac{1}{2}$ with centre (8, 12)
iii Rotate 180° about (6, 8)

Exercise 18.7

1 A_1: (5, 4), (5, 2), (1, 4)
A_2: (7, 4), (7, 0), (5, 0)
A_3: (5, 6), (9, 6), (9, 4)
A_4: (3, 4), (3, 8), (5, 8)

2 **a** B_1: $(\frac{1}{2}, \frac{1}{2})$, $(2, \frac{1}{2})$, $(2, 1\frac{1}{2})$, $(\frac{1}{2}, 2)$
b i The lengths in B_1 are half those in B.
ii 7.5 $units^2$
iii 1.875 $units^2$
iv lengths are divided by 2, areas are divided by 4

3 **b i** the lengths are divided by 4
ii the area is divided 16
iii the angles are the same

4 **b i** $\frac{1}{3}$ **ii** $\frac{2}{3}$ **iii** $\frac{1}{2}$ **c** 9 $units^2$ **d** 9 times
e The perimeter of C_1 is 3 times the perimeter of A.
f The angles are all the same.

5 **b i** (0, 0) **ii** (-0.5, -0.5) **c** $\frac{2}{3}$ **d** 27 $units^2$ **e** 6.75 $units^2$ **f** 12 units

End points

A1 **a** C, D **b** B, C, D **c** B, C, D

A2 **b** F: (-1, -1), (-1, -3), (-2, -3), (-2, -2), (-4, -2), (-4, -1).
G: (-4, 1), (-4, 2), (-2, 2), (-2, 3), (-1, 3), (-1, 1).
H: (3, 1), (3, -1), (6, -1), (6, 0), (4, 0), (4, 1)
c i rotation 180° about (0, 0)
ii reflection in $x = 0$
iii translation $\begin{pmatrix}-2\\2\end{pmatrix}$
d i reflection in $y = 0$
ii rotation 180° about (1, -1)

A3 B: (1, 1), (1, 3), (2, 2), (3, 2), (3, 1)
C: (1, -7), (1, -3), (5, -3), (5, -5), (3, -5)
D: (-5, -2), (-3, -4), (-1, -4), (-1, -6), (-5, -6)
b 10 $units^2$ **c** 2.5 $units^2$ **d** $\frac{1}{4}$ **e** lengths in B are half those in A
g same area, perimeter, but mirror images **j** $\frac{1}{9}$
k lengths in E are a third of those in A

B1 **a** K: (-1, -3), (-4, -3), (-4, -4), (-2, -4), (-2, -5), (-1, -5)
L: (-1, 1), (-3, 1), (-3, 2), (-2, 2), (-2, 4), (-1, 4)
M: (1, -1), (2, -1), (2, -2), (4, -2), (4, -3), (1, -3)
b i rotation 180° about (0, -1)
ii rotation 90° anticlockwise about (0, 0)
iii translation $\begin{pmatrix}0\\-4\end{pmatrix}$
c i translation $\begin{pmatrix}0\\4\end{pmatrix}$ **ii** reflect in $y = -3$
iii rotation 180° about (1, 0)
d i rotation 90° clockwise about (1, -1)
ii rotation 180° about (0, -3)
iii rotation 180° about (0, -1)

SKILLS BREAK 3A

1 a 22 : 40 b 02 : 20
2 3 hour 40 minutes
3 a 1045 crew b 1600 people
4 61 people per boat
5 £287 970
6 £294 930
7 700 feet
8 a 14 knots b 3 hours 19 minutes c $1\frac{1}{2}$ hours
9 a $C = \frac{5(F-32)}{9}$ b $^-2.2$ °C
10 a 1251 b 3800
11 a 0.61 b 0.43 c 0.25
12 38%
13 70 000 tons
14 £(3.0×10^7)
15 1°
16 200.97 ft
17 58°, because the cable is parallel to the tower.
18 14°, because angles in a triangle sum to 180°.
19 a 2 000 000 b 2 180 000 c 2 180 000 d 2 179 600
20 11 people

SKILLS BREAK 3B

1 3.53 m
2 5.88×102^{21}
3 a 40 080 km b 40 100 km
4 12 756 km
5 a 0.2 m b 3600 mm c 7.9 cm
6 2561 days
7 0.6 m
8 307.7 m^2
9 185 m^3
10 a 14.8 mph b 23.7 km/h
11 18 minutes
12 5.5×10^3
13 2200 cm^3
14 a 20.2 cm b 26 800 cm^3 c 27 kg
15 1.75 m
16 1 : 3
17 a i 107 inches ii 267.5 cm iii 2.68 m
18 a 192 b 2277 c 8 digits per second
19 56 days
20 a 51.3° b 242 m c i 5 ms ii 18 km/h
21 54 seconds
22 a 451.6 cm^3 b 2.2 m^3
23 a i 4 cans ii 9 cans b 576
24 after nearly 10 years
25 a 67666 b 2235962
26 Once, 'a basin plant' is an anagram of 'banana split'.

SECTION 19 Probability B

Starting points

A1 0.23
A2 a 0.77 b 0.16 c 0.55 d 0.44
B1 $\frac{1}{6}$
B2 $\frac{1}{3}$
B3 b $\frac{1}{4}$

Thinking ahead to ...

A a $\frac{5}{9}$ b $\frac{1}{9}$ c $\frac{4}{9}$ d $\frac{1}{3}$

Exercise 19.1

1 a 0.11 b 0.50
2 0.89
3 a 8 b 2 c 10
4 a 0.17
b you don't know how many red, green or blue cubes there are.

Exercise 19.3

1 1 – 0.24 – 0.36 = 0.4
3 a 0.11 b 0.09 c 0.15 d 0.17
4 a 0.12 b 0.10 c 0.22
5 a 195 b 147 c 86 d 53 e 39 f 76

Exercise 19.4

1 a A b A c B d B e D f C g A h B i A j D
2 a The outcomes may not be equally likely, perhaps more people are born in the summer, for instance.
b Conduct a survey of people's birth dates.
3 c
4 a the sample is too small
b people leaving a butcher's shop are unlikely to be vegetarian
c not representative of the population
d not representative of the population
e wouldn't be able to answer
5 a not everybody plays soccer
6 True

Exercise 19.5

1 a $\frac{1}{4}$ b $\frac{7}{16}$ c $\frac{11}{16}$
2 c is the sum of a and b
3 a $\frac{5}{16}$ b $\frac{7}{16}$
4 b $\frac{1}{2}$ c Owen has a moustache and an earring
5 a $\frac{1}{4}$ b $\frac{1}{4}$ c $\frac{3}{8}$

Exercise 19.6

1 b No, some criminals have fair hair and a necklace.
2 a $\frac{11}{16}$ b $\frac{5}{16}$ c $\frac{5}{16}$ d $\frac{1}{8}$ e $\frac{1}{4}$
3 a $\frac{7}{16}$ b $\frac{3}{4}$ c $\frac{3}{4}$ d $\frac{9}{16}$
4 a No, some girls could play hockey and basketball.
b Yes, boys and girls don't overlap.
c Yes, you can't be a driver and a passenger.
d No, tall women can drive buses.
e No, some mechanics can dance.

Exercise 19.7

1 a 0.15 b 0.19 c 0.53
2 0.44
3 a a number can't be less than 5 and greater than 7
b 0.55 c 0.45

End points

A1 a 0.17 b 0.07
A2 0.75
B1 b 0.42,below
C1 a theoretical b relative c relative
D1 a 0.36 b a tree cannot be an oak and a lime
D2 Some trees may have canker and deer damage.

Section 20 Inequalities

Starting points

A1 **a** $34 > 12$ **b** $2 < 15$ **c** $^-12 > {}^-56$ **d** $5 > {}^-2$ **e** $^-2 < 0$ **f** $^-17 < {}^-5$

A2 **a** $56 > 28 > 9 > 5 > {}^-4 > {}^-24 > {}^-30$ **b** $^-30 < {}^-24 < {}^-4 < 5 < 9 < 28 < 56$

A3 12, 5, 6.2

A4 **a** $^-5, {}^-4, {}^-3, {}^-2, {}^-1, 0, 1, 2$ **b** $^-4, {}^-3, {}^-2$

A5 There are no integers between $^-2$ and $^-1$

B1 **a** Straight line passing through (0, $^-2$) and (3, 4).
b Straight line passing through (0, 12) and (4, 8).
c Straight line passing through (0, 0) and (4, 12).
d Straight line passing through (0, $^-4$) and (4, $^-4$).

Exercise 20.1

1 $45 \leqslant T \leqslant 80$

2 **a** $17 \leqslant P \leqslant 52$ **b** $10 < P < 64$ **c** $4 < P < 164$

3 **a** $5 \leqslant A \leqslant 65$ **b** $5 < A \leqslant 65$
c 5 year olds can use Fermatol but not Hypaticain.

4 $^-25 < T < {}^-18$

Exercise 20.2

1 B, C

2 **b** 2, 3, 4

3 **b** $^-1$, 0, 1, 2, 3

4 **a** 5, 6, 7, 8, 9 **b** $^-2, {}^-1, 0, 1, 2, 3$
c $^-3, {}^-2, {}^-1, 0, 1, 2, 3, 4, 5$ **d** 68, 69, 70
e $^-11, {}^-10, {}^-9, {}^-8, {}^-7, {}^-6, {}^-5, {}^-4$ **f** $^-5, {}^-4, {}^-3, {}^-2, {}^-1, 0$
g $^-3, {}^-2, {}^-1, 0, 1, 2, 3, 4, 5, 6$ **h** 2164, 2165, 2166
i $^-2, {}^-1, 0, 1, 2, 3$

5 $^-4 < h < 6$

6 $^-4 < g < 2$, $^-4 < g \leqslant 1$, $^-3 \leqslant g < 2$, $^-3 \leqslant g \leqslant 1$

7 **a** 6, 7 **b** $^-4, {}^-3, {}^-2, {}^-1, 0, 1$ **c** 13, 14 **d** $^-1$, 0
e $^-2, {}^-1, 0, 1, 2, 3, 4$ **f** $^-2, {}^-1$

Exercise 20.3

1 **a** A – too large, E – too heavy **c** B

Exercise 20.4

2 **a** $x < 7$ **b** $5 < y < 12$ **c** $^-1 < x < 9$

4 $^-2 < x < 6$ and $1 < y < 5$

Exercise 20.5

1 **a** $x > 4$ **b** $x \leqslant 4$ **c** $x \leqslant 3$ **d** $x \geqslant {}^-4$ **e** $x \geqslant \frac{1}{2}$ **f** $x \geqslant 9$ **g** $x \leqslant \frac{1}{2}$
h $x < 5$ **i** $x \leqslant 16$ **j** $x \geqslant 15$ **k** $x \geqslant 7$ **l** $x \leqslant 4$

2 **a** $y < 7$ **b** $s \leqslant 8$ **c** $k < 29$ **d** $p \geqslant 6$ **e** $b > \frac{7}{2}$ **f** $x > 3$ **g** $a \leqslant {}^-8$
h $x > 8$ **i** $x < 5$ **j** $t \geqslant 4$ **k** $a \geqslant 7$ **l** $c \geqslant 8$ **m** $d < {}^-5$ **n** $a \leqslant {}^-13$
o $x > 6$

3 **a** $a \geqslant 11$ **b** $k < {}^-2$ **c** $x > {}^-9$ **d** $n \geqslant \frac{7}{2}$

End points

A1 **a** $^-1$ **b** $^-5, {}^-4, {}^-3, {}^-2, {}^-1, 0, 1, 2, 3$ **c** 4, 5, 6, 7
d $^-56, {}^-57, {}^-58, {}^-59$ **e** $^-76, {}^-75, {}^-74, {}^-73, {}^-72$ **f** $^-2, {}^-1, 0, 1, 2, 3$

B4 **a** $2 < x < 7$ **b** $y < 2$, $x > 4$

C1 **a** $x \leqslant 6$ **b** $f > 3$ **c** $a \geqslant 12$ **d** $k > {}^-5$ **e** $a \geqslant 6$ **f** $d > 1.5$

Section 21 Working in 3D

Starting points

A1 **a** cube **b** sphere **c** cylinder **d** cuboid **e** cylinder

A3 They are easy to stack.

B1 **a** **ii** A, C, F **iii** B, D, E
b A cuboid, B tetrahedron, C cylinder, D pentagonal pyramid, E octagonal pyramid, F hexagonal prism

C1 **a** 5, 7 **b** 8, 12 **c** 5, 7

D1 64 cm²

D2 85.5 cm²

D3 **a** 1 cm³ **b** 200 cm³

D4 2 m³

D5 13.5 cm³

Exercise 21.1

1 View 1 is from B, view 2 from A and view 3 from C.

2 B, C, D, G, J

3 A

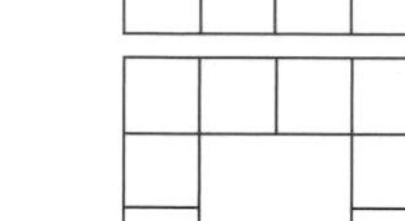

B

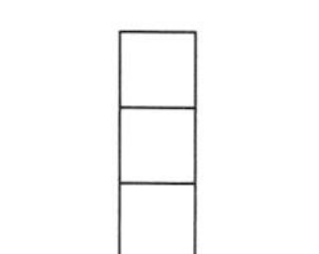

C

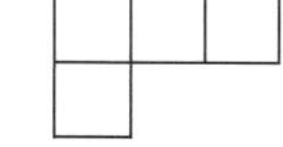

4 **a** 7
b A

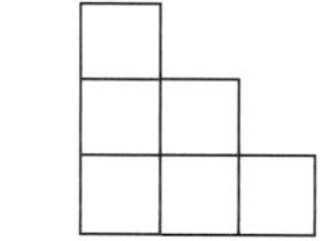

B

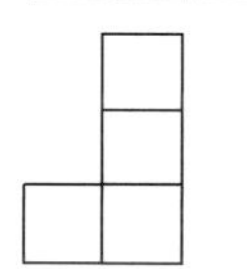

C

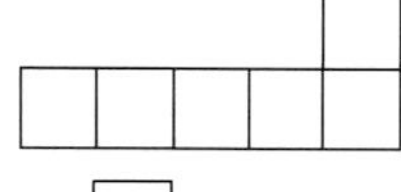

5 **a** 7
b A

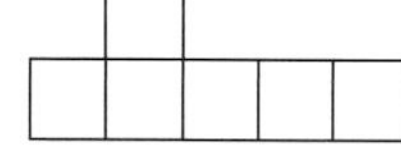

B

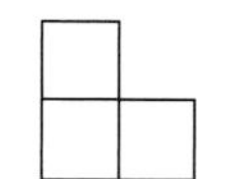

C

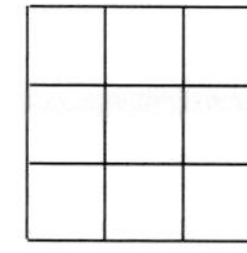

6 A

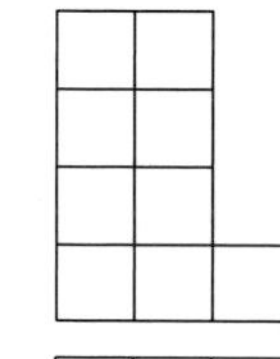

B

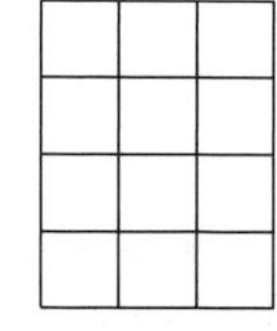

C

Exercise 21.2

1 **a** B and C **b** 1 **c** A

6

Shape	Number of faces	Number of edges	Number of vertices
P	5	9	6
Q	6	12	8
R	7	15	10
S	8	18	12

7 102 faces, 300 edges, 200 vertices, 101 planes of symmetry

8 $n + 2$ faces, $3n$ edges, $2n$ vertices, $n + 1$ planes of symmetry

9 **a i** C, D **ii** A, B
b A is a octagonal pyramid, B is a square pyramid, C is a pentagonal prism, D is a triangular prism.

10 **b** X4, Y2, Z1

11 **a** There is only one more possible net, with the 4 triangles joined together in a line.
b 3

Exercise 21.3

1 **a** 192 cm^3 **b** 216 cm^3 **c** 396 cm^3 **d** 1323 cm^3 **e** 490 cm^3
f 871 cm^3 **g** 180 cm^3 **h** 558 cm^3 **i** 2070 cm^3

2 **a** 200 cm^3 **b** 87.5 cm^3

4 **b** 5 cm × 5 cm × 4 cm

5 **a** 63 cm^2 **b** a = 14, b = 9

6 **a** 23 cm^2 **b** $\sqrt{(23 \div \pi)}$

Exercise 21.4

1 16.5 cm

3 **c** With a radius of 6 cm and a height of 6 cm, the maximum volume is 679 cm^3.

Exercise 21.5

1 **a** 17.0 m^3 **b** 12.7 m^3 **c** 15.3 tonnes

2 **a** 27.2 m^3 **b i** 390 bags **ii** £1996.80

3 **a** 46.25 m^2 **b** 555 m^3 **c** 435 m^3 **d** 435 cm^3

Exercise 21.6

1 **a** 348 cm^2 **b** 120 cm^2 **c** 344 cm^2 **d** 526 cm^3 **e** 2974.5 cm^3

2 7 cm

3 The surface area of B will be 4 times that of A.

Exercise 21.7

1 **a** 300 cm^3 **b** 75 ml **c** 170 g

2 about 75 ml

3 B

4 1.37 litres

5 **a** 21 714 688 cm^3 **b** 21 715 litres

6 217 cm^3, 1.07 g/cm^3; 649 cm^3, 0.952 g/cm^3

7 **A** 1500 cm^3 **B** 720 cm^3 **C** 990 cm^3 **d** 490 cm^3 **E** 440 cm^3

8 By comparing densities: P contains crackers, Q contains Fruit and Fibre, R contains Fruit and Fibre, S contains Trichoc, T contains drinking chocolate

Exercise 21.8

1 **a** length **b** volume **c** area

2 **a** $\frac{1}{2}lz(x + y)$ **b** It is the only expression that represents a volume.

3 **a** $\pi(a + b)$ **b** πab

4 **a** area **b** length **c** length **d** volume **e** volume **f** area

5 The expressions xy and $y(x - z)$ represent areas, but xyz represents a volume.

End points

A1 4

A2 8 faces, 18 edges, 12 vertices

B1 110 cm^3

C1 **a** 6 433 982 cm^3 **b** 6434 litres

C2 19.3g/cm^3

D1 **a** $\pi r(l + r)$ **b** $\frac{1}{3}\pi r^2 h$

SECTION 22 Quadratics

Starting points

A1 **a** $c^2 + 3c - 28$ **b** $a^2 + 8a + 15$ **c** $y^2 - 3y - 40$ **d** $h^2 - 5h - 6$
e $t^2 - 25$ **f** $y^2 - 6y + 9$ **g** $p^2 + 10p + 25$ **h** $k^2 - 11k + 30$

B1 **a** $(n - 1)(n + 5)$ **b** $(y + 2)(y - 5)$ **c** $(k - 1)(k - 6)$ **d** $(h + 4)(h + 5)$
e $(w - 4)(w - 5)$ **f** $(v + 7)(v - 4)$ **g** $(d - 2)(d - 5)$ **h** $(a + 13)(a - 1)$

B2 **a** $(p + 3)^2$ **b** $(m - 5)^2$ **c** $(s - 2)^2$ **d** $(g + 4)^2$

Exercise 22.1

1 **a** $(n + 14)(n - 1) = 0, n = {}^-14$ or 1 **b** $(n + 2)(n - 7) = 0, n = {}^-2$ or 7
c $(n + 1)(n - 14) = 0, n = {}^-1$ or 14 **d** $(b + 5)(b - 3) = 0, b = {}^-5$ or 3
e $(b + 15)(b - 1) = 0, b = {}^-15$ or 1 **f** $(b + 3)(b - 5) = 0, b = {}^-3$ or 5
g $(a + 10)(a - 3) = 0, a = {}^-10$ or 3 **h** $(a + 3)(a + 10) = 0, a = {}^-3$ or ${}^-10$
i $(a + 6)(a - 5) = 0, a = {}^-6$ or 5 **j** $(x + 5)(x - 1) = 0, x = {}^-5$ or 1
k $(x + 1)(x + 5) = 0, x = {}^-1$ or ${}^-5$ **l** $(x - 1)(x - 5) = 0, x = 1$ or 5
m $(y - 3)(y - 5) = 0, y = 3$ or 5 **n** $(y - 1)(y - 15) = 0, y = 1$ or 15
o $(y + 1)(y - 15) = 0, y = {}^-1$ or 15 **p** $(k + 9)(k - 7) = 0, k = {}^-9$ or 7
q $(k - 7)(k - 9) = 0, k = 7$ or 9 **r** $(k + 63)(k - 1) = 0, k = {}^-63$ or 1

2 **a** $(d - 2)(d + 5), d = 2$ or ${}^-5$ **b** $(d + 1)(d - 6), d = {}^-1$ or 6
c $(d + 8)(d - 5), d = {}^-8$ or 5 **d** $(d - 11)(d + 3), d = 11$ or ${}^-3$
e $(d - 11)(d - 3), d = 11$ or 3

3 **a**$(x - 4)^2$ **b** The two factors are the same.

4 **a** $c^2 + 5c - 6 = 0, (c + 6)(c - 1) = 0, c = {}^-6$ or 1
b $y^2 + 6y - 7 = 0, (y + 7)(y - 1) = 0, y = {}^-7$ or 1
c $p^2 - 6p + 9 = 0, (p - 3)(p - 3) = 0, p = 3$ (twice)
d $x^2 - 3x - 10 = 0, (x - 5)(x + 2) = 0, x = 5$ or ${}^-2$
e $w^2 - 4w + 4 = 0, (w - 2)(w - 2) = 0, w = 2$ (twice)
f $b^2 - 15b + 14 = 0, (b - 14)(b - 1) = 0, b = 14$ or 1
g $k^2 - k - 70 = 0, (k + 7)(k - 10) = 0, k = {}^-7$ or 10
h $v^2 + 6v + 9 = 0, (v + 3)(v + 3) = 0, v = {}^-3$ (twice)
i $t^2 - 3t - 28 = 0, (t - 7)(t + 4) = 0, t = 7$ or ${}^-4$
j $y^2 - 7y - 18 = 0, (y - 9)(y + 2) = 0, y = 9$ or ${}^-2$
k $g^2 - 8g + 16 = 0, (g - 4)(g - 4) = 0, g = 4$ (twice)
l $a^2 + 4a - 21 = 0, (a + 7)(a - 3) = 0, a = {}^-7$ or 3
m $p^2 - 3p - 4 = 0, (p - 4)(p + 1) = 0, p = 4$ or ${}^-1$
n $w^2 - 2w - 24 = 0, (w - 6)(w + 4) = 0, w = 6$ or ${}^-4$
o $u^2 + 7u + 10 = 0, (u + 2)(u + 5) = 0, u = {}^-2$ or ${}^-5$

Exercise 22.2

1 **a** ⁻3.4 or 1.4 **b** ⁻1.6 or 3.6 **c** ⁻4.2 or 1.2 **d** ⁻4.4 or 0.4
e ⁻0.3 or 3.3 **f** ⁻5.5 or 0.5 **g** ⁻1.3 or 2.3 **h** ⁻6.3 or 0.3
i ⁻5.5 or 1.5

2 ⁻5.3 or 1.3

3 ⁻0.4 or 5.4

4 **a** ⁻1.35 or 1.85 **b** ⁻1.7 or 1.4 **c** ⁻0.85 or 2.35

5 It can be solved by factorisation

Exercise 22.3

1 **a**

x	⁻5	⁻4	⁻3	⁻2	⁻1	0	1	2	3	4	5
y	25	16	9	4	1	0	1	4	9	16	25

b 25, 0

2 **a**

x	⁻4	⁻3	⁻2	⁻1	0	1	2	3	4
x^2	16	9	4	1	0	1	4	9	16
+1	1	1	1	1	1	1	1	1	1
y	17	10	5	2	1	2	5	10	17

End points

A1 **a** 54° **b** 61°

A2 **a** angle EBD **b** angle CAB

B1 **a** 126° **b** 54°

C1 62°

D1 54°

D2 72°

D3 36°

E1 The four vertices of the quadrilateral lie on the circumference of the same circle.
The angles at opposite vertices are supplementary.

E2 **a** $3k$ must be opposite $6k$, $4k$ must be opposite $5k$.
b $9k = 180$, $k = 20°$

Skills break 4A

1 UK side

2 **a** 2.69 m **b** 4.18 m

3 November

4 73 minutes

5 No

6 9 minutes

7 160 km/h

8 5:27 pm

9 **a** 1:57 pm **b** Calais-Frethun, Paris Nord

10 11:43 am

11 12:44 pm

12 **a** 1:50 pm **b** 6 : 34 pm

13 London

19 July 1995

20 23 383

22 **a** 1.5×10^5 **b** 23513

23 27355

24 5.2×10^6

25 35

26 No, the number of second-class seats cannot be divided exactly by the number of second-class carriages.

27 50

28 **a** 140 km/h **b** 115 km/h

29 105 : 292

30 0.788%

31 19 minutes

32 **a** 6% **b** 18%

33 883

Skills break 4B

1 £9.59

2 £49.41

3 **a** 60 tiles (least amount of boxes) **b** £18.73

4 **a** 6 **b** 2 **c** £20.97

5 28p

6 **b** 1120 **c** 1176

8 **a** hexagon **b** 120° **c** 10 cm

9 **a** He assumes that the tiles will fit exactly.
b $18 \times 11 = 198$ **c** £332.48

10 **a** 3.5 m **b** 2.25 m

11 2.5 litre

12 **a** 4.11 m, 3.61 m **b** £43.47

13 £7.45

14 Forret

15 **a** £16.59 **b** £2.90

16 £1.65

17 £29.96

18 Floral design £6.31 per roll, £5.72 each for 12 rolls or more, Antique embossed £9.19 per roll.

In focus 1

1 B, C, D, E, H

2 **a** $^{-}6 \times 3 = {}^{-}18$ **b** $^{-}4 \times {}^{-}2 = {}^{-}8$ **c** $^{-}14 \div 7 = {}^{-}2$ **d** $^{-}18 \div 6 = {}^{-}3$
e $5 \times {}^{-}4 = {}^{-}20$ **f** $^{-}5 \times {}^{-}7 = 35$ **g** $10 \div {}^{-}2 = {}^{-}5$ **h** $8 \times {}^{-}3 = {}^{-}24$
i $^{-}15 \div {}^{-}3 = 5$ **j** $^{-}4 \times 7 = {}^{-}28$ **k** $^{-}16 \div {}^{-}2 = 8$ **l** $^{-}36 \div 9 = {}^{-}4$

3 **a** 8.54 **b** 9.71 **c** 80.40 **d** 3.62

4 **a** 512 **b** 19.683 **c** 0.125 **d** $^{-}64$ **e** $^{-}5.832$ **f** $^{-}0.027$
g 30 **h** 9 **i** $^{-}1$ **j** $^{-}1.2$

5 **a** 0.5 **b** 0.1 **c** 2.5 **d** $1.\dot{6}$ **e** 3 **f** 6.25 **g** 100 **h** 0.2 **i** $0.08\dot{3}$
j 4.5 **k** $0.\dot{0}\dot{9}$ **l** $3.\dot{6}$

6 **a** $1\frac{1}{2}$ **b** $1\frac{1}{4}$ **c** $\frac{6}{11}$ **d** $\frac{8}{13}$ **e** $\frac{3}{4}$ **f** $\frac{5}{12}$ **g** $\frac{5}{7}$ **h** $\frac{5}{8}$

7

5^{-3}	5^{-2}	5^{-1}	5^0	5^1	5^2	5^3
$\frac{1}{125}$	$\frac{1}{25}$	$\frac{1}{5}$	1	5	25	125

8

6^{-3}	6^{-2}	6^{-1}	6^0	6^1	6^2	6^3
$\frac{1}{216}$	$\frac{1}{36}$	$\frac{1}{6}$	1	6	36	216

9 $7^0 = 1$

10 **a** $\frac{1}{4}$ **b** $\frac{1}{8}$ **c** $\frac{1}{9}$ **d** $\frac{1}{36}$ **e** $\frac{1}{1000}$ **f** $\frac{1}{10000}$
g 1 **h** $\frac{1}{49}$ **i** $\frac{1}{64}$ **j** $\frac{1}{64}$ **k** $\frac{1}{81}$ **l** 1

11 **a** 33 **b** 12.5 **c** 15.75 **d** 0.027 **e** 104 **f** 180 **g** 72.9 **h** 6200

12 **a** 0.2 **b** 0.25 **c** 0.125 **d** 0.01 **e** 0.015625 **f** $0.\dot{1}$
g 0.015626 **h** 0.008 **i** 0.001 **j** 0.01 **k** $0.02\dot{7}$ **l** 0.00001

13 **a** 6.17×10^6 **b** 9.2×10^{10} **c** 3.07×10^5
d 2.5×10^{-5} **e** 2.603×10^{-4} **f** 1.0×10^{-7}

14 **a** 450 000 000 **b** 3 606 000 000 000 **c** 0.000 000 001 24
d 70 000 000 **e** 0.000 000 000 051 **f** 0.000 006 104 7

15 **a** 1.305×10^{11} **b** 1.2384×10^{-3} **c** 8.7×10^{-8} **d** 8.55×10^{10}
e 1.904×10^7 **f** 7.3×10^{-14} **g** 3.47×10^{15} **h** 4.1×10^9

16 **a** 4^{-4} **b** 3^6 **c** 5^4 **d** 2^8 **e** 5^7 **f** 7^{-6} **g** 4^{-9} **h** 6^5

17 **a** $3^8 \times 3^{-2} = 3^6$ **b** $6^5 \div 6^3 = 6^2$ **c** $2^{-4} \div 2^3 = 2^{-7}$ **d** $8^5 \times 8^{-3} = 8^2$
e $3^{-2} \div 3^{-7} = 3^5$ **f** $7^{-5} \times 7^{-3} = 7^{-8}$

18 **a** 2×10^3 **b** 4×10^5 **c** 7×10^{-8} **d** 5×10^{-7}

19 **a** $\frac{11}{15}$ **b** $1\frac{7}{30}$ **c** $1\frac{7}{12}$ **d** $\frac{1}{15}$ **e** $\frac{7}{20}$ **f** $\frac{1}{2}$ **g** $4\frac{5}{6}$ **h** $2\frac{1}{2}$ **i** $4\frac{3}{10}$ **j** $\frac{2}{15}$

20 **a** $\frac{2}{15}$ **b** $\frac{1}{4}$ **c** $\frac{1}{3}$ **d** $\frac{8}{15}$ **e** $\frac{2}{5}$ **f** $1\frac{1}{9}$ **g** $\frac{18}{25}$
h $1\frac{7}{18}$ **i** $4\frac{1}{2}$ **j** $\frac{3}{10}$ **k** $1\frac{9}{10}$ **l** $3\frac{4}{5}$

21 **a** $2\frac{1}{4}$ **b** $6\frac{2}{3}$ **c** 4 **c** 24 **e** 16 **f** $7\frac{1}{2}$
g $3\frac{1}{3}$ **h** 38 **i** $7\frac{1}{5}$ **j** 3 **k** 20 **l** $6\frac{2}{5}$

22 **a** 36 **b** 24 **c** 240 **d** 6300 **e** 700

23 **a** 9 **b** 14 **c** 5 **d** 12 **e** 7

In focus 2

1 **b** $m = 2n + 1$, $m = 4n$, $m = 12n - 8$
d 21 matches, 40 matches, 112 matches

2 **a** $20 \rightarrow 23$
$n \rightarrow n + 3$
b $40 \rightarrow 79$
$t \rightarrow 2t - 1$
c $2 \rightarrow 11$
$3 \rightarrow 15$
$4 \rightarrow 19$
$25 \rightarrow 103$
d $100 \rightarrow 902$ $p \rightarrow 9p + 2$

3 **a** $n \rightarrow 9n - 5$

4 $25 \rightarrow 629$ $n \rightarrow n^2 + 4$

5 **a** A $3n - 2$ **B** $3n + 2$ **C** $6n + 1$ **D** $5n - 1$ **E** $5n + 2$
F $4n - 2$ **G** $6n - 3$ **H** $8n + 1$
b A 88 **b** 92 **c** 181 **d** 149 **e** 152 **f** 118 **g** 177 **h** 241

6 **a** $5p + 10$ **b** $8n - 24$ **c** $4c + 20$ **d** $6r + 2$ **e** $35 + 5x$ **f** $14s - 21$
g $3 - 15t$ **h** $6m - 12$ **i** $100k - 20$ **j** $32 - 8z$ **k** $50 - 20t$ **l** $4a + 4b$
m $6g + 3h$ **n** $5x - 5y$ **o** $21c - 14d$

IN FOCUS 3

1 **a** 47°, 133°, 47°, 133° **b** 51.5°, 128.5°, 51.5°, 128.5°
c 11°, 11°, 169°, 11°

2 **a** 34°, 73°, 73° **b** isosceles
c **i** 34°, 146°, 34°, 146° **ii** 73°, 107°, 73°, 107°
d All sides are of equal length; opposite sides are parallel.
e Only pairs of opposite sides are equal in length.
f **i** 90° **ii** 73° **iii** 73° **iv** 17° **v** 17°

3 **a** For an n-sided polygon, the sum of the exterior angles is $360° \div n$.

4 $360° \div 8 = 45°$

5 $a = 73°$, $b = 46°$, $c = 92°$, $d = 120°$, $e = 17°$, $f = 72°$, $g = 198°$, $h = 275°$,
$i = 120°$, $j = 60°$, $k = 60°$, $l = 60°$, $m = 36°$

IN FOCUS 4

1 A: 4, B: $\frac{3}{2}$, C: $^{-}2$, D: 1

3 **a** $y = 2x + 1$

x	$^{-}1$	0	1	2	3	4
y	$^{-}1$	1	3	5	7	9

$y = 3x - 2$

x	$^{-}1$	0	1	2	3	4
y	$^{-}5$	$^{-}2$	1	4	7	10

$y = x - 2$

x	$^{-}1$	0	1	2	3	4
y	$^{-}3$	$^{-}2$	$^{-}1$	0	1	2

$y = 2x - 3$

x	$^{-}1$	0	1	2	3	4
y	$^{-}5$	$^{-}3$	$^{-}1$	1	3	5

$y = x + 2$

x	$^{-}1$	0	1	2	3	4
y	1	2	3	4	5	6

$y = x$

x	$^{-}1$	0	1	2	3	4
y	$^{-}1$	0	1	2	3	4

4 **a** 4, $^{-}3$ **b** 2, 1 **c** $^{-}3$, 2 **d** 1, 0 **e** $^{-}1$, 0 **f** 5, 2 **g** $^{-}2$, $^{-}1$ **h** 1, 1
i 1.5, 2 **j** 1, 0.5 **k** $^{-}0.5$, 3 **l** 0, 1 **m** $\frac{1}{2}$, 0 **n** $\frac{3}{4}$, $^{-}1$ **o** 1, $\frac{^{-}3}{5}$
p $^{-}1$, 1.6 **q** 1, 1.2 **r** $^{-}3$, 0

5 A: $y = 2x - 8$, B: $y = 3x + 4$, C: $y = \frac{3}{4}x - 1$,
D: $y = 3 - 5x$, E: $y = {}^{-}7x$

6 **a** $\frac{^{-}1}{2}$ **b** $\frac{^{-}1}{3}$ **c** $\frac{1}{3}$ **d** $^{-}1$ **e** 2 **f** $\frac{^{-}1}{5}$

7 **a** $y = 2x - 4$ **b** $y = 2x + 3$ **c** $y = 2x - 3$ **d** $y = 2x + \frac{1}{2}$
e $y = \frac{2}{3}x + 2$ **f** $y = \frac{5}{2}x + \frac{7}{2}$ **g** $y = \frac{3}{4}x + \frac{1}{4}$ **h** $y = 2x - 1$

11 A: $y = \frac{4}{3}x + 1$, B: $y = 3x - 2$, C: $y = \frac{1}{2}x + 2$,
D: $y = \frac{2}{3}x$, E: $y = \frac{1}{2}x - 1$, F: $y = {}^{-}2$

12 a, b, c, d, e, f, g

IN FOCUS 5

1 Children in Care: 222°, 84°, 36°, 18°
Child Safety: 234°, 72°, 36°, 18°

2 Chris: 8, Dani: 7, Sam: 4

3 **b** Bella: 6, Emily: 7, Gavin: 7.5

4 Bella: 5.4, Emily: 6.7, Gavin: 7.2

5 Asif: mode = 8, range = 6, Suki: mode = 5, range = 4

6 Kurt: median = 5, range = 6, Jean: median = 7, range = 5

7 Sally: mean = 5.9, range = 7, Gavin: mean = 6.4, range = 7

8 Peggy: mean = 5.7, median = 6, range = 6
Toby: mean = 5.5, median = 5.5, range = 6
Sue: mean = 6.2, median = 7, range = 7

9 **a** P: 36, Q: 38, R: 38.5
b P: 35, 38, Q: 35.5, 39, R: 37, 40

IN FOCUS 6

1 39.4 cm², 96.4 cm², 12.6 cm², 77.2 cm², 109.6 cm²

2 **a** 3000 mm² **b** 154 905 mm² **c** 161 345 mm²

3 10.2 cm

4 **a** 20.4 cm, 33.2 cm² **b** 30.2 cm, 72.4 cm² **c** 31.9 cm, 60.4 cm²

5 26.25π

6 **a** 73.5 cm² **b** 35.8 cm

7 **a** 126.6 mm **b** 12 655 mm² **c** 395 **d** 5 251 275 mm²

IN FOCUS 7

1 **a** $6k$ **b** $6l + m$ **c** $5n - 7p$ **d** $4q + 6r$ **e** s **f** $9 + 5u + 7v$
g $6w + 2x$ **h** $11 - 3y$ **i** $5z + 2$

2 **a & b**

31	18	35
32	28	24
21	41	25

c no **e** Change bottom middle to $5g + h$

3 **a** $x = 5$ **b** $x = \frac{9}{2}$ **c** $x = \frac{19}{5}$ **d** $x = {}^{-}2$ **e** $x = 4$ **f** $x = {}^{-}3$ **g** $x = 2$
h $x = \frac{3}{2}$ **i** $x = \frac{7}{2}$ **j** $x = {}^{-}1$ **k** $x = 6$ **l** $x = 5$ **m** $x = -\frac{5}{2}$ **n** $x = 9$
o $x = {}^{-}1$ **p** $x = 2$

4 **a** A: 56 cm, B: 54 cm, C: 106 cm, D: 56 cm
b A: $8p$ cm, B: $(6p + 12)$ cm, C: $(16p - 6)$ cm, D: $8p$ cm
c 3 cm **d** 2.5 cm **e** 1.5 cm **f** 6 cm
g 0.75 cm **h** 1.8 cm **i** They are both $8p$ cm

5 A $2n - 3 = 12 - n$, $n = 5$
B $2(n + 1) = 3(n - 5)$, $n = 17$

6 A $x = 7$, $y = 1$ B $x = 1.5$, $y = 3$
C $x = {}^{-}2$, $y = 5$ D $x = 2.8$, $y = 6.8$

7 A $x = 1.5$, $y = 3$ B $x = 12$, $y = 1$
C $x = 10$, $y = {}^{-}2$ D $x = {}^{-}1$, $y = 2.5$
E $x = 2$, $y = 12$ F $x = 4.5$, $y = 2$
G $x = 5$, $y = {}^{-}1$ H $x = 1$, $y = 1$
I $x = 3$, $y = 2.5$ J $x = 1$, $y = {}^{-}5$

IN FOCUS 8

1 **a** 5.67 **b** 12.7 **c** 2140 **d** 534.69 **e** 35 **f** 3000 **g** 13 468.28
h 0.07 **i** 60 **j** 60.0 **k** 9000 **l** 68 **m** 56.290

2 **a** 56.8 **b** 16 400 **c** 2.5 **d** 50 **e** 15.78 **f** 730 000 **g** 90
h 93 748 000 **i** 564.2 **j** 200 **k** 6.8 **l** 15.68 **m** 673 500 **n** 0.036
o 0.0280 **p** 3.00

3 **a** 40 000 **b** 200 **c** 54 000 **d** 250 000 **e** 350 **f** 7.5 **g** 120 000
h 120 **i** 5000 **j** 150 **k** 350 **l** 3500 **m** 350

4 **a** 1200, 1216.9 **b** 20 000, 18 218.8 **c** 2000, 2381.721
d 1000, 172.232 **e** 40 000, 32 421.57 **f** 0.15, 0.1863
g 120, 110.88 **h** 1000, 965.0772

5 **a** 12 000 m² (400 × 30) **b** 200 m² (20 × 10)
c 2 100 000 m² (7000 × 300)

IN FOCUS 9

1 **a** $\frac{1}{6}$ **b** $\frac{1}{2}$ **c** $\frac{1}{3}$ **d** $\frac{1}{2}$

2 **a** $\frac{2}{5}$ **b** $\frac{3}{10}$ **c** $\frac{1}{2}$ **d** $\frac{1}{10}$ **e** $\frac{9}{10}$ **f** 0 **g** $\frac{7}{10}$

3 probability of not B, for example

4

	1	2	3	4	5	6	7	8
1	1, 1	1, 2	1, 3	1, 4	1, 5	1, 6	1, 7	1, 8
2	2, 1	2, 2	2, 3	2, 4	2, 5	2, 6	2, 7	2, 8
3	3, 1	3, 2	3, 3	3, 4	3, 5	3, 6	3, 7	3, 8
4	4, 1	4, 2	4, 3	4, 4	4, 5	4, 6	4, 7	4, 8

5 **a** $\frac{11}{32}$ **b** $\frac{1}{8}$ **c** $\frac{1}{8}$ **d** $\frac{1}{32}$ **e** $\frac{1}{16}$ **f** $\frac{1}{4}$ **g** $\frac{1}{4}$ **h** $\frac{5}{16}$ **i** $\frac{9}{16}$ **j** $\frac{11}{32}$ **k** $\frac{1}{8}$ **l** $\frac{7}{8}$

7 **a** $\frac{1}{8}$ **b** $\frac{3}{4}$ **c** $\frac{1}{2}$ **d** $\frac{1}{8}$ **e** $\frac{1}{4}$ **f** 0

8 **a** $\frac{7}{15}$ **b** $\frac{1}{30}$ **c** $\frac{2}{15}$ **d** $\frac{7}{30}$ **e** $\frac{1}{15}$ **f** $\frac{1}{15}$

9 **a** PQRS, PQSR, PRQS, PRSQ, PSQR, PSRQ, QPRS, QPSR, QRPS, QRSP, QSPR, QSRP, RPQS, RPSQ, RQPS, RQSP, RSPQ, RSQP, SPQR, SPRQ, SQPR, SQRP, SRPQ, SRQP

10 **a** PQ, PR, PS, PT, QR, QS, QT, RS, RT, ST

IN FOCUS 10

1 **a** $(2b + 3.8)$m **b** $(2a + 4b + 17.4)$m **c** 37.4 m

2 **a** $(8p + 20)\text{cm}^2$ **b** 71.2 cm^2 **c** 4.25 cm

3 A and D, B and I, C and H

4 **a** $3a + 18$ **b** $5z + 30$ **c** $3n - 12$ **d** $6n + 12$ **e** $x^2 + 2x$ **f** $4p^2 - 12p$ **g** $3b^2 + 9b$ **h** $2a^2 - 8a$ **i** $12p^2 - 16p$

5 **a** $(w + 8)$cm **b** $(4w + 16)$cm **c** 14 cm

6 300 cm^2

7 **a** $48pq$ **b** $9y^2$ **c** m^2n^2 **d** $4p^2q$ **e** $8m^2n$ **f** $18a^2b$ **g** a^3g^3 **h** $20p^3$ **i** $28a^3b$

8 **a** $24a - 18b$ **b** $pn - p^2$ **c** $s^2 + st$ **d** $6mn - 6m^2$ **e** $5x^2 + 5xy$ **f** $4u^2 + 3u$ **g** $5pq + 3pr$ **h** $12mn + 21n^2$ **i** $27ab + 9a^2$

9 D

10 **a** $7m + 2n$ **b** $21a^2 - a$ **c** $2b$ **d** $2xy + 8x - 2y$ **e** $4p + 3q$ **f** $8p^2 + 5q^2 + 2pq$ **g** $4m^2 + 2mn - 6n$ **h** $2x - 2xy$

11 **a** $56a - 5b$ **b** $19x + 13y$ **c** $16x^2 + 4x$ **d** $30x^2 + 13xy$

12 A: $2c + 3$, B: $3 + 4b$

13 **a** $p(7 + 3q)$ **b** $m(1 + 3n)$ **c** $4q(2p + 1)$ **d** $3y(1 - 4x)$ **e** $2b(b + 5)$ **f** $6b(3a + 4c)$ **g** $xy(x - 2)$ **h** $3ab(3a + 2b)$ **i** $3mn(4m - 3)$

IN FOCUS 12

1

Dye	Blue	Yellow	Total
Grass	120 ml	360 ml	480 ml
Lime	450 ml	1050 ml	1500 ml
Pea	200 ml	360 ml	560 ml
Lime	75 ml	175 ml	250 ml
Pea	600 ml	1080 ml	1680 ml
Grass	170 ml	510 ml	680 ml
Lime	645 ml	1505 ml	2150 ml
Pea	275 ml	495 ml	770 ml
Grass	240 ml	720 ml	960 ml
Lime	360 ml	840 ml	1200 ml
Grass	215 ml	645 ml	860 ml
Pea	250 ml	450 ml	700 ml
Grass	320 ml	960 ml	1280 ml
Lime	267 ml	623 ml	890 ml
Pea	375 ml	675 ml	1050 ml

2 **a** £24, £60 **b** £155, £93 **c** 135 g, 60 g **d** £40, £20, £60 **e** 75 g, 100 g, 25 g **f** 80 ml, 120 ml, 160 ml **g** 120 g, 180 g, 120 g **h** £55, £55, £385 **i** 36 cm, 45 cm, 18 cm **j** 1.25 m, 0.5 m, 0.25 m **k** 0.2 kg, 0.4 kg, 0.8 kg **l** 0.3 cm, 1.8 cm, 0.6 cm

3 **a** 90 g icing sugar, 405 g margarine, 180 g plain chocolate, 405 g flour
b 20 g icing sugar, 90 g margarine, 40 g plain chocolate, 90 g flour

4 1 : 2, 2 : 3, 2 : 1

5 $\frac{2}{7}$, $\frac{3}{4}$, $\frac{4}{7}$, $\frac{1}{4}$, $\frac{3}{8}$

6 **a** 1.4 **b** 7.7 cm

7 **a** 0.75 **b** 5.7 cm

8 **a** £52 770, £79 155, £105 540, £158 310
b They should add up to the amount won

9 **a** 10.5 cm **b** 17.5 cm **c** 183.75 cm^2

10 **a** 5 : 11 : 4
b 281.25 ml, 618.75 ml
c 437.5 ml, 962.5 ml, 350 ml

IN FOCUS 13

1 **a** 5.2 m **b** 13 m **c** 24.18 m **d** 156 m **e** 566.8 m

2 604 mph

3 43.7 seconds

4 0.7

5 **a** 2.15 m/s **b** 2000 m/s
c 0.12 m/s **d** 5.6 m/s

6 **a** 120 km/h **b** 40 km/h **c** 65.1 km/h **d** 21.6 km/h

7 **a** 1 mile per minute **b** 0.5 miles per minute
c 1.5 miles per minute **d** 0.75 miles per minute
e 0.42 miles per minute **f** 0.7 miles per minute

8 **a** 36 mph **b** 72 mph **c** 27 mph **d** 360 mph

9 3120 km

10 **a** 76.5 miles **b** 38.25 miles **c** 12.75 miles **d** 68 miles
e 136 miles **f** 191.25 miles

11 **a** 30 minutes **b** 2 hours **c** 48 minutes **d** 6 hours 40 minutes
e 1 hour 5 minutes **f** 1 hour 13 minutes **g** 1 hour 12 minutes
h 24 minutes

12 The less steep line as the truck would travel slower than the motorcycle.

13 **a** 45 km/h **b** 38.33 km/h **c** 48 mph **d** 85.71 km/h
e 52.5 mph **f** 83.33 km/h **g** 53.33 mph **h** 30 km/h

14 **b** 12 noon

15 **b** 6:15 pm **c** 66.7 mph **d** 60 mph

IN FOCUS 14

1 **a** 56° **b** 68° **c** 61° **d** 47° **e** 13° **f** 51°

2 **a** 4.7 cm **b** 10.5 cm **c** 5.1 cm **d** 16.0 cm **e** 1.6 cm **f** 7.4 cm **g** 1.6 cm

3 **a** 61° **b** 37° **c** 68° **d** 13° **e** 45° **f** 34°

4 **a** **i** 64.3° **ii** 61.2° **iii** 32.1°
b **i** 38.6° **ii** 35.0°
c **i** 2.93 cm **ii** 145 cm **iii** 2.49 cm
d **i** 21.0 cm **ii** 4.95 cm

5 **A** 27.30° **B** 30.47° **C** 25.62°

6 **a** 189 mm **b** 433 mm

IN FOCUS 15

1 **a** 6 **b** $p = t - 1$ **c** 2

2 **a** $x = \frac{P}{4}$ **b** 10.25 cm

3 **a** 270 mins **b** $W = \frac{(T - 20)}{20}$ **c** 5 pounds

4 **a** 8.75 pints **b** $l = \frac{4}{7p}$ **c** 4.86 litres

5 **a** $w = \frac{A}{l}$ **b** 6.2 cm

6 **a** $w = \frac{(p-2l)}{2}$ **b** 4.2 cm

7 **a** $k = \frac{(V+25)}{4}$ **b** $k = \frac{y}{3}+1$ **c** $k=7j$ **d** $k=\frac{4}{3}w$ **e** $k=5-2v$
f $k=b+7$ **g** $k=\frac{5}{2}(d-1)$ **h** $k=8h-3$ **i** $k=p-w$ **j** $k=\frac{(20-q)}{5}$
k $k=3(13-v)$ **l** $k=\frac{(v-s)}{t}$ **m** $k=bh-a$ **n** $k=\frac{h}{(2g)}$ **o** $k=\frac{(3-2y)}{5}$
p $k=\frac{(9+2h)}{3}$

8 **a** $p=\sqrt{A}$ **b** $p=\sqrt{(\frac{y}{5})}$ **c** $p=\sqrt{(g-1)}$ **d** $p=\sqrt{\left(\frac{(q+5)}{3}\right)}$ **e** $p=\sqrt{(4j)}$
f $p=\sqrt{(\frac{3F}{2})}$ **g** $p=\sqrt{(2-z)}$ **h** $p=\sqrt{(\frac{K}{2\pi})}$ **i** $p=\sqrt{\left(\frac{(A-1)}{2}\right)}$
j $p=\sqrt{\left(\frac{(H+8)}{3}\right)}$

9 **a** 80, 45, 85, 28 **b** 35, 12, 22, 4 **c** 19.25, 2.25, 4, $^-1.25$

10 **a** 35 **b** 16 **c** ±7 **d** 38 **e** 225 **f** 11 **g** 1.1 **h** $\frac{7}{10}$

11 **a** $n^2+7n+10$ **b** $f^2+11f+10$ **c** $w^2+8w+12$ **d** $2n^2+7n+3$
e $4d^2+19d+21$ **f** $4y^2+28y+45$ **g** $x^2+3x-18$ **h** v^2+v-6
i $b^2-2b-15$ **j** m^2-8m+7 **k** h^2-6h+5 **l** $t^2-10t+21$
m $6p^2+p-1$ **n** $15x^2+8x-12$ **o** $12y^2-24y-15$
p $12g^2-43g+35$

12 **a** $(x+3)(x+4)$ **b** $(x+2)(x+6)$ **c** $(x+1)(x+6)$ **d** $(x+2)(x+3)$
e $(x+1)(x+12)$ **f** $(x+2)(x+4)$

13 **a** $(x+1)(x+11)$ **b** $(x+3)(x+5)$ **c** $(x+2)(x+2)$ **d** $(x+4)(x+5)$
e $(x+4)(x-1)$ **f** $(x+5)(x-3)$ **g** $(x+6)(x-2)$ **h** $(x+2)(x-3)$
i $(x-1)(x-10)$ **j** $(x+2)(x-4)$ **k** $(x-1)(x-2)$ **l** $(x-2)(x-2)$

IN FOCUS 16

1 **a** bars with heights 1, 4, 1, 4, 5, 4, 2, 3
b bars with heights 1, 2, 2, 2, 5, 7, 5

2 **a** 0.160– **b** 0.170–

4 **a** 0.07 seconds **b** 0.06 seconds

5 **a** 0.163 **b** 0.165

6 There are the same amount of men as women.

8 **a** 3 **b** 2

9 **a** 7 **b** 8

10 **a** 0.165 **b** 0.17

11 **a** 0.027 **b** 0.023

12 The men are on average faster, but the women were closer together

13 **a** bars with heights 7, 27, 20, 3, 2
b bars with heights 2, 10, 7, 2, 2

14 Different totals of athletes

15 **a** 27.1 years **b** 28.3 years

16 **b** 20 minutes (C), 40 minutes (D), 7 minutes (E)
c 86.2 minutes (C), 239 minutes (D), 45.1 minutes (E)
e 85.5 minutes (C), 238 minutes (D), 45.4 minutes (E)
f 5 minutes (C), 15 minutes (D), 2.6 minutes (E)

17 **a** 34, 33, 29, 28, 25, 22, 24, 27, 26

IN FOCUS 17

1 **a** 63.33% **b** 25.45% **c** 283.33% **d** 62% **e** 4.67% **f** 31.25%
g 83.67% **h** 15.21% **i** 16.62% **j** 1.04%

2 **a** 29.5 kg **b** 2310 km **c** 3702.4 miles **d** 1822.5 yards **e** £27 183.70
f 145 062.5 tonnes **g** 1.05 cm **h** 47.25 mm **i** £38.51 **j** 419.75 ml

3 **a** 426.8 ml **b** £36.6 **c** 22.75 mm **d** 0.74 cm **e** 3968.75 tonnes
f £10 523.04 **g** 1267.2 yards **h** 2500.08 miles **i** 2322 km
j 24.96 kg

4 **a** £218.37 **b** £13.74 **c** £52.71 **d** £25.56 **e** £315.55 **f** £187.99
g £159.07 **h** £10.86 **i** £58.74 **j** £1761.33

5 £155.10

6 **a** Price busters **b** They are cheaper by 25p

7 **a** £405 **b** £4216 **c** £20.40 **d** £23.40 **e** £810 **f** £2300 **g** £3.75
h £40.50 **i** £91.60 **j** £94.50

8 **a** £95.51 **b** £582.54 **c** £38.91 **d** £13.18 **e** £7011.2 **f** £3.96
g £578.67 **h** £1355.71 **i** £32.45 **j** £70.58

9 **a** £255.31 **b** £11.06 **c** £38.29 **d** £8.50 **e** £63.83 **f** £76.55
g £22.97 **h** £110.63 **i** £16.16 **j** £119.14

10 **a** £100.53 **b** £15.08

11 £3 157 895

IN FOCUS 18

2 **a** shape with vertices at $(^-2, 3)$, $(2, 1)$, $(2, ^-1)$, $(^-4, ^-1)$, $(^-4, 1)$
b shape with vertices at $(4\frac{1}{2}, 0)$, $(5\frac{1}{2}, ^-\frac{1}{2})$, $(5\frac{1}{2}, ^-1)$, $(4, ^-1)$, $(4, ^-\frac{1}{2})$.

3 **b** **i** Rotation 90° anticlockwise about (1, 1)
ii Reflection in $x = 5$
c **i** C **ii** F **iii** E
d Rotation 180° about $(4, ^-1)$

4 **c** **i** Rotation 180° about (4, 0)
ii Rotation 90° clockwise about (3, 0)
iii Translation $\begin{pmatrix}2\\2\end{pmatrix}$
d Rotation 180° about (1, 0)

5 **b** **i** Rotation 180° about $(^-0.5, 1.5)$
ii Translation $\begin{pmatrix}^-14\\0\end{pmatrix}$

IN FOCUS 19

1 **a** 0.08 **b** 0.78 **c** 0.03 **d** 0.11

2 They represent all the outcomes.

3 **a** 0.92 **b** 0.18 **c** 0.11

4 No, peg down isn't as likely as peg up.

5 **a** 12 **b** 117 **c** 5

7 **b** 0.08 **c** 0.48 **d** 0.52

8 **a** Relative
b Relative
c Theoretical
d Relative
e Relative
f Theoretical
g Theoretical
h Relative

9 **a** 0.19
b No two outcomes can happen at the same time

10 **b** 15 and 30 are both multiples of 3 and 5

11 **a** 0.28 **b** 0.82 **c** 0.18 **d** 0.4

12 0.67

IN FOCUS 20

1 **a** $9 > ^-4$ **b** $^-8 < ^-3$ **c** $4 > ^-2$ **d** $^-2 < 2$ **e** $^-5.6 < ^-4.23$
f $^-8 > ^-11$

2 4, 71, $^-3.2$, $^-0.33$

3 **a** 4, 5, 6, 7, 8 **b** $^-3$, $^-2$, $^-1$ **c** $^-5$, $^-4$, $^-3$, $^-2$, $^-1$, 0 **d** $^-6$, $^-5$, $^-4$
e none **f** 24 **g** $^-14$, $^-15$ **h** 1, 2, 3, 4, 5

5 The number of waiters has to be an integer, the weights do not

7 **a** $x \geqslant 9$ **b** $t < ^-7$ **c** $p \leqslant 10.5$ **d** $^-11 \leqslant k \leqslant 11$ **e** $w < 2$ **f** $q > 9.3$
g $t \geqslant 8$ **h** $h \leqslant 2$ **i** $f < 3$ **j** $^-9 < c < 9$ **k** $g \geqslant 4.5$ **l** $x > ^-3$
m $j < 19$ **n** $d \geqslant ^-2$ **o** $v < 4\frac{1}{2}$ **p** $x \leqslant ^-13$ **q** $s < 0$ **r** $u \leqslant ^-3$

8 **a** 0 **b** 16

11 $x + y \leqslant 7, x < 2$

IN FOCUS 21

1 A: 2, B: 5

3 **a** 3 **b** 9

4 **b** 48 cm^3 **c** 88 cm^2

5 X: 78.6 cm^3, Y: 18.4 cm^3

6 a = 19.8 cm, b = 59.0 cm

7 **a** R: 3927.8 ml, S: 1247.7 ml, T: 74.8 ml, U: 1 231 504.3 ml
b R: 3.9 litres, S: 1.2 litres, T: 0.7 litres, U: 1231.5 litres

8 **a** length **b** area **c** area **d** volume **e** length **f** area
g none **h** volume **i** length **j** area **k** none **l** volume
m area **n** volume **o** area **p** length

In Focus 22

1 **a** -3, -2 **b** -5, 4 **c** -2, 8 **d** -9, 6 **e** -5, 9 **f** -8, 5 **g** 2, 5
h -12, 3 **i** -5, -3 **j** 3, 6 **k** -12, 6 **l** -3, 10 **m** -9, 4 **n** -12, 4
o 3, 9 **p** -11, -7 **q** -8, 9 **r** -7, 6 **s** -8, -7 **t** -5, 12

2 **a** 3.4 or –2.4 **b** 1.5 **c** 3.2 or –1.2 **d** 1.8 **e** 1.9 or –1.4 **f** 1.6

3 **a**

x	-2	-1	0	1	2	3	4	5
y	12	5	0	-3	-4	-3	0	5

5 **a**

x	0	1	2	3	4	5
y	4	0	-2	-2	0	4

6 **a**

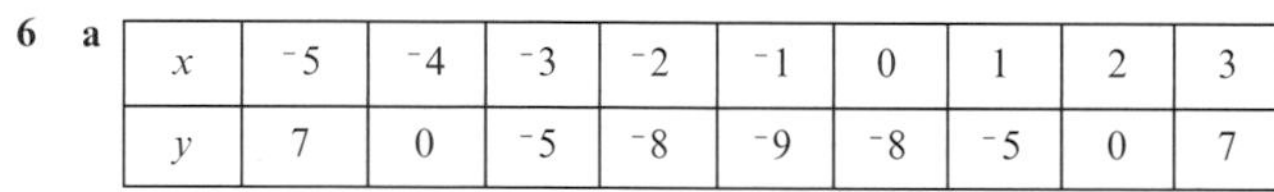

x	-5	-4	-3	-2	-1	0	1	2	3
y	7	0	-5	-8	-9	-8	-5	0	7

c -4, 2

7 **b** -6, 3

8 **b** 1, 5 **d** 0.6, 5.4

9 **b** -7, 3 **d** -6.5, 2.5 **e** -7.4, 3.4

14 **c & d** -0.4, 2.4

17 1.82

In Focus 23

1 Not clear, leading, ambiguous.

7 **d** Moderate positive correlation

8 **a** £7300 **b** £4700

9 **a** 350 cc **b** 680 cc

10 **a** £8900 **b** 1160 cc

11 The vertical scale doesn't start at zero

12 The area of the large video box is about twice as large as the small one

13 **a** The number of injuries reduced steadily each year
d There were 17 or 18

In Focus 24

1 **a** 3.04 cm **b** 30.1° **c** 5.73 cm **d** 6.07 cm

2 A: 6.51 cm, 2.04 cm², B: 14.52 cm, 12.3 cm²

3 **a** **i** 4.20 m **ii** 6.02 m **b** 19°

4 **a** **i** 5 km **ii** 2 km **b** **i** 135° **ii** 068° **iii** 277°
c **i** 5.39 km **ii** 22° **d** 4.24 km **e** 8.06 km

5 **a** 1.1 m
b 27.9°

In Focus 25

1 **b** CD is a diameter

2 **b** 68°

3 22°

4 **a** 109° **b** 33°

5 **b** 96°

6 **b** 110°

7 **b** 75°

Exam Questions

N1.1 £547.50

N1.2 Medium

N1.3 The small jar

N1.4 25 rock cakes

N1.6 The small bottle.

N1.7 **a** 375 g **b** 360 buns

N2.1 **a** 108 **b** 1, 3, 37, 111 **c** 103 **d** 2

N2.2 **a** 48 **b** $2^2 \times 3^2$ **c** 4 and 9

N2.3 **a** **i** 2, 3, 4, 6, 8, 12 **ii** 11, 13, 17 **iii** $2^3 \times 3$
b **i** 31 **iii** 21 + 10, the increase goes up by 2 each time

N2.4 **a** 9, 25, 100 **b** 20, 25, 100 **c** **c** 3, 29

N2.5 **a** **i** 5 **ii** 30 **b** x = 14 or 21, y = 77 or 91

N2.6 3^4

N2.7 **a** 125 **b** 64

N2.8 **a** $4\frac{4}{5}$ **b** 13

N2.9 **a** 4 **b** 13.2

N3.1 £780

N3.2 **a** $\frac{1}{4}$ **b** $12\frac{1}{2}$%

N3.3 **a** £180 **b** £270

N3.4 25%

N3.5 **a** 0.4 for example **b** $\frac{3}{4}$

N3.6 $\frac{5}{12}$

N3.7 **a** 2.4 m **b** 1.92 m **c** 4 bounces

N3.8 £36

N3.9 £85

N4.1 -2, -8

N4.2 **a** 19 **b** **i** $x + 4$ **ii** $x - 4$

N4.3 **a** $10^2 - 9^2 = 19$ **b** $r^2 - (r - 1)^2 = 2r - 1$

N4.4 **a** 33, 45
b **i** 65
ii The difference between terms keeps doubling, so add 32 to 33.
c $2n + 1$.

N4.5 **a** $4^2 + 8 = 6^2 - 12$
b **i** $4^2 + 8 = 4 \times 6$ **ii** $n^2 + 2n = n \times (n + 2)$ **c** $n^2 + 2n$

N4.6 **a** $2n + 3$ **b** $n^2 + 2$

N4.7 **a** $4n + 1$ **b** $n(n + 1)$

N5.1 **a** **i** 3.9×10^5 **ii** 390 000 **b** **i** 6.7×10^{-3} **ii** 0.0067

N5.2 **a** 3.45×10^{10} **b** 5.43×10^{-7} **c** 1.125×10^2

N5.3 **a** 0.00037 **b** **i** 1.012×10^6 **ii** 3.28×10^{-1}

N5.4 **a** 2.44018×10^5 **b** 0.05%

N5.5 9.47×10^{12} km

N6.1 250 g

N6.2 £4

N6.3 60 million pounds

N6.4 125 sweets

N6.5 **a** 35 **b** 40

N6.6 Max £262.50, friend £157.50

N7.1 **a** 140 **b** 49

N7.2 **a** 4 minutes 24 seconds **b** 294 words

N7.3 **a** 22% **b** 1 : 7

N7.4 50 days

N7.5 **a** 5 **b** £77.35

N7.6 **a** **i** $25 \times 20 = 500$ **ii** $120 \div 15 = 8$
b $40 \times 0.03 = 1.2$, so his answer is too big by a factor of 10.

N7.7 **a** 62.4 kg **b** 10% **c** **i** 53.5 kg **ii** 119 pounds

N7.8 **a i** 5 pounds **ii** 6900 kg **b** 24.5 kg

N7.9 £2289.80

N7.10 **a** £184 **b** $2M + 40$ **c i** $2M + 40 = 124$ **ii** 42 miles

A1.1 **a i** £32.16 **ii** 150 miles **b** £153.48

A1.2 **a** 135° **b i** 30 **ii** $n = 360 \div (180 - p)$

A1.3 66

A1.4 2.2

A1.5 **a** 9.26 **b** $\sqrt{(PR)}$

A1.6 **a i** 27 **ii** 7 **iii** $\sqrt{(\frac{y}{3})}$

A1.7 5

A1.8 **a**

2	−3	4
3	1	−1
−2	5	0

A2.1 29

A2.2 $x = 3$

A2.3 **a** $x = 3$ **b** $x = 2$

A2.4 **a** $x = 1.5$ **b** $x = {}^-1$

A2.5 **a** $x = 6$ **b** $x = 0.5$ **c** $x = 23$

A2.6 **a** $x = 4$ **b** $x = 5, y = {}^-1$

A2.7 $x = 1, y = {}^-2$

A2.8 **a** Straight line passing through (1, 2) and (2, 4) **b** $({}^-1, {}^-2)$

A2.9 $x = {}^-0.5, y = 6.5$

A3.1 **a** $(x + 3)(x + 5)$ **b** $x = {}^-3$ or ${}^-5$

A3.2 $x = {}^-5$ or 2

A3.3 **a** $(x + 3)(x + 2) = 42$ **b** 16 cm^2

A3.4 **a** $4x^2 + 4x = 48$ **b** $x = 3$

A3.5 **a**

x	$^-2$	$^-1$	0	1	2	3
y	5	1	$^-1$	$^-1$	1	5

c $x = -0.6, 1.6$

A3.6 $x = 27$

A3.7 $x = 7.6$

A4.1 **b** The solution is the x-coordinate where the two lines intersect.

A4.2 **a** $y = 0.4x + 18$ **b** £58

A4.3 $a = {}^-0.6, b = 3$

A4.4 **a i** C **ii** A **iii** D

A4.5 $y = 6 - 1.5x$

A4.6 **a i**

x	$^-3$	$^-2$	$^-1$	0	1	2	3
y	2	$^-3$	$^-6$	$^-7$	$^-6$	$^-3$	2

b $x \perp 2.6$ **c** $y - {}^-7$

A4.7 **a**

x	$^-3$	$^-2$	$^-1$	0	1	2	3
y	$^-4$	1	4	5	4	1	$^-4$

c $x = \pm 2.2$

A5.1 $^-1$, 0, 1

A5.2 $1 < x < 2$

A5.3 2, 3

A5.4 **a** $x \leqslant 1$ **b i** $x \geqslant {}^-1.4$ **ii** $^-1$

A5.5 $x < {}^-2$

A6.1 **a** $2x - 1$ **b** $2a(2 - b)$ **c** $x = 5$

A6.2 $\frac{a^4}{c}$

A6.3 **a** t^8 **b** p^4 **c** a^4

A6.4 **a** $3x^3 - 5x$ **b** $6x^2 - x - 2$

A6.5 **a** p^4 **b** $24abc$ **c** 1

A6.6 **a** $x^2 - x - 8$ **b** $3a(a - 2)$ **c** $2x^2 - 7x + 3$

A6.7 **a i** a^8 **ii** a^2 **iii** a^{15} **b i** $(a^5)^3$ **ii** $a^5 \div a^3$

A6.8 **a i** $6a^5$ **ii** $2a^3$ **b** $2x(x + 2)$

A6.9 **a** $x - 6$ **b** $x = 2$

A7.1 **a** $3x(x - 2)$ **b** $3x^2 - 10x - 8$ **c** $t = (4W - 3) \div 5$

A7.2 **a** $x + 1$ **b** 3

A7.3 **a** $6x^2 + 7x - 20$ **b** $5xy(x + 3y^2)$

A7.4 **a** $x = 6$ **b** $x = 3.5, y = 0.5$

A7.5 **a** $5x + 1$ **b i** $5x + 1 = 16, x = 3$ **ii** 5 sides

A7.6 **a** $G(0, 5)$, $H(10, 0)$ **b ii** $x = 2$ **c** $x = (y - b) \div a$

S1.1 **a** 36° **b** 108°

S1.2 **a** 45° **b** 67.5°

S1.3 **a** 65°, because alternate angles are equal.
b 70°, because this angle and 110° are supplementary.

S1.4 $w = 40°, x = 119°, y = z = 61°$

S1.5 **a** 180° **b** $5x$ **c** 72°, 72°, 36° **d** isosceles

S1.6 36°

S1.7 **a i** $x°$ **ii** $180 - 2x°$ **b** 64°

S2.1 506.25 cm^2

S2.2 **a** 75 cm^2 **b** 42 cm

S2.3 0.63 m^2

S2.4 **a** 4000 cm^2 **b** 10 000 cm^3

S2.5 14 000 cm^3

S2.6 **b** $\frac{1}{2}(a + b)c$

S2.7 **a** $2\pi a(a + b)$, $\frac{1}{2}(a + b)c$ **b** $\pi a^2 b$

S2.8 **a** $2(v + 2w + x + y + z)$ **b** $\frac{1}{2}z(x + y)w$

S2.9 **a** 440 cm **b** 770 000 cm^3

S2.10 **a** 220 cm **b** 30 800 cm^3

S3.1 **a** 50 mph **b** 1 hour 20 minutes

S3.2 **a** 90 mph **b** 4 hours 30 minutes

S3.3 **a** 47 km **b** 1207

S3.4 **a** 200 miles **b** 30 minutes

S3.5 18 mph

S3.6 **b** 96 mph

S4.1 **a i** 037° **ii** 250° **b** 3.6 km

S4.2 **a** 6.3 miles

S4.3 299°

S4.4 296°

S5.1 **a** 49° **b** 5.4 cm

S5.2 **a** 9.4 m **b** 66°

S5.3 **a** 44° **b** 632 cm^2

S5.4 **a** $\frac{3}{5}$ **b i** $\frac{4}{5}$ **ii** 9 cm

S5.5 2.08 m

S5.6 $41^2 = 40^2 - 9^2$

S5.7 **a** 2.1 m **b** 4.8 m

S7.1 **a** 6 squares, 0 triangles, 0 rectangles
b 1 square, 4 triangles, 0 rectangles
c 0 squares, 2 triangles, 3 rectangles

S7.2 **a** P **b** 54 cm^2 **c** 300

S7.3 **a** 12 cm^3

S7.4 **b** 3.5 cm

S8.2 **a** reflection in $x = 3$ **b** rotation of 180° about (3, 0)

S8.3 **a i** H **ii** G **ii** F
b reflection in the y-axis

S8.4 **a i** triangle DEF **ii** triangle CLF
b reflection about the line JD, rotation of 180° about O

S8.5 **a** $x = 8, y = 3.5$
b **i** ($^-4$, 1) **ii** translation of 9 units to the right and 4 units up

S8.6 **a** rotation of 90° about (0, 0) **b** reflection in $y = x$
c Triangle with vertices at (3, 2), (5, 0), (9, 4)

S9.1 3.6 cm

S9.2 14.4 cm

S9.3 **a** angle p = 130°, angle q = 20° **b** 8 cm

S10.1 **a** $\frac{1}{3}$

S10.2 **a** (3, 6) **b** (3, 2)

S10.3 **a** $x = 63°$, $y = 27°$ **b** **i** 13 cm **ii** 63 cm^2

S10.4 **a** 6.4 cm **b** 8.7 cm

S10.5 **a** 333° **b** 22 km^2

D1.1 **a**

	1	2	3	4	5
2	3	4	5	6	7
3	4	5	6	7	8
4	5	6	7	8	9
5	6	7	8	9	10
6	7	8	9	10	11

b **i** $\frac{1}{25}$ **ii** $\frac{6}{25}$

D1.2 **a**

	5	6	7	8
1	4	5	6	7
2	3	4	5	6
3	2	3	4	5
4	1	2	3	4

b $\frac{1}{2}$

D1.3 **a** 1H, 1T, 2H, 2T, 3H, 3T, 4H, 4T, 5H, 5T, 6H, 6T
b $\frac{1}{4}$ **c** $\frac{3}{4}$

D1.4 **a** $\frac{1}{3}$ **b** $\frac{1}{6}$ **c** $\frac{1}{2}$

D1.5 **a** $\frac{1}{25}$ **b** $\frac{1}{5}$ **c** $\frac{2}{5}$

D1.6 **a** green jacket and red skirt, green jacket and yellow skirt, black jacket and red skirt, black jacket and yellow skirt, white jacket and red skirt, white jacket and yellow skirt
b **i** $\frac{1}{6}$ **ii** $\frac{1}{3}$

D1.7 **a** 0.72 **b** 0.26

D1.8 **a** $\frac{5}{19}$ **b** 0.75

D1.9 **b** **i** 0.16 **ii** 0.84

D1.10 **b** 0.42

D1.11 **a** 14 **b** 0.33
c Yes, the probability of obtaining a tail should be close to 0.5.

D1.12 **a** He has not thrown the dice enough times to check the relative frequency.
b $\frac{19}{100}$ **c** 190 times

D1.13 **a** $\frac{1}{4}$ **b** $\frac{1}{5}$ **c** 200 times

D2.1 **a** Pie chart with angles 80°, 60°, 120°, 100°
b 30 children

D2.2 **a** pie chart with angle 170°, 120°, 70°.
b **i** 130
iii Sales of national daily newspapers continue all day, but are highest in the morning; sales of the Echo occur only in the afternoon, and increase as the afternoon goes on.

D2.3 **a** pie chart with angles 108°, 72°, 120°, 60°
b **i**

Main use	Number of people	Percentage of people asked
e-mail	11	44
Internet	5	20
Word-processing	3	12
Games	6	24
Total	25	

ii bar chart with heights 44, 20, 12, 24

D2.4 pie chart with angles 70°, 190°, 100°

D2.5 **a** **i** Britain **ii** 18 people **b** 12.5%

D3.1 **a** Tall fathers produce tall sons. **c** 175 cm

D3.2 **b** Older patients react more slowly.
d **i** 25 **ii** 52 hundredths of a second
e The 30 year olds have higher reaction times on average. The spread of reaction times is similar for each age group (relative to the median).

D3.3 **b** strong positive correlation **c** 6 years

D3.4 **b** The 100 page book costing £9.
c There is a strong positive correlation.

D3.5 **a** **i** C **ii** B **iii** D
b 'I got a poor result because I didn't spend much time revising'.
c The more time they spent revising, the better their test result.

D4.1 $^-1$°

D4.2 **a** **i** 14 minutes **ii** 5 minutes
b Either of:
The buses are on average later than the trains.
There is less variation in the lateness of the buses.

D4.3 **a** 2, 3, 10, 10 for example
b 6, 10, 10, 10 for example
c 4, 10, 10, 12

D4.4 2.2 goals per pupil

D4.5 **a** **i** £3 **ii** £3.49
b The 16 year olds have the larger range, because their maximum amount is higher.

D4.6 36 years

D4.7 33 minutes

D5.1 **a** **i**

Age, A, years	Number of people (millions)	Cumulative frequency (millions)
$0 \leq A < 10$	8	8
$10 \leq A < 20$	7	15
$20 \leq A < 30$	8	23
$30 \leq A < 40$	9	32
$40 \leq A < 50$	8	40
$50 \leq A < 60$	7	47
$60 \leq A < 70$	6	53
$70 \leq A < 80$	4	57
$80 \leq A < 90$	2	59
$90 \leq A < 100$	1	60
	60	

b **i** 38 years **ii** 37 years **c** 10 million people

D5.2 **b** 8 **c i** 25 hours **ii** 5 hours
d No, the interquartile range is higher.

D5.3 **a i**

Age, A, years	Number of people	Cumulative frequency
$0 \leqslant A < 10$	20	20
$10 \leqslant A < 20$	130	150
$20 \leqslant A < 30$	152	302
$30 \leqslant A < 40$	92	394
$40 \leqslant A < 60$	86	480
$60 \leqslant A < 80$	18	498
$80 \leqslant A < 100$	2	500

b i 27 years **ii** 19 years **c** 17%

D6.1 **a** There are more people on the roads at that time.
b i 240 **ii** 30

D6.2 Question 1: It is unclear which box you tick if you get up at 7.00.
Question 2: There is no 'other' category for people that have none of these.
Question 3: Respondents should be given a choice of time bands.

D6.3 **b** No, he will miss all the people who do not borrow books.

D6.4 **a** 219
c He will not be asking anyone who does not take any exercise.
He will not be asking any boys.
d i It is a leading question. **ii** There is no 'no' choice.

D7.2 **a** 0.01 **b** 0.001

D7.3 **a** More than 4 hours but less than or equal to 5 hours
b Ann could have finished just under 5 hours and Ben just over 5 hours.
c polygon joining heights 10, 15, 40, 57, 23

PRACTICE NON-CALCULATOR PAPER

1 **a** £3.25 **b** $1.5 + 0.25x$
2 **a** 23 **b** $4x - 3$
4 **a** angles 75°, 129°, 108°, 48° **b** 10%
5 **a** 55° **b** 125°
6 **a** 13 **b** 6 **c** 6.5
7 **a** 33.3 mph **b** 5h
8 **b** 21
9 2550 g
10 50
11 **a** $2 \times 2 \times 2 \times 3 \times 3$ **b** 8 **c** 24
13 **a** x^5 **b** x^2 **c** x^3
14 5
15 Savings Bond better
16 $10\pi + 44$
17 140
18 $x = 3$, $y = 1$
19 $y = \frac{1}{2}x + 2$
20 **a** 6×10^8 **b** 3×10^9
21 English results are less spreadout. Averages are the same.
22 **a** 65 **b** 115

PRACTICE CALCULATOR PAPER

1 £2.76
2 **a** $a + 6b$ **b** 7 **c** 4.8
3 **a** 444 **b** £139.86
4 **a** $4x$ **b** $16x$ **c** 9
6 Yes, 63 is bigger than 60
7 **b** 13.3 km/h
8 **b** positive correlation
9 130 cm^2
10 **a** $5 \times 6 = 6^2 - 6$ **b** $n \times (n+1) = (n+1)^2 - (n+1)$
11 **a** 1.33 **b** 4.472
12 **a** $5 < T \leqslant 10$ **b** 11.9 minutes
13 6000
14 9
15 3.3
16 3400 m^2
17 **a** 5 **b** 1 **c** 3 **d** $\frac{1}{8}$ or 0.125
18 110°
19 23.3°
20 **a** $x(x+4) = 45$ **b** $x^2 + 4x - 45 = 0$
c $x = 5$, perimeter = 28 **d** 89.25
21 **a** 0.6 **b i** 0.6, 0.3, 0.7, 0.3, 0.7 **ii** 0.12
22 **a** 6 **b** 6
23 $x = \frac{c}{a - b}$